The Performance and Design of Direct Current Machines

The Performance and Design of Direct Current Machines

Albert E Clayton

DSc (Engineering) Lond; Associate of King's College London, FIEE

Formerly Senior Lecturer in Electrical Machinery and Design in the Victoria University of Manchester, at the College of Technology, Manchester

Revised by the author in collaboration with

N N Hancock

PhD BSc (Special) Hons Lond; BSc (Engineering) Hons Lond.
MSc Tech Manchester; MIEE, Mem AIEE

Senior Lecturer in Electrical Engineering in the University of Manchester Institute of Science and Technology

Third Edition

CBS Publishers & Distributors Pvt. Ltd.

New Delhi • Bengaluru • Chennai • Kochi • Kolkata • Mumbai
Hyderabad • Nagpur • Patna • Pune • Vijayawada

ISBN: 81-239-0927-6

First Indian Reprint: 1986
Reprint: 2000, 2001, 2002, 2003, 2004

This edition has been reprinted in India by arrangement with
Pitman, London

Published by **Satish Kumar Jain** and produced by **Varun Jain** for
CBS Publishers & Distributors Pvt. Ltd.,
4819/XI Prahlad Street, 24 Ansari Road, Daryaganj, New Delhi - 110002
delhi@cbspd.com, cbspubs@airtelmail.in • www.cbspd.com
Ph.: 23289259, 23266861, 23266867 • Fax: 011-23243014

Corporate Office: 204 FIE, Industrial Area, Patparganj, Delhi - 110 092
Ph: 49344934 • Fax: 011-49344935
E-mail: publishing@cbspd.com • publicity@cbspd.com

Branches:

- ***Bengaluru:*** 2975, 17th Cross, K.R. Road, Bansankari 2nd Stage, Bengaluru - 70 • Ph: +91-80-26771678/79 • Fax: +91-80-26771680 E-mail: cbsbng@gmail.com, bangalore@cbspd.com
- ***Chennai:*** No. 7, Subbaraya Street, Shenoy Nagar, Chennai - 600030 Ph: +91-44-26681266, 26680620 • Fax: +91-44-42032115 E-mail: chennai@cbspd.com
- ***Kochi:*** Ashana House, 39/1904, A.M. Thomas Road, Valanjambalam, Ernakulum, Kochi • Ph: +91-484-4059061-65 Fax: +91-484-4059065 • E-mail: cochin@cbspd.com
- ***Kolkata:*** 6-B, Ground Floor, Rameshwar Shaw Road, Kolkata - 700014 Ph: +91-33-22891126/7/8 • E-mail: kolkata@cbspd.com
- ***Mumbai:*** 83-C, Dr. E. Moses Road, Worli, Mumbai - 400018 Ph: +91-9833017933, 022-24902340/41 • E-mail: mumbai@cbspd.com

Representatives:

- Hyderabad: 0-9885175004
- Patna: 0-9334159340
- Jharkhand: 0-9811541605
- Nagpur: 0-9021734563
- Pune: 0-9623451994
- Uttarakhand: 0-9716462459

Printed at:
Neekunj Print Process, Delhi (India)

PREFACE

TO THE THIRD EDITION

THE main changes introduced in this edition are the following—

The general theoretical treatment is now based on the rationalized m.k.s. system of units.

The sections on control and special machines have been extended considerably.

The chapter on insulation has been re-written and brought up to date.

The illustrations of machines have largely been replaced by more modern examples.

A supplementary chapter on eddy-current losses in deep armature conductors has been added.

As far as possible, symbols used in the previous edition have been modified where necessary to conform with official recommendations.

In addition the author has generally revised the whole of the subject-matter and made such modifications and additions as seemed to be desirable. For example, duplex lap windings now receive much more attention in view of their application to machines of very large output. The section on temperature-rise curves has been extended and should be new to many. The section dealing with the physical limits to the output of d.c. machines has been extended, and there are various other modifications throughout the book. The supplementary chapter on eddy-current losses in armature conductors was included, after much thought, in the hope that the simplified treatment given therein will be helpful to the young designer.

This book deals with principles and with the practice of electrical design. In the actual design of *any* apparatus the designer will naturally work in terms of units which in his view are best suited *from all considerations*. Thus some leading designers of electrical machinery have used the inch; others, including the author, have used metric units, expressing dimensions chiefly in centimetres, but using the larger and smaller units—the metre, kilometre, and millimetre—freely where circumstances warranted their use. It is simplicity itself to modify fundamental formulae and data developed *either* with m.k.s. *or* c.g.s. units, so that they can be used directly with the inch or any other unit of length. It is probable therefore that designers will continue to use the same units of length as hitherto: those who have used the inch will most likely continue to do so; others, like the author, will continue to use the centimetre where it is called for, still finding that the metre is too large, and the square metre altogether too large for most design details. Further, as hitherto, the absolute unit for any particular quantity may not be suited to all actual problems; hence the use of the kilowatt-hour, the microfarad, and so on. In design, the c.g.s. unit of magnetizing force was never favoured, designers using the ampere-turn per centimetre (or per inch). The m.k.s. unit of ampere-turn per metre is inconveniently small, so that designers will probably continue to use the same unit as hitherto. Again for example, it will probably be found more practical to express current density in a conductor as 4 A/mm² or 400 A/cm² rather than as 4,000,000 A/m². In the design chapters of the book, the author has endeavoured therefore to bear all these matters in mind.

In the preparation of this edition the author invited the assistance of Mr. N. N. Hancock—a man of wide practical experience in the performance and design of electrical machinery. Mr. Hancock has not only collaborated in such matters as reading the MS and proofs, but also by providing the new matter on insulation, and on special machines and control. The author would like to express his appreciation of all this valuable collaboration.

The author's thanks are also due to the firms who provided so generously material from which the new illustrations have been selected: The British Thomson-Houston Co., Ltd.; The English Electric Co., Ltd.; The General Electric Co., Ltd.; Laurance, Scott and Electromotors, Ltd.; The Metropolitan-Vickers Electrical Co., Ltd.; Messrs. Veritys, Ltd.; Watford Electric and Manufacturing Co., Ltd.

It is possible in a revision such as this, involving so many modifications of formulae, that there are some sins of omission or commission. The author hopes they are few and of minor importance.

A. E. CLAYTON

CHEADLE HULME
October, 1956

COLLABORATOR'S NOTE

It is most regretfully necessary to record the sudden death of the author during the proof stage of this third edition. Since only minor corrections have been required in the page proofs, however, the subject matter remains in the exact form of which Dr. Clayton approved.

N. N. H.

THE MANCHESTER COLLEGE OF SCIENCE AND TECHNOLOGY
July, 1958

PREFACE

TO THE FIRST EDITION

THIS book is intended as a textbook on direct current machinery for use primarily by students pursuing a course of study in Electrical Engineering at universities and technical colleges. It is hoped at a later date to publish a companion volume dealing with Alternating Current Machinery, so that the two volumes will cover the needs of the more important examinations in Electrical Machinery.

For most examination purposes, the subject of design is linked with that of performance; in writing the present volume the author has endeavoured to strike a correct balance between the requirements of the two branches of the subject.

The young designer, and the student who is making a very special study of the design of electrical machinery, must of necessity have recourse to standard works on electrical design, such as *The Design and Specification of Electrical Machinery*, by Professor Miles Walker, and *The Dynamo*, by C. C. Hawkins, to mention but two. The author hopes, however, that the present volume may also be of service to the young designer in helping him to obtain a thorough knowledge of the performance of direct current machines and of the principles governing their design.

In general, it is assumed that the reader will possess a certain small knowledge of electrical engineering principles, the knowledge assumed corresponding roughly to that required for the Ordinary National Certificate in Electrical Engineering. The ordinary second-year student at our universities should be possessed of sufficient knowledge to enable him to follow the treatment adopted. At the discretion of the reader, certain paragraphs and sections of the book may well be left to a second reading. For example, most students in their second year will not find it necessary to deal in full detail with the calculation of the magnetic circuit, or with detailed calculations relating to the inductance of the coils undergoing commutation. These matters may well be left until an actual design is being attempted.

The method of presenting the subject-matter will in certain cases, it is hoped, be found to possess certain advantages over methods which have become rather stereotyped. The treatment of armature windings is based at once on the use of slotted armatures, and not upon an equivalent smooth armature, as is most often the case; the result is a very considerable simplification of a subject which often is rather puzzling to the student.

In the sections relating to actual design calculations, no attempt has been given to reduce to the very minimum the actual labour involved, but the reader is taken step by step through the various calculations in order that he may fully appreciate the various points that arise. In practice, the design of a machine in most cases results in filling in a few figures on a printed design sheet, so arranged that the minimum time is expended. Such design sheets, of universal application, have already been published by Professor Miles Walker in his paper on "The Predetermination of the Performance of Electrical Machinery," and in the above-mentioned book. The student who is making a very special study of the subject of design, involving as it does the working out of a large number of complete designs, must prepare and employ such sheets to avoid unnecessary waste of time. For the purpose of the general student it will suffice to carry out calculations on the lines of those tabulated in the last chapter in the book. Limitations of space have made it necessary to prescribe the amount of written matter relating to mechanical construction; the author has, however, endeavoured to make up for this deficiency by the inclusion of a relatively large number of drawings, which afford ample information regarding mechanical details. Since the average engineer will have to deal with electric motors of relatively modest dimensions, rather than with large direct current generators, the drawings given relate chiefly to comparatively small machines.

Most of the illustrations have been specially prepared for the book; those illustrating the actual construction of machines have been very kindly provided by the various firms concerned: Messrs. The British Thomson-Houston Co., Ltd.; Electromotors, Ltd.; The English Electric Co., Ltd.; The Lancashire Dynamo and Motor Co., Ltd.; The Metropolitan-Vickers Electrical Co., Ltd.; Messrs. Sankey, Ltd.; and The Standard Brush-Holder Co. The thanks of the author are also due to The Lancashire Dynamo and Motor Co., and to Messrs. Electromotors, Ltd., for the design data included in the last chapter of the book.

The author has also to acknowledge, with gratitude, help he has received from his colleague, Mr. Bernard Thomsett, B.Sc. (Eng.), and his former student, Mr. N. M. Barratt, B.Sc.Tech.

A. E. C.

College of Technology
Manchester
1927

CONTENTS

CHAPTER V

CHAPTER VI

CHAPTER VII

CHAPTER VIII

CHAPTER IX

CHAPTER X

CHAPTER XI

CHAPTER XII

CHAPTER XIII

CHAPTER XIV

CHAPTER XV

CHAPTER XVI

CHAPTER XVII

CHAPTER XVIII

CHAPTER XIX

CHAPTER XX

LIST OF SYMBOLS

When distinction is necessary, a dash is used to indicate that a symbol represents a quantity expressed in the c.g.s. system of units. Symbols without dashes indicate quantities in general, or more specifically quantities in m.k.s. units.

a	constant; number of pairs of parallel paths in armature winding; ratio of *max* to *min* flux during starting; $\sqrt{\omega\mu_0 b_c/\rho b_s}$
b	breadth; width; periphery of copper section in slot; a/k for starter
b_c	breadth of copper in slot
b_{cp}	breadth of compole at air-gap
b_p	breadth of pole limb
b_{ps}	breadth of pole shoe
b_s	slot width
b_t	tooth width at armature surface
$b_{t_{ef}}$	effective tooth width (1/3 from root)
b_v	breadth of radial ventilating ducts
d	diameter; diameter of uninsulated wire; depth, winding depth
d_c	radial depth of core below slots
d_i	diameter of insulated wire
d_y	radial depth of yoke
e	induced e.m.f. per conductor
e_{r_a}	average reactance e.m.f. per turn
f	frequency; mechanical (normal) stress
h	height; radial height of armature conductor
h_c	total radial height of copper in slot
h_p	radial height of pole limb
h_{ps}	radial height of pole shoe
h_s	radial height of slot $= l_t$
h_t	radial height of tooth-tip above copper in slot
i	instantaneous value of current
i_a	instantaneous value of current in armature conductor

k	constant; ratio of *max* to *min* current during starting
k_i	space factor of core
k_s	ratio of gross area of air-gap surface to surface area of teeth
	length
l_a	length of armature conductor
l_c	length per pole of flux path in armature core
l_s	length of mean turn of shunt or separately-excited field winding
l_{fr}	length of armature conductor outside core at one end
l_g	minimum radial air-gap under main pole (on centre line)
$l_{g_{cp}}$	radial air-gap under compole
l_{gx}	radial air-gap at any point under main pole
l_p	length per pole of flux path in pole
l_{se}	length of mean turn of series field winding
l_t	length per pole of flux path in teeth ($= h_s$)
l_y	length per pole of flux path in yoke
l_G	radial air-gap at pole tip
m	*multiplexity* of armature winding
n	number of; rev/sec; resistance sections of starter
n_v	number of radial ventilating ducts
p	number of pairs of poles; pressure; dissipation per unit surface area
p_a	dissipation per unit surface of armature
p_s	dissipation per unit surface of shunt or separately-excited field winding
$_B$	bearing pressure
q	armature strength = ampere-conductors per unit armature periphery
r	resistance
r_a	resistance of armature winding
r_b	internal resistance of battery
r_{cp}	resistance of compole winding
r_s	resistance of shunt or separately-excited field winding
r_k	resistance of compensating winding
r_m	resistance of machine armature circuit

r_{se}	resistance of series field winding
$r_1, r_2 \ldots$	resistances of sections of starting resistor
s	distance
t	time; thickness; thickness of laminations
t_b	thickness of brush
t_m	thickness of mica separators of commutator
v	voltage; velocity
v_b	brush contact volt drop
v_s	voltage per coil of shunt field winding
v_A	peripheral speed of armature
v_B	booster voltage; velocity of shaft in bearing
v_C	peripheral speed of commutator
y_C	commutator pitch of armature winding in segments
A	area
A_a	cross-sectional area of armature conductor
A_b	contact area of *all* brushes
A_c	cross-sectional area of armature core below teeth
A_s	cross-sectional area of shunt-winding conductor
A_f	area of pole face
A_p	cross-sectional area of pole limb
A_s	cross-sectional area of slot
A_{se}	cross-sectional area of series-winding conductor
A_t	cross-sectional area of teeth per pole
A_y	cross-sectional area of yoke
B	flux density; bending moment
B_{appt}	apparent flux density in teeth
B_{av}	average flux density in air-gap over a pole-pitch
B_c	flux density in armature core below teeth
B_{cp}	flux density in compole air-gap
B_g	peak flux density in air-gap
B_o	flux density in free space
B_p	effective flux density in pole limb
B_{real}	actual flux density in teeth
$B_{t_{av}}$	average flux density in teeth
B_y	flux density in yoke

C	number of commutator segments = number of armature coils; thermal dissipation rate (W/°C)
C_l	lead of brushes in segments
C_n	half distance between pole-tips
C_o	output coefficient
C_1C_2	see Fig. 27, p. 43
D	diameter; external diameter of armature
D_i	internal diameter of armature core
D_m	diameter of armature 1/3 up from root of tooth
D_{min}	minimum possible armature diameter for given rating
D_t	diameter of armature at bottom of slots
D_y	mean diameter of yoke
D_B	bearing diameter
D_C	commutator diameter
E	e.m.f.; total induced e.m.f. per circuit; modulus of elasticity
E_c	average induced e.m.f. per conductor
$E_{r_{av}}$	average reactance e.m.f. per coil, i.e. between segments
E_A	armature e.m.f.
E_C	commutating e.m.f.
F	m.m.f.; force
F_a	armature m.m.f. per pole
F_{av}	average force
F_c	m.m.f. on armature coil
F_{cp}	compole winding m.m.f. per pole
F_d	demagnetizing component of armature m.m.f. per pole
F_s	shunt or separately-excited winding m.m.f. per pole
F_f	total field m.m.f. per pole
F_g	m.m.f. on air-gap
F_i	m.m.f. per pole for iron part of magnetic circuit
F_k	compensating winding m.m.f. per pole
F_p	m.m.f. drop in pole itself
F_q	cross-magnetizing component of armature m.m.f. per pole
F_{se}	series field winding m.m.f. per pole
F_t	m.m.f. on teeth
F_{t+g+c}	m.m.f. on teeth gap and core
F_B	slot m.m.f. at bottom of conductor
F_T	slot m.m.f. at top of conductor

H	magnetizing force; thermal capacity
H_c	magnetizing force in armature core
H_g	magnetizing force in air-gap
H_p	magnetizing force in pole limb
H_t	magnetizing force in armature teeth
H_y	magnetizing force in yoke
I	current; moment of inertia
I_a	current in armature conductor
I_c	short-circuit current
I_s	current in shunt or separately-excited field winding
I_m	nominal slot m.m.f. at mid-section of conductor
I_o	total no-load current
I_{se}	current in series field winding
I_A	armature current
I_A	no-load armature current
$I_1 I_2$	*max* and *min* armature currents during starting
J	moment of inertia
K	constant
K_a	Carter's permeance factor
K_e	ratio of eddy loss in conductor to nominal I^2R loss, both in general and for lower layer of two-layer winding
K_f	field form factor
K_g	Carter's contraction coefficient
$K_{g_{cp}}$	Carter's contraction coefficient for compole gap
K_{gs}	Carter's contraction coefficient for slots
K_{gv}	Carter's contraction ceofficient for ventilating ducts
K_o	total loss factor for conductor = ratio of total loss to nominal I^2R loss $= K_e + 1$
K_t	total loss factor for whole armature winding
K_u	ratio of eddy loss in conductor to nominal I^2R loss for upper conductor of two-layer winding
L	length, gross axial length of armature core; length between bearings
L_a	total length of all armature conductors
L_{cp}	axial length of compole
L_i	net axial length of iron in armature core
L_n	net axial length of armature core (excluding ventilating ducts)

L_o axial length of coil overhang for *one* end
L_p axial length of main poles
L_A axial length of armature over windings
L_B axial length of journal

M intensity of magnetization; equiv. total bending moment

M_r function of *ah, see* p. 477

N number of turns per coil; number of rev/min
N_a number of turns per armature coil
N_e number of turns per coil of shunt field winding
N_r function of *ah, see* p. 477
N_{se} number of turns per coil of shunt field winding

P power; power converted from elec. to mech. form or vice versa; rate of heat dissipation
P^* power converted in *kilowatts*
P_c copper loss
P_e power loss in shunt winding; eddy current power loss in conductor
P_i iron loss
P_o iron, friction and windage loss
P_{se} power loss in series field winding
P_B total bearing load
P_E eddy loss in iron
P_F fixed losses
P_H hysteresis loss in iron
P_L total power losses

R resistance; reluctance
R_x resistance of external circuit
$R_1 R_2 \ldots$ total circuit resistance on steps 1, 2 . . . of starter

T torque; elec. or mech. time constant; accelerating time
T_C time of commutation of armature coil
T_{Cx} time of commutation of x coil-sides in one layer

V terminal voltage; volume
V_a volume of copper in armature winding
V_{av} average voltage (between commutator segments)
V_e voltage available for excitation; volume of copper in shunt field winding

V_{max}	maximum voltage (between commutator segments)
V_o	no-load terminal voltage
V_{se}	terminal voltage of series winding
V_A	armature terminal voltage
V_G	generator terminal voltage
V_M	motor terminal voltage
W	work, energy; weight; weight of armature
W_a	weight of copper in armature winding
W_f	energy stored in magnetic field
W_r	work done per revolution of armature
W_H	hysteresis loss per cycle
Z	number of armature conductors
Z_a	total number of armature conductors
Z_c	number of armature conductors in series per circuit
Z_s	number of armature conductors per slot
α	pole arc/pole pitch; current density
β	1/2 angle moved through during commutation; armature core length/pole-shoe breadth
β_x	distance moved through by commutator surface during commutation of x coil-sides of one layer
β_{xA}	distance moved through by armature surface during the commutation of the x coil-sides of one layer
δ	gap contraction coefficient; shaft deflexion
ε	base of natural logs
η	efficiency; Steinmetz coefficient
η_M	motor efficiency
η_G	generator efficiency
θ	temperature; angle turned through by armature; temperature rise
θ_{max}	maximum temperature rise
θ_t	temperature rise at time t
λ	total permeance per unit length of core
λ_{fr}	permeance around free portion of armature conductor per unit length of core

λ_i	total permeance around embedded conductor per unit length
λ_{ps}	leakage permeance per unit height of pole-shoe
λ_{pl}	leakage permeance per unit height of pole-limb
λ_s	leakage permeance per unit length of slot
λ_t	leakage permeance per unit length of tooth-tip
μ	permeability; coefficient of friction
μ_o	permeability of free space
μ_r	relative permeability
ν	leakage coefficient
ρ	resistivity
σ	space factor of copper
τ	pitch
τ_p	pole pitch
τ_s	slot pitch
τ_{sef}	effective slot-pitch ($l_t/3$ from root of tooth)
τ_C	commutator segment pitch
ϕ	instantaneous mean flux per pole; 1/2 angle of demagnetizing belt
ψ	angle of pole-tip (*see* Fig. 11, p. 23)
Λ	permeance
Λ_{pl}	permeance of leakage path from pole-limb
Λ_{ps}	permeance of leakage path from pole-shoe
Φ	useful flux per pole
Φ_c	$2 \times$ (flux in armature core)
Φ_{cp}	compole flux
Φ_i	flux in iron over one slot pitch
Φ_n	flux in air over one slot pitch
Φ_p	mean flux per pole
Φ_{pl}	leakage flux from pole-limb
Φ_{ps}	leakage flux from pole-shoe
Φ_s	flux over one slot pitch
Φ_y	$2 \times$ (flux in yoke)
Φ_L	leakage flux from pole
$\Phi_1, \Phi_2 \ldots$	*max* and *min* flux per pole during starting

CHAPTER I

GENERATION OF E.M.F.: WORK; POWER; TORQUE

THERE are two main types of direct current machine: 1. homopolar; 2. heteropolar. In the former type there is only one active pole, so arranged that the active conductors move through a field of constant intensity. This type is employed only in very rare instances. In the heteropolar type—the type commonly employed—the active field system consists of a number of poles, arranged alternately N and S, so that the e.m.f. induced in each active conductor alternates with a frequency determined by the product of the speed and the number of pairs of poles.

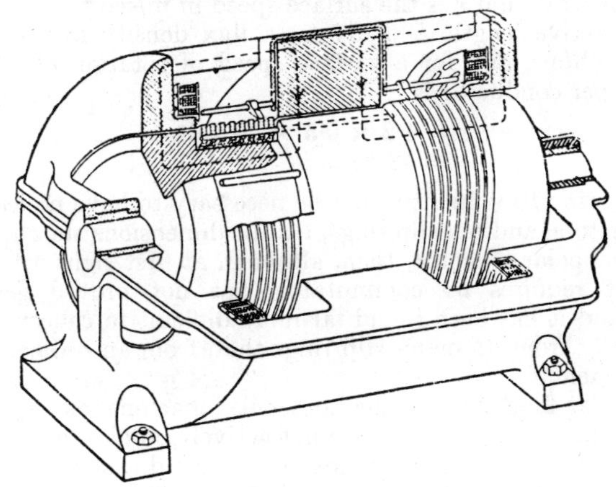

FIG. 1. HOMOPOLAR DYNAMO

(Reproduced, by kind permission, from "The Electrician" 21st August, 1908)

Homopolar Machine. Various arrangements of magnet systems have been used, one of which is indicated in Fig. 1. In this type there is a divided magnetic circuit. The stationary portion is a steel casting consisting in essentials of three rings, magnetically connected together by a suitable number of yoke pieces. The armature consists of a solid steel cylinder, and is surrounded by two exciting coils, as shown, carrying currents in opposite directions. These stationary coils establish a magnetic flux as indicated, the flux density in the air gap under the central pole being quite uniform.

The active conductors are carried in slots, arranged axially, and are each connected to a slip ring at either end. With homopolar d.c. machines, the only method of connecting armature conductors in series is by thus having two slip rings for each active conductor, and connecting the conductors in series by means of brushes bearing upon the rings. For six conductors in series, no less than twelve slip rings would be required, from each of which the full current output of the machine would have to be collected. It is largely this difficulty of satisfactory collection of the current that has limited the use of the homopolar machine to extremely rare cases; it is best suited for low voltage, very heavy current machines, where one active conductor is sufficient. For some purposes it suffices to have brushes bearing upon the rotor core itself. As the flux is constant, there are no iron losses, but this fact is more than counterbalanced by the heavy losses at the slip rings, if many are used.

The value of the e.m.f. induced in each active conductor is at once obtained from—

$$e = B_g L v \quad . \quad . \quad . \quad (1)$$

For the e.m.f. to be given directly in volts, m.k.s. units must be used, i.e. B_g is the air-gap flux density in weber/m², L is the active length in metres, and v is the surface speed in m/sec.*

For an active length L of 0·5 m, a flux density in the gap of $B_g = 1{\cdot}2$ Wb/m², and for a surface speed of rotation of 70 m/sec the e.m.f. per conductor is

$$e = 1{\cdot}2 \times 0{\cdot}5 \times 70 \text{ volts}$$
$$= 42 \text{ volts}$$

To generate 210 volts, it would be necessary to have no less than five conductors, and ten slip rings, for the dimensions stated.

The homopolar machine, then, although at first sight attractive, because it requires no commutator, has not proved generally suitable, and it has been found far more difficult to collect current satisfactorily from its many slip rings than from an ordinary commutator machine.

Heteropolar Machine. In the homopolar machine, as depicted in in Fig. 1, the two return poles are not actively employed. With the normal arrangement, adopted universally for d.c. machines, of N and S poles arranged alternately around the circumference as in Fig. 2, the poles are all fully employed. In consequence of this heteropolar arrangement it of course becomes necessary to employ a commutator, but this apparent disadvantage is completely outweighed by the fact that it becomes possible to group any desired number of conductors in series.

Since the homopolar machine is rarely met with, it will not further be discussed here, and when reference is made to d.c. machines it will be assumed henceforth that they are of the standard heteropolar type.

E.m.f. Induced in the D.C. Machine. Taking first the now obsolete smooth armature and surface winding, as indicated in Fig. 2. If a

* With c.g.s. units, B'_g is in gauss, i.e. maxwells/cm², L' is in cm, v' is in sm/sec, and e' is in c.g.s. units of 10^{-8} volt.

coil is placed as shown, so as to span an exact pole-pitch, it is at once apparent that, by rotating the armature, the two axial sides of the coil will cut through the magnetic lines of force. Since they are a pole-pitch apart, the e.m.f. induced in the two conductors will always act in conjunction, either clockwise or counterclockwise around the coil, and the total e.m.f. will at any instant be given by

$$e = 2B_g Lv^* \quad . \quad . \quad . \quad . \quad (2)$$

where B_g is the value of the flux density in the air gap through which the conductor is passing, L is the active length of the machine, and v is the surface speed of the conductor. At the same time it is apparent that, by rotating the coil, the number of lines linked with

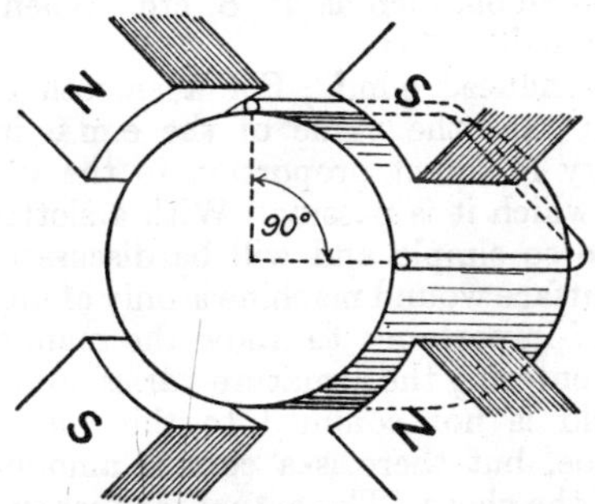

FIG. 2. HETEROPOLAR ARRANGEMENT

the coil will change periodically, and it is obvious that the rate at which the flux changes is identical with the value given above, i.e. $2B_g Lv$. In practice, then, it is immaterial whether the value of the e.m.f. induced in the turn is expressed in the form $2B_g Lv$, or in the form $-\frac{d\Phi}{dt}$; both forms amount to the same. If now the conductors are housed in slots—as in all modern machines—with the machine at standstill, practically the whole of the flux will pass through the teeth, and but little down the slots, so that it is not at once apparent that formula (2) can be applied. On the other hand, it is still quite apparent that, by rotating the armature, the value of the flux linked with the coil will change periodically, and so e.m.f. will be induced, just as with the smooth core machine.

Units of Flux and Flux Density. *Unit flux* is *one weber* in the m.k.s. system, and one maxwell in the c.g.s. system. The e.m.f. induced in a single turn is one volt when the flux linked with the turn is changing at the rate of one weber/sec.†

* The e.m.f. in all cases is given directly in volts if m.k.s. units are used for Φ, B, L and v. If c.g.s. units Φ', B' etc. are used, the e.m.f. is in c.g.s. units of 10^{-8} volts.

† 10^{-8} volt when the rate of change of the flux is one maxwell/sec.

Other units of flux are used in practice; the relation between them is—

One maxwell, or c.g.s. line $= 10^{-8}$ weber
One kiloline $= 10^3$ maxwells $= 10^{-5}$ weber
One megaline $= 10^6$ maxwells $= 10^{-2}$ weber
One weber $= 10^8$ maxwells

For *flux density* the m.k.s. unit is one Wb/m² and the c.g.s. unit is one maxwell/cm², i.e. one gauss

so that one gauss $= 10^{-4}$ Wb/m²
and one kilogauss $= 10^{-1}$ Wb/m²

In this book, unless otherwise stated, m.k.s. units are used and denoted by plain symbols such as Φ, B, etc. When c.g.s. units are used the symbols are Φ', B' etc.

Wave Shape of Induced e.m.f. For a smooth core machine it is at once evident that the value of the e.m.f. induced in each conductor will vary in direct proportion to the value of the field intensity through which it is passing. With a slotted armature the action is not quite so simple and will be discussed later, but the difference from a surface wound machine is only of minor importance.

It is therefore of importance to know the manner in which the density of the flux entering the armature varies around the machine. Naturally, the field is not confined to the portion immediately under the pole shoe, but there is a certain amount of "*fringing*" outside the tips of the shoes. The actual manner in which the field is distributed may be determined experimentally or may be calculated mathematically, and may be represented graphically by means of a *Flux Distribution Curve*, or, as it is sometimes called, a *Field Form.*

Flux Distribution Curve, and Fringing. The actual flux distribution will be dependent to a considerable extent upon the shape of the pole shoe, the relative magnitude of the arc subtended by the shoe, and upon the relative value of the air gap and the clearance between the tips of adjacent pole shoes.

In Fig. 3 is represented the flux distribution obtaining for a rectangular pole, without pole shoe, and with a uniform clearance between the face of the pole and the armature. The flux distribution curve is also shown, the ordinates of this latter giving the value of the flux density *entering* the armature from point to point around the machine.

The total value Φ of the useful flux per pole is then given by the product of the average flux density over the whole pole-pitch, and the area of the armature surface over one pole-pitch. Thus

$$\Phi = B_{av} \times \tau_p L \quad . \quad . \quad . \quad . \quad (3)$$

where τ_p represents the pole-pitch and L the axial length. This mean flux density is given by the average ordinate of the flux distribution curve.

Alternatively, the total value of the flux is sometimes stated as the product of the *average flux density under the pole shoe* and *the area under the pole shoe.* The drawback to the latter method lies in the fact that owing to fringing the value of the actual flux density under the pole shoe is always somewhat indefinite. In this book, then, we shall deal with the average flux density over the whole pole-pitch. As will be discussed later, the value of the iron losses in the armature teeth is determined largely by the *maximum* value

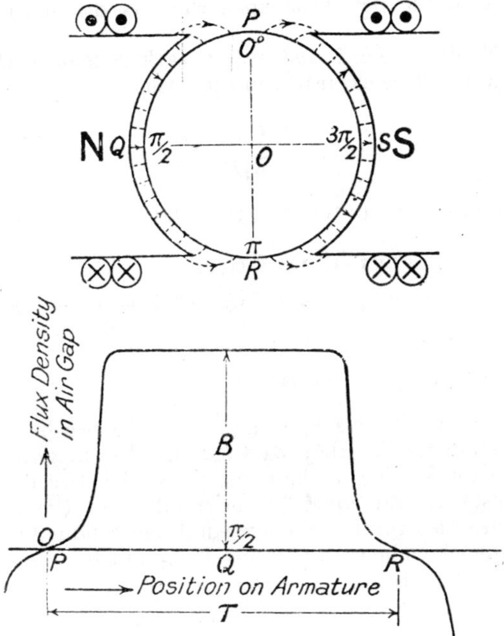

FIG. 3. FLUX DISTRIBUTION CURVE FOR 2-POLE MACHINE

of the flux density at which they are worked. This density normally obtains with an unloaded machine at the centre of the pole shoe, but with a loaded machine, owing to effects discussed later, it occurs towards the tip of the pole shoe. It is desirable, then, to have some means of expressing this maximum air-gap density in terms of the average density. For this purpose the term "field form factor" is used, defined by

$$\text{Field form factor} = \frac{\text{Mean air gap density over the whole pole-pitch}}{\text{Maximum density in air gap}}$$

$$= \frac{B_{av}}{B_g}$$

Denoting this factor by K_f, the value of the useful flux per pole is then

$$\Phi = K_f B_g \tau_p L \quad . \quad . \quad . \quad (4)$$

and

$$B_g = \frac{\Phi}{\tau_p L} \times \frac{1}{K_f} = \frac{B_{av}}{K_f} \quad . \quad . \quad . \quad (5)$$

The method of determining the value of the field form factor is discussed in the chapter on the magnetic circuit, where the influence of the shape of the pole shoe upon the flux distribution is also discussed.

Average Value of e.m.f. Induced in each Armature Conductor. This may variously be expressed in the forms—

$$E_c = B_{av} L v \quad . \quad . \quad . \quad (6)$$

$$E_c = K_f B_g L v \quad . \quad . \quad . \quad . \quad (7)$$

or it may be expressed in terms of the total flux per pole, its value being

$$E_c = \text{total flux cut per second}$$

$$= \text{flux per pole} \times \text{number of poles} \times \text{rev/sec}$$

$$= 2p\Phi n \quad . \quad . \quad . \quad . \quad . \quad . \quad . \quad (8)$$

where $2p$ is the number of poles.

With a commutator machine, and with the brushes placed to affect the conductors as they pass through the magnetic neutral position, the total value of the e.m.f. induced between brushes is sensibly constant, and equal to the product of the number of conductors in series and the average e.m.f. per conductor.

Total Value of Induced e.m.f. for Z_c conductors connected in series per circuit is then

$$E = Z_c E_c$$

$$= Z_c \times 2p\Phi n \quad . \quad . \quad . \quad . \quad . \quad (9)$$

or, in terms of the maximum flux density in the gap

$$E = K_f Z_c B_g L v$$

The value of Z_c is determined by dividing the total number of conductors on the armature, Z_a, by the number of paths in parallel, i.e. $2a$. That is,

$$Z_c = \frac{Z_a}{2a} \quad . \quad . \quad . \quad . \quad (10)$$

The number of parallel paths is determined by the type of winding employed, as discussed in Chapter III.

The total e.m.f. can therefore be expressed in terms of the total number of armature conductors as

$$E = \frac{2p}{2a} Z_a\Phi n = \frac{p}{a} Z_a\Phi n \quad . \quad . \quad . \quad (11)$$

or

$$E = \frac{1}{2a} K_f Z_a B_g L v \quad . \quad . \quad . \quad . \quad (12)$$

The expressions given for the e.m.f. induced in the d.c. machine hold equally for generators and motors, as it is quite immaterial whether the motion of the armature is due to a mechanical prime mover such as, for example, a steam engine, or due to the action of the machine as an electric motor.

Direction of Induced e.m.f. In a generator, the e.m.f. induced is in such a direction as to tend to establish a current which will give rise to a force opposing the prime mover.

In a motor, the direction of the e.m.f. is such as to oppose the flow of the current, these two statements being in accordance with a principle first enunciated by Lenz.

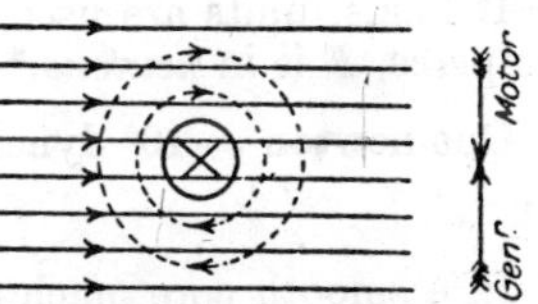

Fig. 4. Motion of Conductor carrying Current in Magnetic Field

Thus, if a conductor is carrying current in a magnetic field as indicated in Fig. 4, the current tends to establish a magnetic field in the direction shown by the dotted circles, the direction being clockwise when the current is flowing away and counter-clockwise when flowing towards the observer. (The corkscrew rule is a convenient memory aid.)

As a result the magnetic field, for the case shown, is strengthened at the top and weakened below the conductor, and a force is set up tending to move the conductor downwards. For a motor carrying current in the direction indicated, the conductor would then be moved downwards. Conversely to generate e.m.f. acting in the direction shown, the motion would have to be reversed as the force set up must now oppose and not create the motion.

As examples of e.m.f. calculations, the following may be taken—

1. Calculate the maximum value of the e.m.f. induced in the armature conductors of a machine driven at 300 rev/min, the external diameter of the armature being 180 cm and the core length 35 cm. The maximum flux density in the air gap is 1 Wb/m².

The surface speed $v = \pi \times 1{\cdot}8 \times 300/60 = 28{\cdot}3$ m/sec
Hence $e = 1{\cdot}0 \times 0{\cdot}35 \times 28{\cdot}3 = 9{\cdot}9$ volts

2. An 8-pole machine has a flux of 0·09 weber per pole. There are 768 conductors on the armature, and 8 circuits in parallel. Calculate the e.m.f. induced at 375 rev/min.

Flux cut per rev	$= 8 \times 0{\cdot}09 = 0{\cdot}72$ Wb
Flux cut per sec	$= 0{\cdot}72 \times 375/60$ Wb
	$= 4{\cdot}5$ Wb
Av. e.m.f. per conductor	$= 4{\cdot}5$ volts
No. of condrs. in series	$= 768/8 = 96$
Total induced e.m.f.	$= 96 \times 4{\cdot}5 = 432$ volts

3. Calculate the number of conductors required on the armature of a 4-pole motor, generating an e.m.f. of 420 volts at a speed of 950 rev/min. The flux per pole is 0·012 Wb, and there are two circuits in parallel.

Av. e.m.f. per conductor $= 4 \times 0{\cdot}012 \times 950/60 = 0{\cdot}76$ volt
No. of condrs. in series $= 420/0{\cdot}76 = 552$
Total No. of condrs., for
two paths in parallel $= 2 \times 552 = 1{,}104$.

Force on a Conductor carrying Current in a Magnetic Field. When a conductor carries current in, and at right angles to, a uniform magnetic field, a force is set up between the conductor and the field. This force is given by the product: length of conductor × density of the field × current is

$$F = BLI_a \quad . \quad . \quad . \quad . \quad (13)$$

If m.k.s. units are used, i.e. B in Wb/m², L in metres, and I_a in amperes, F is in *newtons*.*

One newton $= 10^5$ dynes $= 0{\cdot}1019$ kg wt
$= 0{\cdot}224$ lb wt

In a smooth core machine, the lines of force at the centre of the pole are practically radial, and therefore perpendicular to the armature conductors. The above formula will therefore hold for such a machine.

As an example, take the conductors on an armature 35 cm long, each carrying a current of 200 A, the flux density being 0·9 Wb/m². In this

$$F = 0{\cdot}9 \times 0{\cdot}35 \times 200 = 63 \text{ newtons}$$
$$= 0{\cdot}1019 \times 63 \text{ kg} = 6{\cdot}42 \text{ kg}$$
$$= 0{\cdot}224 \times 63 \text{ lb} = 14{\cdot}1 \text{ lb}$$

Work Done on Moving a Current Through a Magnetic Field. Due to the force set up between a conductor carrying current and the field, work is done if the conductor be moved in the direction opposed to the force. The work done is given by the product of the distance moved and the opposing force. If then the conductor be moved through a distance s in the uniform magnetic field the work done is

$$W = sF$$
$$= s \times BLI_a$$

In moving the distance s, the conductor cuts through a magnetic flux of $\Phi = sBL$. The work done in moving the conductor can therefore also be expressed in the form

$$W = \Phi I_a \quad . \quad . \quad . \quad . \quad (14)$$

* With c.g.s. units—gauss, centimetre and I_a in c.g.s. units of 10 A—the force is in dynes.

This expression holds not only for a field of uniform density, but for *all* cases, when a conductor carrying a current of I_a amperes is moved so as to cut a flux Φ, the work done is ΦI_a.

If m.k.s. units are used, i.e. Φ in webers, and I_a in amperes, W is in joules, i.e. watt sec.*

This may at once be applied to a d.c. armature. In one complete revolution of the armature, each conductor moves through a total flux of $2p\Phi$, Φ being the flux per pole, and $2p$ the number of poles. When the conductor passes from the influence of one pole to that of the next, the current in the conductor is reversed by the commutator, so that the total work done in one complete revolution for each conductor on the armature is

$$W_r = 2p\Phi I_a \quad . \quad . \quad . \quad . \quad (15)$$

Taking, as an example, an 8-pole machine having a flux per pole of 9×10^{-2} Wb, and a current of 200 A in each conductor, the work done for each conductor during one complete revolution is

$$W_r = 8 \times 0{\cdot}09 \times 200 = 144 \text{ watt sec}$$
$$= 550/746 \times 144 = 106 \text{ ft lb}$$

The total number of conductors on the armature being Z_a, the work done by the complete winding during each revolution is

$$W_r = 2p\Phi \times I_a Z_a \text{ watt sec} \quad . \quad . \quad . \quad (16)$$

The term $2p\Phi$ is simply the total flux entering and leaving the armature, and is often called the "*total magnetic loading*" of the machine.

On the other hand, $I_a Z_a$ represents the sum of the currents in all the conductors on the armature, and is often called the "*total electric loading*," or the "*total current volume*," or the "*total ampere conductors on the armature.*"

The work done for one complete revolution of the armature is then

$$W_r = (\text{total magnetic loading in webers} \times \text{total electric loading}) \text{ watt sec} \quad . \quad (17)$$

The work done in each complete revolution of the armature is then

$$W_r = 0{\cdot}737 \times 2p\Phi \times I_a Z_a \text{ ft lb}\dagger \quad . \quad . \quad (18)$$

It should be noted that these expressions hold equally for generator and motor action. In a motor, electrical energy is supplied to the armature and, as a result of the force set up between

* If Φ' is in maxwells and I'_a is in c.g.s. units W' is in ergs, i.e. in c.g.s. units $W' = s'B'L'I_a' = \Phi' I_a'$ ergs.

† One joule = one watt sec = 1/9·81 kg m = 0·1019 kg m
= 1/1·356 ft lb = 0·737 ft lb

the conductors and the magnetic field, actual mechanical work is done. In a generator, the force set up when the conductors carry current opposes the motion of the armature, and the prime mover does mechanical work in overcoming this force. This work, neglecting losses, is all converted into electrical energy.

If then the 8-pole machine discussed above (on page 9) had, in all, 768 conductors on the armature, the work done, neglecting losses, when the current in each conductor is 200 A aggregates during each complete revolution

$$W_r = 0{\cdot}737 \times 0{\cdot}72 \times (200 \times 768) \text{ ft lb}$$
$$= 81{,}500 \text{ ft lb}$$

Power Developed by the Armature. Power being defined as the rate of doing work, it follows at once that the power is determined by the product of the work done per revolution and n the number of revolutions of the armature per second, i.e.

$$P = n \times 2p\Phi \times I_a Z_a \text{ watts}$$
$$= 1{\cdot}34n \times 2p\Phi \times I_a Z_a \times 10^{-3} \text{ h.p.} \quad . \quad . \quad (19)$$

This expression gives the total value of the power developed—including losses—and does not, in a motor, represent the actual value of the power output.

To a rough approximation, the value of the total flux, and the total electric loading, is independent of the actual value of the working voltage of the machine, and not largely dependent upon the actual value of the speed of operation. To obtain a relatively large value for the power output of a machine it is then desirable that the angular speed shall be as high as possible. The armature is capable of doing a certain amount of work in each complete revolution; if it rotates very quickly it accomplishes this work quickly and is therefore relatively powerful.

It will be seen later that it is the value of the work done per revolution which determines the dimensions of a machine, and not solely the power output. Thus, for example, a 50 h.p. motor operating at 1,000 rev/min is of relatively small dimensions, but a 50 h.p. motor to operate at, say 60 rev/min, is relatively large.

Thus, for the 8-pole machine discussed above, the power developed at 375 rev/min, i.e. 6·25 rev/sec from equation (19) is

$$P = 1{\cdot}34 \times 6{\cdot}25 \times 0{\cdot}72 \times 153{,}600 \times 10^{-3} \text{ h.p.} = 925 \text{ h.p.}$$

At 500 rev/min, with the same flux and current, the power would be 1,234 h.p., and at 300 rev/min, no more than 740 h.p.

Torque Developed by the Armature. Since the work done per revolution is given by

$$W_r = 2\pi T \quad . \quad . \quad . \quad . \quad (20)$$

where T is the value of the torque, the expression for the torque follows at once from those already given for the work done in one complete revolution. Thus

$$T = W_r/2\pi = 2p\Phi \times I_a Z_a/2\pi \text{ newton at one metre} \quad (21)$$
$$= 0{\cdot}737 \times 2p\Phi \times I_a Z_a/2\pi \text{ lb ft}$$
$$= 0{\cdot}1173 \times 2p\Phi \times I_a Z_a \text{ lb ft} \quad (22)$$

Alternatively, the expression for the torque may be developed from that for the average force on each conductor. Thus

$$F_{av} = B_{av} I_a L$$

This acting at a radius of $D/2$ corresponds to a torque

$$T = 0{\cdot}5\, I_a B_{av} DL \text{ newton at one metre}$$

so that the total torque developed by the Z_a conductors is

$$T = 0{\cdot}5\, I_a Z_a \times B_{av} DL \text{ newton at one metre}$$

The total magnetic loading $2p\Phi$ is obviously equal to $B_{av} \times \pi DL$ whence

$$T = 2p\Phi \times I_a Z_a/2\pi$$

as already obtained. The numerical value for the torque is thus obtained by dividing that for the work done in one revolution by 2π, and it follows that the dimensions of an armature are determined by the value of the torque it has to develop rather than its output in horse power.

For the 8-pole machine previously discussed, the torque is then

$$T = 0{\cdot}1173 \times 2p\Phi \times I_a Z_a$$
$$= 0{\cdot}1173 \times 0{\cdot}72 \times (200 \times 768)$$
$$= 13{,}000 \text{ lb ft}$$

As already deduced, the power developed at 375 rev/min is 925 h.p. This value could also have been obtained from the torque. Thus

$$P = n \times 2\pi\, T \text{ ft lb/sec}$$
$$= 2\pi \times 13{,}000 \times 6{\cdot}25/550 \text{ h.p.} = 925 \text{ h.p.}$$

(In Ex. 2, page 7, the induced e.m.f. under these conditions is calculated to be 432 V. With 8 circuits in parallel, and 200 A per circuit the total current is 1,600 A. The total power is then 432 × 1,600 = 691 × 10^3 watts = 925 h.p.)

D.C. Machine as a Generator or Receiver of Electrical Energy. Consider, first, a voltaic battery of e.m.f. E connected to a supply system the voltage of which is maintained at the value V, the positive terminal of the battery being connected to the positive supply main, so that the e.m.f. of the battery acts in opposition to the supply voltage, as in Fig. 5. Unless the e.m.f. E is exactly equal to the applied voltage V, current will flow, the unbalanced voltage sending current out of the positive terminal of the battery being $E - V$. If r_b is the "internal" resistance of the battery, the current actually flowing out of the positive terminal of the battery is

$$I = \frac{E - V}{r_b} \quad . \quad . \quad . \quad . \quad (23)$$

If E is greater than V, the current actually flows in the direction in which E acts, and as a result the battery generates electrical energy, or "discharges."

If $E = V$, no current flows in the battery, and the battery is said to "float" on the supply system.

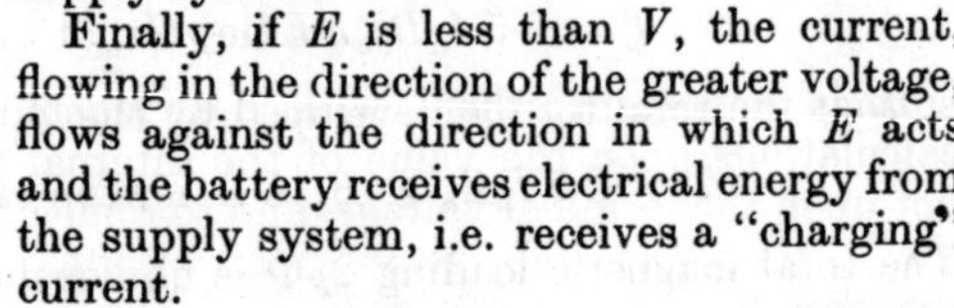

Finally, if E is less than V, the current, flowing in the direction of the greater voltage, flows against the direction in which E acts and the battery receives electrical energy from the supply system, i.e. receives a "charging" current.

FIG. 5. GENERATOR AND BATTERY IN PARALLEL

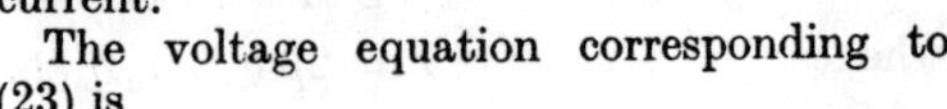

The voltage equation corresponding to (23) is

$$V = E - Ir_b \quad . \quad . \quad . \quad . \quad (24)$$

When I is positive, the current is discharging, and when negative, charging the battery.

By multiplying each term in (24) by the value of the current, the power equation is obtained

$$VI = EI - I^2r_b \quad . \quad . \quad . \quad . \quad (25)$$

Of these three terms, EI denotes the total power developed by the battery, I^2r_b is the power wasted in internal ohmic losses, and VI is the power output of the battery. When $V > E$, I is negative and the products EI and VI are both negative, indicating that the battery receives energy from the supply system.

Similar considerations hold for a d.c. machine. If the machine is connected to a supply system, and the voltage between its terminals is less than the induced e.m.f., the machine acts as a *generator* of electrical energy, and converts mechanical energy supplied by the prime mover into electrical energy. If, on the other hand, the voltage between terminals is maintained at a higher value than that of the induced e.m.f., the machine *receives* electrical energy and converts it into mechanical energy, and, in short, functions as a motor. The pressure equation for the d.c. machine is then

$$V = E - Ir_m$$

where r_m is the total resistance in series between the terminals. Again, when I is positive, the machine generates electrical energy, and when negative, acts as a receiver, i.e. as an electric motor. It is important to notice that the conversion of electrical into mechanical energy involves the flow of current against an opposing e.m.f., the e.m.f. being commonly called the "back e.m.f." An inspection of (25) at once establishes that if $E = 0$, the whole of the power corresponding to VI is transformed into ohmic losses.

In ordinary electrical practice it is a convenience to deal solely with numerical values and to express the voltage equation for the d.c. machine in two forms—

1. For *a generator*, numerically,

$$V = E - Ir_m \quad . \quad . \quad . \quad . \quad (26)$$

2. For *a motor*, numerically,

$$V = E + Ir_m \quad . \quad . \quad . \quad . \quad (27)$$

It is important, then, with d.c. machines, to base magnetic calculations upon the value of the internal or induced e.m.f., and not upon the terminal voltage, and to remember that in a generator, the induced e.m.f. is greater, and for a motor less than the value of the terminal voltage, the difference increasing in direct proportion to the value of the load current.

CHAPTER II

MAGNETIC CIRCUIT OF THE D.C. MACHINE

Laws of the Magnetic Circuit. A very brief revision of the laws of the magnetic circuit may be desirable.

It may then be stated that the laws governing the steady flow of electricity in a circuit may readily be modified so as to be at once applicable to the magnetic circuit. Thus

Magnetomotive force = Flux × Reluctance

$$F = \Phi \times R \qquad (28)$$

corresponding exactly to

Electromotive force = Current × Resistance

$$E = I \times R$$

Further

$$\text{Reluctance } R = \frac{\text{length}}{\text{area}} \times \frac{1}{\text{permeability}} = \frac{l}{A\mu} \qquad (29)$$

for a material of uniform sectional area corresponding exactly to

$$\text{Resistance} = \frac{\text{length}}{\text{area}} \times \frac{1}{\text{conductivity}}$$

(It is somewhat anomalous that, although generally we deal with the reluctance of a magnetic path, the specific term invariably used is "permeability.")

It often is convenient to calculate in terms of unit dimensions. We then have

$$\text{m.m.f. per unit length} = \text{Flux} \times \text{Reluctance per unit length}$$

$$= \frac{\text{flux}}{\text{area}} \times \frac{1}{\text{permeability}}$$

$$= \text{flux density} \times \frac{1}{\text{permeability}}$$

i.e.

$$H = \frac{B}{\mu} \qquad (30)$$

This corresponds exactly to

Volt drop per unit length = current density × resistivity

For a material carrying a uniform distributed flux, and of length l, the total m.m.f. is then equal to the m.m.f. unit length × l, i.e.

$$F = Hl \qquad (31)$$

For dealing with magnetic circuits in which the flux has to permeate a number of different portions "in series," the methods employed for dealing with series electric circuits can at once be applied, the total reluctance being the sum of the values for the various portions. Normally, the value of the reluctance is only of moment in so far as it permits of a value being determined for the m.m.f. required to establish a given flux in the circuit. It is often the simplest method to determine this value of the total m.m.f. by summing the values of the m.m.f. necessary to establish the flux through the various component parts of the circuit. This is exactly in line with the method of calculating the total voltage drop in an electric circuit by summing the values of the voltage drop in the various parts.

Thus the total value of the m.m.f. acting around a complete magnetic circuit is given by

$$F = \int H dl \quad . \quad . \quad . \quad . \quad (32)$$

or, where the circuit consists of a number of homogeneous parts, each of uniform cross section, and of lengths l_1, l_2, etc.

Total m.m.f.

$$F = \Sigma Hl = H_1 l_1 + H_2 l_2 + \ldots \quad . \quad . \quad (33)$$

If the values for the area and permeability of the various portions of the circuit are, respectively, A_1, μ_1, etc., the total m.m.f. becomes

$$F = \frac{B_1}{\mu_1} l_1 + \frac{B_2}{\mu_2} l_2 + \ldots \text{ where } B_1 = \frac{\Phi}{A_1}, \text{ etc.} \quad . \quad (34)$$

Occasionally it is convenient to express the fundamental law of the magnetic circuit in the form

Flux = magnetomotive force × *permeance* . (35)

Permeance is then the reciprocal of reluctance; for dealing with paths in parallel, the total permeance is the sum of the values for the individual paths.

Units. Magnetomotive Force. The practical unit of m.m.f. F, has been one ampere-turn since the beginning of this century, and little use has been made in practice of the c.g.s. unit. In the rationalized m.k.s. system unit m.m.f. is also one ampere-turn.

For magnetizing force, or m.m.f. per unit length H, the practical unit has similarly been one ampere-turn per centimetre. In the m.k.s. system the unit is one ampere-turn per metre—a very small unit, in contrast with the large unit of flux in this system.

The following relations hold—

In the rationalized m.k.s. system, and for many years in general use

$F^* = IN$ ampere-turns

In the m.k.s. system $H = IN/l$ ampere-turns/m

but much used in practice is $H'' = IN/l'$ ampere-turns/cm

* In the c.g.s. system $F' = 0{\cdot}4\pi IN$ gilberts,
and $H' = 0{\cdot}4\pi IN/l'$ oersteds.

The main difference between electric and magnetic calculations arises from the fact that whereas the resistance of an electric circuit is not directly dependent upon the value of the current flowing, and may be regarded as constant for normal purposes, the value of the reluctance of a magnetic substance is dependent to an enormous extent upon the value of the flux permeating it.

As a result, equation (34) has little, if any, practical application, reference being but very rarely made to actual permeability values for magnetic materials. On the other hand, equation (33) is very widely used indeed, in conjunction with magnetization curves such as Fig. 6, which give at once the magnetizing force necessary for any normal value of the flux density.

The various expressions for the magnetic circuit given above, equations (28) to (35), are of quite general application, and hold for all systems of units. Since $\mu = B/H$, its numerical value will depend on the units in which B and H are expressed. Thus

Using m.k.s. units (rationalized) $\Phi = F/R = IN\mu A/l$

Using c.g.s. units $\Phi' = F'/R' = 0{\cdot}4\pi IN\mu' A'/l'$

$$= 0{\cdot}4\pi IN\mu' \times 10{,}000A/100l$$

Since $\Phi'/\Phi = 10^8$, it follows that $\mu = (4\pi \times 10^{-7})\mu'$

Relative Permeability. The relative permeability μ_r is the ratio of the flux density B obtaining for a given value of H in a material, to that B_o obtaining in a vacuum—or free space—for the same value of H. For free space the relation is $B_o = \mu_o H$, so that the relative permeability $\mu_r = \mu/\mu_o$, and $\mu = \mu_r\mu_o$.

In the c.g.s. system μ_o' has been taken to be unity. It follows from the previous paragraph that in the rationalized m.k.s. system $\mu_o = 4\pi \times 10^{-7}$.

For all non-magnetic materials the relative permeability is, effectively, unity so that for all non-magnetic materials and, in particular, for air

$$\mu^* = 1{\cdot}0 \times \mu_o = 4\pi \times 10^{-7} \text{ in m.k.s. units}$$

Magnetization Curves for Iron. These curves, although in general not representing actual permeability values, are frequently referred to as permeability curves. The magnetic properties of iron and steel depend to a very large extent upon the chemical composition and previous history of the material. Slight changes in the composition may effect very marked changes in the magnetic properties, so that the graphs given in Fig. 6 must be regarded simply as representing average conditions, although manufacturers of dynamo sheet steel are prepared to guarantee, within narrow limits, the magnetic properties of the materials they supply.

* $\mu' = 1{\cdot}0\mu_o' = 1{\cdot}0$ in c.g.s. units.

In the design of dynamo electric machinery, magnetization curves are required to permit of the predetermination of the value of the m.m.f. necessary to establish a given flux density in various parts of a magnetic circuit. Since the value of the permeability is of use only in that it enables the value H to be calculated, in practice it is more convenient to have data plotted at once in terms of this magnetizing force and the flux density, and to omit all reference to actual permeability values. Magnetization curves may then be plotted in terms of the flux density in webers per square metre and

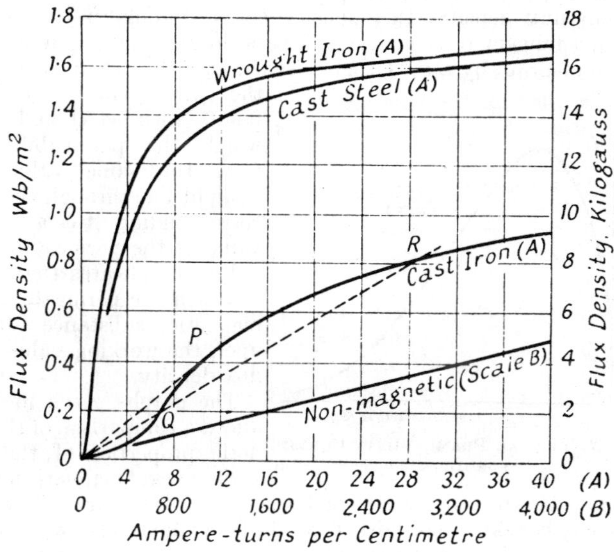

FIG. 6. TYPICAL MAGNETIZATION CURVES

the magnetizing force H in ampere-turns per metre, or ampere-turns per centimetre, as in Figs. 6 and 8. For use in the design of actual machines it may be more convenient to have the curves plotted in terms of other units. It is a very simple matter, as in Figs. 6 and 8, to provide duplicate scales so that the same set of curves can be used for flux densities in either webers per square metre or in gauss, and either ampere-turns per metre or ampere-turns per centimetre. (Some designers use inch units, and find it convenient to use magnetization curves based on flux densities in "lines per square inch," and magnetizing force in ampere-turns per inch.)

Since the value of the permeability is numerically equal to B/H, it follows that the magnetization curve for a substance having constant permeability is a straight line passing through the origin.

Non-magnetic substances have constant permeability, so that for such a substance*

$$H = B/\mu_o$$

hence $$H = B/4\pi \times 10^7 = 0{\cdot}796\, B \times 10^6 \text{ AT/m} \quad . \quad . \quad (36)$$

The magnetization curve for a non-magnetic substance is also included in Fig. 6, but it will be noted that for this the scale for the m.m.f. is such that 1 cm represents 100 times as much as for the magnetic substances. The shape of the magnetization curves given in Fig. 6 indicates that the permeability reaches a maximum at a certain flux density, this density corresponding to the point at which a tangent to the curve will pass through the origin, as shown in the diagram by OP. Other lines through the origin, and cutting the curve, will cut it at two points such as Q and R, for which the permeability will have the same value, each straight line through the origin corresponding to a definite value of the permeability.

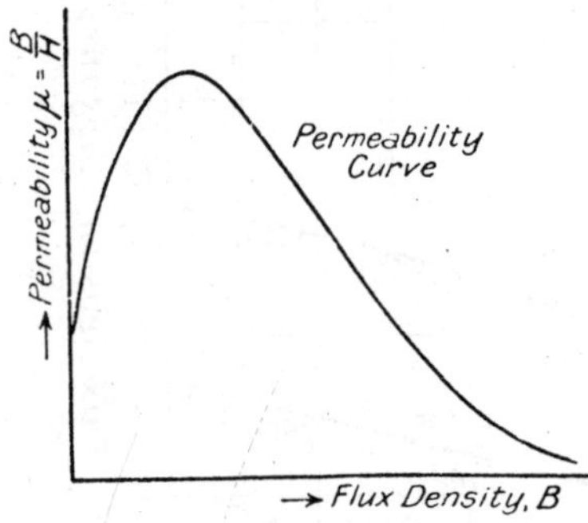

FIG. 7. TYPICAL PERMEABILITY CURVE FOR IRON

Fig. 7 indicates the manner in which the permeability of a magnetic substance depends upon the working value of the flux density.

The graphs given in Fig. 6 afford a comparison of the magnetic properties of the three main types of magnetic material employed in the construction of d.c. machines. The curve for wrought iron may be taken to be fairly typical where mild steel or wrought iron is used. For the other two substances it must be remembered that owing to differences in chemical composition, heat treatment, etc., appreciable variations may be expected with different castings. The curves for cast steel and cast iron indicate that to a rough approximation cast steel can be worked at double the flux density that is possible with cast iron. The use of cast iron for the frames of d.c. machines is therefore almost entirely abandoned in favour of steel, mainly on account of the saving in weight resulting thereby.

Intensity of Magnetization. For a non-magnetic substance $B_n = \mu_o H$, and for a magnetic substance the flux density exceeds this value by an amount M, such that

$$B = M + \mu_o H$$

* In c.g.s. units:

$$H' = B'/1{\cdot}0 = B' \text{ oersteds}$$

and $$H'' = 0{\cdot}796\, B' \text{ AT/cm}$$

where B' is in gauss.

The *intensity of magnetization* M is therefore given by $M = B - \mu_o H$, and differs but little from B until magnetic saturation is approached. Once the maximum intensity of magnetization M_{max} is reached, any further increase in the magnetizing force H simply produces the same *increase* in the flux density as would obtain for

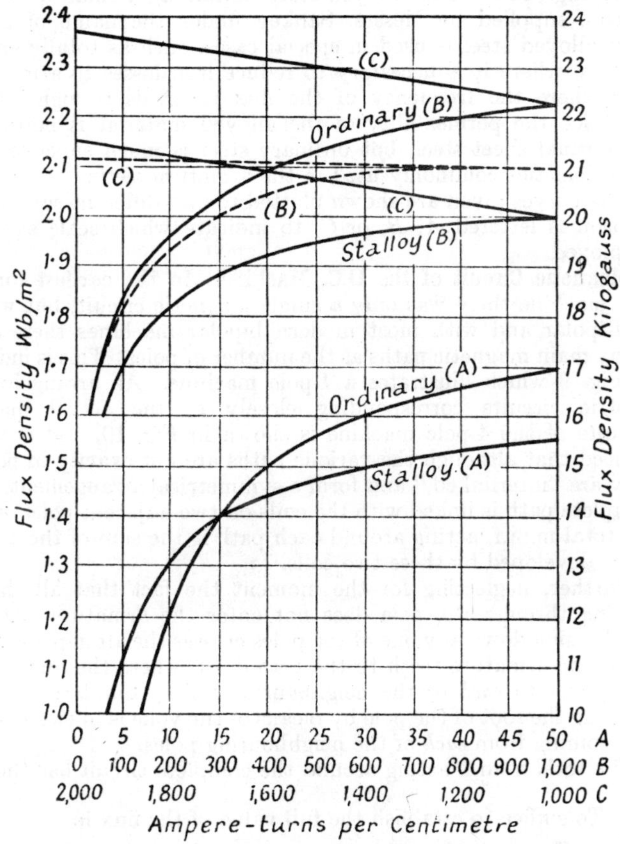

FIG. 8. TYPICAL MAGNETIZATION CURVES FOR ORDINARY AND STALLOY SHEET STEEL AS USED FOR ARMATURE CORES

(*Note.* Full lines refer to flux density. Dotted line refers to the intensity of magnetization for ordinary sheet steel, *see above*.)

a non-magnetic substance, i.e. at very high values of H, $B = M_{max} + \mu_o H$.*

In Fig. 8 the chain-line curve shows the manner in which M varies with the magnetizing force for ordinary dynamo sheet steel. The saturation value is about $M_{max} = 2{\cdot}1$ Wb/m^2.

* In c.g.s. units $M' = B' - 1{\cdot}0\, H'$, and at very high flux densities for ordinary sheet steel $B' \simeq 21{,}000 + 1{\cdot}25 \times H'$.

Magnetization Curves for Dynamo Sheet Steel. The normal working values of the flux density in the various parts of the magnetic circuit of d.c. machines range from about 1 to 2·2 webers per square metre. The highest values occur in the armature teeth.

Fig. 8 gives typical magnetization curves for ordinary dynamo stampings, and for an alloyed steel containing a small quantity of silicon, supplied by Messrs. Sankey under the name of Stalloy. This alloyed steel is used in special cases, such as totally enclosed motors, where it is necessary to reduce iron losses to a minimum, and where the frequency of the flux reversals is high. At low densities the permeability of the alloyed material is better than for normal sheet steel, but ordinary steel is much superior at the high densities commonly used in the armature teeth.

The curves given are shown plotted to two different scales; each portion is lettered *A*, *B*, or *C*, to indicate which scale should be employed.

Magnetic Circuit of the D.C. Machine. In the earliest forms of d.c. machine there was only a single magnetic circuit, but with all multipolar and with most modern bipolar machines there are as many main magnetic paths as the number of poles. This is indicated in Fig. 9 which illustrates a 4-pole machine. An arrangement of electric circuits corresponding closely to the actual magnetic circuits of the 4-pole machine is shown in Fig. 10, and it will be noticed that although the various paths are not exactly in parallel, they are "interlinked," and form a symmetrical arrangement. Each complete path is linked with the coils on two adjacent poles, so that the total m.m.f. acting around each path is the sum of the ampere-turns developed by these two coils.

Further, neglecting for the moment the fact that all the flux passing through the pole does not enter the armature, the total flux passing down any one of the poles crosses the air gap and passes down the armature teeth to the core itself, where the flux divides, half going to *each* of the neighbouring poles. Similarly, the flux entering the root of the pole by means of the yoke is in two portions, half coming from *each* of the neighbouring poles.

The total m.m.f. acting around the complete circuit has then—

1. To suffice to establish the full value of the flux in
   (*a*) Two poles in series;
   (*b*) Two air gaps in series;
   (*c*) Two sets of armature teeth in series; and

2. To suffice to establish one half the full value of the flux in
   (*d*) One section of the armature core, and
   (*e*) One section of the yoke.

Since this total m.m.f. includes the effect of two field coils, the

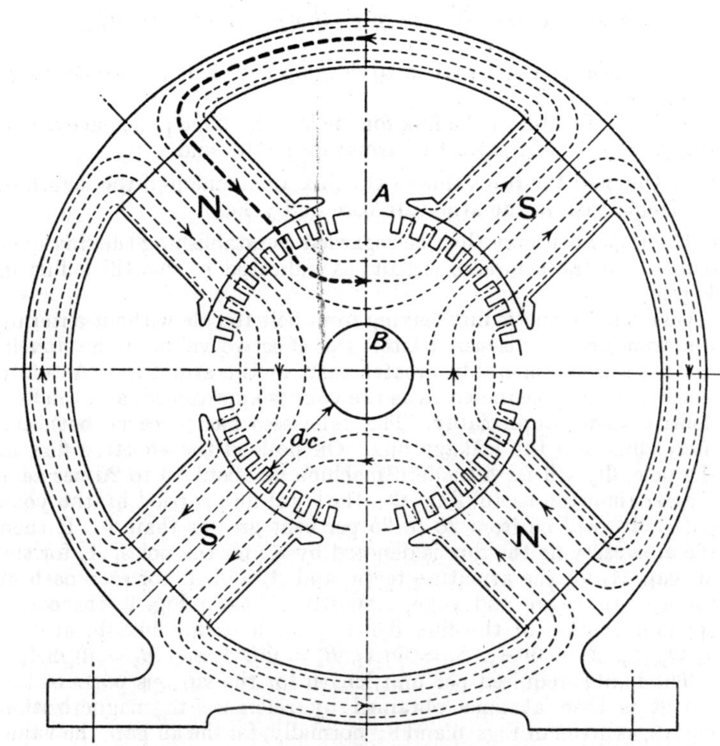

FIG. 9. MAGNETIC CIRCUIT OF 4-POLE MACHINE

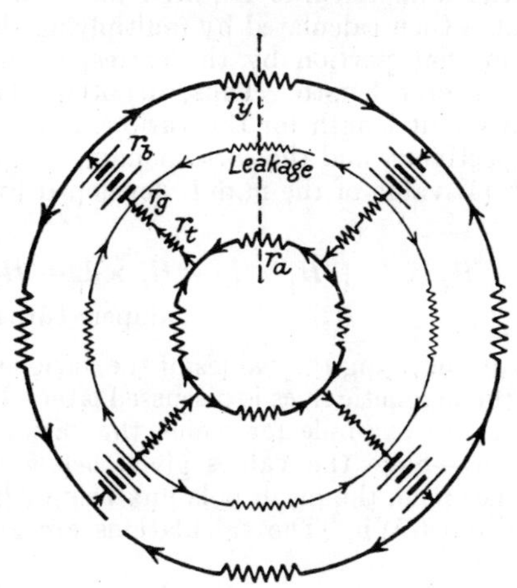

FIG. 10. ELECTRIC CIRCUIT ANALOGOUS TO MAGNETIC CIRCUIT OF 4-POLE MACHINE

duty of ***each*** field coil is then to develop sufficient ampere-turns to establish—

1. The full value of the flux for the length of one pole, once across the air gap, and in one set of armature teeth; and

2. One half tho full value of the flux along one half the length of the single path in the armature core and yoke.

Each field coil may then be regarded as establishing the flux over one half of the complete circuit, as indicated by the thick line in Fig. 9.

Owing to magnetic flux leaking from pole to pole without entering the armature, as discussed later, the effective value of the flux in the poles is considerably greater than in the armature. Actually the flux in the pole increases as the yoke is approached, and the flux density varies appreciably. The yoke also has to carry both the useful flux and the leakage flux. On no-load the effective flux in the pole, Φ_p will for a modern machine be about 15 to 20 per cent greater than the useful flux, Φ; the total flux carried by the yoke paths, Φ_y, will be from 20 to 25 per cent greater than Φ. If then the area carrying the flux is denoted by A_p for the poles, A_g for the air gap, A_t for the armature teeth, and A_c and A_y for *each* path in the armature core and yoke, respectively, as in Fig. 9, the corresponding values of the flux density, for a useful flux Φ, are $B_p = \Phi_p/A_p$, $B_g = \Phi/A_g$, $B_t = \Phi/A_t$, $B_c = \Phi/2A_c$ and $B_y = \Phi_y/2A_y$.

The m.m.f. required per unit length for the various parts of the circuit is then at once obtained by reference to magnetization curves, as given in Figs. 6 and 8; normally, for the air gap, the value is calculated as already indicated, i.e., $= 0{\cdot}796 \times 10^6 B_g$ AT/m. The value of the ampere-turns required for each portion of the magnetic circuit is then calculated by multiplying the length of the magnetic line in that portion by the corresponding value of the ampere-turns per unit length. Thus, denoting the value of the ampere-turns per unit length for the various parts by H_p, H_g, H_t, H_c, and H_y respectively, and the corresponding lengths, by l_p, l_g, l_t, l_c, and l_y, the total value of the m.m.f. developed by each field coil is then

$$F_f = H_p \times l_p + H_g \times l_g + H_t \times l_t + H_c \times l_c + H_y \times l_y \quad \text{ampere-turns per pole} \qquad (37)$$

The method of obtaining the values of the various sectional areas and lengths of the magnetic lines is discussed later. For the moment we will work out an example for which the various dimensions—areas and lengths—have the values given below, the useful flux being 0·08 Wb per pole, the mean pole flux being 0·092 Wb and the total yoke flux 0·096 Wb. The calculations are given in tabular form.

The values of the densities, and the dimensions generally, in this table are fairly typical of a machine of medium size. The factor

Part	Area cm² m² × 10⁻⁴	Flux density Wb/m² gauss × 10⁴	H'' AT/cm	Length cm	F
Air gap	842	0·95	7,560	1·15 × 0·5	4,350
*Armature teeth (sheet steel)	381	2·1	400	4·8	1,920
*Armature core (sheet steel)	2 × 333	1·2	10	16	160
*Pole (sheet steel)	575	1·6	30	20	600
†Yoke (cast steel)	2 × 400	1·2	7	30	210
		Total ampere-turns per pole F_f =			7,240

* cf. Fig. 8. † cf. Fig. 6.

1·15, by which the air gap dimension, 0·5 cm, is multiplied, is an approximate allowance for the increased reluctance of the air gap due to the armature slots. Once the values of the flux densities in the various parts of the magnetic circuit are known, the method of computing the total m.m.f. is thus extremely simple.

Dealing now in greater detail with each portion of the magnetic circuit.

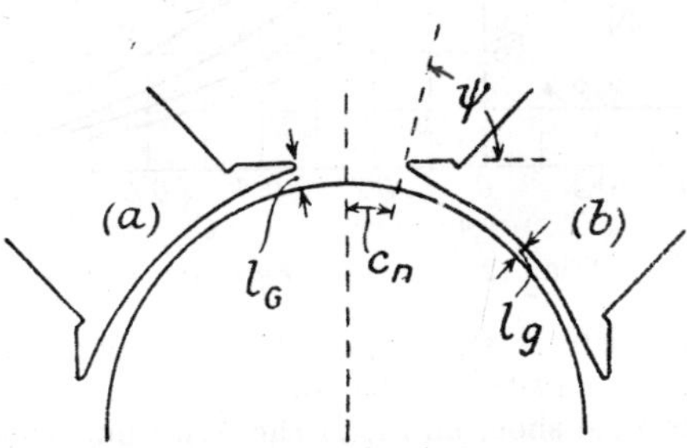

FIG. 11. POLE PROFILES

Air Gap. The exact calculation of the reluctance of the air gap is a matter of difficulty owing to—

(1) The gap dimension not being constant over the whole pole-pitch, and

(2) The fact that modern armatures are toothed, so that the flux tends to concentrate upon the teeth.

In (1) not only is the effect of the pole fringe included, but, in addition, the dimension of the gap under the pole shoe is not constant. With d.c. machines the pole shoe is generally of the form shown in Fig. 11 (*a*) but may have the form shown in Fig. 11 (*b*). In either case it will be noted that the dimension of the gap at the tips is greater than that at the centre of the shoe, 50 per cent greater being a common figure.

Flux Distribution Curve, or Field-form. To obtain the value of the reluctance of the air gap it is necessary to know the distribution of the magnetic field in the air gap.

It is necessary therefore to determine the shape of the flux distribution curve, or field form.

For a smooth core armature, in which the iron parts are unsaturated, there is no difficulty in determining the manner in which the flux density varies at any portion of the armature surface lying well under the pole shoe, for it may be taken that the flux density

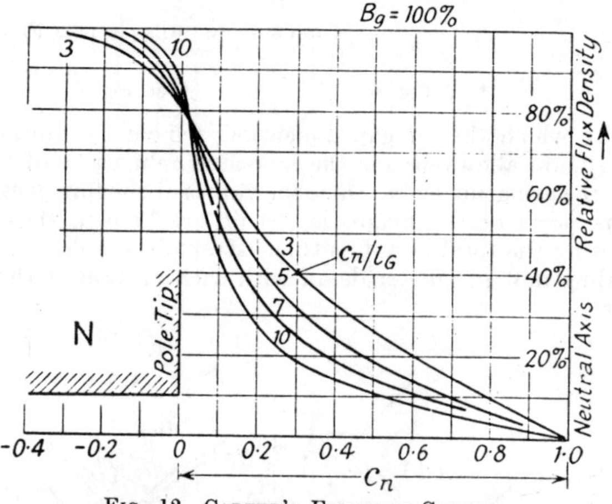

FIG. 12. CARTER'S FRINGING CURVES

will vary inversely as the length of the air gap. Thus if B_g is the maximum flux density in the air gap, and occurring under the centre of the pole shoe, and l_g is the gap dimension at that point, the flux density at any other point where the gap dimension is l_{gx} will be $B_g \times l_g/l_{gx}$.

When a toothed armature is considered, it becomes necessary to take magnetic saturation into consideration, as shown later.

For the flux distribution in the inter-polar region, no such simple calculation is possible. To determine the magnitude of this "fringing" effect, much experimental work has been done, and F. W. Carter attacked the problem mathematically. Carter's mathematics have been amply verified by experimental results, and most designers make use of the results published by him.

These results have been given in graphical form and indicate the manner in which the flux density at the armature surface varies over the whole pole-pitch, and in particular outside the pole shoe. The actual shape of the fringing portion of the flux distribution curve is dependent upon the proportions of the machine, i.e. upon the ratio that the distance from pole tip to neutral axis bears to the air gap dimension at the pole tip—this ratio being C_n/l_G, as in Fig. 11. Actually, the density in the air gap begins to fall off under the pole shoe, a little before the actual tip is reached, as indicated by the curves given in Fig. 12. For values of C_n/l_G not covered by the curves plotted in Fig. 12, the method of interpolation may be used, and the appropriate "fringe" can thus be fitted to any particular field form, and the flux distribution curve readily obtained for the case of an unsaturated machine.

This may be illustrated by considering a pole shoe of the form indicated in developed view in Fig. 13, there being a slight bevel to

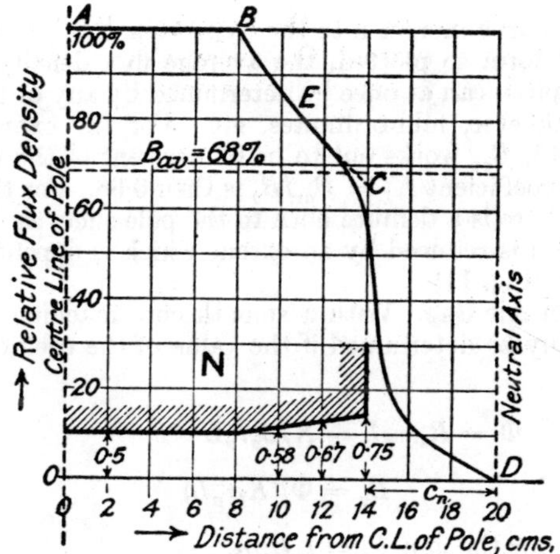

FIG. 13. FIELD FORM (NO SATURATION)

the shoe. The figure refers to a machine having a pole-pitch of 40 cm, a minimum air gap of 0·5 cm, a bevel of 6 cm and a maximum gap of 0·75 cm. The pole arc is 70 per cent of the pole-pitch, so that the distance from pole tip to neutral axis is 6 cm. The value C_n/l_G is thus $6/0{\cdot}75 = 8$, and the corresponding fringing curve will be intermediate between those marked 7 and 10 in Fig. 12, and actually just slightly below the 7 curve.

The relative flux density at various points outside the shoe can then be deduced by multiplying the values as obtained from Fig. 12, by the ratio $l_g/l_G = 0{\cdot}5/0{\cdot}75 = 0{\cdot}67$.

Neglecting the effect of the fringe, the density at the tip will be 0·67 of the maximum value.

At 4 cm from the tip the gap is $0{\cdot}5 + (2{\cdot}5 \times 1/3) = 0{\cdot}583$ cm, and at 2 cm from the tip, the gap is 6·67 mm. The corresponding values of the flux density are then 5/5·83 and 5/6·67, i.e. 86 per cent and 75 per cent, respectively, of the maximum value. The corresponding portion of the flux distribution curve is then represented by the hyperbola *BEC*. The values of the fringing field density can conveniently be found at 5 equidistant points between the pole tip and neutral axis i.e. at 14, 15·2, 16·4, 17·6, and 18·8 cm, respectively, from the centre line of the pole, as indicated by the following table—

Distance from pole tip, cm.	0	1·2	2·4	3·6	4·8	6·0
Relative density, from Fig. 12 for $C_n/l_g = 8$	0·83	0·33	0·175	0·10	0·05	0
Actual density	55%	22%	11·8%	6·7%	3·3%	0

The fringing curve merges into the hyperbola *BEC*, as indicated. From the field form so plotted, the average flux density over the complete pole-pitch can at once be determined by any of the known methods—planimeter, mid-ordinates, etc. For the example illustrated in Fig. 13, B_{av} works out to be 68 per cent of the maximum value, and the coefficient $K_f = B_{av}/B_g$ is then 0·68. For the normal case in which there is a definite horn to the pole shoe, as in Fig. 11, the fringing field is reduced by an extent which is dependent upon the angle ψ (see Fig. 11).

Reluctance of Air Gap. With a smooth core machine the reluctance can readily be determined if the value of the coefficient K_f is known,

for $$\Phi = B_{av}\tau_p L = K_f B_g \tau_p L$$

hence $$B_g = \Phi/(K_f \tau_p L)$$

and the m.m.f. for the gap $F_g = l_g B_g/\mu_0$

whence the reluctance

$$R_g = F_g/\Phi = \frac{1}{\mu_0} \times \frac{l_g}{K_f \tau_p L} \quad . \quad . \quad (38)$$

That is, the actual reluctance of the air gap is obtained by dividing K_f into the value of the reluctance of a uniform gap of length l_g, extending over the complete area of the pole-pitch, $\tau_p L$.

Calculations for the reluctance of a smooth armature can therefore be based upon the value of the air gap at the centre of the pole shoe, and upon an equivalent area given by

$$A_g = K_f \times \tau_p L \quad . \quad . \quad . \quad (39)$$

Naturally, the designer need not determine the value of K_f for each individual machine, as the value will differ but little for machines designed on the same standard lines.

In d.c. machines, it is not essential to know the actual wave shape of the fringing field, so long as its total effect upon the reluctance of the gap is known. Carter has therefore expressed his results in a form which readily gives the actual permeance of the paths of this field. John Hopkinson, in his classic paper on dynamo electric machinery, allowed for fringing by adding to the dimensions of the pole arc and the axial length a constant amount of 1·6 times the gap dimension. Carter, from his fringing curves has expressed the allowance which should thus be made as a function of the gap length l_G and of the ratio C_n/l_G.

Fig. 14 gives Carter's "permeance factor" K_a for each fringe of the pole arc, in terms of the ratio C_n/l_G, where l_G is the radial air

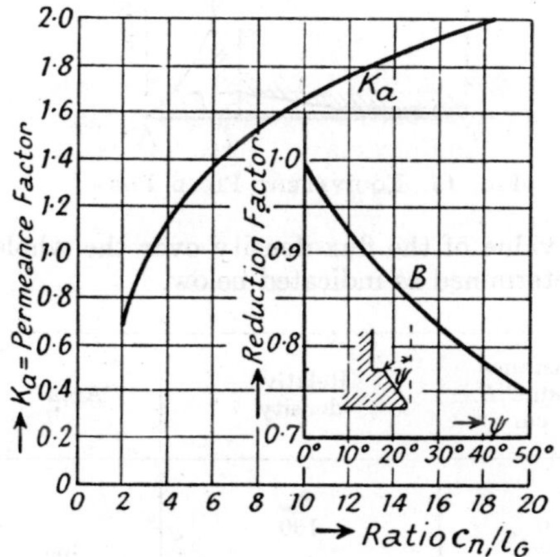

FIG. 14. CARTER'S CURVE FOR THE PERMEANCE FACTOR FOR THE FRINGE

gap at the pole tip, the effect of the fringe being equivalent to extending the pole arc by an amount $K_a l_G$ at each pole tip.* The actual field form may then be regarded as equivalent to the field form *ABCDE* shown in Fig. 15.

Fig. 14 gives the value of the fringe permeance factor for the type of pole indicated; when, as is usual, there is a definite polar horn the permeance is reduced, roughly as indicated in Fig. 14 by curve *B*, which shows the reduction factor for various values of the angle ψ. As an example of the use of the curves given in Fig. 14 the coefficient K_f will be determined for the case indicated by Fig. 13.

* Actually K_a is the permeance determined by Carter of the fringing field from pole tip to neutral point on the armature, expressed in c.g.s. units per cm of axial length.

The permeance per metre axial length for each fringe is $K_a/l_0 = 4\pi \times 10^{-7} K_a$ m.k.s. units.

The ratio $C_n/l_G = 8$, hence, from Fig. 14, $K_a = 1{\cdot}53$. The amount to be added to each end of the pole arc is then, as in Fig. 15, $K_a \times l_G = 1{\cdot}53 \times 0{\cdot}75 = 1{\cdot}15$ cm.

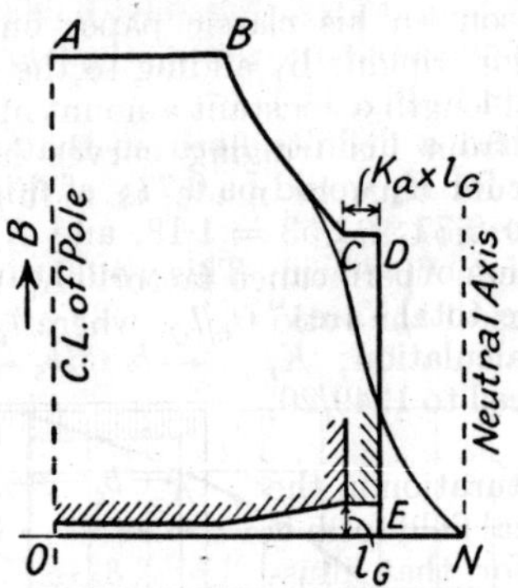

FIG. 15. EQUIVALENT FIELD FORM

The average value of the flux density over the whole pole-pitch may then be determined as indicated below.

Distance from *C.L.* cm	Relative density	"Area"
0	% 100	
		200
	100	
		200
	100	
		200
	100	
		200
8	100	
		186
10	86	
		161
12	75	
		142
14	67	
		77
15·15	67	
15·15	0	
		0
20	0	
	Total "area"	1,366

Av. Value of Relative density = 1,366/20 = 68·3%.

The "area" represents, for the major part, the area of a strip of the flux distibution curve 2 cm wide; for the "equivalent fringe" the width of the strip is 1·15 cm.

The field form coefficient K_f is therefore 0·683. (This value accords well with the 0·68 found by the method of mid-ordinates from the actual field form in Fig. 13.)

If now the pole tip had a definite horn so that the angle ψ is 45°, the reduction factor for K_a would be 0·77; cf. Fig. 14, curve *B*.

K_a is thus reduced to $0{\cdot}77 \times 1{\cdot}53 = 1{\cdot}18$, and $K_a \times l_G$ to $1{\cdot}18 \times 0{\cdot}75 = 0{\cdot}89$, a reduction of 0·26 cm. This would cause a reduction of $0{\cdot}26 \times 67 = 17$ in the total "area" given in the above calculation; K_f would therefore be reduced to 1349/20, i.e. to 67·5 per cent.

Owing to magnetic saturation in the armature teeth, the actual field form is rather "flatter" than for that illustrated in Fig. 13, as discussed later on page 48.

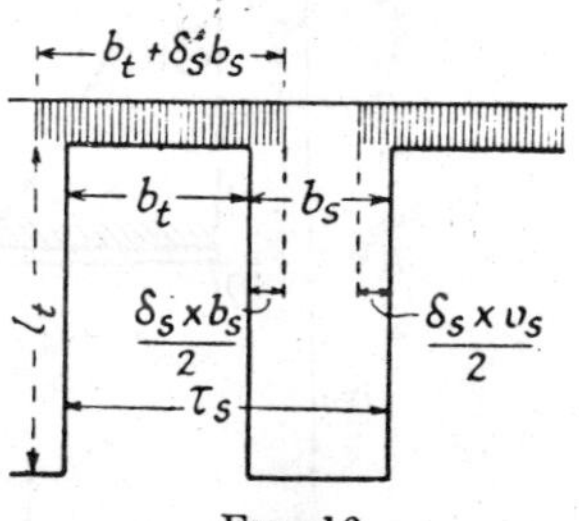

FIG. 16

Reluctance of Air Gap with Slotted Armatures. It is obvious that the air gap reluctance with slotted armatures will exceed that of corresponding smooth core machines. Much experimental and analytical work has been published on this subject, and again most designers utilize results developed by F. W. Carter.

A convenient method is to assume that the flux is distributed quite uniformly over the whole of each slot pitch, excepting a fraction of the slot width, as indicated in Fig. 16. This fraction will depend upon the ratio of the slot width to the air-gap width.

Over each slot pitch, then, the flux may be assumed to be carried by a width $b_t + \delta_s b_s$, where b_t and b_s are the tooth width and slot width respectively, and δ_s is a "gap coefficient" as deduced by Carter. The manner in which this coefficient depends upon the ratio b_s/l_g is indicated in Fig. 17.

As compared with a smooth core machine, then, the reluctance of the air gap for a slotted armature will be increased in the ratio

$$\frac{\text{Reluctance with slotted armature}}{\text{Reluctance with smooth core}} = K_{gs} = \frac{b_t + b_s}{b_t + \delta_s b_s}$$

The coefficient K_{gs} may then be defined as the "air-gap coefficient" for the slots.

If τ_s denotes the slot pitch, the coefficient may also be expressed

$$K_{gs} = \frac{\tau_s}{\tau_s - b_s(1 - \delta_s)} \qquad . \qquad . \qquad . \qquad (40)$$

Fig. 17 is constructed so that the value $(1 - \delta)$ can at once be read off.

Effect of Ventilating Ducts upon Reluctance. The effect of the ventilating ducts* can be allowed for in a manner precisely similar to that of the slots. Fig. 17 can then be used for various values of the ratio: Width of vent duct/width of gap.

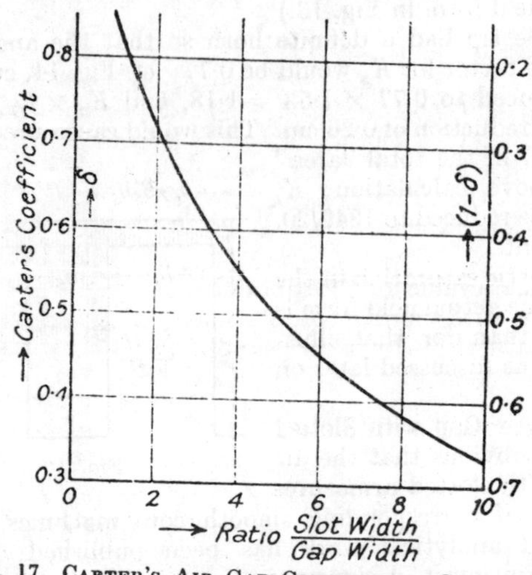

FIG. 17. CARTER'S AIR GAP CONTRACTION COEFFICIENT

The reluctance of the air gap for a machine with vent ducts will then be greater than that for a machine without ducts in the ratio

$$K_{gv} = \frac{L}{L - n_v b_v (1 - \delta_v)} \quad . \quad . \quad . \quad (41)$$

where L is the total, or gross length of core, n_v is the number of ventilating ducts, b_v is the width of each duct and δ_v is the coefficient obtained from Fig. 17.

Since the axial armature length is usually made rather greater than that of the poles, there is no need to allow further for fringing at the ends of the core.

The total value of the air-gap coefficient, including the effects of both the armature slots and the ventilating ducts is then

$$K_g = K_{gs} \times K_{gv} \quad . \quad . \quad . \quad . \quad (42)$$

A fairly normal value of K_g is from 1·15 to 1·18 for fairly large d.c. machines with open slots, and up to 1·25 for machines with a very small air gap. In practice it is a convenience for the designer to have the value of the coefficient K_{gs} calculated and plotted for various ratios of the slot pitch/slot width, and slot width/gap width.

* See Fig. 18, p. 31.

The total reluctance of the air gap for a slotted armature is then

$$R_g = K_g \times \frac{l_g}{\mu_o K_f \tau_p L} \quad . \quad . \quad . \quad (43)$$

In this expression, $l_g/\mu_o\tau_p L$ is the reluctance that would obtain for a smooth core machine having a uniform air gap l_g over the whole pole-pitch; K_f takes account of the fact that a definite or salient pole is used, and allows for fringing at the pole tips and bevel of the shoes; K_f is numerically equal to the ratio of the average density over the whole pole-pitch to the maximum density under the centre of the shoe; K_g takes account of the fact that the armature is slotted, and allows also for the effect of the ventilating ducts.

An approximate value for K_f is obtained, for a machine with normal proportions and degree of saturation, by taking, simply, K_f = pole arc/pole-pitch. For precise calculations, the actual value of the coefficient should be determined.

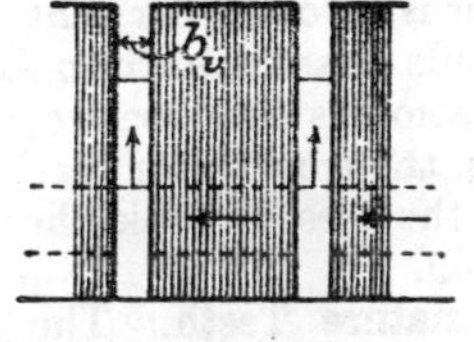

FIG. 18. VENTILATION DUCTS

Air-gap Ampere-turns. There is no need in practice to work out the actual reluctance of the air gap, the air gap correction coefficient K_g being used as a simple "correcting" factor to the calculation based on a smooth-core machine. Thus

$$F_g{}^* = K_g \times B_g l_g/\mu_o$$
$$= 0{\cdot}796\, K_g B_g l_g \times 10^6 \quad . \quad . \quad . \quad (44)$$

with m.k.s. units for B_g and l_g.

Active Iron Length. The armature core of a d.c. machine is built up of a number of plates, or laminations, lightly insulated from each other by paper, varnish or some other suitable material. These steps are taken to reduce eddy currents to reasonable proportions. Further, to keep down the temperature of the active parts, it is necessary to ventilate the core, and this usually is effected by subdividing the core into packets of 5 to 7 cm in width, separated at intervals by distance pieces called "vent spacers" so as to leave a ventilating duct through which air can pass. These ducts are radial, as indicated in Fig. 18, and are from 8 mm to 10 mm in width.

Of the total gross core length, then, a certain proportion is taken up by the ventilating ducts, and of the other part only a portion is iron, the remainder being the insulation on the core plates and an amount due to the springiness and irregularities in the thickness of the core plates. It is usual to define the space factor of the iron in

* $F_g = 0{\cdot}796\, K_g B_g' l_g'$ with B_g' in gauss and l_g' in cm (cf. page 18).
In these expressions B_g and B_g' denote the maximum flux density in the air gap, for a smooth-core armature.

the core as the "iron factor"; this "iron factor" is the ratio of the axial length of actual iron in a bunch of assembled core plates to the total axial length of the bunch. So that the total axial length of actual iron in the core is given by

$$L_i = \text{iron factor} \times (\text{gross length} - \text{total width of all the ventilating ducts})$$
$$= k_i\,(L - n_v b_v) \qquad . \quad . \quad . \quad (45)$$

The value of the iron factor depends largely upon the thickness of the core plates used and upon the type of insulating material; for normal stampings 0·5 mm thick and paper insulation, the iron factor is 88 or 89 per cent. With the plates insulated by means of a suitable enamel, an iron factor of 0·92 can be obtained. Where no insulation is used between plates, as for example with laminated poles, the iron factor reaches the value of 95 per cent.

In the present book, the iron factor in armature cores will be taken as 0·89.

Armature Teeth. The exact calculation of the ampere-turns necessary to establish the flux in the teeth is a matter of difficulty owing to

(1) The fact that the teeth are wedge-shaped or tapered; and

(2) The fact that the teeth are frequently worked at such a high flux density that their permeability is very small, and, as a result, an appreciable portion of the flux is diverted and passes down the slot itself.

Calculations for Tapered Teeth. The total m.m.f. required for the teeth is given by $\int H dl$, the integration being carried out for the complete height of the teeth, l_t. To obtain correctly the value of the required m.m.f. it is therefore necessary to construct a graph showing the manner in which the magnetizing force varies over the height of the tooth. The mean ordinate of such a graph gives the equivalent magnetizing force for the whole of the tooth.

To determine this magnetizing force graph it is necessary to construct first a graph showing the manner in which the flux density varies from the root to the tip of the tooth. Since the flux density varies inversely with the sectional area, for a tooth having a simple taper as in Fig. 19, the density will follow a hyperbolic and not a linear law. Thus, for the case of a 2 : 1 taper, and a maximum density of 2 Wb/m², the flux density midway up the tooth is *not* 1·5 but 2/1·5 = 1·333 Wb/m².

In general, if b_{min} is the minimum tooth width, and b is the width at a height h from the root, $b = b_{min}\,(1 + ah)$ where a is some constant.

The flux density at the root being B_{max}, its value at any height h is then: $B = \dfrac{B_{max}}{1 + ah}$.

The flux density graph being plotted, that representing the magnetizing force is at once obtained from the magnetization curve for the stampings in use. Fig. 19 represents a tooth having a taper of 1·4 and a height of 4 cm, the minimum width being 1 cm and the maximum 1·4; B_{max} is taken as 2·1 Wb/m². The necessary calculations for obtaining the magnetizing force graph are shown in tabular form. The values for the m.m.f. per cm corresponding

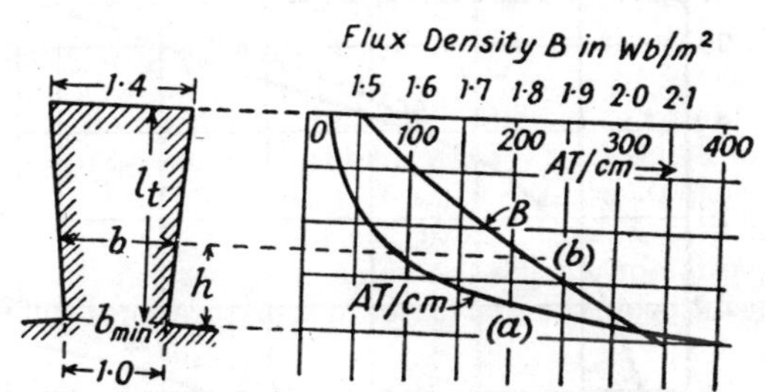

Fig. 19. Calculation of m.m.f. for Tapered Tooth

to the various values of *B* are read off from the magnetization curve in Fig. 8, and the average value of this m.m.f. per cm is calculated as shown.

CALCULATION OF M.M.F. FOR TAPERED TOOTH

Height from root cm	Width cm	Flux density Wb/m²	Magnetizing Force AT/cm	Mean Magnetizing Force AT/cm
0	1·0	2·1	400	
				275
1	1·1	1·91	150	
				108
2	1·2	1·75	65	
				48
3	1·3	1·615	30	
				25
4	1·4	1·5	20	
		Total ampere-turns for tooth		**456**

Equivalent: ampere-turns per cm = 456/4 = 114.

As an alternative method, Simpson's rule can be used to determine the average magnetizing force. Thus, in the above example, the limiting flux densities correspond to magnetizing forces of 400 and 20 AT/cm respectively, and for the mid-section the value is 65 AT/cm. Hence the mean magnetizing force is 1/6 (400 + 20 + 4 × 65) = 680/6 = 113 AT/cm.

To simplify such calculations Fig. 20 is given in which the two scales given for the magnetizing force represent respectively $H'/6$ and $4\,H'/6$, so that the three readings taken from the curve have simply to be added.

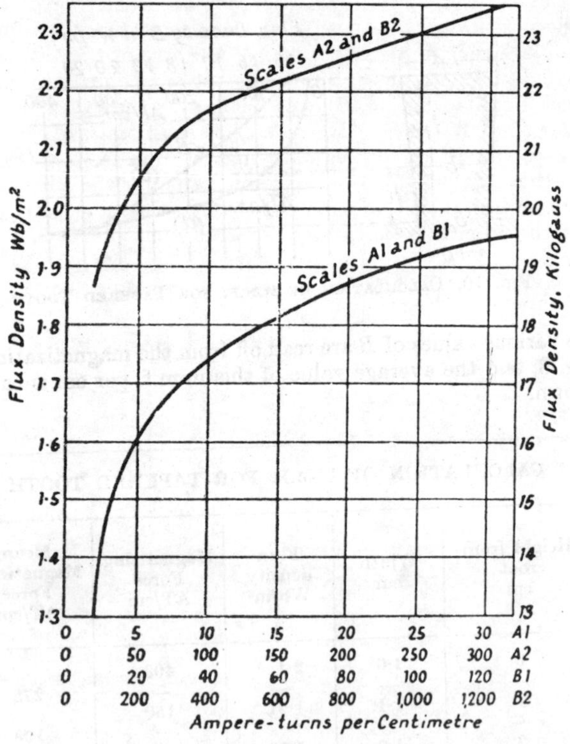

FIG. 20. MODIFIED MAGNETIZATION CURVE FOR ORDINARY SHEET STEEL FOR SIMPSON'S RULE

Thus, using scale A at $B = 2{\cdot}1$ and $1{\cdot}5$ Wb/m², respectively we read off 67 and 3, and using scale B we read off 43 at $B = 1{\cdot}75$ Wb/m². The mean value of the magnetizing force is then $67 + 3 + 43 = 113$ AT/cm.

The magnetization curve Fig. 8 shows that 114 ampere-turns per cm correspond to a flux density of 1·85 Wb/m². This occurs at a width of $1 \times 2{\cdot}1/1{\cdot}85$, i.e. 1·135 cm, which is approximately at a point one-third of the way up the tooth, as measured from the root.

For ordinary purposes, in d.c. machine design, it is then customary to base all calculations upon the density obtaining one-third of the way from the root of the tooth, and not, as might casually be thought, upon the density at the mid-tooth section, or even the average of the minimum and maximum densities. This method is, of course, not exact, but is generally sufficiently accurate for direct current machines.

With the teeth, as with the air gap, the flux density varies according to the position of the tooth with relation to the pole shoe, and is maximum when the tooth is under the centre of the shoe.

The average density in all the teeth over one complete pole-pitch is obtained by dividing the value of the flux by the section of all the teeth over the pole-pitch. Owing to the reasons already given for the air gap, the effective area carrying the flux is, however, only K_f times this total area. By taking this value for the effective area carrying the flux, it will be apparent that, just as in the air gap, the value of the m.m.f. is calculated for the point of maximum flux density. It should be remembered that it is only necessary to calculate the m.m.f. required for the whole of any one particular flux line, and it is merely a matter of convenience whether we take this line as passing down the tooth at the centre of the pole or at some other point. So far as the teeth are concerned it is desirable to work in terms of this maximum density, since it will be seen later that the value of the iron loss in the teeth is determined directly by this maximum value.

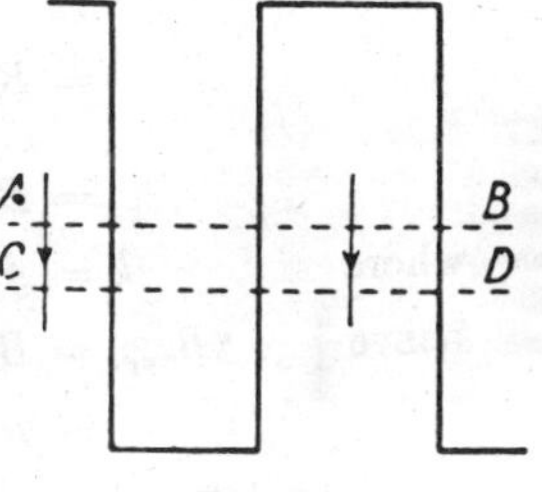

FIG. 21

Real and Apparent Flux Density in Teeth. Consider deep teeth with parallel sides as in Fig. 21, and take two planes AB and CD each normal to the magnetic field and 1 cm apart. For very deep teeth the planes AB and CD will represent equipotential planes. When the teeth are highly saturated, the difference of magnetic potential between the two planes will suffice to establish an appreciable flux in the slot itself. If H represents the magnetizing force necessary to establish the flux density in the teeth, the flux density established in the "air path" in the slot will also correspond to the magnetizing force of H, and, the material in the slot being of permeability μ_0, the flux density in the slot will be $\mu_0 H$.

As a result, the real value of the flux carried by the teeth is always less than the total flux. In consequence, the real flux density in the teeth is always less than the "apparent flux density."

The apparent flux density is defined

$$B_{appt} = \frac{\text{Total flux}}{\text{Tooth section}} \quad . \quad . \quad . \quad . \quad (46)$$

and the real density

$$B_{real} = \frac{\text{Actual flux in tooth}}{\text{Tooth section}} \quad . \quad . \quad . \quad (47)$$

In an actual machine, taking the flux over one slot pitch, there are two paths in parallel—

(1) The iron path, equal in area to

$$A_i = \text{Tooth width} \times \text{net iron length},$$
$$= b_t \times L_i \quad . \quad . \quad . \quad . \quad . \quad (48)$$

and (2) The non-magnetic path, sometimes called the "air path," of an area given by

$$A_n = (\text{tooth pitch} \times \text{gross length}) - (\text{tooth width} \times \text{iron length})$$
$$= (\tau_s \times L) - (b_t \times L_i) \quad . \quad . \quad . \quad (49)$$

If Φ_s represents the total value of the flux over one slot pitch, of which Φ_i is carried by the iron, and Φ by the non-magnetic paths,

$$B_{appt} = \frac{\Phi_s}{A_i} = \frac{\Phi_i + \Phi_n}{A_i}$$
$$= B_{real} + \frac{\Phi_n}{A_i} = B_{real} + \frac{\Phi_n}{A_n} \times \frac{A_n}{A_i}$$
$$= B_{real} + kB_n = B_{real} + (k_s - 1)B_n \quad . \quad (50)$$

where $k = A_n/A_i$, and $k_s = k + 1 = \tau_s L/b_t L_i$

hence $*B_{appt} = B_{real} + \mu_o H(k_s - 1) \quad . \quad . \quad . \quad (51)$

$$= B_{real} + 1{\cdot}257H\,(k_s - 1) \times 10^{-6} \quad . \quad (52)$$

Take as a definite example a real flux density $B_{real} = 2{\cdot}2$ Wb/m². From Fig. 8 this corresponds to 800 AT/cm, or 80,000 AT/m.

If the area A_n were equal to the area A_i, i.e. $k_s = 2$, the apparent flux density would therefore be $2{\cdot}2 + 0{\cdot}1 = 2{\cdot}3$ Wb/m², for $k_s = 3$, the apparent flux density would be $2{\cdot}2 + 0{\cdot}2 = 2{\cdot}4$ Wb/m² and so on.

If, in the first case, calculations were based upon the apparent density, the ampere-turns per centimetre would be taken as about 1,500, instead of the actual value of about 800. For the second case the error would be enormous.

To allow for the effect of the flux which passes down the slots, it is then desirable to have data available which will at once give the magnetizing force for various values of B_{appt} and k_s, and at the same time which will give the value of B_{real} for various values of

* In c.g.s. units: $B'_{appt} = B'_{real} + 1{\cdot}257H'' \times (k_s - 1)$.

the two above quantities. Designers usually employ a family of such curves, calculated for various values of k_s, such as 1·5, 2·0, 2·5, 3·0, interpolation being used for intermediate values of k_s. In Fig. 22 the lowest curve, $k_s = 1{\cdot}0$, is the ordinary magnetization curve for iron stampings. The second curve, $k_s = 1{\cdot}5$, is obtained by stepping up from various points on the lowest curve $0{\cdot}628 \times 10^{-6}$ of the corresponding value of H at these points, that is $0{\cdot}628 \times 10^{-4}$ times the value of the ampere-turns per cm. The other curves are constructed quite simply by setting off with dividers, for example,

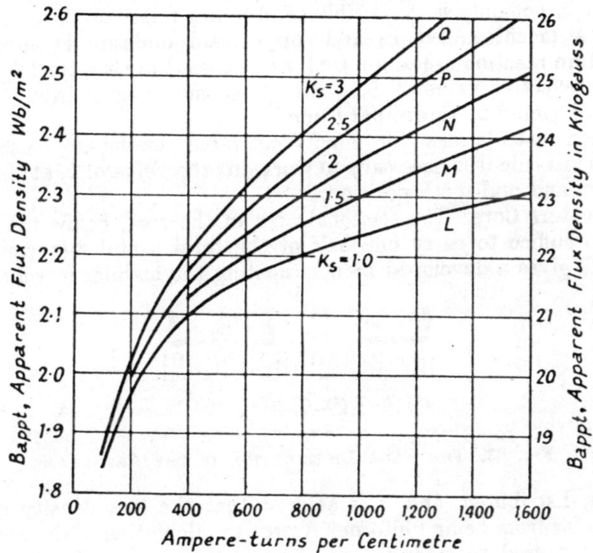

FIG. 22. MAGNETIZATION CURVE FOR ARMATURE TEETH FOR VARIOUS VALUES OF k_s

$LM = MN = NP = PQ$, etc., and constructing similarly at various other points.

As an example of the use of such curves, the following may be taken, the dimensions given being measured one-third of the way up the teeth, as already indicated.

$B_{appt} = 2{\cdot}25$; active iron length $L_i = 25$ cm; gross length, $L = 32$ cm; slot pitch, 3·0 cm; slot width, 1·2 cm.

$$\text{Then } k_s = \frac{3 \times 32}{1{\cdot}8 \times 25} = \frac{96}{45} = 2{\cdot}13.$$

From Fig. 22 by interpolating between the curves $k_s = 2{\cdot}0$ and $k_s = 2{\cdot}5$, the value of B_{real} is 2·165, and the ampere turns per cm, 620.

Where greater accuracy is required, the real flux density should be calculated at the top, bottom and mid-section of the tooth and the effective value of the ampere-turns per cm determined by the aid of Simpson's rule. The use of Fig. 20 in this connexion has already been given.

It is only at very high values of the flux density that the slots, etc., carry an appreciable portion of the flux, and for values of B_{appt} not exceeding 1·9* it is generally negligible, and B_{real} may be taken as equal to B_{appt}. By having the teeth highly saturated, however, not only is it possible to obtain a relatively large output from a machine, but in addition certain undesirable effects of armature reaction are minimized, as discussed in Chapter IV.

The effective value of B_{real} with large machines is from 2·05 to 2·15, or even 2·2, for ordinary iron.

When calculations for highly-saturated teeth are based on Simpson's rule it is necessary to work out the value of k_s at the top, bottom and mid-tooth position.

Armature Core. The sectional area of the core, below the slots, has to suffice to carry one half of the total useful flux per pole. Fig. 23 gives a developed view indicating the manner in which the

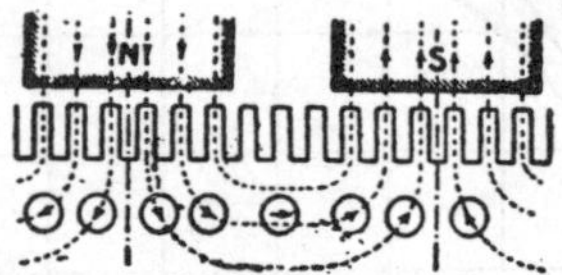

FIG. 23. THE FLUX DISTRIBUTION IN THE ARMATURE

flux is distributed, and it is evident that the flux density in the core is far from being uniform. A precise calculation of the ampere-turns required to establish the flux is then not to be expected; fortunately, the reluctance is normally so very small that an approximate value is sufficient.

An approximate, and sufficiently exact, method is to base calculations upon the maximum value of the flux density, which will occur at AB in Fig. 9, p. 21, and to take the total length of the path as being the value of the pole-pitch measured at the mean diameter as indicated by the heavy dotted line in Fig. 9.

Thus, if D_t = the diameter at the root of the teeth, and d_c = the core depth below slots, the length of the magnetic line may be taken

$$= \frac{\pi (D_t - d_c)}{2p}$$

and the value corresponding to each pole being one half of this amount, becomes

$$l_c = \frac{\pi (D_t - d_c)}{4p} \quad . \quad . \quad . \quad (53)$$

* 1·9 for ordinary iron, and 1·7 for Stalloy.

The value for the sectional area

$$A_c = d_c L_i \quad \ldots \quad (54)$$

and the flux density

$$B_c = \frac{\Phi}{2A_c}$$

The flux density at which the core is worked is determined almost entirely from considerations of iron losses and temperature rise. In general, except with small machines, d.c. machines have a number of poles which at normal values of the speed will cause the frequency of the flux reversals in the armature to be of the order of 25 to 50 c/s. When the frequency is about 25 c/s the flux density in the core may be 1·2 to 1·3 Wb/m². When the frequency is high it becomes necessary to work at slightly lower values of the density, 1·0 Wb/m² or less.

Poles and Pole Shoes. The poles may either be solid, or built up of laminations, as discussed later. If solid poles are used, the pole shoes will generally be laminated, although, very rarely, the shoes may be of cast iron. In any case it is necessary that steps should be taken to keep down the losses that take place in the shoes owing to *flux swinging*. In the main body of the pole, the direction and magnitude of the flux lines is not influenced by the relative position of the armature teeth, but in the shoes the flux is not uniformly distributed, tending to bunch in the direction of the teeth. As the armature moves, the flux thus swings, each portion going forward for a short time with one tooth until the arrival of another tooth presents another path and the flux swings back across the slot into the next tooth, and, incidentally, establishes the working e.m.f. in the armature conductors placed in the slot.

The frequency at which the flux swings is determined by the product of the number of teeth on the armature and the speed in rev/sec. This flux swinging gives rise to hysteresis and eddy current losses in the poles shoes—the eddy losses predominating—the losses often being called "pulsation losses," and at the same time sets the teeth vibrating, causing them to emit the characteristic note or "hum" often experienced to a disagreeable extent. The value of the pulsation losses may be kept low by having a small value for the ratio: slot width/gap·width.

When the poles are built up of laminations, the shoe can form part of the main punching, and there is no need to take further steps to reduce the pulsation losses.

Form of Pole Section. With solid poles, a circular, oval, rectangular or square section could be employed.

The circular section represents the one having the smallest periphery for a given sectional area, and *hence will require the smallest weight of magnet coil*, since this weight is largely determined by the length of each turn of the winding, and thus obviously is dependent upon the periphery of the pole. In addition, cylindrical coils can be wound very compactly, so that the heat developed can easily travel out from the centre. With rectangular coils, there is a certain amount of slackness at the centre of each side which impairs the heat conduction, owing to the presence of small air spaces.

A drawback to the use of circular poles is that they are not suited to relatively long armature cores, but are obviously best suited where the pole arc and the axial length of the machine have about the same dimension.

In practice it is necessary, in the interests of economy, to arrange for a standard frame to work with various values of the core length. For this the laminated pole is at an advantage, as the axial pole length can easily be adjusted to any desired value. Further, owing to the removal of any doubt as to the homogeneity of the pole material, it is possible to work laminated poles at rather a higher density than is possible with solid cast steel poles.

When cast steel poles were used, the frame and the poles all formed part of the same casting.

With the wide use of fabricated steel yokes, solid poles when used are of forged steel, and not cast steel; but for ordinary industrial machines laminated poles are used as a general rule.

With the laminated pole it must be remembered that the active iron length is only 95 per cent of the total axial length.

Magnetic Leakage. A considerable proportion of the total flux passing down the poles does not enter the armature, but completes its path through the air. John Hopkinson, to allow for this, introduced the term "leakage coefficient," the value of this coefficient being defined as the ratio of the total flux in the pole to the total useful flux entering the armature. In old bi-polar machines the value of this coefficient is sometimes as high as 1·5, and in modern multipolar machines the value is about 1·15 to 1·20. The coefficient gives an indication of the relative values of the reluctance of the main magnetic path and of the leakage paths; for a coefficient of 1·5, the leakage path has twice the reluctance of the main magnetic circuit, and for a coefficient of 1·2, its reluctance is 5 times that of the main path.

Normally, designers rely upon test results on previous machines for the value of the leakage coefficient. At first it might be concluded that magnetic leakage is very serious, but actually it acts mainly to the detriment of the pole and yoke, which must be designed to carry not only the useful flux, but the leakage flux in addition. The increased sectional area of the pole will in addition necessitate an increased weight of copper for the magnet coils.

For rectangular poles, the actual leakage flux may be divided into two parts, i.e. the direct leakage from one pole flank to the other, and the leakage occurring at the ends, as in Fig. 24, which represents a plan taken midway through the poles so as to correspond to average conditions.

The paths of the leakage lines are also shown diagrammatically in Fig. 25, and it will be seen that whereas those lines leaking out of the pole shoes are linked with the whole of the turns on the two poles, the lines leaking near the root of the poles are only linked with a few turns. Half way down the field coils, as shown, the leakage lines only link with one half the complete number of turns. Now the value of the m.m.f. acting around any complete line of force is obtained, simply, by summing the ampere-turns with which the line is linked. This is equivalent to summing, algebraically, the ampere-conductors enclosed by the line.

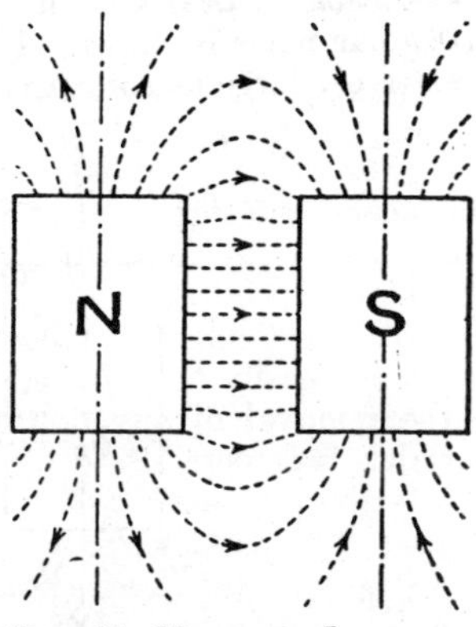

Fig. 24. Magnetic Leakage

For a normal cylindrical field coil, therefore, the m.m.f. driving the leakage flux will increase uniformly from zero towards the roots of the poles, to a maximum value at the other end of the coil. As a result, the density of the leakage flux will increase progressively from the

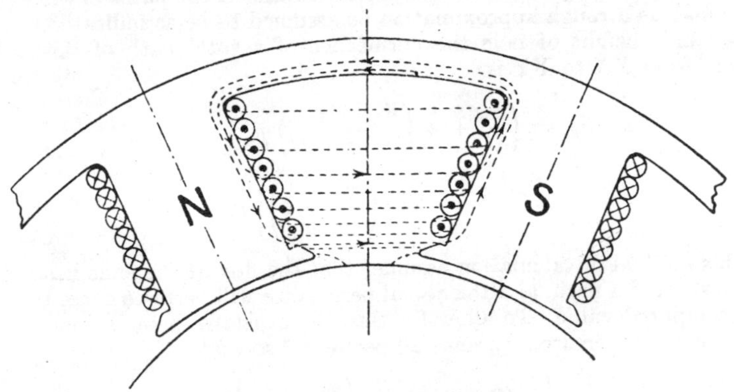

Fig. 25. Magnetic Leakage

root of the pole as the armature is approached. The increase is made more marked by the fact that, with the normal d.c. machine, the distance between poles gets progressively less as the armature is approached; with these machines, a large proportion of the total direct leakage takes place between pole shoes, as these are relatively close together, and the m.m.f. acting along the path is the maximum value.

For rectangular poles having an axial length which is large compared with the clearance between poles, the direct leakage can be calculated with sufficient accuracy, but the calculation of the leakage at the ends of the poles can be attempted only very approximately, on the lines indicated below.

Estimation of Leakage Flux. In Fig. 26 a plan of two adjacent rectangular poles is shown. If the axial dimension L_p of the poles be relatively large as compared with the distance between them, C,

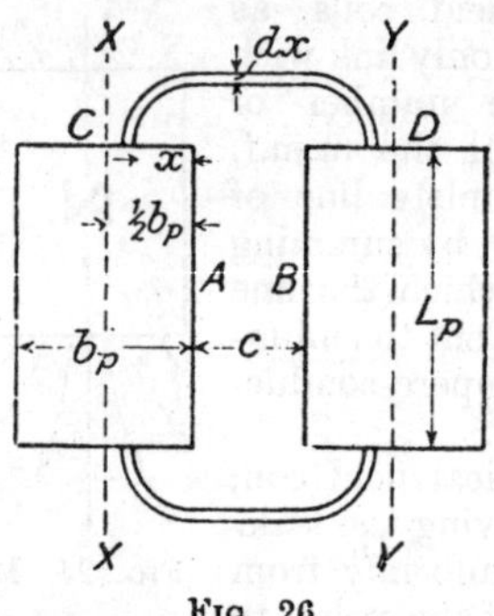

FIG. 26

flux may be assumed to be uniformly distributed between the axial surfaces A and B, and for the end surfaces C and D the paths of the lines may as a rough approximation be assumed to be as indicated.

For unit height of pole the permeance of a total path of this nature from XX to YY is

$$\lambda_{xy} = \left(\frac{L_p}{C} + 2\int_0^{\frac{1}{2}b_p} \frac{dx}{\pi x + C}\right)\mu_o$$

$$= \left\{\frac{L_p}{C} + \frac{2}{\pi}\log_e\left(\frac{\pi b_p}{2C} + 1\right)\right\}\mu_o$$

This method of calculation assumes that the flux at the ends has too restricted a path, and the actual permeance will perforce exceed the computed value. To allow for this, the constant in the second term may be increased by some 20 per cent,* so that

$$\lambda_{xy} \simeq \left\{L_p/C + 1{\cdot}8\log_{10}\left(\frac{\pi b_p}{2C} + 1\right)\right\}\mu_o \quad . \quad . \quad (55)$$

The flux leaking from the axis XX to YY represents one half of the actual flux leaking from the pole, so that the total permeance of the leakage paths per unit height of pole is $\lambda = 2\lambda_{xy}$.

* I.e. $\frac{4{\cdot}6}{\pi}$ is replaced by 1·8.

The leakage flux is produced by the joint action of the coils on the two adjacent poles.

For the flux leaking out at the shoes the total m.m.f. acting is therefore the sum of the ampere-turns $2(F_t + F_g + F_c)$, say $2F_{t+g+c}$.

For the flux leaking out of the pole limbs the m.m.f. decreases steadily from this value at the pole shoe to zero at a point just before the other end of the exciting coils. The mean value acting from about the mid point M of the windings may with sufficient accuracy be taken to be F_{t+g+c}.

Hence dividing the leakage flux into two parts, emanating from the shoe and limb respectively (for dimensions see Fig. 27).

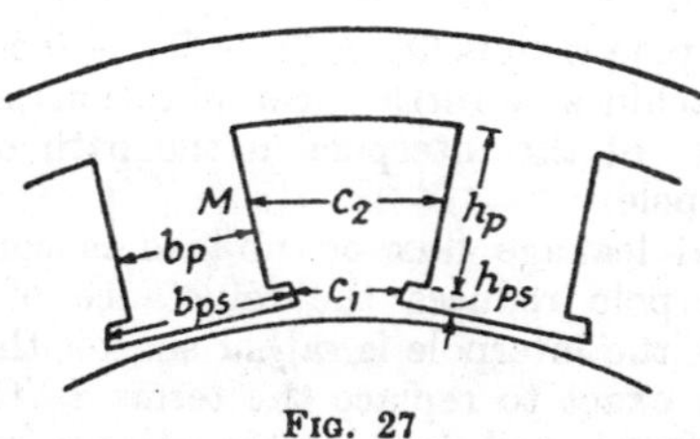

FIG. 27

The permeance of the leakage paths from the shoe

$$\Lambda_{ps} = 2\lambda_{ps}h_{ps} = 0{\cdot}4\pi \times 2h_{ps}\left\{\frac{L_p}{C_1} + 1{\cdot}8 \log\left(\frac{\pi b_{ps}}{2C_1} + 1\right)\right\} 10^{-6} \quad (56)$$

and the total leakage flux from the shoe—

$$\Phi_{ps} = 2F_{t+g+c}\Lambda_{ps} = 0{\cdot}4\pi \times 4F_{t+g+c}h_{ps}\left\{\frac{L_p}{C_1} + 1{\cdot}8 \log\left(\frac{\pi b_{ps}}{2C_1} + 1\right)\right\} 10^{-6} \quad (57)$$

For the pole limbs, the leakage permeance is

$$\Lambda_{pl} = 2\lambda_{pl}h_p = 0{\cdot}4\pi \times 2h_p\left\{\frac{L_p}{C_2} + 1{\cdot}8 \log\left(\frac{\pi b_p}{2C_2} + 1\right)\right\} 10^{-6} \quad (58)$$

and the total leakage flux from the pole limb

$$\Phi_{pl} = F_{t+g+c}\Lambda_{pl} = 0{\cdot}4\pi \times 2F_{t+g+c}h_p\left\{\frac{L_p}{C_2} + 1{\cdot}8 \log\left(\frac{\pi b_p}{2C_2} + 1\right)\right\} 10^{-6} \quad (59)$$

In the above two formulae for the leakage flux, the leakage is given in webers if the dimensions h_{ps} and h_p are expressed in metres, and in megalines if they are in centimetres.*

* The other dimensions, L_p, C, etc. may be expressed in any convenient unit.

For the dimension h_{ps} of the pole shoe a mean figure should be taken to correspond to the average dimension below the coils for the whole of the shoe.

The total leakage flux is then

$$\Phi_L = \Phi_{ps} + \Phi_{pl}$$

The total flux entering the yoke, and the maximum flux in the pole is

$$\Phi_y = \Phi + \Phi_L$$

and the minimum flux in the pole is

$$\Phi + \Phi_{ps}$$

The mean flux in the pole is thus $\Phi_p = \Phi + \Phi_{ps} + 0{\cdot}5\Phi_{pl}$.

With interpole machines a further complication is introduced owing to the presence of the interpole in the path of the direct leakage from pole to pole.

So far as the total leakage flux on no-load is concerned, the presence of the interpole reduces the reluctance of the leakage paths. If, as is usual, the interpole is rather shorter than the main pole, it is sufficiently exact to replace the terms L_p/C_1 and L_p/C_2 in (57) and (59) by $L_p/2d_1$ and $L_p/2d_2$, respectively, where d_1 and d_2 are the distances between the tips and the mean distance between limbs of the main pole and interpole, as in Fig. 119. This is sufficiently exact for most purposes when the axial length of the interpole is about two-thirds of that of the main pole.

With d.c. machines, very precise calculations of the magnetic leakage flux are rarely necessary, and it is in general sufficient to base calculations upon a leakage coefficient determined experimentally from existing machines of similar type, by determining the relative value of the flux midway along the pole and in the armature.

Denoting the leakage coefficient by ν the effective flux in the pole is $\Phi_p = \nu\Phi$, where Φ is the useful flux.

Hence the effective flux density in the pole, $B_p = \dfrac{\nu\Phi}{A_p}$.

The length of the magnetic line in the pole may be taken as the height of the pole above the shoe, the m.m.f. required for the latter being negligibly small.

Effect of Load, Saturation, etc., upon Leakage Coefficient. The main portion of the leakage lines passing through air, the reluctance is nearly constant, and to a close approximation the leakage flux increases in direct proportion to the total ampere-turns absorbed by the armature and air gap. The reluctance of these latter parts of the magnetic circuit is, however, not constant, but varies mainly according to the degree of saturation of the teeth. Hence the leakage coefficient will increase fairly rapidly when the teeth become highly saturated.

In general, if F = m.m.f. for (teeth + gap + core) at normal voltage, V

and F_1 = m.m.f. at V_1

Then if ν = leakage coefficient at V volts.

The coefficient at V_1 is $\nu_1 = \frac{F_1}{F} \times \frac{V}{V_1}(\nu - 1) + 1$ . . (60)

On load, too, to counteract the effect of armature reaction, as explained later, it becomes necessary to increase the excitation in order to maintain the required value of the flux.—In brief the effect of the currents in the armature is to alter the distribution of the main field and to increase the reluctance of the main magnetic circuit.—This increased excitation increases the leakage flux, although only serving to maintain normal value of the useful flux. Hence the leakage coefficient with the machine loaded is greater than on open circuit.

The Yoke. Normally the yoke will be either of forged or cast steel, or cast iron. Cast steel, as used for dynamos, has very good magnetic properties, but cast iron is very inferior from this point of view, and it may be taken that owing to its poor magnetic properties it is necessary to employ about twice as large a section for a cast-iron frame as for one of cast steel. This adds considerably to the weight of the machine, so that the modern practice is to employ forged steel or sometimes cast steel frames, as indicated in Figs. 273 and 290.

It is not desirable, with steel yokes, to work at too high a flux density, and a normal working value for both cast and forged steel is about 1·2–1·3 Wb/m². The flux carried by the yoke is, for reasons already given, one half of the total flux entering the root of the pole. The yoke has then to carry one half the full value of the sum of the useful flux and the leakage flux.

For the total length of magnetic path in the yoke it is sufficient to take the dimension corresponding to the average flux line, and this is approximately equal to the value of the pole-pitch measured at the mean yoke diameter, $\pi D_y/2p$, less one half the width of one pole, plus the depth of the yoke, d_y.

Hence $l_y = \frac{1}{2}\{\pi D_y/2p + d_y - \frac{1}{2}b_p\}$ . . . (61)

The leakage coefficient for the yoke should be taken rather higher than the value taken as representing average conditions for the pole.

Magnetization Curve of D.C. Machine. This curve shows the manner in which the value of the m.m.f. required per pole varies with the value of the flux. Since, for constant speed, the induced e.m.f. is directly proportional to the flux, the same graph can be utilized to show the manner in which the induced e.m.f. varies with the magnetizing current, i.e. for the open-circuit characteristic.

To construct the curve it is then necessary to calculate the ampere-turns per pole required to establish various values of the

OPEN CIRCUIT MAGNETIZATION CURVE FOR 1,000 kW, 500 V GENERATOR

Part	Area	Length	E = 350 V			E = 425 V			E = 500 V, Φ = 0·1093 Wb			E = 550 V			E = 575 V		
	cm²	cm	B Wb/m²	AT/cm	AT	B Wb/m²	AT/cm	AT	B Wb/m²	AT/cm	AT	B Wb/m²	AT/cm	AT	B Wb/m²	AT/cm	AT
Air Gap $K_g = 1·14$	1,130	0·6	0·677	—	3,690	0·822	—	4,480	0·967	8,800	5,280	1·064	—	5,800	1·112	—	6,060
Teeth $K_s = 2·1$	518	5	1·475	18	90	1·79	80	400	2·11	330	1,650	2·32	850	4,250	2·43	1,250	6,250
Core	2 × 454	19	0·84	5·5	100	1·025	7	130	1·205	10	190	1·325	12·5	240	1·385	14	270
F_{t+g+c}					3,880			5,010			7,120			10,390			12,580
Leakage coef. (poles)			1·12			1·125			1·15			1·20			1·23		
Poles	838	30	1·02	7	210	1·25	10·5	320	1·5	20	600	1·72	55	1,650	1·85	110	3,300
Yoke (C.S.)	2 × 583	35	0·76	3	110	0·935	4	140	1·12	6	210	1·29	9	320	1·38	12	420
Total AT on open circuit					4,200			5,470			7,930			12,360			16,300

useful flux. It is normally sufficient to carry out the calculations for values of the flux varying from about 70 per cent of normal, to about 115 per cent. Below 70 per cent, the reluctance of the iron parts is insignificant in comparison with that of the air gap, and the curve merges into a straight line corresponding to the air gap characteristic.

A convenient tabular form for carrying out the calculations is given opposite, calculations being given for a 1,000-kW generator, designed completely later in the book. The ampere-turns are calculated for 70 per cent, 85 per cent, 100 per cent, 110 per cent, and 115 per cent of normal induced e.m.f.

The calculations are based upon a constant value for the flux coefficient K_f, although, as already stated, the value will increase when the iron parts of the armature, and in particular the teeth, become saturated.

The various dimensions are all deduced by the methods indicated, the necessary calculations being given in Chapter XVIII.

In the upper part of the table, calculations are given for the air gap and armature. The leakage coefficient for the main poles is taken as 1·15 at 500 V. For the yoke the figure taken is 1·20, and at all other voltages these coefficients are taken to have the same ratio, i.e. 1·15/1·2.

The leakage coefficient at voltages other than 500 is deduced before the calculations for the pole and yoke are carried out. For example, at 500 V the leakage flux is due to 7,120 ampere-turns and amounts to 15 per cent of normal flux.

At 575 V the useful flux has increased in the ratio 575/500 and the leakage flux in the ratio of the ampere turns F_{t+g+c}, i.e. 12,580/7,120. The leakage flux is therefore 12,580/7,120 = 177 per cent of the value at 500 V, and the useful flux 575/500 = 115 per cent. The leakage coefficient is then

$$(0{\cdot}15 \times 177/115) + 1 = 0{\cdot}23 + 1 = 1{\cdot}23.$$

At 350 V the leakage coefficient is 1·12, and rises to 1·23 at 575 V, the rise being rapid above normal voltage, owing to the high degree of saturation then obtaining in the teeth.

Owing to the increased leakage at high voltages, the poles tend to saturate, and as a result the magnetization curve flattens out very rapidly due to this effect adding to that of the teeth.

The flux densities shown for the teeth are the apparent densities, and for both the teeth and the air gap the densities really correspond to those under the mid-point of the pole shoe.

It will be observed that the m.m.f. required for the armature core and the yoke is insignificant. Normally the designer finds it sufficient to allow a liberal amount and does not calculate this m.m.f.

If in the poles, the leakage coefficient had been assumed to retain the value 1·15, at 575 volts the m.m.f. for the poles would have been under-calculated to the extent of about 1,000 ampere-turns.

Fig. 28 gives the actual magnetization curve plotted from the data in the table.

Effect of Magnetic Saturation upon Field-form and Flux Coefficient. It may be taken that the sum $F_t + F_g$ remains constant

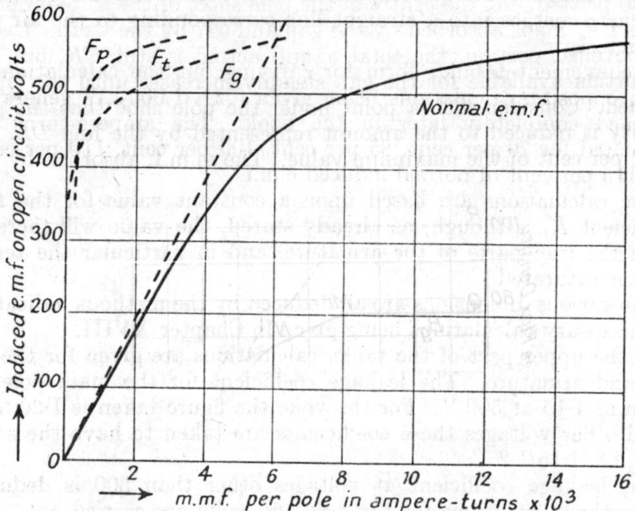

FIG. 28. THE OPEN-CIRCUIT MAGNETIZATION CURVE OF A 1,000-KW GENERATOR; 500 V, 300 REV/MIN

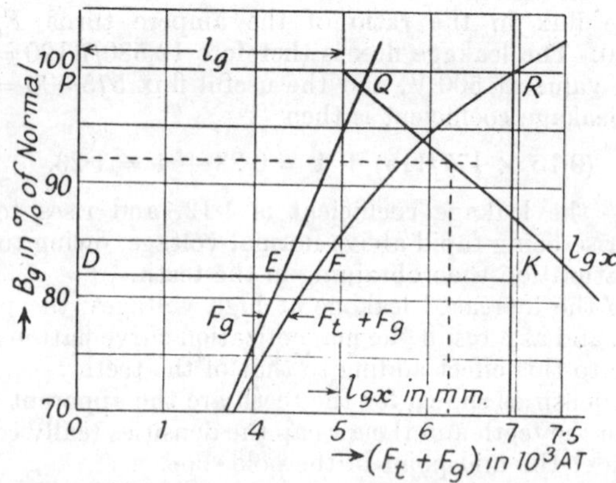

FIG. 29. MAGNETIZATION CURVE FOR AIR GAP AND TEETH

over the whole of the pole face. The gap dimension is, however, not constant and as a result the degree of saturation in the teeth is greater at the centre of the pole than at the tips. In Fig. 29 is plotted the "teeth + gap" magnetization curve for a typical generator, the ordinates being expressed as a percentage of normal value, and in terms of B_g instead of Φ. The air gap characteristic is also plotted. At the centre of the pole shoe, of the total ampere-turns PR, PQ is absorbed by the gap and QR by the teeth. Under the bevelled portion, the total m.m.f. acting is still PR, but the proportion available for the gap steadily increases until the tip is reached. Consider such a point under the pole shoe that the gap density is reduced to the amount represented by the line DE, i.e. to 82 per cent of the maximum value. The m.m.f. absorbed by the

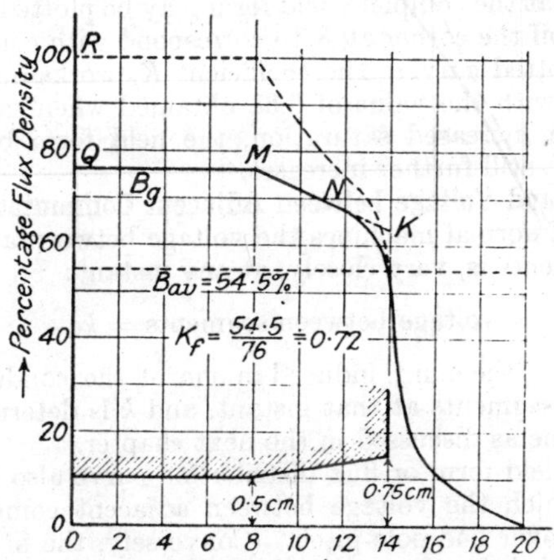

Fig. 30. Flux Distribution Curve for Generator on No-load, the Teeth being Fairly Saturated

teeth is now given by EF, so that $DE + FK$ represents the total ampere-turns acting along the air gap. Since DE represents the amount necessary to produce the flux density of 82 per cent for normal length of gap, the length of the actual gap at this point is therefore given by

$$l_{gx} = l_g\left(1 + \frac{FK}{DE}\right)$$

Hence a graph can readily be constructed showing the manner in which the flux density varies with the gap dimensions as shown in Fig. 29.

Having deduced this latter curve, the actual field form can readily be plotted. Thus, in Fig. 30 the field form shown by the dotted line corresponds exactly to that given in Fig. 13 for a machine in which magnetic saturation is entirely absent, the bevel of the pole shoe being as indicated.

The "teeth + gap" magnetization curve for this machine is as given in Fig. 29 above, the point R corresponding to normal e.m.f.

on open circuit. The excitation PR only produces the same value of B_g that would be produced by PQ, if saturation were absent, hence the value of B_g may be represented by OQ in Fig. 30 where OQ/OR represents the ratio $F_g/(F_t + F_g)$. At the pole tip the gap is 7·5 mm and from Fig. 29 it appears that the flux density will be 83·5 per cent of that obtaining at the centre of the shoe, i.e. 83·5 per cent of B_g. Similarly, 11 cm from the centre line of the pole, the gap dimension is 6·25 mm, and from Fig. 29 the relative density is 92 per cent. The curve MNK is thus obtained. Except in the immediate neighbourhood of the pole tips, the field form at the fringe will be the same with the teeth saturated as without, so that the complete field form may be plotted as indicated, by rounding off the corner at K to correspond with conditions holding for the dotted curve. The coefficient K_f works out to be 0·72, as compared with the value of 0·68 obtained when saturation was absent. With increased saturation, the field form becomes even flatter, and K_f still further increases.

Field-form and Voltage between Adjacent Commutator Segments. In all cases of normal machines the voltage between adjacent commutator segments is, very closely, at any instant:

$$\text{voltage between segments} = ke$$

where $e = BLv$, the e.m.f. induced in one of the conductors joined to one of the segments at that instant, and k is determined by the winding scheme as discussed in the next chapter.

Hence the field form or flux distribution curve also indicates the manner in which the voltage between adjacent commutator segments varies over one pole-pitch. (Conversely, the field form may be determined by measuring the voltage between commutator segments, over one pole-pitch, under normal no-load conditions, exploratory brushes spaced one commutator segment pitch apart being used in conjunction with a suitable brush rocker and graduated scale.)

The ratio between V_{av}, the average value, and V_{max}, the maximum value of this voltage between segments is evidently

$$V_{av}/V_{max} = B_{av}/B_g = K_f$$

Since V_{av} = (total e.m.f. between main brushes)/(no. of commutator segments per pole), the maximum voltage between segments is:

$$V_{max} = \frac{\text{Voltage between main brushes}}{K_f \times \text{no. of commutator segments per pole}}$$

The ratio V_{max}/V_{av} on no-load ranges from about 1·4 to 1·5 the smaller value being for the larger ratio of α, (i.e. pole arc/pole-pitch) $= 0{\cdot}7$ with reasonable magnetic saturation, and the value 1·5 being more normal for $\alpha = 0{\cdot}67$ and little saturation.

Flux Plotting. For such purposes as estimating magnetic leakage and for pre-determining the field form two dimensional maps of the magnetic field are sometimes made. The main methods are—

1. By freehand sketching, employing the method of curvilinear squares.

2. By experiment on a kindred problem of the flow of current in an electrolyte in a suitable electrolytic tank.

3. By calculation, using Southwell's relaxation method.

In all the methods it is assumed that the iron has infinite permeability, so that the magnetic lines of force enter and leave the iron at right angles to its surface. Further, full use is made of any axes of symmetry, to reduce the labour involved.

1. Freehand Sketching, the Method of Curvilinear Squares. This method is used to map both the lines of force and the magnetic equipotential lines, and with care can give relatively good results.

In broad outline the method is based on the following—

Both lines of force and equipotential lines are drawn.

There are no discontinuities, i.e. abrupt changes in direction, in either lines of force or equipotentials. As a result, equipotential lines meet axes of symmetry at right angles.

The more nearly the equipotential lines approach the equipotential boundaries, the more nearly they approximate to the shape of the boundaries.

Equipotentials and lines of force intersect at right angles, thus forming *curvilinear rectangles*. In flux plotting, the spacing of the lines of force and of the equipotentials are made the same, and thus form *curvilinear squares*. These squares thus enclose areas bounded by four lines, some or all of which may be curved, but which intersect at right angles at the corners. When the areas are repeatedly subdivided into further smaller curvilinear squares, these latter more and more nearly approach to true squares.

Having first drawn the boundaries and axes of symmetry the procedure is then to sketch in pencil a relatively small number of lines of force and equipotentials, approximating to the above requirements. The areas so formed are then repeatedly subdivided by adding more lines of force and equipotentials, and correcting any errors which this process reveals as it progresses.

2. Electrolytic Tank Method. The analogy between the laws of the electric and magnetic circuits have already been discussed. Others also exist, two important ones analogous to that of magnetic field distributions are the stream-line flow of liquids, and electric current flow. Some 50 years ago Hele Shaw published work relating to tooth fringing—with beautiful photographs—employing the former analogy. More recently* the electrolytic tank has been

* Cf. H. Diggle and E. R. Hartill *Some applications of the Electrolytic Tank to Engineering Design Problems.* J.I.E.E. 1954, Vol. 101, pp. 349–364.

further developed and widely used for magnetic and other problems. In this, a shallow tank made of insulating material contains the electrolyte in which model electrodes represent, to scale, the boundary conditions. A constant alternating potential difference is maintained between electrodes, and the equipotentials determined by an exploratory probe in the electrolyte, the probe being joined through a detector to a voltage divider connected in parallel with the electrodes. Using a low voltage 1,000 c/s supply, a telephone head-set forms a suitable detector.

The equipotentials so determined have the same pattern as those for the analogous magnetic problem.

3. SOUTHWELL'S RELAXATION METHOD. This method enables the magnetic potential distribution to be determined to any normal desired degree of accuracy if sufficient simple computations are made, but the total process may become lengthy. For a full description of the method the reader is advised to consult the paper by Motz and Worthy.*

Relation between Flux Wave and Flux Map. If it is required to obtain the field form from the flux plot, it is not necessary to count the lines of force (which do not necessarily appear on maps obtained by the electrolytic tank or relaxation methods). Since flux density, for a constant difference of magnetic potential, is inversely proportional to the length of the lines of force, the flux density at any point is inversely proportional to the spacing of the equipotentials at that point. It is merely necessary to measure the distance between the boundary, i.e. the armature surface, and the nearest equipotential at intervals around the armature and to plot the inverse of this length. The scale of the field form thus obtained can be determined from the density calculated for the pole centre-line where the lines of force cross the gap in straight lines.

* Motz and Worthy: *Calculation of the Magnetic Field in Dynamo-Electric Machines by Southwell's Relaxation Methods.* J.I.E.E. 1945, Vol. 92, p. 522

CHAPTER III

ARMATURE WINDINGS

MODERN d.c. armatures invariably have *closed windings*. Although ring windings, as developed independently by Gramme and Pacinotti, are obsolete, it may be well to consider first the ring winding, as shown in Fig. 31. In this winding there are a number of coils,

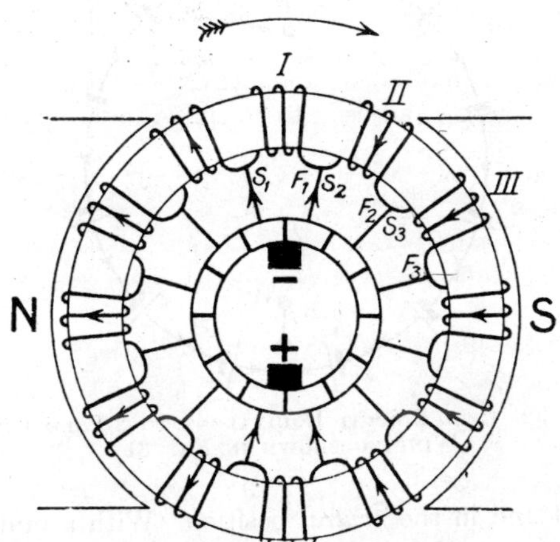

FIG. 31. 2-POLE RING WINDING
(12 segments; 3-turn coils)

1, 2, 3, etc., wound radially around the ring formed by the core plates. Each of these coils may consist of any desired number of turns, i.e. may either be a "single-turn coil" or a "multi-turn coil." The coils are joined consecutively, as it were in series, so as to form a completely closed circuit. Thus, the "finish" F_1 of coil 1 is joined to the "start" S_2 of coil 2; F_2 is joined to S_3; and so on, until ultimately the "finish" of the last coil is joined to the "start" of the first. The junctions between neighbouring coils are connected to the commutator segments as in Fig. 31, so that any particular commutator segment is connected to the "finish" of one coil and to the "start" of the next coil. When such a closed winding is rotated between the poles of an ordinary magnet system, the e.m.f. induced in the coils under the north pole of the field system balances exactly that induced in the coils under the south pole. The total value of the e.m.f. acting around the closed armature winding is thus zero, the arrangement corresponding to the cell circuit shown in Fig. 32. However, when brushes are placed on the commutator as indicated in Fig. 31, there will always be two circuits in parallel between them. A little consideration suffices to establish the fact that, for all positions of the brushes, the total e.m.f. acting in the one of these circuits will always be exactly equal to that induced in the other. Maximum e.m.f. will be induced between brushes, when they are placed so as to divide the winding into two equal sections, one of which is entirely under the north pole, and the other entirely under the south pole. That is, the brushes must be placed one pole-pitch, or 180 electrical

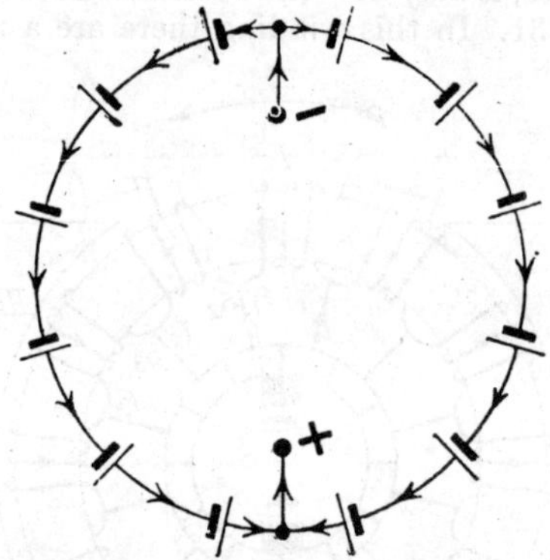

Fig. 32. Cell Circuit Equivalent to 2-Pole Ring Winding shown in Fig. 31

degrees, apart and in the *neutral* position. With a multipolar ring winding the arrangement corresponds to that of a number of similar batteries joined back to back to form a closed circuit, as in Fig. 34. To obtain full advantage of the winding it then becomes necessary to use as many, equally spaced, brush sets as there are poles, brushes of like polarity being connected in parallel as indicated in Fig. 33. Thus with the simple ring winding there are as many circuits in parallel as there are poles, and the number of coils in series in each of these circuits is equal, simply, to the number of commutator segments between adjacent brush sets.

Drum Windings. The ring winding has so long been obsolete that the reasons why it has been supplanted by the drum winding are merely of historical interest. It will, then, suffice to state that the ring winding was very inferior from both the mechanical and constructional point of view, it was wasteful of copper, and gave rise to unsatisfactory commutation conditions.

In the ring winding each coil may be regarded as having only one active side, the two ends and the side passing through the armature tunnel being inactive. In the drum winding, on the other hand, each coil may be regarded as having *two* active sides, the turns being no longer wound radially as with the ring winding, but entirely upon the outer periphery of the armature.

The simplest form of drum-wound slotted armature is the two-slot or shuttle armature indicated in Fig. 35, in which there is a single

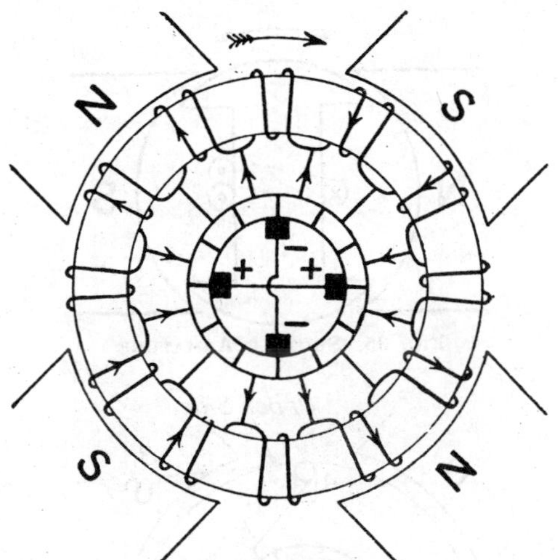

FIG. 33. 4-POLE RING WINDING
(12 segments; 2-turn coils)

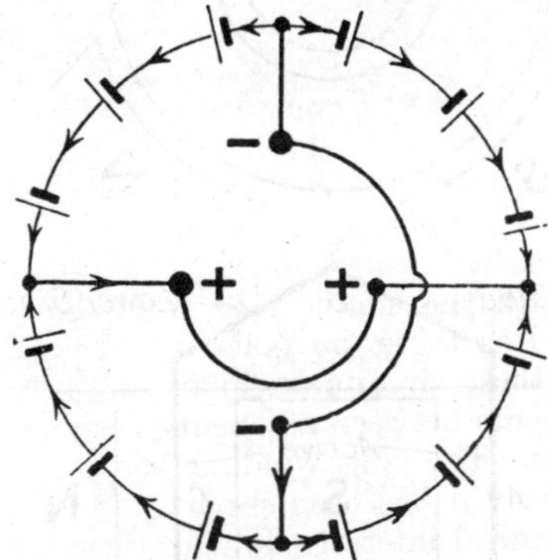

FIG. 34. CELL CIRCUIT EQUIVALENT TO 4-POLE RING WINDING SHOWN IN FIG. 33

armature coil consisting normally of very many turns, the active sides of which are housed in two diametrically opposite slots. The coil thus embraces, as a maximum, the whole of the useful flux.

In a normal slotted armature the coil is wound, as indicated diagrammatically in Fig. 36, to embrace a certain number of teeth and will thus consist of two active sides in two different slots together with two inactive ends.

Alternatively, it may be stated that whereas with a ring armature

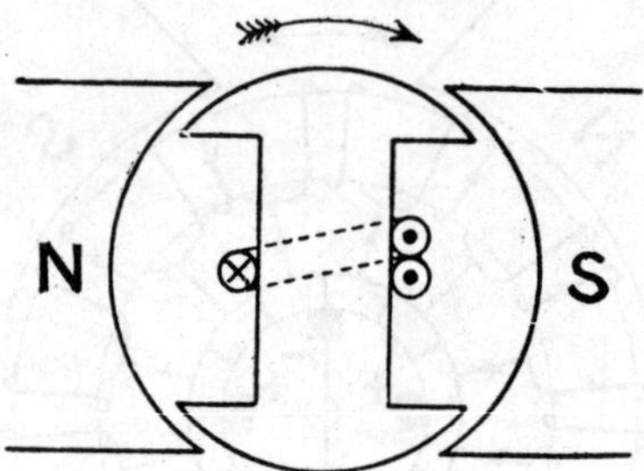

Fig. 35. Shuttle Armature

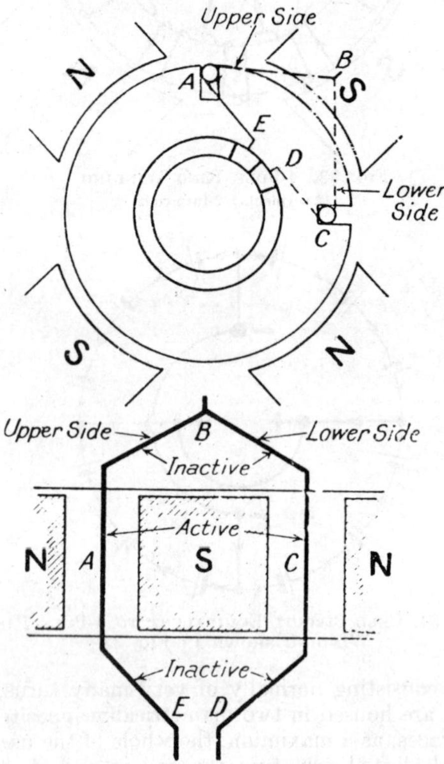

Fig. 36. Drum-type Armature Coil

each coil can embrace as a maximum only one half of the useful flux per pole, with a drum winding the coil may be given such a span that it embraces as a maximum the whole of the useful flux per pole, as shown in Fig. 36. In the drum winding, just as with the ring type, each coil may have any desired number of turns, the single-turn coil always being preferred to the multi-turn coil so far as commutation is concerned.

Two-layer Windings. To obtain a satisfactory arrangement of the end-connexions, d.c. armature windings are universally of the two-layer type. Each coil thus consists, as in Fig. 36, of an "upper side"

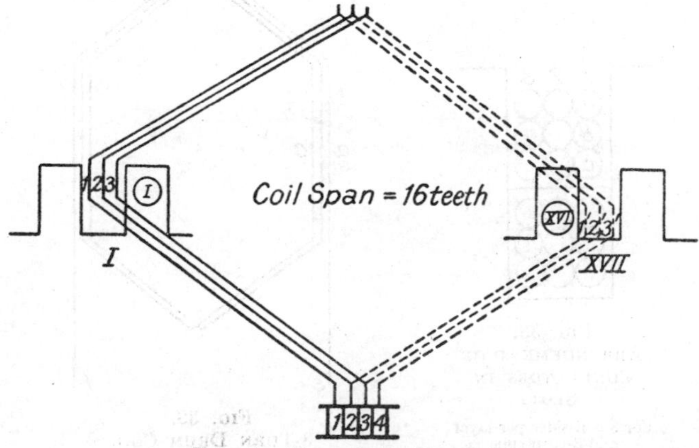

FIG. 37. TWO-LAYER WINDING

placed at the top of one slot, and a "lower side" situated at the bottom of another slot, normally distant by about one pole-pitch from the "upper side," as discussed later. Generally the complete coil is made out of one length of conductor, so that there are no joints at the back end of the machine, the change from the upper to the lower portion of the coil being effected by a bend introduced at the back end of the coil.

With a two-layer winding, then, each slot will contain the upper sides of one or more coils, together with the lower sides of a corresponding number of other coils.*

Number of Coil-sides per Layer. If each slot contained only one upper coil-side and one lower coil-side, the number of slots required would be equal to the number of coils. This is not possible, except very rarely, the number of slots being limited for reasons that are discussed on page 412. It therefore becomes necessary to have 2, 3, 4 or sometimes even more coil-sides per layer. The coils are invariably arranged side by side, as indicated in Fig. 37, for three

* That is to say, the upper conductors in a slot do not form part of the same coils as the lower conductors.

coil-sides per layer, and single-turn coils. Fig. 38 gives the normal arrangement of the conductors for a winding having three coil-sides per layer and multi-turn coils, the number of turns per coil shown being three. In Fig. 37 the upper coil-sides are denoted, respectively, 1, 2, and 3, and it is obvious that by arranging, as shown, that the lower sides of these three coils, denoted 1′, 2′, and 3′, all come in one slot, it is possible to shape the three coils in succession upon the same former, and to insulate them *en bloc* so as to form one unit or

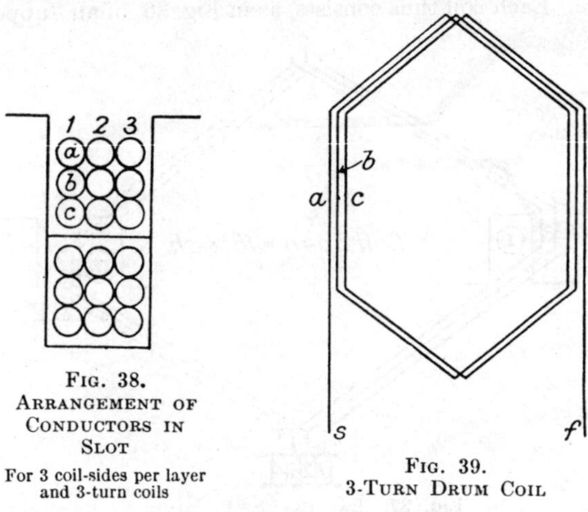

Fig. 38.
Arrangement of Conductors in Slot
For 3 coil-sides per layer and 3-turn coils

Fig. 39.
3-Turn Drum Coil

cell of the winding. Examples of such units or cells are shown in Fig. 40.

Coil Span. If a drum-type coil spans an exact pole-pitch, at every instant the e.m.f.'s induced in the two sides will act in conjunction around the coil. If the span is rather less, or rather more, than the exact pole-pitch, there will be brief instants during which the e.m.f.'s will oppose, but the conductors during these instants will be in the region of the neutral axis—i.e. midway between the poles—and the e.m.f. induced will be very small in any case.

Put another way, to obtain maximum possible e.m.f. it is essential to adopt such a span that the coil embraces as a maximum the whole of the useful flux per pole. For a coil span of an exact pole-pitch this condition is realized; for a smaller span, the coil will not embrace, as a maximum, quite the whole of the flux, but the reduction will not be very marked unless the coil span is so much reduced that it spans but little more than an amount corresponding to the pole shoe.

With modern interpole machines it is not practicable to employ a coil span of much less than one pole-pitch, and it may in general be taken that with d.c. machines the number of teeth embraced by the coils is in most cases as nearly as possible equal to the number of teeth per pole on the armature. Occasionally, the span is deliberately reduced by an amount corresponding to one slot pitch, as this facilitates commutation.

If the coil spans an exact pole-pitch it is called a "full-pitch" coil. If its span is less than a pole-pitch it is said to be a "fractional-pitch" coil, or, alternatively, it is said to be chorded. For example, in an 8-pole armature having 128 slots, a normal full-pitch coil would span or embrace 128/8, i.e. 16 teeth. A coil spanning 15 teeth

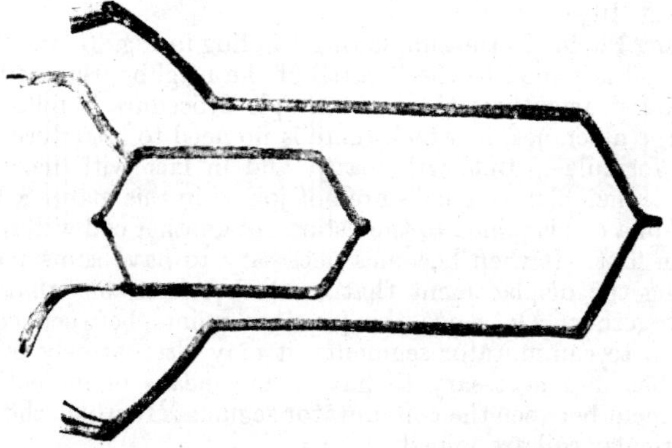

FIG. 40. STRAP COIL AND MULTI-TURN COIL

would constitute a fractional-pitch or chorded coil, and is said to be short chorded by one slot pitch.

With certain types of winding, however, the number of slots is not exactly divisible by the number of poles. For example, a 4-pole wave winding may well have 39 slots. The coil may then be taken to embrace either 9 or 10 teeth. Nearly always it will be arranged that the upper and lower sides of any coil shall occupy the same relative position in the slots, as indicated in Fig. 37 already given.

It may then be taken that for all types of d.c. windings, whether lap or wave is immaterial, the number of teeth embraced by the complete slot group or cell, i.e. the span of the coil, is at once obtained by dividing the total number of teeth by the number of poles. If the quotient is an integer, then by taking this number for the span of the coil, as measured by the teeth embraced, the normal case of a full pitch winding is obtained. For special cases, as already stated, the span may be taken as one less than the quotient and a slightly chorded winding obtained. When the quotient is not an integer, the nearest whole number usually is taken, although it is of no great moment whether the number taken is the nearest integer greater or less than the actual quotient.

Sometimes, with large machines, single armature bars are used and joints made at the back end of the armature, in addition to the joints at the commutator. With this arrangement it is possible to arrange that the upper and lower sides of the completed coil shall not occupy the same relative position in the slot. Thus, in a winding having 3 coil sides per layer, the upper side may be against the left wall of the slot, and the lower side in the central position. This arrangement is used by some designers to obtain more uniform commutation conditions for all the coils. The coil span in these rare cases can conveniently be expressed as a fractional number of teeth—e.g. $10\frac{1}{3}$.

Winding Pitch. In the simple ring winding in Fig. 31, the "finish" of each coil is joined to the "start" of the neighbouring coil; with drum-wound armatures the same simple procedure is followed for most large machines, in which there is no need to introduce mathematical formulae. But with many, and in fact with most, small armatures, neighbouring coils are not joined in this manner, but the "finish" of a coil is joined to the "start" of another coil well removed from the first. It then becomes necessary to have some means of indicating the displacement that exists between coils that are so joined together. Or, since the junction points between coils are connected to commutator segments, it may alternatively be stated that it becomes necessary to have some means of indicating the displacement between the commutator segments to which the ends of one particular coil are joined.

The displacement between the coils is sometimes called the total winding pitch, or total pitch. It will be referred to here as, simply, the winding pitch.

In a similar manner the displacement between the commutator segments to which the two ends of a coil are joined is called the commutator pitch.

Commutator Pitch. This is expressed in terms of the number of commutator segments. Thus, if the commutator segments are numbered consecutively, as in Fig. 41, and the ends of a coil are connected, respectively, to segments 1 and 11, the commutator pitch is $11 - 1$, i.e. 10. For the simple ring winding already discussed the commutator pitch is thus unity. Actually, then, the commutator pitch is equal to the number of insulation segments between the two ends of the coil.

The commutator pitch will be denoted by the symbol y_C.

Method of Numbering Armature Conductors. Various methods of numbering the conductors are in use. The simplest method is to number only the coils and not the individual coil-sides. This really resolves itself into numbering, simply, the upper coil-sides, as shown in Fig. 41. If it is required to number the lower coil-sides the following notation is often used: the upper side of coil 1 is denoted simply as 1, and the lower side as 1′. Since the number of coils is equal to the number of commutator segments, it becomes possible to arrange that the number of the coil shall be the same as that of the commutator segment to which its upper side is joined, as in Fig. 41. The convenience arising from this is obvious.

The *winding pitch* is expressed in terms of the number of coils. Thus, if the "finish" of coil 1 is connected, as in Fig. 41, to the "start" of coil 11, the winding pitch is 11 − 1, i.e. 10.

It will be noted that with the method adopted for numbering the coils the winding pitch has exactly the same value as the com-

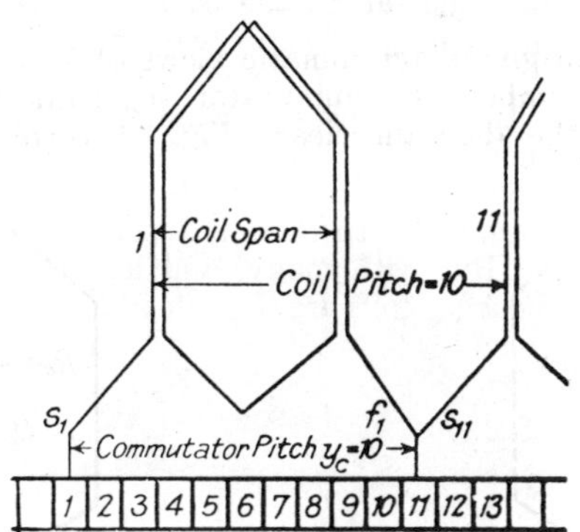

FIG. 41. COIL SPAN AND COMMUTATOR PITCH

mutator pitch, so that, in practice, it simply becomes necessary to deal with the commutator pitch.

With multi-turn coils, the coils only are numbered, as in Figs. 38 and 41, and not the individual turns; the treatment is thus identical with that for a single-turn coil.

Difference between Coil Span and Winding Pitch. It is still the custom in some books to make use of the term "back pitch" to denote what the author has called coil span. The *coil span* has no influence upon the type of winding, and will have no influence whatever upon the number of parallel circuits in the complete winding. On the other hand, the *winding pitch*—i.e. the commutator pitch—is vitally important in determining the type of winding and the number of parallel circuits. Put another way, a slight change in the value of the coil span would have but a minor effect upon the performance of a machine, whereas a slight change in the value of the winding pitch would completely alter the performance, as will appear later. The author, therefore, finds it desirable to use the term "coil span," and in general, as already stated, to express this span in terms of the number of teeth embraced by the coil. The winding or commutator pitch, on the other hand, is best expressed as the number of "micas" on the commutator between the two legs of the coil.

Simple Lap Winding. This winding differs from the simple ring winding only in the method of forming the individual coils. The winding scheme is identical with that for the ring winding, the "finish" of each coil being connected to the "start" of the next coil, so that the winding or commutator pitch is unity.

The winding may be "progressive" or "retrogressive." That is, it may progress in the direction in which the coils are wound, as in Fig. 42, or in the opposite direction as in Fig. 43.

Thus for a simple lap winding the commutator pitch is

$$y_C = \pm 1 \quad . \quad . \quad . \quad . \quad (62)$$

Figs. 44 and 45 give diagrammatic views of a 4-pole simple lap winding. The number of commutator segments taken is very small to make the diagram clear. Fig. 44 corresponds to the

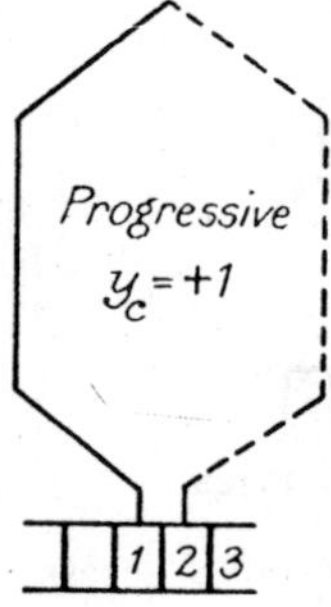

FIG. 42. COIL OF PROGRESSIVE LAP WINDING

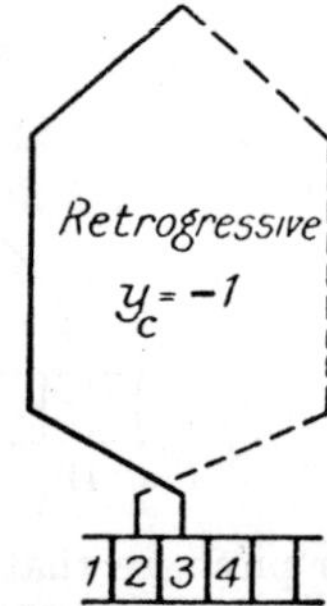

FIG. 43. COIL OF RETROGRESSIVE LAP WINDING

case of a disc-type winding rather than to the drum type. A definite attempt has been made to indicate that the conductors are grouped in slots; the lower sides of the coils are denoted by broken lines. The winding is progressive and the coil spans 3 teeth, i.e. a full pole-pitch, and corresponds exactly to the ring winding illustrated in Fig. 33. With this limited number of commutator segments, the total resulting e.m.f. induced between brushes would not be quite constant. Thus, an inspection of the developed view given in Fig. 45 indicates that for the armature position shown there are two coils in series between adjacent brush sets and that each brush "short-circuits" one coil. When the commutator has rotated through a distance corresponding to one half the width of one segment these coils will no longer be short-circuited and there will then be three coils in series between adjacent brush sets. However, one of them is almost inactive. With the number of segments normally used, the variations in the induced e.m.f. are very small, although still present, and in general it may be taken that the number of coils in series between brushes is equal to the number of coils per pole.

Hence, as in the simple ring winding, the number of parallel circuits, for a simple lap winding, is equal to the number of poles i.e.

$$2a = 2p \quad . \quad . \quad . \quad . \quad . \quad (63)$$

and the number of conductors in series per circuit is

$$Z_c = Z_a/2p \quad . \quad . \quad . \quad . \quad . \quad (64)$$

The coil span is determined, for this and all other types of winding, by the method already indicated. For the normal full-pitch coil,

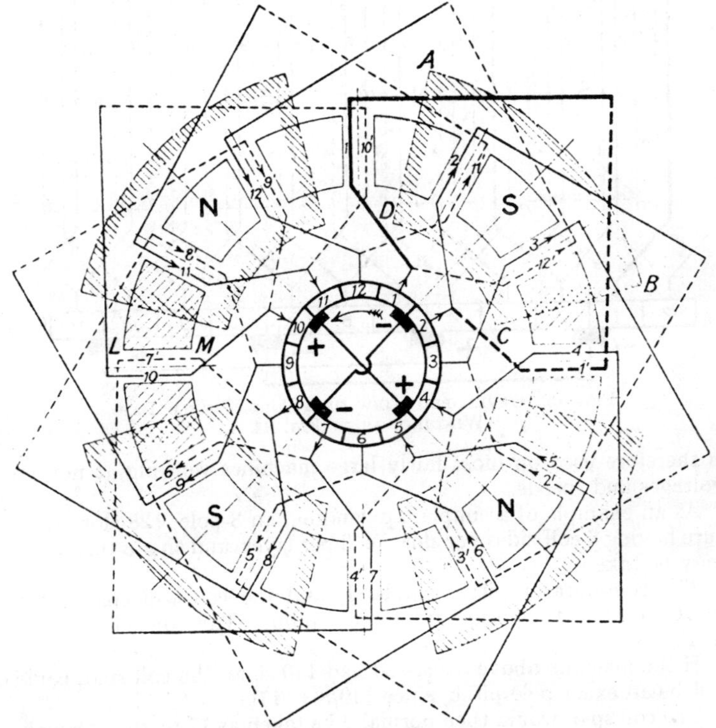

FIG. 44. DIAGRAMMATIC REPRESENTATION OF A 4-POLE, SIMPLE LAP WINDING

12 segments; single-turn coils; 12 slots; 2 coil-sides per slot

the number of teeth spanned by the coil is simply equal to the number of teeth per pole on the armature, i.e. 3, in Fig. 44. When the number of teeth does not divide exactly by the number of poles, the nearest integer to the actual quotient will in general be taken, although the winding would work perfectly well with either of the next integers.

Because of the number of parallel circuits obtained with this winding, it is often called the "Multiple Winding." The simple lap winding is well suited where the current to be handled is large, as it is then convenient to have a fair number of circuits in parallel in the armature. The current per brush set, and that carried by each conductor is thus reduced to a reasonable amount. The winding

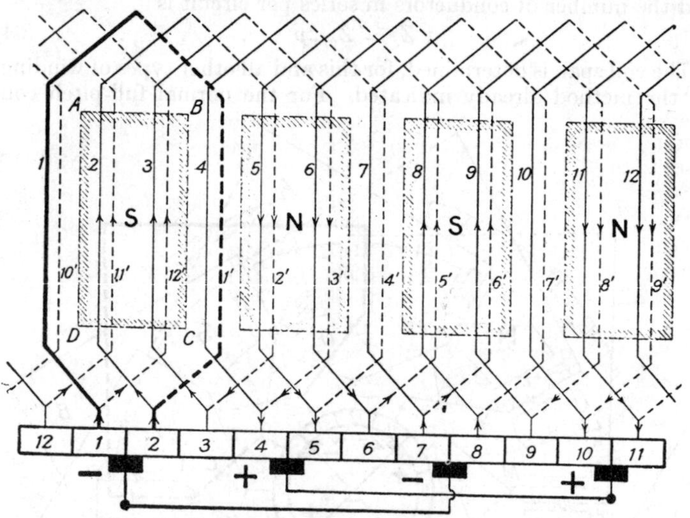

FIG. 45. DEVELOPED VIEW OF 4-POLE, SIMPLE LAP WINDING, AS IN FIG. 44

is therefore used on most fairly large machines working at normal voltages and speeds.

As an example of a simple lap winding, an 8-pole, 128-slot armature having 6 coil-sides per slot, i.e. 3 per layer, and single-turn coils, may be taken.

The commutator pitch will be 1, and the coil will span 128/8 = 16 teeth, for the normal full pitch winding. This coil is illustrated in Fig. 37.

If the machine above discussed had 140 slots, the coil span could not be an exact pole-pitch, since 140/8 = 17·5.

The coil span would then normally be taken as 17 teeth, although, as already stated, the nature of the winding would not be affected if 18 were taken.

Equalizing Rings. If batteries are joined in parallel, circulating currents will flow unless the e.m.f.'s of the batteries have exactly the same value. Similarly with a multipolar lap-wound armature, circulating currents will flow unless the values of the e.m.f. in the various circuits are identical. Further, reference to Fig. 45, will establish the fact that these circulating currents will have to flow through the brushes, and, when the machine is loaded, the tendency will be to overload certain of the brushes.

In practice, it is probable that the e.m.f.'s induced in the various circuits of a multipolar armature will not have the same value, as it is not to be expected that the values of the reluctance of the various magnetic circuits will be identical. The air gap may vary for different poles, and the iron parts may have considerable variation. The unequal loading of the circuits will result however in increased armature reaction by the more heavily loaded circuits tending to reduce the flux linked with the circuits concerned and so tending to equalize the e.m.f.'s.

Steps must be taken to prevent circulating currents flowing through the brushes; these steps consist in providing other permanent paths for the circulation of such currents, by connecting together certain points on the winding which should be at the same potential. Thus, for example, with an 8-pole armature having 400 commutator segments and coils, coils 1, 101, 201 and 301 are always exactly similarly situated with regard to the magnet system and should always be at exactly the same potential. These coils may therefore permanently be connected to an equalizing ring, the connexion to which may be taken either from the corresponding commutator segments or from the back of the coils as indicated in Fig. 46.

It is of course not possible to equalize the potential of every coil with that of correspondingly situated coils, but it generally suffices to employ from 10 to 20 equalizing connectors or rings, each having as many coils connected to it as the number of pole pairs. Normally it is arranged that the tappings to the various equalizing rings shall be equally spaced.

Thus, in the case above considered, 10 rings might be used. As there are 400 coils in all, i.e. 100 per pair of poles, the "pitch" of the tapping points would be 100/10 = 10. The connexions to the rings would then be as indicated in tabular form below.

Ring No.	1	2	3	4	5	6	7	8	9	10
Connected to coils numbered	1	11	21	31	41	51	61	71	81	91
Do. do.	101	111	121	131	141	151	161	171	181	191
Do. do.	201	211	221	231	241	251	261	271	281	291
Do. do.	301	311	321	331	341	351	361	371	381	391

The actual sectional area of the equalizing connexions is largely a matter of the designer's fancy. In general the rings are given a sectional area from 50 per cent to 100 per cent of that of the armature conductors.

To illustrate the use of equalizing rings, Fig. 46 is given, for the case of a 6-pole simple lap winding having 18 slots and 4 coil-sides per slot. There are 36 coils and commutator segments. Three equalizing rings are shown, the tappings on any particular ring are displaced by 120°, and for ring 1 are from coils 1, 13, and 25, as indicated.

Equalizing connexions may take one of two common forms.

The first type uses wires to join the appropriate commutator segments. These wires are formed to a shape similar to that of the

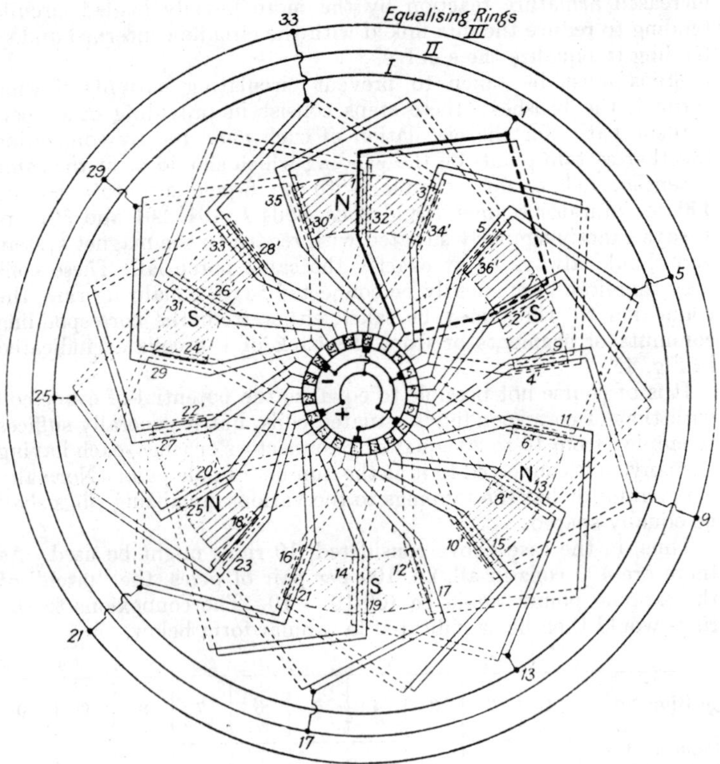

FIG. 46. 6-POLE SIMPLE LAP WINDING
4 coil-sides per slot; 36 segments; $y_C = 1$; coil span = 3 teeth (progressive winding)

coil ends and may be mounted at the back of the commutator under or over the coil ends themselves. This type has the disadvantage that they either obstruct the coil ends or are themselves obstructed by the coil ends.

The second type is fitted at the opposite end to the commutator and connected to tappings on the noses of the coils. It takes the form of rings made of copper strap. Each ring is insulated from the others and is connected to a group of equipotential coils. This type is usually mounted on the armature spider under the coil ends and is the type normally used on large machines.

Choice of Number of Slots for Lap Winding Limited. A scheme for a simple lap winding can be worked out for any number whatever of coils. If equalizing connexions are to be used, however, the winding must be symmetrically arranged with regard to the poles, that is, the number of slots must be a multiple of the number of pairs of poles. Generally, the number of slots used is a multiple of the number of poles so that full-pitch windings are possible. Occasionally, the number of slots is an odd multiple of the number of pairs of poles, e.g. 27 slots per pair of poles.

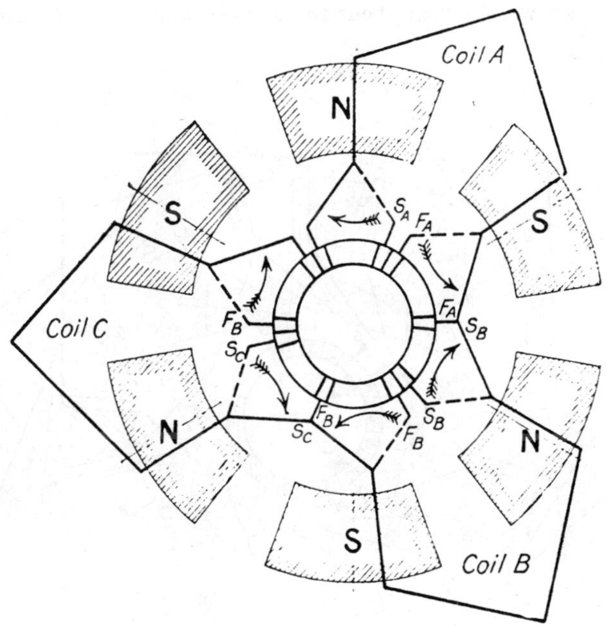

FIG. 47. PRINCIPLE OF SIMPLE WAVE WINDING

Simple Wave, or Series Winding. In the simple lap winding all the pole-groups of coils that generate e.m.f. in the same direction at any instant are connected in parallel by the brushes. In the simple wave winding, the arrangement is such that all coils carrying current in the same direction are connected in series. Thus all the coils carrying current in a clockwise direction are connected in series, and those carrying current in the opposite direction are connected in series. There are thus only two parallel circuits in the winding, quite independent of the number of poles.

Consider, for example, a 6-pole armature, as indicated in Fig. 47. The coils A, B, and C, two pole-pitches apart, are shown in their fully active position. In Fig. 47 the front connexions are drawn in broken lines for a lap winding, so that there is simply one coil between each pair of commutator segments. Since the three coils shown will always be generating e.m.f.'s in the same direction they may be connected in series to form a group of three coils, and the ends of the group joined to a pair of adjacent segments. To do this, it is of course not necessary to interfere with the span of the coils but simply to bend the front ends of the coils outwards as indicated by the arrows, F_A being bent to meet S_B, and so on. Connexions

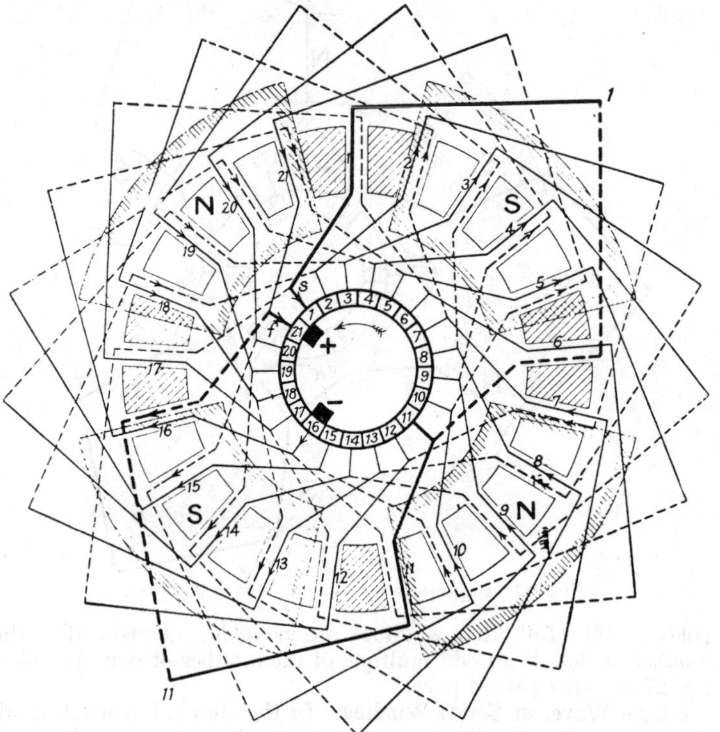

FIG. 48. 4-POLE SIMPLE WAVE WINDING
2 coil-sides per slot; 21 segments; $y_C = 10$; coil span = 5 teeth
(retrogressive winding)

to the commutator are taken from the junction points of all the coils, so that, for the 6-pole case indicated, three times the value of the commutator pitch must correspond to one more, or to one less, than the total number of commutator segments.

In general, the number of coils thus joined in series between adjacent commutator segments is equal to the number of *pairs of poles*. So that the relation between C, the number of coils or commutator segments, and the commutator pitch, for a machine having p pole pairs is

$$p \times y_C = C \pm 1 \quad . \quad . \quad . \quad . \quad (65)$$

If $+ 1$, the winding is "progressive," and if $- 1$, it is "retrogressive." In practice it is quite immaterial whether a progressive or a retrogressive winding is used.

A simple example of this type of winding is a 4-pole 21-slot armature, having single-turn coils, the number of commutator segments and of coils thus being $C = 21$ as in Figs. 48 and 49. For

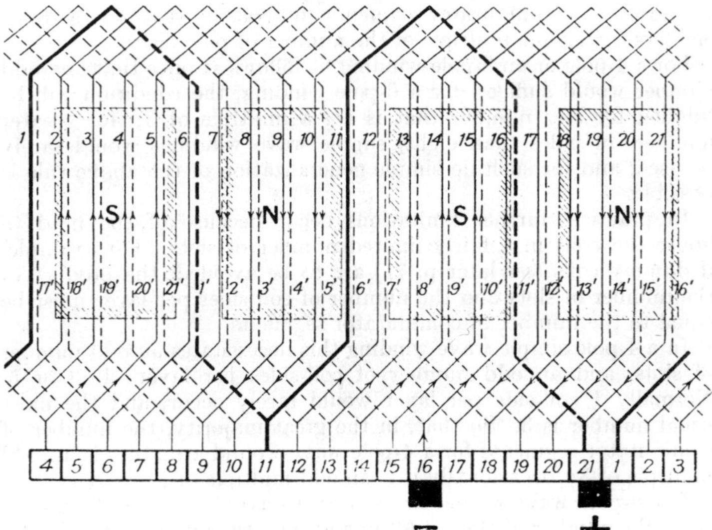

FIG. 49. DEVELOPED VIEW OF WINDING SHOWN IN FIG. 48

a retrogressive winding, as shown, the commutator pitch is $y_C = (21 - 1)/2 = 10$. The coils shown in heavy type indicate clearly that two coils are connected in series between each pair of adjacent commutator segments, the coils being practically diametrically opposite. The coil span has been taken to be 5 teeth, as 5 is the nearest integer to 21/4.

Further examples of simple wave windings may be—

1. A 4-pole armature having 129 commutator segments. Here either a commutator pitch of 64 or a pitch of 65 may be used. If the lower figure is employed the winding is retrogressive, and if 65, progressive.

In general, it may be taken that it is only with 4-pole armatures that the same number of segments will be suitable for either a progressive or a retrogressive winding.

2. A 6-pole armature having 128 commutator segments. Here, $3 \times 43 = 128 + 1$, so that with a commutator pitch of 43 a progressive winding can be obtained.

For a simple 2-circuit wave winding and the given number of segments a pitch of 42 is *not* suited, for the condition given in (65) above must be fulfilled *exactly*.

3. An 8-pole armature having 129 commutator segments. Here a commutator pitch of 32 is the only possibility ($4 \times 32 = 129 - 1$).

Choice of Number of Slots for Wave Winding Limited. It has been seen that for a simple wave winding the number of coils and commutator segments must conform to the expression $C = py_C \pm 1$. Since the commutator pitch of necessity is an integer, it follows that the number of commutator segments and coils must be one more or one less than some multiple of the number of pole pairs.

For a 4-pole or an 8-pole winding, it follows at once that any odd number would suffice; for a 6-pole winding, the condition will be fulfilled for any number that is *not* a multiple of 3; for greater numbers of poles than eight, simple wave windings would rarely be used, and for such no simple generalization of the above rule is possible.

In practice, further limitations occur because of the need to house the coils in a fairly limited number of slots. For example, if dummy coils (*see* later p. 71) are to be avoided, the product of the number of slots and the number of coil sides per layer must be equal to the number of commutator segments.

In a 4-pole simple wave winding this necessitates an odd number of slots, and an odd number of coil-sides per layer—1, 3 or 5. Normally 1 coil-side per layer would rarely occur, and the most usual number is 3. So that, in the great majority, the number of commutator segments for a 4-pole wave-wound armature is an odd multiple of 3; occasionally it is an odd multiple of 5.

For 6-pole wave-wound armatures, to avoid the use of dummy coils, the number of slots, and, in addition, the number of coil-sides per layer, must not be a multiple of 3. It therefore becomes necessary to employ either 2 or 4 coil-sides per layer.

For an 8-pole machine the same conditions hold as with a 4-pole.

The student should have no difficulty in determining a satisfactory arrangement for other numbers of poles, should occasion arise.

The windings so far illustrated have but one coil-side per layer. As further illustrations of simple wave windings, therefore, Fig. 50 is given illustrating a 4-pole winding having 3 coil-sides per layer, i.e. 6 coil-sides per slot—a very common value in 4-pole machines. The number of slots, 13, is of course very much lower than normal. There are 39 coils and commutator segments. That illustrated is for a progressive winding, so that the commutator pitch $y_C = (39 + 1)/2 = 20$. A retrogressive winding might equally well have been taken—the pitch would then be $y_C = (39 - 1)/2 = 19$.

The coil span has been taken to be 3 teeth (13/4 = $3\frac{1}{4}$).

Fig. 51 illustrates a 6-pole simple wave winding having 16 slots and 2 coil-sides per layer, single-turn coils. There are 32 coils and commutator segments.

In this, a progressive winding is the only type possible, the

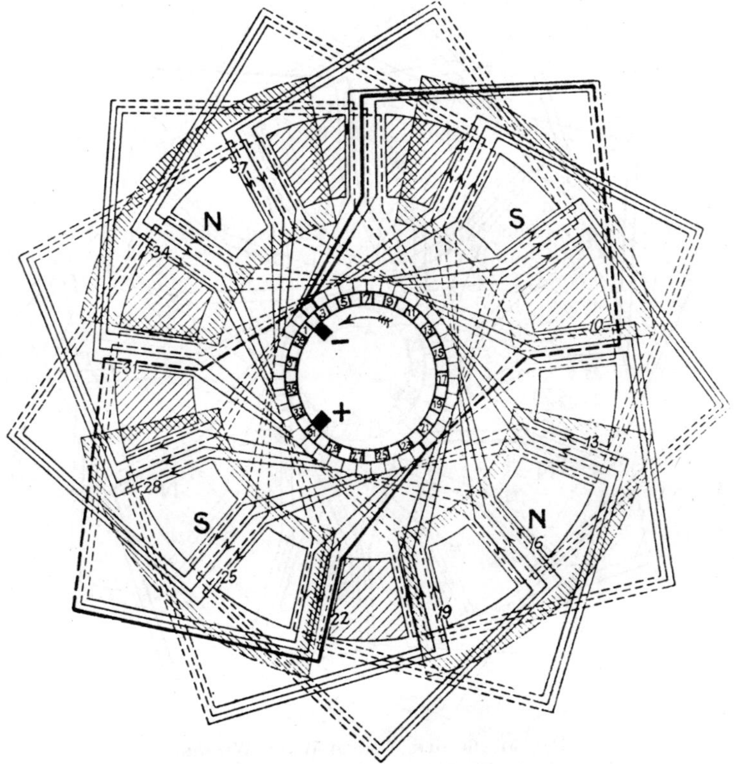

FIG. 50. 4-POLE, SIMPLE WAVE WINDING

6 coil-sides per slot; 39 segments; y_C = 20; coil span = 3 teeth (progressive winding)

pitch being $y_C = (32 + 1)/3 = 11$. A retrogressive winding is not possible, since $(32 - 1)$ is not a multiple of 3.

The coils in heavy type indicate clearly that between adjacent segments there are now three coils connected in series, the coils being mutually displaced, by approximately 120°. A span of 3 teeth has been chosen, corresponding to rather more than a full pole-pitch.

Dummy Coils. Occasionally, to make use of a standard armature punching, it becomes necessary to employ for a simple wave winding a number of slots that does not conform with the above requirements, so that more coils—usually only one more—are provided than can be utilized. The extra coil, called a dummy coil, is put in position to maintain the mechanical balance of the armature, but is completely insulated from the remainder of the

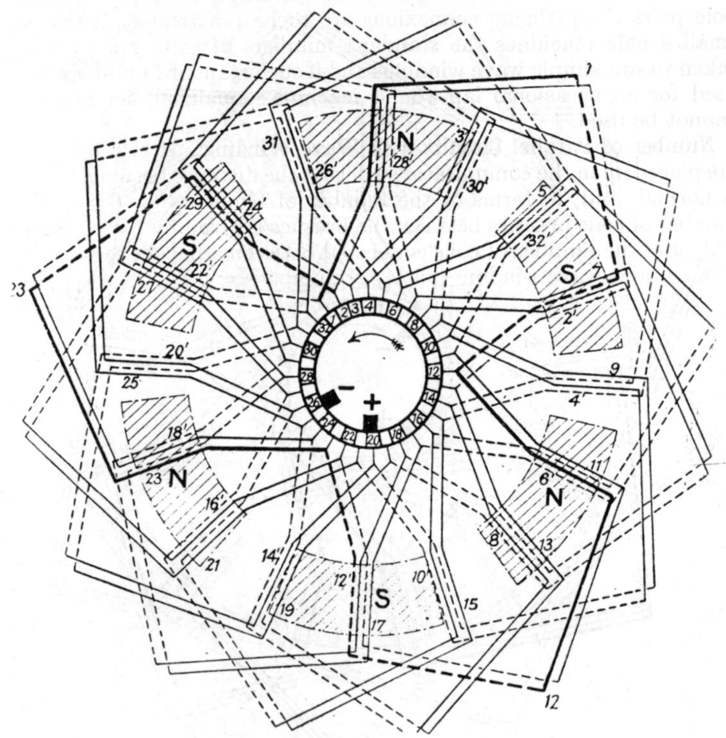

FIG. 51. 6-POLE, SIMPLE WAVE WINDING
4 coil-sides per slot; 32 segments; $y_C = 11$; coil span = 3 teeth (progressive winding)

winding. Dummy coils are avoided if at all possible, owing to the slight dissymmetry they introduce.

An example of the use of a dummy coil is an armature having 72 slots, 4 coil-sides per slot—i.e. 2 per layer—and wound with a 6-pole simple wave winding. The total number of coils, including the dummy, is $2 \times 72 = 144$. The nearest possible number of commutator segments for a 6-pole winding is 143. A commutator pitch of 48 would give a progressive winding since $3 \times 48 = 143 + 1$.

A dummy coil is therefore necessary. In numbering the coils, no number is given to the dummy, and the best plan is to give the number 1 to the coil next the dummy.

Occasionally, to employ a standard commutator for a special purpose, it may be necessary to use a dummy commutator segment, in which case the segment is connected to one of its neighbours.

It will be noted that, if dummy coils are to be avoided, with simple wave-wound machines the number of slots cannot be a multiple of the number of pole pairs, whereas with lap-wound machines, the number of slots must be a multiple of the number of pole pairs if equalizing connexions are to be practicable. In very small 4-pole machines the standard numbers of slots are usually taken to suit simple wave windings and if such standard numbers are used for an occasional lap-wound machine, equalizing connexions cannot be used.

Number of Parallel Circuits with Wave Winding. If two brushes are placed upon the commutator such that the distance between them is normal, and, in terms of the number of segments, $= C/2p$, the number of coils in series between the brushes is given by the product $C/2p$ and the number of coils connected between adjacent segments. For a simple wave winding this latter number, as in Fig. 47, is equal to the number of pole pairs, i.e. to p. Hence the number of coils in series for a simple wave winding is $p . C/2p = C/2$. The total number of coils being C, the number of parallel circuits, quite independent of the number of poles, is two; i.e. for a simple wave winding.

$$2a = 2 \quad . \quad . \quad . \quad . \quad . \quad (66)$$

and the number of conductors in series per circuit is

$$Z_c = Z_a/2 \quad . \quad . \quad . \quad . \quad . \quad (67)$$

To illustrate this fact the table below has been prepared for a 6-pole, 50-commutator segment, simple wave armature. (The number of commutator segments has been taken much smaller than in practice, to prevent the table from being unwieldy.) In this table, the two circuits through the winding, starting from coil 1, are given. The portion of the table to the right indicates that from segment 1 we pass through coil 1 to segment 18, thence by coil 18 to segment 35, thence through coil 35 to segment 2. From segment 2 we pass to segment 19, and so on until segment 9 is reached, by which time we have passed through 24 coils. Passing then through coil 9 brings us to segment 26. Travelling now through the winding in the reverse direction from segment 1, as indicated in the portion of the table to the left, from segment 1 we pass through coil 34 to segment 34; thence through coil 17 to segment 17, and so on until segment 26 is reached.

It will be noticed that if brushes are placed upon segments 1 and 9 respectively, although these brushes are not quite a pole-pitch apart, there are 24 coils in the circuit indicated to the right of the table, and a little consideration will establish the fact that the e.m.f.'s induced in these 24 coils always act in conjunction. Similarly for the circuit to the left, between segments 1 and 44 there are 24 coils in series. If we wish to include the full maximum possible number of coils in each circuit, i.e. 25, instead of placing the second brush upon segment 9 it must be placed upon 26. This large displacement of the brush serves simply to include coil 9 in the circuit to the right.

Brush Sets and Brush Position for Simple Wave Winding. In the example just considered it is apparent that it is sufficient to employ

WINDING SCHEME FOR 6-POLE SIMPLE WAVE WINDING HAVING 50 COMMUTATOR SEGMENTS

Progressive winding: $y_c = (50 + 1)/3 = 17.$

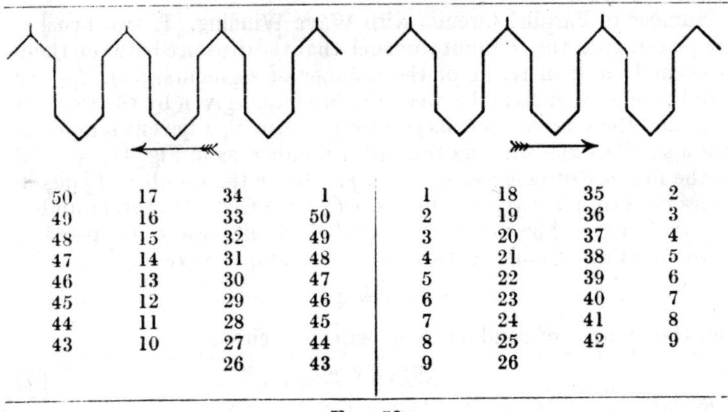

50	17	34	1	1	18	35	2
49	16	33	50	2	19	36	3
48	15	32	49	3	20	37	4
47	14	31	48	4	21	38	5
46	13	30	47	5	22	39	6
45	12	29	46	6	23	40	7
44	11	28	45	7	24	41	8
43	10	27	44	8	25	42	9
		26	43	9	26		

FIG. 52

two brush sets only, placed upon segments corresponding to 1 and 26, i.e. 3 pole-pitches apart. With the brushes in the neutral position the coils joining segments 26 and 9, and 26 and 43 respectively, will be idle, so that it will also suffice if brushes are placed upon either of segments 9 or 43 instead of upon 26. That is, the brush sets may be either 1 pole-pitch apart or 3 pole-pitches.

With simple wave-wound armatures it is then necessary to use only one set of positive brushes and one of negative, quite independent of the number of poles. The brush arms must be displaced by some odd multiple of a pole-pitch. If, however, it is desirable to keep down the length of the commutator, the number of brush sets may be increased up to the normal number in use for a lap-wound machine. Normally, small d.c. 4-pole armatures are wave wound and have only two brush sets, as in Fig. 48. With a 6-pole wave-wound armature 2, 4, or 6 brush sets may be used.

Unless the current to be collected is very small, as many brush sets are used as in lap windings, and advantage taken of the small current per brush arm to employ a very short, and therefore cheap commutator.

Coil Span. This is determined exactly as for any other type of drum winding. Since, if there are no dummy coils, the number of slots will not be a multiple of the number of poles, the coil span cannot be an exact pole-pitch, and the number of teeth embraced by the coil is taken to correspond as closely as possible to the actual

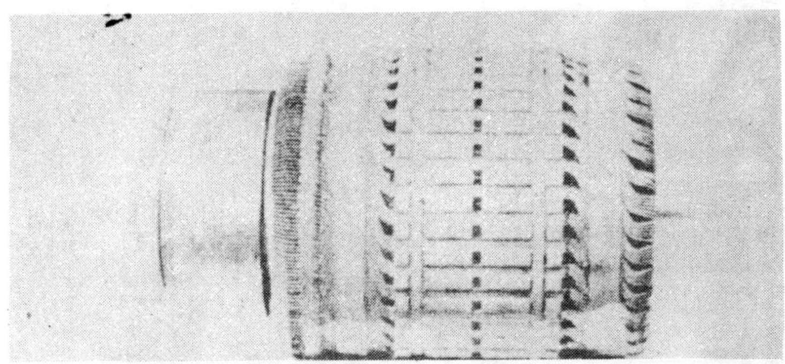

FIG. 53. SMALL CURRENT, STRAP-WOUND ARMATURE
Simple wave winding; 12 bars per slot
(*Lancashire Dynamo and Crypto, Ltd.*)

number of teeth per pole. Thus, for the three examples already considered on page 69,

1. For a 43-slot, 3 coil-sides per layer, 4-pole armature,

$$43/4 = 10{\cdot}75$$

The coil span may be taken as 11 teeth, but satisfactory results would obtain if the value 10 were taken.

2. For a 64-slot, 2 coil-sides per layer, 6-pole armature,

$$64/6 = 10{\cdot}67.$$

The coil span may again be either 11 or 10 teeth.

3. For a 43-slot, 3 coil-sides per layer, 8-pole armature,

$$*43/8 = 5{\cdot}375.$$

The coil span would then be 5 teeth.

Appearance of Lap- and Wave-wound Armatures. A mere inspection of a normal armature suffices to establish whether it is lap wound or wave wound, since the direction in which the ends of the upper portions of the coils are bent can normally be seen. Thus Fig. 54 obviously represents a lap-wound armature, whereas Fig. 53 represents a wave-wound machine, the front ends of the coils being bent "inwards" for the former and "outwards" for the other. Fig. 53 represents a small machine in which the armature is wave wound, so that there are only two paths in parallel; the commutator is short, and there are 12 thin conductors per slot. Fig. 54 represents

* An 8-pole armature would generally have many more than 43 slots.

a large heavy current armature, the rated current being of the order of 8,000 A.

FIG. 54. ARMATURE OF 7,000 H.P. MILL MOTOR; 800 V, 50/120 REV/MIN

(*Metropolitan-Vickers Electrical Co., Ltd.*)

Voltage between Adjacent Commutator Segments for Simple Lap- and Wave-windings.

If N_a = no. of turns in each armature coil

E_c = average e.m.f. induced in each conductor, i.e.

$= B_{av}Lv = 2p\Phi n.$

Then the average value of the voltage between adjacent commutator segments is

$V_{av} = 2N_aE_c$ for the simple lap winding
and $\quad = 2pN_aE_c$ for the simple wave winding.

As already stated on p. 50 the maximum voltage between segments is $V_{max} = V_{av}/K_f$ and is normally 1·4 to 1·5 times V_{av}. As discussed later, the working value of the voltage between adjacent commutator segments is strictly limited, so that for large machines where E_c is relatively large $N_a = 1$ and in addition a simple lap-winding is generally used. For small machines where E_c is small, the simple wave winding is best suited and may be in addition multi-turn coils, with $N_a = 2, 3, 4$, etc. are used. Inevitably there are border-line cases for which either the simple lap- or the simple wave-winding may be used, and other cases where neither of these arrangements is quite suitable and a multiplex winding may be desirable.

Multiplex Windings. Although for the vast majority of cases, either the simple lap- or the simple wave-winding is suited, occasion may arise when the simple wave-winding gives too few parallel circuits, i.e. when it is desirable to have more than two circuits but less than the number given by a simple lap winding. Or it may be that the simple lap winding does not provide enough parallel circuits. Such cases can be met by the use of what, for the moment, we may call a multiplex winding. Here it should be mentioned that, although the various winding schemes that can be devised are many, in practice but few have been found desirable. Further, much confusion has arisen out of the fact that various authors do not agree in their definitions of certain terms. It is not helpful to find the same type of winding described by various authors as "duplex, singly re-entrant," "duplex, doubly re-entrant," and "duplex with two independent windings"; and another type as "simplex, doubly re-entrant," "duplex, singly re-entrant" and "duplex with single helix."

In common with the author who avoided the use of the term "re-entrant," to wit, C. C. Hawkins, the present writer favours the following notation: "A duplex lap, or wave, winding is a lap, or wave, winding which has twice as many parallel paths as the simple lap, or wave, winding respectively."

Similar definitions would hold for triplex or quadruplex, etc., windings, although these latter are not likely to be met in practice.

As will be shown shortly, a duplex winding may consist of a single closed circuit or of two entirely separate closed circuits. Similarly a triplex winding might consist of a single winding or of three independent windings. Normally, the author finds it sufficient to define a winding simply in terms of the number of parallel circuits it contains, and to avoid altogether the use of an elaborate nomenclature.

Duplex Lap Windings. Fig. 55 indicates a lap winding in which the pitch of the coils, instead of being unity, is increased to 2,

i.e. $$y_C = \pm 2 \quad . \quad . \quad . \quad . \quad (68)$$

It is apparent that the number of coils in series between adjacent brush sets is now only one half of the total number of coils per pole; but, on the other hand, by using brushes of sufficient thickness to cover always at least two commutator segments, the number of circuits in parallel is double that obtaining for a simple lap winding.

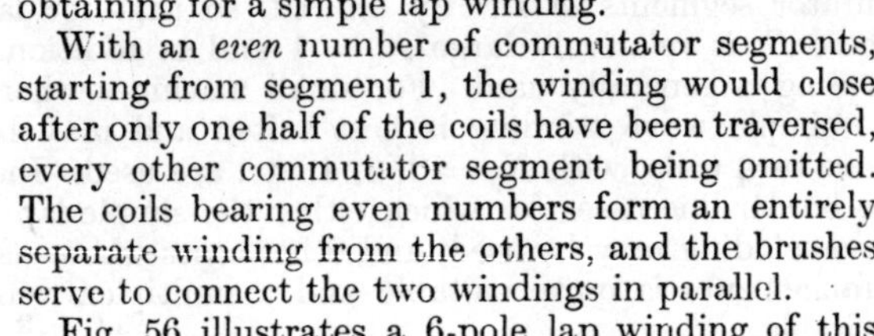

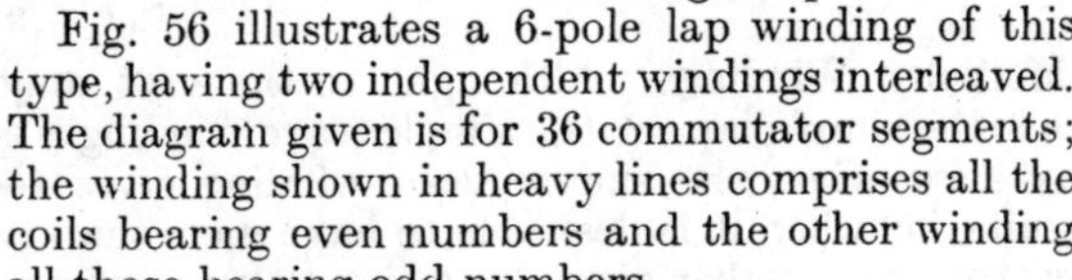

Fig. 55. Coil of Duplex Lap Winding

With an *even* number of commutator segments, starting from segment 1, the winding would close after only one half of the coils have been traversed, every other commutator segment being omitted. The coils bearing even numbers form an entirely separate winding from the others, and the brushes serve to connect the two windings in parallel.

Fig. 56 illustrates a 6-pole lap winding of this type, having two independent windings interleaved. The diagram given is for 36 commutator segments; the winding shown in heavy lines comprises all the coils bearing even numbers and the other winding all those bearing odd numbers.

With an *odd* number of commutator segments, the winding will not close after one complete tour of the commutator, but during

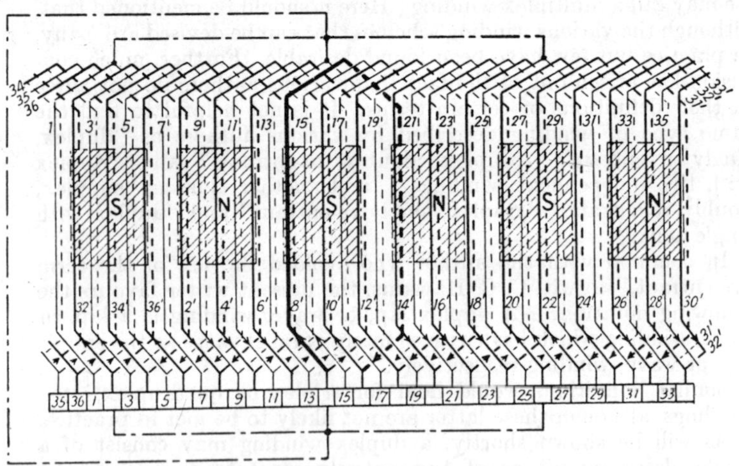

Fig. 56. 6-Pole Duplex Lap Winding

Doubly-closed; 36 segments; 12 circuits; $y_C = +2$

the second tour will include those coils and commutator segments that were passed over during the first tour, and finally will close on the starting point after two complete tours of the commutator.

This duplex winding thus consists of a single closed circuit. (A winding of this type was called by Arnold and S. P. Thompson "doubly re-entrant.")

A singly-closed winding of this type is illustrated in Fig. 57 for a 33-segment 6-pole armature. Starting on a tour through the

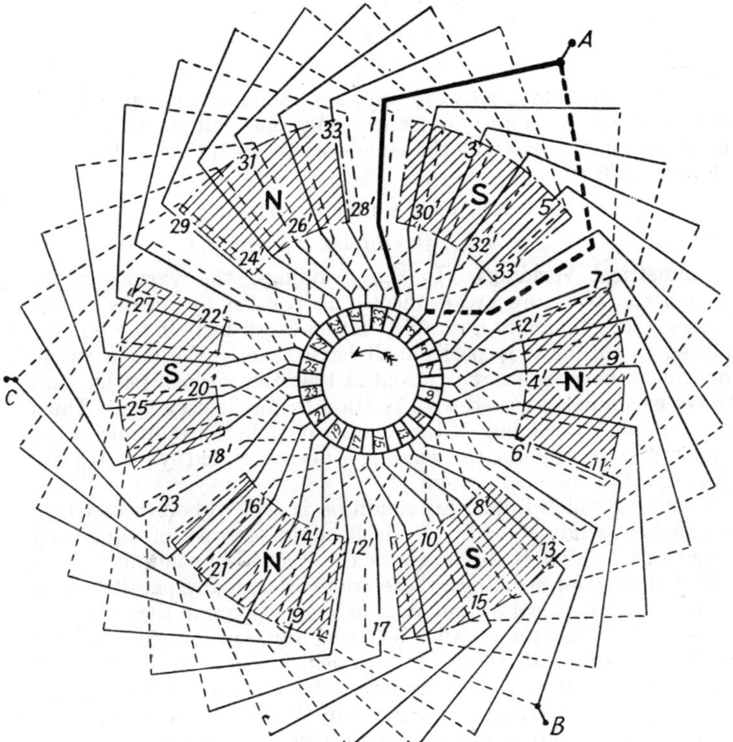

FIG. 57. SINGLY-CLOSED DUPLEX LAP WINDING
6 poles; 12 circuits: $C = 33$; $y_C = 2$

winding from coil 1, first all the odd numbered coils are encountered until coil 33 is reached. Then the coils having even numbers are traversed until, having passed through every coil, the winding finally closes at the starting point.

With this type of duplex lap winding there is thus a circuit between neighbouring segments, consisting of two paths in parallel, each path comprising one half of the total number of armature coils. The resistance between neighbouring segments should then test out at one quarter of the total resistance of the whole of the armature coils assumed connected in series.

With the former type, having two entirely independent windings, there is no circuit between adjacent commutator segments if the brushes are lifted. In both cases, however, when brushes covering always not less than 2 segments are placed normally in position, the number of parallel circuits is twice the number of poles.

For a duplex lap winding then

$$2a = 4p \quad . \quad . \quad . \quad . \quad . \quad (69)$$

and

$$Z_c = Z_a/4p \quad . \quad . \quad . \quad . \quad . \quad (70)$$

The average voltage between adjacent commutator segments for the duplex lap winding is evidently only one half its value for a simple lap winding based on the same working conditions of flux, etc.

For single turn coils—

$V_{av} = 2E_c$ for a simple lap winding

and $V_{av} = E_c$ for the duplex lap winding.

Symmetrical Windings. In d.c. windings, the various paths through the armature are connected in parallel by means of the brushes. It is therefore desirable that at every instant the e.m.f.'s induced in the various paths shall have exactly the same value. For this to be possible it is essential that the various paths shall always be similarly placed with relation to the field system. Thus, in the simple lap winding, if the number of slots is a multiple of the number of pairs of poles, the winding is always placed symmetrically with regard to the field system.

A winding which fulfils this condition is said to be symmetrical, and such a winding readily lends itself to the use of equalizing connexions, as already discussed for the simple lap winding.

With duplex lap windings, there are twice as many parallel paths as the number of poles, and it is at once apparent that the two paths immediately in parallel between adjacent brushes are *not* similarly placed with regard to the field system. It is therefore impossible to obtain a perfectly symmetrical duplex lap winding.*

Equalizing Connexions to Duplex Lap Windings. With suitably-designed duplex lap windings certain coils are precisely two pole-pitches apart. Equalizers can therefore be used, a necessary condition being, as with the simple lap winding, that the number of slots must be a multiple of the number of pairs of poles.

This condition is indeed fulfilled with both the 6-pole duplex windings illustrated, in Figs. 56 and 57, pp. 78, 79, both being for one coil-side per layer in the slots. Thus in Fig. 56 there are 36 coils and 6 poles, and equalizers can be connected to coils 1, 1 + 36/3, i.e. 13 and 25, and to 4, 16 and 28, and so on.

In Fig. 57 there are 33 coils, 11 for each pair of poles and equalizers can be connected to coils 1, 1 + 11, i.e. 12, and 23, at the points indicated A, B, and C and to coils 2, 13, and 24, and so on.

* The singly-closed duplex lap windings already given may be regarded as "hemisymmetrical."

It will be noted that for the singly-closed winding, Fig. 57, connexions are made alternately to odd- and even-numbered coils, that is, to each in turn of the two interleaved paths in the winding.

But for the doubly-closed winding, Fig. 56, each individual equalizer is joined solely to either odd- or to even-numbered coils. However, it is possible to equalize the two paths represented by these odd- and even-numbered coils. Thus Fig. 56 shows an equalizer joined at the back end of the armature to coils 1, 13, and 25, and another equalizer joined at the front end to coils 2, 14, and 26. Evidently segment 2 can be "equalized" with the back end *A* of coil 1, so that the two equalizers can be connected by an inactive conductor passing through the tunnel of the armature, as shown by the dotted line. The need for this can be avoided with certain numbers of poles as discussed later.

(Both the windings illustrated have of course too few commutator segments to be practicable.)

With *singly-closed duplex lap windings* there must be an odd number of coils and of commutator segments. If equalizers are to be used the number of slots per pole pair must be odd, also the number of coil sides per layer must be odd. This limits such windings to machines having an *odd number of pole pairs,* and also to 1, 3, etc. coil-sides per layer.

With *doubly-closed duplex lap windings,* if it is desired that the connexions to each equalizer shall alternate between odd- and even-numbered coils, the number of coils per pair of poles must be odd. Since the total number of coils must now be even, it follows that the number of pole pairs must be even, the number of slots per pole pair and of coil-sides per layer must both be odd.

Thus for both the singly-closed and doubly-closed duplex lap windings there must be an odd number of commutator segments per pair of poles, an odd number of slots per pair of poles, and an odd number of coil-sides per layer. For the singly-closed winding the number of pole pairs *must not* be even, but must be 3, 5, 7, 9 etc.: for the doubly-closed winding the number of pole pairs *must* be even, such as 4, 6, 8, 10 etc.

These windings are of only very limited application, but have been adopted for a few machines of very large rating, as discussed later, in order to use more commutator segments than would be required by the simple lap winding and so keep down the voltage between adjacent commutator segments, as stated in p. 80.

Taking 45 slots per pair of poles, 3 coil-sides per layer, i.e. 135 commutator segments per pair of poles, the connexions to the equalizers would be—

Equalizer 1 joined to coils 1, 136, 271, 406, 541, 676
Equalizer 2 joined to coils 4, 139, 274, 409, 544, 679, etc.

As these windings are used on only very large machines—working towards normal output limits—a large number of equalizers would be used. In the above example there would be 45.

The connexions given above are for a *12-pole doubly-closed*, duplex-lap machine having 270 slots and 810 commutator segments.

They also hold, with the last numbers deleted, 676, 679, etc., for a *10-pole singly-closed* duplex lap winding having $5 \times 45 = 225$ slots and 675 commutator segments.

For a *16-pole doubly-closed* duplex lap winding it is necessary to extend the connexions to coils 946, 949, etc. for equalizers 1, 2, etc. and there would now be 360 slots and 1,080 commutator segments.

With an increase in the number of poles to 18 with $9 \times 45 = 405$ slots and 1,215 commutator segments there would be 9 connexions to each equalizer, further connectors being taken from coils 1,081, 1,084, etc. to equalizers 1, 2, etc.

In a similar manner schemes for various numbers of poles can quickly be worked out for any odd number of slots per pair of poles and 3 coil-sides per layer, e.g. 39 slots, 117 commutator segments per pair of poles. The connexions to one equalizer for a 12-pole armature would be coils 1, 118, 235, 352, 469, and 686.

By conforming to the limits mentioned above, a winding scheme is possible with the connexions to each equalizer made alternately to odd- and even-numbered coils. But if a doubly-closed duplex lap winding having an even number of slots per pair of poles is used, each equalizer will be connected solely to the odd- or to the even-numbered coils, so that the windings would still be quite separate save for the parallelling effect of the brushes.

As already shown in Fig. 56 two sets of equalizers can be used, one at the back-end of the armature joined to the "odd" coils and the other at the front-end joined to the "even" commutator segments. These equalizers can be "equalized" by inactive conductors taken through the tunnel of the armature.

When it is possible, it is probably best to use an odd number of commutator segments per pair of poles together with either a singly-closed or doubly-closed winding, which ever suits the number of poles.

It may be stated, however, that practice is not uniform in this matter—and of course it may not always be possible to use an odd number of coil-sides per layer—and such duplex doubly-closed lap windings having an even number of commutator segments per pair of poles are in use, with equalizers joined to one end only of the winding. In such instances the maximum possible number of equalizers is used, i.e. a connexion is taken from each coil. As already stated, these windings are not widely used at present, and then only for very large machines.

It will be noticed that in all the above duplex-lap windings the number of connexions to each equalizer is equal to the number of pole pairs, just as with the simple lap winding, whereas the number of pairs of parallel paths is now $a = 2p$.

Duplex Wave Windings. Just as in lap windings, it is possible to wind an armature with two entirely independent simple wave windings, interleaved as in Fig. 58. For the duplex lap winding, it will be remembered, the commutator pitch was 2. For the duplex

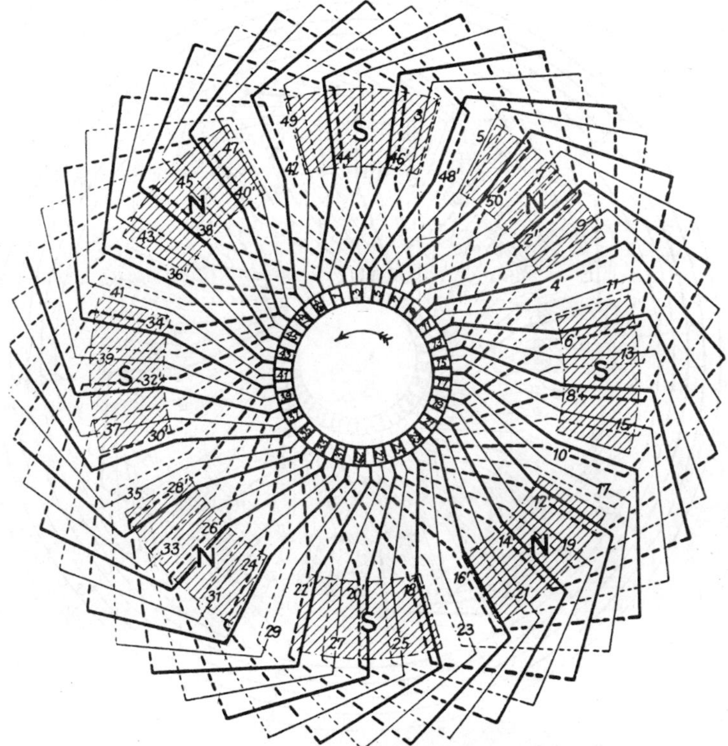

FIG. 58. 4-CIRCUIT, 8-POLE WAVE WINDING WITH TWO SEPARATE WINDINGS

$C = 50$; $y_C = 12$ (retrogressive winding)

wave winding, starting from any given coil and traversing as many coils as there are pairs of poles will take us to the coil *next but one*—either in front or behind—to the starting coil. Thus, for a duplex wave winding

$$py_C = C \pm 2 \quad . \quad . \quad . \quad . \quad (71)$$

as compared with $py_C = C \pm 1$, for the simple wave winding. As in the duplex lap winding, the duplex wave winding may consist of a single closed circuit, or of two entirely separate closed circuits. A little consideration suffices to establish that, again with an even number of commutator segments, there will be two independent windings if the commutator pitch is also even, the one winding comprising all the coils bearing even numbers, and the other all the coils having odd numbers.

The winding will however be singly-closed for all other conditions,

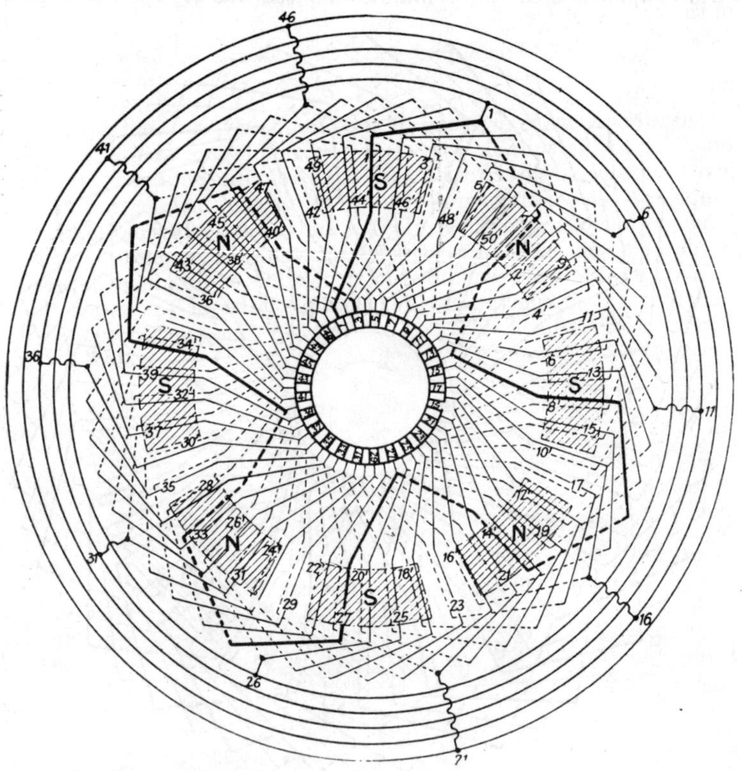

FIG. 59. 4-CIRCUIT, 8-POLE WAVE WINDING, SINGLY-CLOSED
$C = 50$; $y_C = 13$ (progressive winding)

that is, where the number of commutator segments is "odd," or where the commutator pitch is an odd number.

As a simple illustration, take an 8-pole, 50-segment commutator. To obtain 4 circuits in parallel the commutator pitch for a retrogressive winding is $y_C = (50 - 2)/4 = 12$.

The number of segments being even, in addition to the commutator pitch, a double winding results as indicated in Fig. 58. On the other hand, the pitch for a progressive winding becomes $y_C = (50 + 2)/4 = 13$, and a single winding results, as shown in Fig. 59.

With duplex wave windings it is necessary that the number of commutator segments, of coils, and of slots, shall suit the condition represented in (71).

The number of parallel circuits is double that obtaining for the simple wave winding, and is therefore always four, quite independent of the number of poles.

$$\text{For duplex wave } 2a = 4. \quad . \quad . \quad . \quad (72)$$

and

$$Z_c = Z_a/4 \quad . \quad . \quad . \quad (73)$$

The duplex wave winding is used on fairly rare occasions in practice, and is suitable for where the simple lap winding would give too many parallel circuits, and the two circuits obtained from the simple wave winding are insufficient. Such include machines having a fairly large number of poles and a relatively small current output, e.g. low speed or relatively high voltage machines.

With duplex wave windings it is possible to obtain symmetry under certain circumstances. For example, take an 8-pole, 126-slot winding having 6 coil-sides per slot and 378 commutator segments. For this duplex winding the commutator pitch may be (378 + 2) ÷ 4, i.e. 95, or (378 — 2) ÷ 4, i.e. 94. By taking 95—an odd number—the winding will be singly-closed, an arrangement usually preferred. In this machine the number of slots is divisible by 2, and, in consequence, the number of coils and commutator segments is also a multiple of 2. Hence a circuit starting from any given commutator segment is always exactly similarly situated with regard to the field system as another circuit starting from a commutator segment diametrically opposite the first. For example, coil 1 and coil 1 + 378/2, i.e. coil 190 are always similarly situated with regard to the field system. The commutator pitch being 95, the two circuits in parallel in the forward direction from a brush bedding on segments 378 and 1, to the next brush will comprise, respectively, the coils

1 — 96 — 191 — . . .

and 378 — 95 — 190 — 285 . . .

If the magnetic balance of the machine is perfect, the e.m.f.'s induced in coils 1, 96, 191, etc., are at every instant equal respectively to those induced in coils 190, 285, 2, etc., so that the two circuits are perfectly balanced and the winding is symmetrical.

To obtain a symmetrical duplex wave winding it is then necessary that—

1. The number of pole pairs shall be divisible by 2.
2. The number of slots shall also be a multiple of 2.

If the number of slots is a multiple of 2, and of course no dummy coils are used, it follows that the number of coils and the number of commutator segments will also be some multiple of 2. In addition, it is of course necessary that the number of coils shall conform to $py_C = C \pm 2$.

It follows at once from condition 1 that symmetrical duplex wave windings cannot be obtained for machines having an odd number of pairs of poles, and, in particular for 6-pole and 10-pole machines.

For a 4-pole machine, the simple lap winding has the same number of parallel circuits as the duplex wave, hence the duplex wave is not often used for 4-pole d.c. machines.

For an 8-pole machine all the conditions with regard to symmetry and number of coils will be observed if the number of slots is an odd multiple of 2, and there is an odd number of coil sides per layer, e.g. 3 or 5.

For larger numbers of poles—12, 16, etc.—no simple generalization is possible, and each case must be treated on its merits.

Equalizing Connexions to Duplex Wave Windings. As in simple lap windings, it is desirable to equalize the potential of certain points of the winding which, with perfect magnetic balance, etc., should be naturally at the same potential.

With symmetrical duplex wave windings (and normally only symmetrical windings should be used) for all even numbers of pole pairs, diametrically opposite commutator segments should always be at the same potential and therefore may be connected by an equalizing connexion. Independent of the actual number of poles there will be simply two connexions to each equalizer.

Thus, in the example considered above, the pitch of the connectors to the equalizers is $378/2 = 189$, so that coils 1 and 190 can be equalized. The number of equalizing connectors is chosen quite arbitrarily. In this, 12 connectors would suffice, and would be spaced 16 segments apart, with three exceptions, where the spacing would be 15.*

The spacing between every fourth pair of tappings would then be made 15, the connexions to the various connectors being: 1 — 190; 17 — 206; 33 — 222; 48 — 237; 64 — 253; 80 — 269; 96 — 285; 111 — 300; 127 — 316; 143 — 332; 159 — 348; 174 — 363.

In Fig. 59 showing an 8-pole 4-circuit winding, the number of slots is 50, there being 2 coil-sides per slot. Connexions to 5 equalizers are shown. Coil 1 is equalized with the diametrically opposite coil, i.e. coil 25, coil 6 with 31, and so on.

Triplex and Multiplex Windings. In general, the pitch of a lap coil may be m commutator segments, and that for a wave winding may be $y_C = (C \pm m)/p$ commutator segments, where m is any integer.

If $m = 1$ the windings become the common type, referred to here as the "simple lap" and the "simple wave." Actually the very vast majority of windings in use are of these simple types. In general, m is called the "multiplexity" of the winding; if $m = 2$ the winding is "duplex" as already discussed, if $m = 3$, triplex, and so on. Actually most of these special windings exist only on

* $189 = 9 \times 16 + 3 \times 15$.

paper, triplex windings being extremely rare. In fact some firms confine their designs practically entirely to the simple windings above discussed.

The triplex wave winding gives six parallel circuits, a number sometimes suited to low speed machines having a large number of poles. To obtain a symmetrical winding it is now necessary to find three circuits absolutely similarly placed with respect to the field system, and this necessitates the number of pairs of poles, the number of slots and the number of commutator segments being each a multiple of 3. The symmetrical triplex wave winding is thus only possible for 6, 12, or 18, etc., poles. With the 6-pole machine the simple lap winding gives 6 parallel circuits, so that the symmetrical triplex wave winding is limited to 12-pole machines, and possibly, in still rare cases, 18, etc., poles.

The triplex winding may comprise three entirely independent windings, interspaced as in duplex windings, or it may be singly-closed. The condition necessary for three independent windings is that the commutator pitch shall be a multiple of 3 and the number of commutator segments also a multiple of 3. (For a 12-pole triplex wave winding, the number of segments is of necessity a multiple of 3.)

To obtain a singly-closed winding, which is the type preferred by most designers, it is then simply necessary that the commutator pitch shall not be a multiple of 3, for any symmetrical triplex wave winding.

The rules relating to multiplex wave windings may then be summarized

Number of parallel circuits $= 2m$

Commutator pitch $y_C = \left(\dfrac{C \pm m}{p}\right)$

Number of commutator segments $= Ap \pm m$
where A is an integer.

For a symmetrical winding

Number of pole pairs
Number of slots
Number of commutator segments
} must be a multiple of m

For a singly-closed winding—

The number of commutator segments and the commutator pitch must not both be multiples of m. For a symmetrical winding, since the number of pole pairs must be a multiple of m, the condition for a singly closed winding is that the commutator pitch shall not be a multiple of m.

This latter statement holds for values of m up to and including 3. It is not necessary to consider further values, as even triplex windings are extremely rare.

For single-turn coils, the *average voltage between adjacent commutator segments* for a duplex wave winding is one half, and for a triplex wave winding is one third of that for a simple wave winding working with the same value E_c of the average e.m.f. induced in each conductor.

Thus $V_{av} = 2pE_c$ for a simple wave winding,

$V_{av} = pE_c$ for a duplex wave winding,

and $V_{av} = 2/3 \cdot pE_c$ for a triplex wave winding

Examples. As examples of multiplex wave windings the following may be given.

1. Duplex Wave Winding, 8 poles.

8 poles, 90 slots, 4 circuit winding having 6 coil-sides per slot, 270 coils and commutator segments.

The coil span will be 90/8 teeth approximately. In this case 11 teeth would be taken.

The commutator pitch is given by $4y_C = 270 \pm 2$, and y_C may therefore be either 68 or 67.

If the pitch is taken as 68 two independent windings would result. To obtain a singly-closed winding, the type most preferred, the pitch must be odd for a duplex wave winding, and therefore would be taken as 67.

Equipotential Connexions. Equipotential points on the commutator are displaced by $C/2 = 135$ segments. The number of equalizing connectors may be taken as 9, tapping points to adjacent equalizers being $135/9 = 15$ segments apart. The connexions to equalizers are thus 1 — 136; 16 — 151, etc.

2. Duplex Wave Winding, 12 poles.

12 poles, 160 slots, 4 circuit winding having 4 coil-sides per slot, 320 coils and commutator segments.

The coil span would be taken as the nearest whole number to 160/12, i.e. would be taken as 13 teeth.

The commutator pitch is given by $6y_C = 320 - 2$, and is therefore equal to 53. (Note $320 + 2$ is not a multiple of 6, so that there is no choice in the value to take for y_C.) As the commutator pitch is an odd number, the winding is singly closed.

Equipotential Connexions. Equipotential points on the commutator are displaced by $320/2 = 160$ segments. Taking 10 equalizing connexions, the connexions to these are then: 1 — 161; 17 — 177; 33 — 193, etc.

3. Triplex Wave Winding, 12 poles.

123 slots, 12 poles, 6 circuit winding having 5 coil-sides per layer, i.e. 10 per slot; 615 coils and commutator segments.

Coil span = 10 teeth.

The commutator pitch is $y_C = (615 \pm 3)/6$, and may therefore be either 103 or 102. The former is not a multiple of 3 and therefore would give a singly-closed winding. The latter, 102, being a multiple of 3, and the number of segments also being of necessity a multiple of 3 for a symmetrical winding, would result in the winding closing after one-third of the coils are traversed, and would thus necessitate three separate windings. To obtain a singly-closed winding the pitch of 103 would therefore be used.

Equipotential Connexions. There are now 6 parallel circuits and 3 connexions to each equalizer. These connexions are equally spaced around the commutator and are thus $615/3 = 205$ segments apart. Taking 12 equalizers, the average spacing between adjacent tappings to successive equalizers will be 205/12 segments. With one exception the spacing will then be 17 segments, the exception being 18. The connexions to the various equalizing rings are then: 1 – 206 – 411; 18 – 223 – 428, etc.

4. Triplex Wave Winding, 18 poles.

> 213 slots, 18 poles, 6 circuit winding having 4 coil-sides per layer, i.e. 8 per slot; 852 coils and commutator segments. The coil span may be taken to be 12 teeth.

The commutator pitch is given by $9y_C = 852 + 3$, and, hence $y_C = 95$. This is not a multiple of 3 so that the winding will be singly closed.

Equipotential Connexions. There are in all 213 slots, i.e. 3×71. There will be 3 connexions to each equalizer as in the previous example, the connexions being displaced mutually by exactly 71 slots, and $852/3 = 284$ commutator segments. Using 15 equalizers, the average spacing between adjacent tappings to successive equalizers will be 284/15. With one exception, the spacing will, therefore, be 19 segments, the exception being 18. The connexions to the various equalizers are then: 1 – 285 – 569; 20 – 304 – 588; . . . 266 – 550 – 835.

It will be noticed that in all the latter four windings the conditions necessary for symmetry have been observed, i.e. the number of poles is divisible by the "multiplexity" or, in these cases, the number of pairs of parallel circuits. The number of slots, and in consequence the number of coils and commutator segments, has also been taken to be a multiple of the number of pairs of parallel circuits.

CHAPTER IV

ARMATURE REACTION

Magnetizing Action of the Armature Currents. Consider first, the simplest form of drum type, slotted armature, namely, the 2-slot or shuttle armature as used in magneto machines. A consideration of Fig. 60 at once establishes the fact that when the armature carries current it becomes an electro-magnet, the total value of the m.m.f. developed being given by the product of the number of turns and the value of the current, and the axis of magnetization coinciding with the axis of the coil.

In a similar manner, each coil on a normal drum armature develops m.m.f. when the armature carries current and differs from that of the simple shuttle armature only in that, instead of using two large slots per pair of poles, the winding is distributed in a fairly large number of slots. Just as in the shuttle armature, then, any d.c. armature when carrying current acts as an electro-magnet, the *total* value of the m.m.f. developed being given for a 2-pole machine by the product of the number of turns on the armature and the value of the current in each conductor.* The axis of magnetization differs for the various coils, so that the resulting axis of magnetization for the complete winding will be that corresponding to the coil which is centrally placed with regard to the others, that is, to the coil for which the active conductors are midway between the brushes.

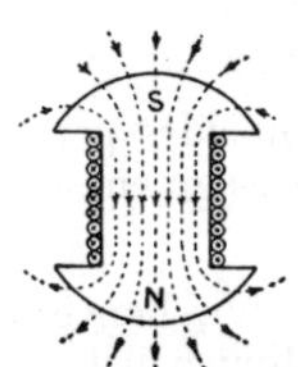

Fig. 60. Magnetic Field due to Shuttle Armature when Carrying Current

The magnetic field set up by an armature can readily be explored by removing the armature from its bearings, and supporting it with the shaft vertical, so that the armature can pass through a horizontal sheet of cardboard, cut so as to receive it. Current can then be passed through the armature by means of temporary leads bound on to suitable commutator segments, and the magnetic field set up around the armature explored with iron filings or a compass needle.†

The whole action of the d.c. machine is dependent upon this magnetizing action of the armature currents, or armature reaction, as it is commonly called. In the d.c. motor it is the mechanical forces set up between the main field magnet and the "armature

* As in the main field circuit, it is a convenience to divide the magnetic circuit of the armature flux into two equal parts, so that the value of the m.m.f. *per pole* becomes one half of this total product.

† It is well to cover the armature surface with paper or cloth, to prevent filings getting in to vital parts.

magnet" that cause the motor to operate, and in the generator it is by rotating the armature against these mechanical forces, set up between the main and armature magnetizing actions, that the steam engine, or other prime mover, is able to supply the energy which the generator changes to the form of electrical energy.

M.m.f. established by a Full-Pitched Drum Coil. In Fig. 61 is represented a drum coil on a bipolar machine, placed in two narrow, diametrically opposite slots, the direction of current in the coil being indicated. The arrangement is clearly that of a simple shuttle

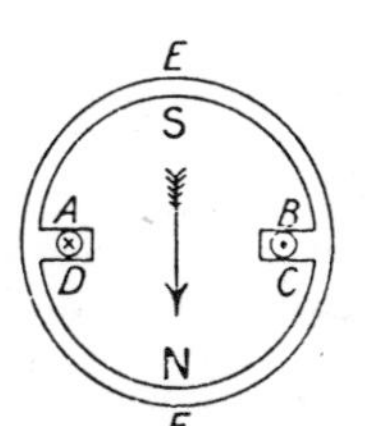

Fig. 61. Full-pitch Drum Coil

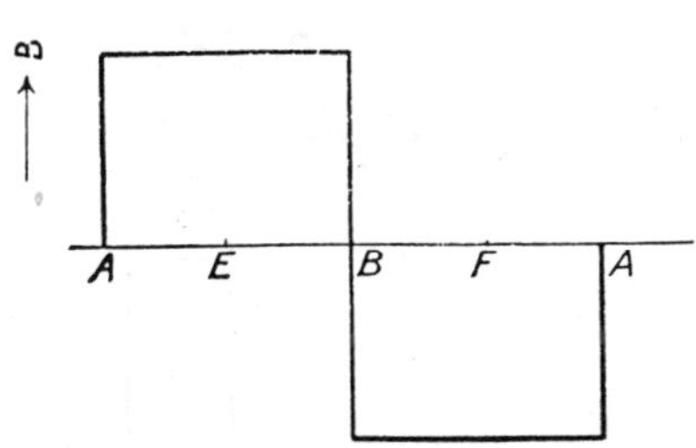

Fig. 62. The Field Form Corresponding to Fig. 61

armature having extremely narrow slots, so that an electromagnet is formed having its S pole uppermost. Every line of magnetic force—neglecting small effects—links completely with the coil, so that the m.m.f. establishing the magnetic field is constant along all the lines and equal to the ampere-turns developed by the coil, excepting only at the slots. If then the armature were placed in a stator having a cylindrical inner surface, so that the air gap is constant, as in most a.c. machines, the flux density in the air gap would be quite constant if the reluctance of the iron were negligible. (The effect of the width of the slot is neglected.) In Fig. 61, the flux would travel from the stator to the armature between the points *A* and *B*, and would return from the armature to the stator between *C* and *D*. Assuming that the width of the slot is negligibly small, the field form or flux distribution produced by this armature coil would then be the rectangular wave form shown in Fig. 62. This wave form may also be taken to represent the manner in which the m.m.f. acting across the air gap varies around the armature, and will therefore be referred to as the m.m.f. distribution curve due to the single coil.

In the present book the convention adopted is that the m.m.f. acting across the air gap and the flux density in the gap are shown as *positive* when they act in a direction *from* the field system to the armature. Thus the portion of the field form corresponding to that part of the armature under a main N pole is shown as positive. A m.m.f. tending to establish N polarity at the *armature surface* is then shown as negative.

Take now an armature having 3 slots per pole, as in Fig. 63. With a normal 2-layer winding there will be 6 sets of coils, divided by the brushes into two sets of 3, carrying currents in opposite directions, as in the diagram. For simplicity, the figure is constructed for one coil-side per layer. Considering the 3 coils whose upper sides are in the slots to the left: each of these coils will give rise to a rectangular m.m.f. distribution curve, but the three curves will be displaced in space by an amount corresponding to the displacement

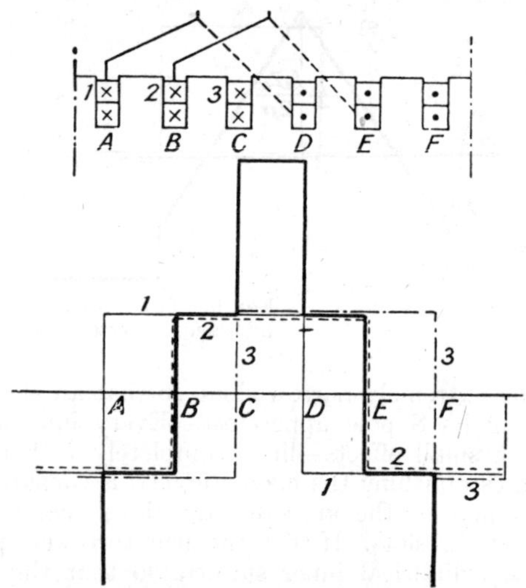

FIGS. 63 AND 64. THE M.M.F. DIAGRAM FOR THE EFFECT OF THE ARMATURE CURRENTS

between the coils, i.e. to one slot pitch. The three graphs are then as shown in Fig. 64, which is lettered to correspond to Fig. 63. The total m.m.f. acting across the gap at any point is obtained by summing the effects due to the three coils, and is thus represented by the "stepped" diagram shown. It will be noticed that the axis of magnetization corresponding to the joint effect of the three coils coincides with that due to coil 2, and thus lies along the brush axis. A little consideration will suffice to establish the fact that the joint effect of the other three coils is exactly equal at every point to that of the 3 just considered, so that the maximum value of the m.m.f. developed around the complete circuit is the product of the current and the total number of turns on the armature, and the value acting across *each air gap* is one half of this product.

A further subdivision of the winding by the use of a larger number of slots would increase the number, but decrease the magnitude, of the individual steps in the resulting m.m.f. distribution.

Finally, if the number of slots per pole is large, or if a surface winding is used, the actual m.m.f. distribution due to the armature currents merges into the triangular form shown in Fig. 65, the amplitude of the triangle being equal to one half of the total product of the number of turns and the armature current, for a 2-pole machine.

A similar m.m.f. distribution is obtained with each pole of multipolar machines, the maximum value of the ampere-turns developed

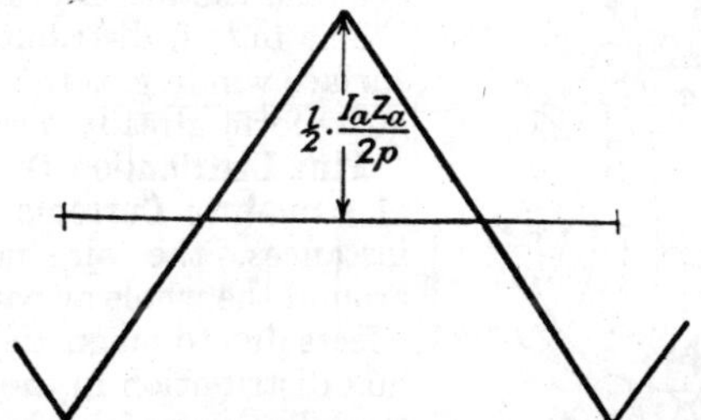

Fig. 65. The m.m.f. Distribution due to Armature Reaction, for a Smooth-core Machine

per pole being in every case given by dividing the total number of ampere-turns developed on the armature, $I_aZ_a/2$, by the number of poles, $2p$. Thus, armature ampere-turns per pole

$$F_a = \tfrac{1}{2} \cdot \frac{I_aZ_a}{2p} \quad . \quad . \quad . \quad . \quad . \quad (74)$$

The joint effect of the armature currents is thus to establish a magnetizing action which differs from that due to normal salient poles in that it is distributed instead of concentrated. The axis of magnetization, i.e. the axis at which maximum armature reaction is experienced, always coincides with the brush axis, that is, at the point at which the current in the conductors undergoes reversal.

As an alternative, the shape of the m.m.f. distribution curve due to the armature currents may be developed as follows. Consider the 2-pole armature shown in Fig. 66 for a surface winding. The distribution of the magnetic field around the armature will obviously be somewhat as indicated, from reasons of symmetry, and, as already stated, this can be verified experimentally by exploring the field around an armature by means of iron filings or compass needles.

The total value of the m.m.f. acting around any complete magnetic path is given by finding tbe algebraic sum of all the "currents" enclosed by the path, that is, the "ampere conductors." Thus the line shown at A embraces $Z_a/2$ conductors, and hence the total m.m.f. acting around the path is equal to the armature ampere conductors per pole. This m.m.f. has to drive the flux completely round the circuit, so that the value acting across each air gap, when the armature is in position, is one half of this total m.m.f., and hence is equal to the armature ampere-*turns* per pole.

The line shown at B embraces two less conductors than that at A; that at C, two less conductors than the one at B; and so on until at X the m.m.f. acting is simply that due to a single conductor. Beyond X, the direction of the flux entering the armature changes, so that the m.m.f. acting across the air gap changes in direction midway between the brush positions.

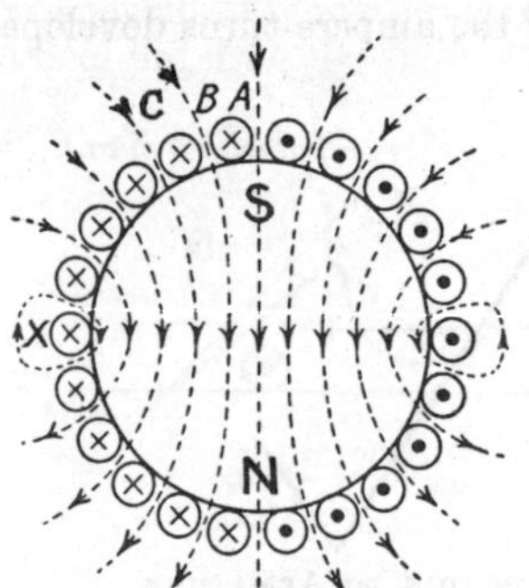

FIG. 66. MAGNETIC FIELD DUE TO SMOOTH ARMATURE WHEN CARRYING CURRENT

The m.m.f. distribution wave, for a surface winding, is therefore the triangular form already discussed.

Flux Distribution Due to Sole Action of Armature Currents. If, as in rare instances, the air gap is constant around the whole armature, except for effects due to magnetic saturation, the flux distribution in the air gap, due to the sole action of the armature currents, would be of exactly the same form as the m.m.f. distribution, i.e. triangular. With the brushes set in the neutral position, i.e. so that they affect the conductors situated midway between poles, the effect of armature reaction is zero at the mid-point of the pole shoe, and maximum at the brush axis.

With normal salient pole machines, there is no particular difficulty in determining, under the pole shoes, the flux distribution due to the armature currents, but in the interpolar space the reluctance of the various paths cannot simply be determined. With modern machines, however, the flux distribution outside the pole arc is not of paramount importance, and it is sufficient for our purpose to concentrate attention mainly upon the happenings under the pole shoe.

If the gap dimension is constant over the whole pole arc, the flux distribution curve due to the sole action of the armature currents will then be of the form indicated in Fig. 67. The part corresponding to interpolar space is, of course, only approximately correct, but it is important to realize that, due to the sole action of the armature currents, a magnetic field of appreciable intensity will be established, entering the armature surface midway between the poles. In spite of the great length of the air path of the magnetic lines at this point, the large value of the m.m.f. developed by the armature suffices to establish, as stated, appreciable flux density.

In normal machines, the pole arc dimension is 70 per cent of the pole-pitch, so that at the tips of the pole shoe the value of the armature m.m.f. is 70 per cent of its maximum value, i.e. 70 per cent of the value of the armature ampere-turns per pole. For a uniform gap dimension under the pole shoe, the flux density in the air gap due to the sole action of full-load current in the armature would reach a considerable percentage of the normal working value. By having a bevel on the pole shoes the gap dimension is increased towards the tips of the shoes, and additional reluctance thus put in the path of the armature flux at the points where the value of the armature ampere-turns begins to be important.

The flux distribution curve under these conditions could be determined experimentally. The machine would be driven at normal speed with no current in the field windings but with full-load current

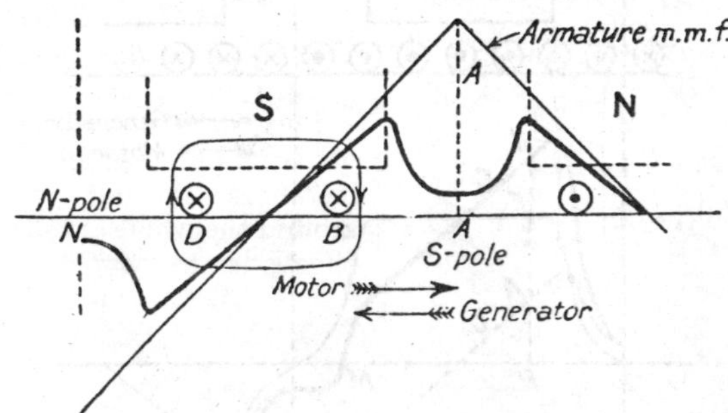

FIG. 67. THE FLUX DISTRIBUTION PRODUCED BY THE SOLE ACTION OF THE ARMATURE CURRENTS

fed into the armature from a suitable low-voltage source. The voltage between two exploratory brushes spaced one commutator pitch apart would be measured at various points over the pole-pitch on the commutator, a suitable brush rocker and graduated scale being used to determine the actual position of the test brushes. The distribution curve can also be obtained oscillographically if a search coil of pitch equal to the pole-pitch is fitted to the armature and its ends connected to a pair of slip-rings.

If, with the armature alone carrying current, brushes were placed at 90° to their normal position as at C and F in Fig. 64 the total e.m.f. at normal speed between these brushes would be considerable. For the case illustrated in Fig. 67 its value would be about one-third of the normal working voltage. This e.m.f., due to the cross-magnetizing action of the armature currents, is made use of in a number of special machines, some of which are discussed later in Chapter XIII.

Effect of Armature Reaction upon Flux Distribution Curve of Loaded Dynamo. There are many "rules" giving the direction in which e.m.f. is induced in a conductor when the conductor is moved through a magnetic field. The author finds it best to apply the principle enunciated in Lenz' Law, and not to trust to memory. For example, in a dynamo, the magnetizing action due to the armature currents will give rise to a force between the armature and field magnet systems. There are only two alternatives, this force may act in the direction in which the armature is being driven, or it may act in the opposite direction. If the former held, a state of affairs would be set up which would give rise to ever increasing acceleration—obviously impossible; so the latter must hold, in conformity with Lenz' Law, and the force set up between the

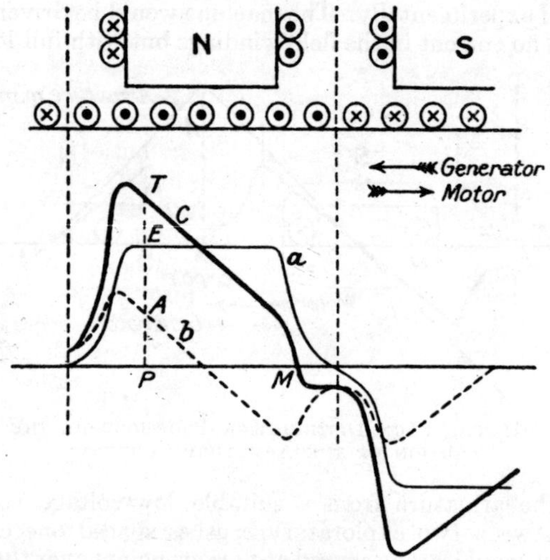

FIGS. 68 AND 69. THE RESULTING FLUX DISTRIBUTION ON LOAD FOR AN UNSATURATED D.C. MACHINE

armature magnet and the field magnet must be in the opposite direction to the driving force.

For example, in the machine indicated in Fig. 67, for the direction of motion indicated, from right to left, the magnetic pole produced by the armature currents at the neutral plane AA must be of the same polarity as the main pole to the left of A. In general the poles *on the armature* due to armature reaction must always, for a generator, have the same sign as the main poles in advance of them.

The direction of the current in the armature conductors is at once deduced from the polarity of the magnet produced by them; for this it is simply necessary to remember the corkscrew rule for the direction of the magnetic field set up around a conductor. For example, in Fig. 67, the m.m.f. of the armature has to establish S polarity at A, and hence the line of force passing through B and D must have a clockwise direction around the conductors it embraces. The current in these conductors must therefore be in the direction indicated, i.e. flowing from the reader.

Dealing now with the effect of armature reaction on a loaded generator. Fig. 68 represents the directions of the current in the armature conductors and field coils of a d.c. generator, the armature being rotated in the direction indicated by the appropriate arrow.

Assuming first that magnetic saturation is entirely absent, the value of the flux density at any point in the air gap may be taken to be proportional to the total ampere-turns acting along the magnetic path passing through that point. It is then permissible to determine separately the flux distribution curves due to the main field coils, and to the armature. The corresponding ordinates of these curves may then be added to obtain the ordinate of the total flux distribution curve, due to the joint action of the main exciting winding and the armature currents.

In the diagram, Fig. 69, the flux distribution curve for the generator on no-load is shown by the thin line a. This diagram has already been discussed. The flux distribution due to the sole action of the armature currents is indicated by the dotted line b, this graph corresponding exactly to Fig. 67, and being constructed for where the value of the armature ampere-turns per pole on full load is equal to that of the ampere-turns developed by the field winding.

The value of the resulting field intensity at any point on the armature is then obtained by summing the values due to the field winding and the armature winding. Thus, in Fig. 69 the ordinate PT of the total flux distribution curve on full load at P is given by $PT = PE + PA$. This latter full-load flux distribution curve is shown by the heavy line c, and it will be noted that the effect of armature reaction is to crowd the magnetic field towards the trailing end of the pole shoe, for the case considered, i.e. a generator. The field form, or flux distribution curve, is thus much distorted and is no longer a symmetrical curve about an axis through the centre of the shoe. That shown is for brushes at the exact geometric neutral axis, i.e. placed so that they commutate the current in each coil when the active conductors are midway between the poles.

For this particular instance of negligible reluctance of the iron parts of the magnetic circuit, and for the brushes in the geometric neutral position, it will be observed that the total value of the useful flux per pole is unaltered from its value on no-load; it is merely the distribution of the flux that is altered.

In particular the field distribution in the interpolar space is altered, and the position of zero field intensity in the air gap at the armature surface—or *magnetic neutral axis*—is shifted forward from the geometric neutral position, by an amount which is mainly dependent upon the value of the armature current, to a position M.

Lead of the Brushes and its Effects. If the brushes are displaced in the direction of rotation from the geometric neutral position, they are said to be given "lead." This is said to be "positive lead" if the brushes are thus moved forward, and "negative lead" if they are moved backward. (Sometimes the terms, "forward lead" and "backward lead" are used.) This lead is sometimes measured in "commutator segments" and sometimes expressed as an angle, the distance between brushes, i.e. the pole-pitch as measured on the commutator, corresponding to 180 electrical degrees.

For reasons that will appear later, generators working without interpoles require positive, or forward, lead to be given to the brushes for full-load working.

Fig. 70 illustrates a generator in which the brushes are given

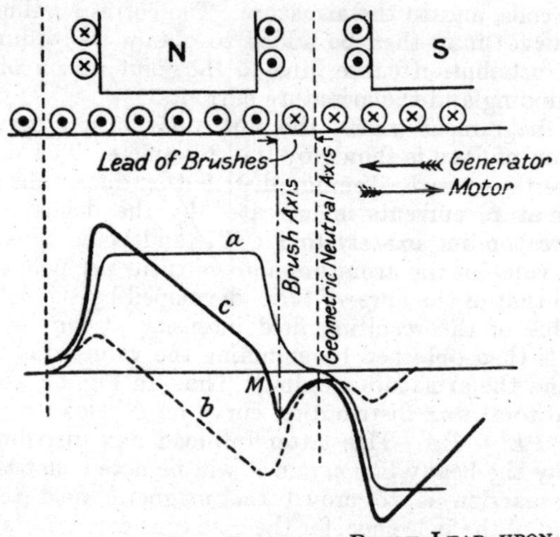

FIGS. 70 AND 71. THE EFFECT OF BRUSH LEAD UPON THE RESULTING FLUX DISTRIBUTION OF A LOADED D.C. MACHINE

positive lead. The effect is to delay the reversal of the current until the conductors are well beyond the geometric neutral position, and thus to shift forward the axis of magnetization of the armature currents. The magnetizing action due to the armature is no longer entirely at right angles to that of the field system, but possesses a component which is in direct opposition to the m.m.f. due to the field coils, with the result that the value of the useful flux per pole is decreased by the action of the armature currents when the machine is loaded. This is illustrated also in Fig. 71, which shows the m.m.f. due to armature reaction displaced by the angle of lead relatively to its position in Fig. 69, and gives also the graphs for the no-load flux distribution as well as those due to the sole action of the armature currents, and the resulting distribution. As in Fig. 69, the reluctance of the iron parts is assumed to be negligible. With forward lead, the result of armature reaction, as shown in Fig. 71, is to decrease the flux density over the greater portion of the air gap, and so to result in an appreciable reduction of the value of the useful flux. At the same time, the magnetic neutral position M is shifted still farther forward.

In a similar manner it may be shown that giving backward lead to the brushes of a d.c. generator will result in an increase in the value of the useful flux when the armature carries current, and in the magnetic neutral being shifted from the position occurring with no lead, back towards or beyond the geometric neutral.

Effect of Magnetic Saturation upon Flux Distribution of Loaded Generator. The value of the e.m.f. induced in each armature conductor is determined at any instant by the flux density in the air gap at the point through which the conductor is passing. If the flux density in the gap is then increased much above the no-load value, the maximum value of the e.m.f. induced in each turn of the armature winding will rise, and as a result the maximum value of the voltage existing between adjacent segments of the commutator may become excessive. By normally working the armature teeth at a high degree of saturation the effect of armature reaction upon the flux density under the trailing end of the pole shoe is very much reduced, although at the same time the actual value of the

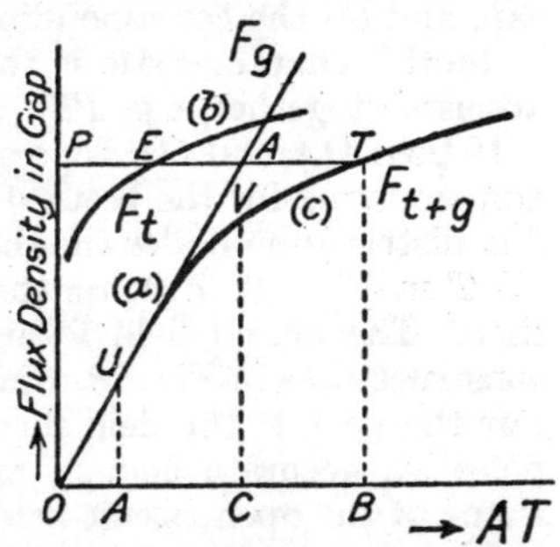

FIG. 72. THE "GAP + TOOTH" MAGNETIZATION CURVE

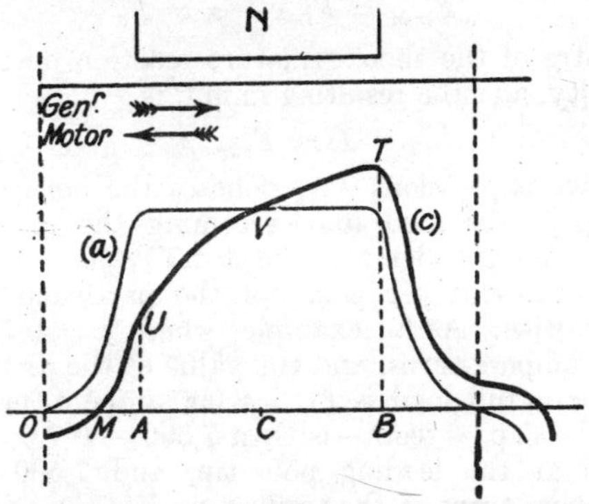

FIG. 73. THE EFFECT OF MAGNETIC SATURATION UPON THE FLUX DISTRIBUTION OF A LOADED D.C. MACHINE

total flux per pole is reduced thereby. This is illustrated by Fig. 73, which is constructed for a uniform air gap under the pole shoe, and brushes in the geometric neutral axis.

Since the value of the resulting m.m.f. due to the combined effects of armature reaction and the main field winding varies uniformly over the pole arc, i.e. according to a linear law, the field form under the pole shoe will be simply a reproduction of a portion of the magnetization characteristic for the (teeth + air gap) portions of the magnetic circuit. Normally, the magnetization curves for the various parts of the magnetic circuit are plotted in terms of the value of the total useful flux per pole. They may, equally, be plotted in terms of the flux density in the air gap, and this latter method is more convenient in dealing with the field form. Thus, in Fig. 72, the graph (*a*) represents the magnetization curve for the single air gap, and (*b*) the corresponding curve for the teeth. The total "gap + tooth" characteristic is then obtained by adding corresponding abscissae together, e.g. $PT = PE + PA$.

If then OA and OB represent respectively the values of the net ampere-turns for the teeth + gap at the two tips of the poles, the flux distribution under the shoe will be an exact replica of the curve UVT in Fig. 72, V being the density under the centre of the pole shoe. The actual field form for the loaded machine can then be constructed as in Fig. 73, which is lettered to correspond to Fig. 71. For the part of the field form corresponding to the space between poles an accurate enough approximation can be made from the shape of the open circuit field form.

If α = the ratio pole arc ÷ pole-pitch, the minimum value of the total m.m.f., occurring at the leading tips of the pole shoes is

$$F_{max} = F_{t+g} - \alpha \times F_a$$

and the maximum value, occurring at the trailing tips

$$F_{min} = F_{t+g} + \alpha \times F_a$$

At the centre of the shoe, armature reaction does not influence the flux density, and the resulting m.m.f. is

$$F = F_{t+g}$$

In the above expressions F_{t+g} denotes the normal value of the m.m.f. per pole absorbed in overcoming the reluctance of the teeth + gap, and F_a denotes the total value of the armature reaction ampere-turns per pole, for the particular value of load current dealt with. As an example, where $\alpha = 0{\cdot}7$, the value of F_{t+g} is 7,500 ampere-turns, and the value of the armature ampere-turns per pole on full load is F_a = 8,000 ampere-turns, the m.m.f. acting—for the gap + teeth—is then 7,500 − 0·7 × 8,000 = 1,900 ampere-turns at the leading pole tip, and 7,500 + 0·7 × 8,000 = 13,100 ampere-turns at the trailing tip.

In a dynamo having an air gap of constant value over the whole of the pole arc, it is evident that, with teeth worked normally at a high degree of saturation, extreme saturation will be attained in the teeth at the trailing tip, and as a result the value of the total flux per pole will be seriously reduced when the machine is heavily loaded. As already stated, by arranging for the value of the gap dimension to increase at the pole tips, by either of the methods indicated in Fig. 11, this reduction in the value of the flux can be kept within reasonable limits.

As a result of magnetic saturation in the teeth, therefore, it becomes necessary to increase the value of the ampere-turns developed by the main poles when the armature is loaded, to maintain the initial value of the flux per pole. The relative amount of this increase is dependent upon the value of the ampere-turns developed by the armature, and is thus dependent upon the value of the load current. In addition, the extent of the increase necessary is dependent upon the degree of saturation at which the teeth are worked, and upon the shape of the pole shoe in so far as it affects the dimension of the air gap.

All these considerations hold even when the brushes are in the geometric neutral position. When the brushes are rocked forward, to maintain the value of the flux it becomes necessary to increase still more the value of the ampere-turns developed by the main field winding owing to the demagnetizing component now introduced into the armature m.m.f.

The majority of modern machines work with brushes fixed at the geometric neutral position, and to allow for the effect of armature reaction upon the value of the main flux per pole, it is the usual practice to assume that the ampere-turns developed by the field winding must be increased by a certain fraction of the value of the armature ampere-turns. This fraction varies from 15 to 25 per cent, according to the degree of saturation in the teeth, etc., and for a normal case 20 per cent may be taken.

The field distortion can be further reduced by increasing the reluctance of the tips of the pole shoes, as indicated later, by having plates with one normal tip and one shortened tip, arranged alternately.

Armature Reaction with Motors. For motor operation, the direction of rotation, corresponding to a given polarity of the main magnets and a given direction for the armature currents, will be opposite to that for generator operation. The diagrams given, indicating the effects of armature reaction, have therefore been provided with two arrows in opposite directions, the one holding for generator and the other for motor operation.

For motor operation, then, the effect of armature reaction differs from that with generator operation, only in that the field is shifted backward against the direction of rotation, instead of forward.

When the brushes are given "lead," as in non-interpole machines, forward lead will result in the field being strengthened, and backward lead—illustrated for motor operation in Fig. 70—will weaken the field.

Demagnetizing and Cross-magnetizing Ampere-turns. With the brushes in the geometric neutral position, the effect of armature reaction is solely of the nature of cross magnetization, although owing to saturation in the armature teeth, the main flux is reduced. When the brushes are given lead, the total effect of the armature conductors may conveniently be resolved into two parts, as indicated in Fig. 74. The belt of conductors (on either side of the geometric neutral axis) extending by an angle corresponding to the brush lead, directly demagnetizes, acting in dead opposition to the main

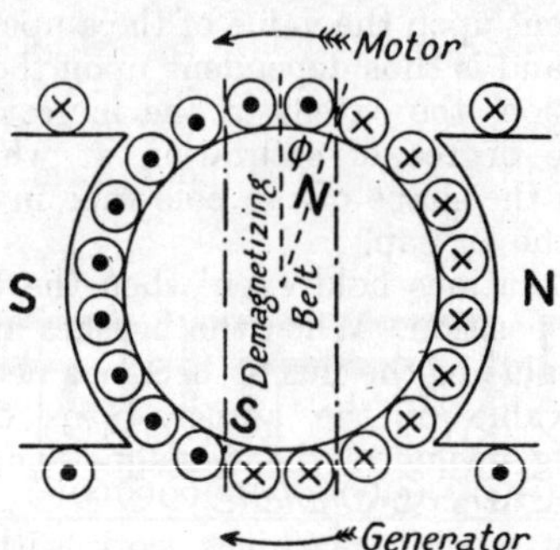

FIG. 74. DEMAGNETIZING BELT

field winding, as indicated. The axis of magnetization due to the remaining conductors is at right angles to the main field.

If then ϕ is the equivalent angle of lead of the brushes, expressed in electrical measure in order to correspond to the actual number of poles—

The demagnetizing ampere-turns per pole

$$F_d = \frac{2\phi}{\pi} \times \text{armature m.m.f. per pole} \quad . \quad . \quad . \quad (75)$$

and the cross-magnetizing ampere-turns

$$F_q = \left(\frac{\pi - 2\phi}{\pi}\right) \times \text{armature m.m.f. per pole} \quad . \quad (76)$$

where the armature m.m.f. per pole is given by $I_a Z_a / 4p$ as in (74).

In terms of the number of commutator segments, etc.—

Let C_l = the number of segments lead

C = number of commutator segments.

Then, the demagnetizing ampere-turns per pole

$$F_d = \frac{C_l}{\frac{C}{4p}} \times \frac{I_a Z_a}{4p}$$

$$= \frac{C_l}{C} \times I_a Z_a \quad . \quad . \quad . \quad (77)$$

Owing to magnetic saturation, however, any treatment based upon a consideration of the separate effects of these two components will be incorrect, and it is necessary to consider the problem on the lines previously indicated.

Effect of Armature Reaction upon Iron Losses. The value of the iron losses in the teeth and in the pole shoes is determined by the maximum value of the flux density at which they work. Since armature reaction, by distorting the field form, increases this

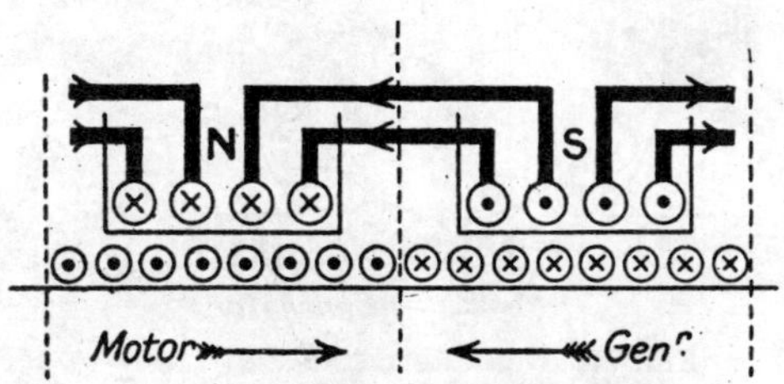

FIG. 75. COMPENSATING WINDINGS

maximum value considerably above that corresponding to no-load conditions, the iron losses, and in particular those in the teeth, are greater on load than on no-load. Further, owing to the very high degree of saturation of the teeth, some of the flux finds its way into the core end plates, and these being fairly massive, the eddy currents induced therein cause appreciable losses.

To a rough approximation the "extra losses" that thus occur in the iron when the machine is loaded may be taken to vary as the square of the armature current, and occasionally in the aggregate they amount on full load to about 50 per cent of the no-load value of the iron losses.

Compensating Windings. It is sometimes desirable to prevent the distortion of the field form that is established by armature reaction. To do this, it is necessary to neutralize the magnetizing action of the armature currents, as far as is possible over the whole of the pole-pitch.

Compensating windings are used for this purpose, and generally are of the concentric coil type as indicated in Fig. 75.

The coils span the interpolar space so that their axis of magnetization coincides with the neutral axis. The coil-sides are housed in axial slots in the pole shoes, and carry current in the opposite direction to the adjacent armature conductors.

Complete neutralization of the armature m.m.f. is only possible in machines having non-salient poles. To obtain perfect compensation under the pole shoe it is necessary that the total ampere-conductors per pole for the compensating winding shall be exactly equal to the total ampere-conductors on the portion of the armature under the shoe.

Thus, to prevent distortion of the flux, under the pole shoe only, the m.m.f. developed per pole by the compensating windings will be

$$F_k = \frac{I_a Z_a}{4p} \times \frac{\text{pole arc}}{\text{pole pitch}} \quad . \quad . \quad . \quad (78)$$

Occasionally, to neutralize armature reaction exactly at the geometric neutral axis, the m.m.f. developed by the compensating

FIG. 76. FIELD SYSTEM OF LARGE D.C. MOTOR SHOWING INTERPOLES AND COMPENSATING WINDING
(British Thomson-Houston Co., Ltd.)

windings is made exactly equal to the armature m.m.f. per pole. This will result in a slight degree of over-compensation for all points under the pole shoe.

To obtain correct compensation at all loads, it is essential that the compensating windings shall be series wound. Naturally these windings add considerably to the weight of copper required for the machine. They are thus used only where field distortion would give rise to an excessive value for the maximum voltage induced in the armature coils, which in turn would give too high a value of the voltage between commutator segments, and cause flashing over between segments and even from brush to brush. These include very high speed machines such as turbo-generators, and machines working at high speed with relatively weak fields, such as shunt motors working over a very wide range of speed. The windings are also desirable in certain other applications.

Generally, it is unnecessary to compensate armature reaction over the whole of the pole-pitch, and it is sufficient to neutralize the reaction in the region of the geometric neutral axis. For this purpose a narrow coil wound on an interpole may be used, but, as is discussed in the next chapter, the function of the interpole is not merely to compensate armature reaction in the geometric neutral zone.

Fig. 76 illustrates the field system of a large motor, fitted with compensating windings.

CHAPTER V

GROWTH OF CURRENT IN A CIRCUIT; IRON LOSSES

Inductance. The flow of current in a circuit usually causes an appreciable magnetic field to be established, linking with the circuit. In this magnetic field, as is shown later, energy is stored. As a result, in establishing the current, energy has to be supplied over and above that corresponding to ohmic losses. The problem of building up a current is therefore somewhat analogous to that of accelerating a body, where, in addition to the energy absorbed by friction, energy has also to be supplied to provide the kinetic energy associated with the motion of the body.

If friction effects are negligible, the connexion between the accelerating force and the rate at which it causes the velocity to increase, i.e. the acceleration, is

$$\text{Acceleration} = \frac{dv}{dt} = \frac{\text{Force}}{\text{Mass}} = \frac{F}{M}$$

In the electric circuit the property corresponding to mass in the above expression is known as "inductance"—denoted by L.

When voltage is applied to a circuit the rate at which the current increases is given by—

$$\frac{di}{dt} = \frac{\text{Voltage}}{\text{Inductance}} = \frac{V}{L} \quad . \quad . \quad (79)$$

if resistance effects are negligible.

In the electric circuit it is apparent that any change in the value of the current gives rise to a corresponding change in the value of the magnetic flux linked with the circuit. This change in the magnetic flux induces an e.m.f. in the circuit, the e.m.f. being in such a direction as to oppose the change in the flux, and hence also to oppose the change in the current. The e.m.f. is then called an "e.m.f. of self induction", or sometimes a "back e.m.f.," and if resistance effects are negligible the current will increase at such a rate that this e.m.f. of self induction just balances the applied voltage.

The value of the e.m.f. of self induction, expressed in terms of the flux Φ, and number of turns N with which that flux is linked is

$$N\frac{d\Phi}{dt} \quad . \quad . \quad . \quad . \quad (80)$$

It follows then from (79) that the inductance must have a value such that

$$L\frac{di}{dt} = N\frac{d\Phi}{dt} \quad . \quad . \quad . \quad (81)$$

When the flux varies in direct proportion to the current, the value of the inductance is then

$$L = \frac{N \times \Phi}{I} \quad . \quad . \quad . \quad . \quad (82)$$

The product $N\Phi$ of the flux and the number of turns with which it is linked is called the number of flux linkages.

A simple definition of the inductance of a circuit is then: the inductance is equal to the number of flux linkages, in webers, established by a current of one ampere.

The unit of inductance is called one henry. For a circuit possessing unit inductance, and negligible resistance, an applied voltage of one volt would cause the current to build up at the rate of 1 A/sec.

Where the value of the flux does not vary in direct proportion to that of the current, such as, for example, coils wound on an iron core, the value of the inductance will vary according to that of the current. Many writers, even for this, define the inductance as in (82). The value so obtained is, however, not of much use. A better general definition is obtained direct from (81), i.e.

$$L = N\frac{d\Phi}{di} \quad . \quad . \quad . \quad . \quad (83)$$

When dealing with acceleration problems in actual practice, frictional effects must be taken into account; these reduce the force actually producing acceleration to $F - F_r$, where F is the total force and F_r that absorbed in overcoming friction.

Similarly with the electric circuit. Of the total applied voltage a certain portion, iR is absorbed in overcoming the resistance of the circuit, so that the total voltage available for increasing the current,—by the time the current attains the value i—reduces to $V - iR$. The current will then at any instant be increasing at a rate given by

$$\frac{di}{dt} = \frac{V - iR}{L} \quad . \quad . \quad . \quad . \quad (84)$$

This differential equation is of a type repeatedly occurring in electrical engineering problems. Its solution may be obtained as follows—

From (84) $$\frac{di}{V - iR} = \frac{1}{L}\,dt$$

integrating $$\int \frac{di}{V - iR} = \int \frac{1}{L}\, dt + A$$

Then $$-\frac{1}{R}\log_e (V - iR) = \frac{t}{L} + A$$

where A is the integration constant, the value of A being determined

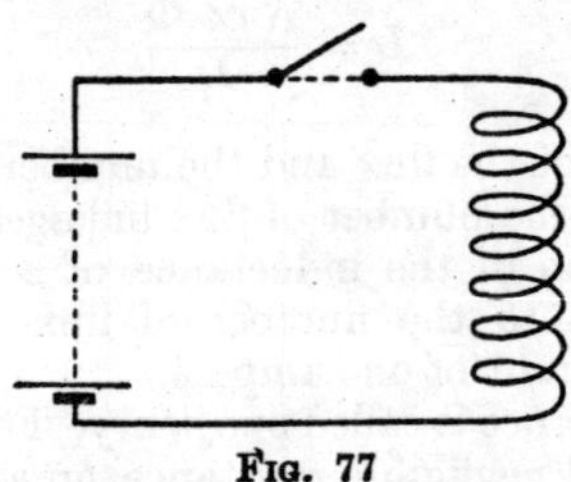

Fig. 77

by the initial conditions; ε is the base of the system of natural logarithms and $= 2{\cdot}718$ approximately.

Hence $$\log_e (V - iR) = -\frac{R}{L}t - RA$$

and $$V - iR = \varepsilon^{-\frac{R}{L}t} \times \varepsilon^{-RA} = B\varepsilon^{-\frac{R}{L}t}$$

and $$i = \frac{1}{R}(V - B\varepsilon^{-\frac{R}{L}t}) \quad . \quad . \quad (85)$$

where $B = \varepsilon^{-RA}$, and is a constant determined by the given starting conditions.

Growth of Current on Applying a Constant Voltage. Here no current is assumed to be flowing at the instant at which the circuit is first completed, i.e. the circuit is first open as in Fig. 77

The initial conditions are then, obviously, that when $= 0$, $i = 0$.

Substituting these values in the general equation (85) above

$$0 = \frac{1}{R}(V - B\varepsilon^{o}) = \frac{1}{R}(V - B)$$

whence $B = V$.

The value of the current at any time t after closing the switch is then—

$$i = \frac{V}{R}(1 - \varepsilon^{-\frac{R}{L}t}) \quad . \quad . \quad (86)$$

This equation is represented by the graph shown in Fig. 78, the current growing according to an exponential law.

The initial rate of growth of the current is quite independent of the resistance of the circuit, since when $t = 0$, i also $= 0$, and $\frac{di}{dt} = V/L$.

The value ultimately attained by the current is independent of the inductance, and, in accordance with Ohm's Law is $I = V/R$.

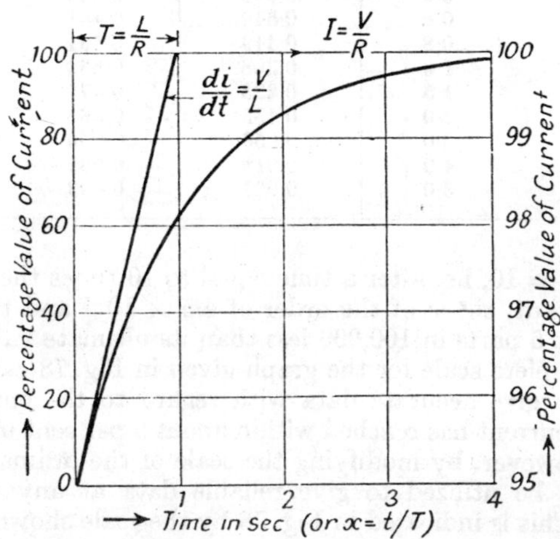

FIG. 78. THE GROWTH OF A DIRECT CURRENT

If then the initial rate of growth of the current could be maintained the time taken to attain the value $I = \frac{V}{R}$ would be $\frac{V}{R} \div \frac{V}{L}$, i.e. L/R sec, as indicated in Fig. 78. The ratio of the inductance of the circuit to its resistance is then called the *Time Constant*, T.

The time constant is measured in seconds, and for a highly inductive circuit its value may reach, or exceed, two or three seconds. The value of the current, in terms of its ultimate value, is then $i = I\,(1 - \varepsilon^{-\frac{t}{T}})$, where T is the time constant, or if the time be expressed in terms of the time constant, the current may be expressed: $i = I\,(1 - \varepsilon^{-x})$, where $x = t/T$.

Values of ε^{-x} are tabulated in many books of mathematical tables or can readily be calculated.

Fig. 78 is plotted for a time constant of unity, and, as indicated, will hold for the general case if the time is plotted in terms of the time constant, i.e. $x = t/T$ is plotted instead of t.

So that the reader may be able to plot this curve to a larger scale for future reference, data relating to it are given in tabular form below.

$x = \frac{t}{T}$	ε^{-x}	$1 - \varepsilon^{-x}$
0	1·0	0
0·2	0·819	0·181
0·4	0·670	0·330
0·6	0·549	0·451
0·8	0·449	0·551
1·0	0·368	0·632
1·5	0·223	0·777
2·0	0·135	0·865
3·0	0·050	0·950
4·0	0·018	0·982
5·0	0·007	0·993

When $x = 10$, i.e. after a time equal to 10 times the value of the time constant, ε^{-x} is of the order of $4{\cdot}5 \times 10^{-5}$ and the current is only about 5 parts in 100,000 less than its ultimate value.

The complete scale for the graph given in Fig. 78 cannot be used directly to give accurate data with regard to the current growth when the current has reached within about 5 per cent of its ultimate value. However, by modifying the scale of the ordinates the same graph can be utilized to give reliable data at any value of the current. This is indicated in Fig. 78 by the scale shown to the right of the diagram, starting from 95 per cent and reading up to 100 per cent. Using this scale, the diagram indicates the time taken for the current to increase from 95 per cent of its maximum value to any higher value. Thus, for the current to reach 95 per cent of its final value for a circuit having a time constant of unity, the time taken is practically 3 sec. By using the scale to the right of the diagram it is seen that the time taken for the current to rise from 95 per cent to 99·75 per cent is a further 3 sec, the total time taken to reach 99·75 per cent being thus 6 sec.

The range chosen of 95 to 100 per cent is quite arbitrary. The graph can be utilized for any scale whatever, ending of course at 100 per cent. This is due to the fact that to increase the value of the current from i by an amount* $k(I - i)$ requires a perfectly definite time which depends solely upon k and not at all upon the initial value i. Thus to increase the current from 0 to 90 per cent requires exactly the same time as is necessary to raise it from 50 per cent to 95 per cent, from 90 per cent to 99 per cent, or from 95 per cent to 99·5 per cent, etc. Similarly to increase the current from 0 to 50 per cent requires exactly the same time as that necessary for it to grow from 98 to 99 per cent; and so on.

* Where k has any value not greater than 1, and $I = V/R$.

So far as d.c. machines are concerned, a direct application of the particular problem just discussed occurs in the building up of the field current of a shunt motor on the application of a steady voltage. A slight complication is introduced owing to the fact that it is usual to work with a fair degree of magnetic saturation in some of the iron parts of the magnetic circuit, so that the value of the inductance is not constant. Thus, in Fig. 79 is shown the magnetization curve of an 8-pole machine, the working value of the magnetic flux in each pole being 0·075 Wb, the number of turns on the shunt coils being 600, and the total resistance of the shunt circuit being 100 ohms, the current necessary for 0·075 Wb being 6·25 A at 625 V. The inductance of the complete shunt circuit as normally expressed would then be

$$L = \frac{8 \times 600 \times 0{\cdot}075}{6{\cdot}25} = 57{\cdot}6 \text{ henrys}$$

since all the shunt coils are connected in series.

As shown in the diagram, until a flux of 0·055 Wb is reached the magnetization curve is practically straight. The excitation necessary for 0·055 Wb is 3·6 A. The corresponding value of the inductance is then

$$L_o = \frac{8 \times 600 \times 0{\cdot}055}{3{\cdot}6} = 72{\cdot}3 \text{ henrys.}$$

To increase the flux from 0·055 to 0·075 Wb an increase of 6·25 − 3·6 = 2·65 A is necessary, corresponding to a mean value for the inductance during this period of $\frac{8 \times 600 \times 0{\cdot}02}{2{\cdot}65} = 36{\cdot}2$ henrys. This value holds exactly for the straight line *RP*. A close approximation to the actual conditions may then be obtained by assuming the magnetization to consist of the two straight line portions *OR* and *RP*.

For the first portion the corresponding time constant is 72·3/100 = 0·72 sec, and for the latter portion 36·2/100 = 0·36 sec. The time constant corresponding to 57·6 henrys is 0·576 sec.

In Fig. 80, curve *A* indicates the manner in which the current would rise following the application of 625 V if the inductance maintained its initial value of 72·3 henrys. Curve *B* indicates the manner in which the current would grow if *L* were constant at 36·2 henrys, and *F* is the corresponding curve for an inductance of 57·6 henrys, the resistance in each case being 100 ohms. Until the point *R* of the saturation curve, Fig. 79, is reached—i.e. until the current is 3·6 A—the current will build up according to curve *A*. If from this point the magnetization curve were the straight line *RP*, Fig. 79, the current would then build up as in curve *D*, which is drawn at a fixed horizontal displacement, *QR*, from curve *B*.

Actually the transition from the one curve to the other will be gradual as indicated by the dotted line.

For greater accuracy, the portion RP of the magnetization curve

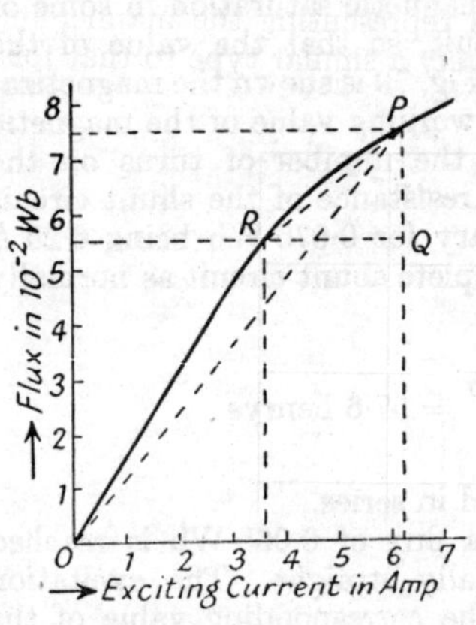

Fig. 79. Magnetization Curve

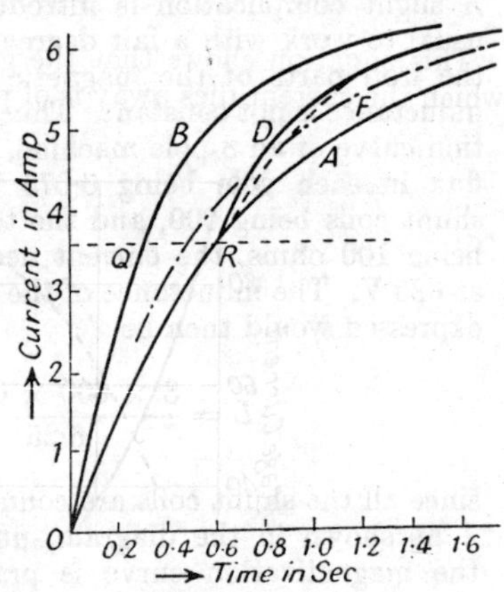

Fig. 80. The Growth of Current in Generator Field Circuit

may be divided into two or more sections, and the method described above extended.

It will be noticed that curve F does not represent at all truly the manner in which the current builds up. (With solid poles, the actual building-up curve will be profoundly modified, owing to the effect of eddy currents.)

Fig. 81

Decay of Direct Current in a Closed Circuit. This corresponds to short-circuiting a circuit in which current is flowing, as, for example, would occur if the switch S in Fig. 81 were closed. The current in the coil would then decay at a rate determined by the time constant of the circuit.

Applying the general equation (85): $i = \frac{1}{R}(V - B\varepsilon^{-\frac{R}{L}t})$, to this case.

Let the initial value of the current be I.

The value of the applied voltage after closing the switch S is $V = 0$. Hence $i = -\frac{1}{R}B\varepsilon^{-\frac{R}{L}t}$.

When $t = 0$, $i = I$, hence $I = -\frac{1}{R}B\varepsilon^{0}$ and $B = -IR$.

The current at any time t after closing the short circuiting switch is then

$$i = I\varepsilon^{-\frac{R}{L}t} \quad . \quad . \quad . \quad . \quad (87)$$

This equation shows that the graph representing the manner in which the current dies away is of precisely a similar type to that for

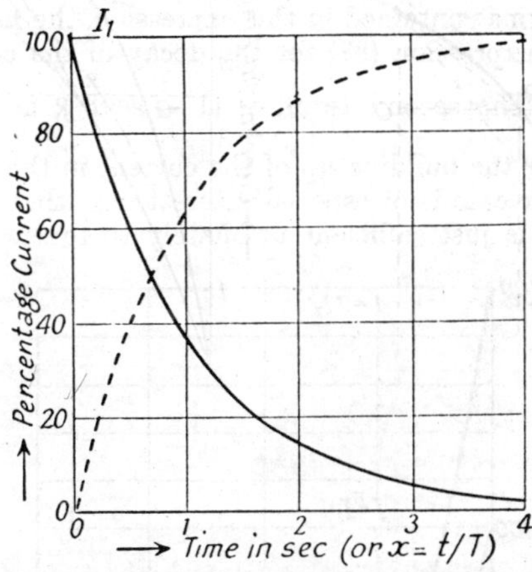

FIG. 82. THE DECAY OF CURRENT IN A CLOSED CIRCUIT

the building up of the current, but relatively "inverted," as in Fig. 82. In this diagram both the building up and the die-away curves are shown.

As with the *growth* curve, the *decay* curve can be used with any desired scale, ending in zero in this case. Thus equal times are taken for the current to fall from 100 per cent to 80 per cent, from 10 per cent to 8 per cent, and from 1 per cent to 0·8 per cent, and so on.

Reversal of Direct Current. Dealing finally with the application of a reversing voltage to a circuit in which a d.c. is flowing, the circuit remaining closed.

Let the initial value of the current be I
and that of the applied voltage be $-V$

That is, positive directions of current are taken to be those obtaining when $t = 0$.

Substituting $-V$ for V in the general equation (85),

$$i = \frac{1}{R}(-V + B\varepsilon^{-\frac{R}{L}t})$$

When $t = 0$, $i = I$, hence $I = -\frac{V}{R} + B$

from which $B = IR + V.$

The value of the current at any time t is then

$$i = \frac{1}{R}\{-V + (IR + V)\varepsilon^{-\frac{R}{L}t}\}$$

$$= I\varepsilon^{-\frac{R}{L}} - \frac{V}{R}(1 - \varepsilon^{-\frac{R}{L}t}) \quad . \quad . \quad . \quad (88)$$

Of the two terms contained in this expression, the former, $I\varepsilon^{-\frac{R}{L}t}$ is the ordinary expression (87) for the decay of the current I in a closed circuit. The second term, $\frac{V}{R}(1 - \varepsilon^{-\frac{R}{L}t})$, is the ordinary equation (86) for the building up of the current in the circuit.

One particular case is of especial interest, e.g. that in which the voltage applied is just sufficient ultimately to reverse the current

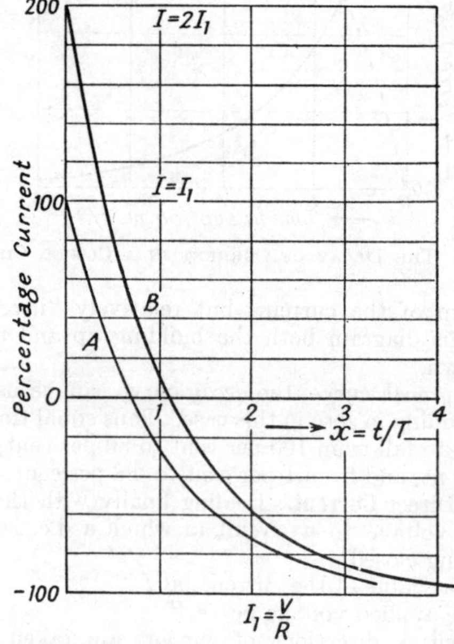

FIG. 83. THE REVERSAL OF A DIRECT CURRENT

already flowing, i.e. in which $V = IR$. The current equation is then

$$i = -\frac{V}{R}(1 - 2\varepsilon^{-\frac{R}{L}t}) \quad . \quad . \quad . \quad (89)$$

This is illustrated by curve A in Fig. 83; curve B is for when the initial value of the current is twice that obtaining after reversal. Actually this curve is simply an extension of the other.

It will be noted that for curve A the time taken for the current to be reduced to zero is identical with that taken to build it up from zero to one half of its ultimate maximum value.

Voltage Necessary to Reverse Current in a Given Time. As in the previous section, the value attained by the current after t sec under the action of a steady reversing voltage V, is, for an initial current of I amperes,

$$i = I\varepsilon^{-\frac{R}{L}t} - \frac{V}{R}(1 - \varepsilon^{-\frac{R}{L}t})$$

Hence if the current is reversed by the time T_C

$$i = -I = I\varepsilon^{-\frac{R}{L}T_C} - \frac{V}{R}(1 - \varepsilon^{-\frac{R}{L}T_C})$$

and
$$V = IR\frac{1 + \varepsilon^{-\frac{R}{L}T_C}}{1 - \varepsilon^{-\frac{R}{L}T_C}} \quad . \quad . \quad . \quad . \quad . \quad (90)$$

Where the time constant $\frac{L}{R}$ is large in terms of the time T_C,

$$V \simeq \frac{2LI}{T_C} \quad . \quad . \quad . \quad . \quad . \quad (91)$$

This problem has an important bearing upon commutation in d.c. machines.

Energy of the Magnetic Field. When the current in a circuit is changing, the applied voltage at every instant is utilized partly in overcoming resistance, and partly in changing the current, the voltage equation being

$$v = iR + N\frac{d\Phi}{dt} \ldots = iR + L\frac{di}{dt} \quad . \quad . \quad (92)$$

The power supplied to the circuit at any instant is then

$$vi = i^2R + Ni\frac{d\Phi}{dt} \ldots = i^2R + Li\frac{di}{dt} \quad . \quad . \quad (93)$$

The power absorbed in overcoming resistance is i^2R, the remainder of the input to the circuit is then utilized in increasing the energy stored in the magnetic field.

If $\delta\Phi$ and δi respectively represent the change in the flux and current that take place in a very short time δt, so short that the values of the various quantities, Φ, i, and v, may be assumed to be nearly constant, the energy supplied to the circuit during this small time is

$$\delta w = vi\delta t = i^2R\delta t + Ni\delta\Phi \ldots = i^2R\delta t + Li\delta i \quad . \quad (94)$$

The total energy supplied to the circuit in a given time can then be obtained by integrating the above expression, and is given by

$$W = \int_0^t vi\,dt = \int_0^t i^2R\,dt + \int_0^\Phi Ni\,.\,d\Phi$$

or

$$= \int_0^t i^2R\,dt + \int_0^I Li\,.\,di \quad . \quad . \quad . \quad . \quad (95)$$

In these expressions, Φ and I denote the values of the flux and current respectively at time t, and both flux and current are assumed to be zero at $t = 0$. (The same expressions can be used for other limits of integration, if necessary.) The energy degraded into heat is given by $\int_0^t i^2R\,dt$, so that the value of the energy supplied to build up the magnetic field.

$$W_f = \int_0^\Phi Ni\,.\,d\Phi \quad . \quad . \quad . \quad . \quad (96)$$

or, in terms of the current,

$$W_f = \int_0^I Li\,.\,di \quad . \quad . \quad . \quad . \quad (97)$$

The unit in which this energy is expressed is the joule or watt-second. The product Ni of the number of turns and the current gives the m.m.f. in ampere-turns.

If then the magnetization curve for the magnetic circuit which links with the current is represented by Fig. 84, and flux is plotted in webers, and m.m.f. in ampere-turns, the horizontal strip shown will represent $Ni\delta\Phi$, and its area is a direct measure of the energy that must be put into the magnetic field to change its value by the amount $\delta\Phi$. Similarly, to change the value of the flux from Φ_1 to Φ_2, the energy that must be supplied to increase the value of the flux is represented by the area shown shaded. If the scale of flux is 1 cm = A webers, and of m.m.f. 1 cm = B ampere-turns, 1 cm^2 will correspond to AB watt-sec.

The total value of the energy stored in the magnetic field established by the flow of current in a circuit can then at once be deduced from the magnetization curve, the total energy for a flux Φ_2, as in Fig. 84, being given by the area $OABCDO$ between the corresponding portion of the magnetization curve and the vertical line through the origin.

This principle can be applied not only to a complete magnetic circuit but to each individual part of the circuit. Thus for a portion of the circuit in which the uniform flux density is B, the area A, and the length of the path l, the energy stored in the volume lA is

$$\int_0^\Phi Ni\,.\,d\Phi = A\int_0^B Ni\,.\,dB = lA\int_0^B \frac{Ni}{l}\,dB$$

The energy stored per unit volume of the material is then

$$W_f = \int_0^B \frac{Ni}{l} dB \text{ watt-sec} \quad . \quad . \quad . \quad (98)$$

$$= \int_0^B H \, . \, dB \text{ watt-sec} \quad . \quad . \quad . \quad (99)$$

If then the magnetization curve of a material be plotted in terms of the flux density, and the ampere-turns per metre, $\frac{Ni}{l}$, as the unit of magnetizing force, the scales being 1 cm = a Wb/m², and 1 cm = b

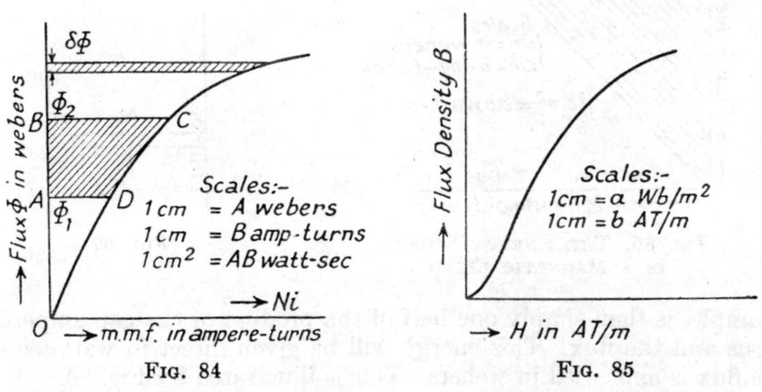

FIG. 84 FIG. 85

ampere-turns per metre, each cm² of area will then correspond to ab watt-sec, per metre³ of the material, as indicated in Fig. 85.*

Dealing now with the other expression for the value of the energy stored, i.e. $\int_0^I Li \, . \, di$. For the general case where the field established does not vary in direct proportion to the current, this expression will only hold if the inductance L is defined $L = N\frac{d\Phi}{di}$. For non-magnetic substances, and in particular for "air-cored" coils, this definition accords with the usual form, $L = \frac{N\Phi}{I}$. For such, the

* Occasionally magnetization curves are still plotted using the c.g.s. unit of magnetizing force, i.e. in terms of H', instead of the ampere-turns per metre. Since $H' = \frac{4\pi}{10} \times$ AT/cm, if the scales to which the curve is plotted are 1 cm = a gauss, and 1 cm = c gilbert (i.e. c.c.g.s. units of magnetizing force), 1 cm² of area will then represent $\frac{ac}{4\pi}$ ergs per cm³ of the material.

energy stored is then given by

$$\int_0^I Li \,.\, di = L\int_0^I i \,.\, di = \tfrac{1}{2}LI^2 \text{ watt-sec} \quad . \quad . \ (100)$$

since L is constant.

This energy may also be expressed in the form

$$\int_0^\Phi Ni \,.\, d\Phi = \frac{NI \times \Phi}{2} \quad . \quad . \quad . \ (101)$$

The total energy stored in the air gap field of a dynamo, for

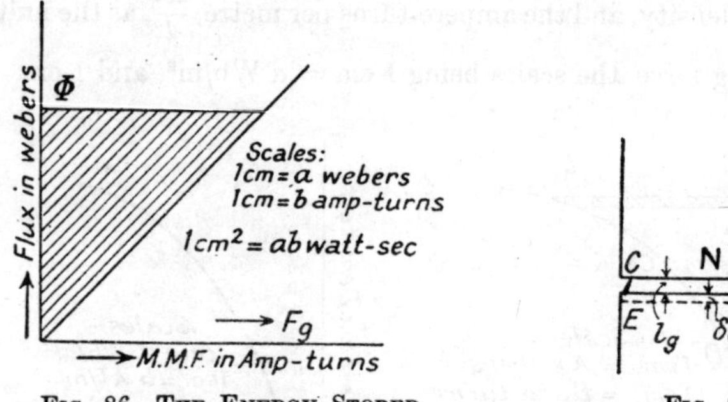

FIG. 86. THE ENERGY STORED IN A MAGNETIC FIELD

FIG. 87

example, is then simply one half of the product of the gap ampere-turns and the flux. This energy will be given direct in watt-sec if the flux is expressed in webers. This is illustrated by Fig. 86.

For a non-magnetic material the energy stored per unit volume, when the flux density is B, becomes

$$\int_0^B HdB = \int_0^B \frac{B}{\mu_o} dB = \frac{B^2}{8\pi} \times 10^7 \text{ watt-sec/m}^3 \quad . \ (102)$$

For a fairly large d.c. machine the flux density in the air gap is of the order of $B = 1{\cdot}0$ Wb/m², i.e. $B' = 10{,}000$ gauss, and at this density the energy stored per cubic metre is $10^7/8\pi = 398{,}000$ watt-sec, or 0·398 watt-sec/cm³.

Thus, for a pole face 30 × 30 cm², and an air gap of 5 mm the energy stored at this density is 900 × 0·5 × 0·398 = 179 watt-sec.

Magnetic Pull between Poles and Armature. In Fig. 87 let CD represent one pole of a magnet and EF its armature, the air gap being l_g and the surface area of the pole face A_g. Then if B is the flux density in the gap, neglecting all fringing effects the energy stored in the magnetic field is

$$W = \frac{1}{2}\frac{B^2}{\mu_o}A_g l_g$$

If the armature is moved so as to increase the air gap by the very small amount δl_g the work done will be given by the product of the force of attraction F between pole and armature and the distance δl_g. This work is entirely utilized in changing the value of the energy stored in the air gap, hence

$$F \cdot \delta l_g = \frac{1}{2}\frac{B^2}{\mu_0} A_g \delta l_g$$

and

$$F = \frac{1}{2}\frac{B^2}{\mu_0} A_g \quad . \quad . \quad . \quad . \quad . \quad . \quad . \quad (103)$$

$$= \frac{B^2}{8\pi} A_g \times 10^7 \text{ newtons, using m.k.s. values}$$

$$= 4{\cdot}06\, B^2 \times 10^4 \text{ kg/m}^2 \quad . \quad . \quad . \quad . \quad (104)$$

As an example, the force of attraction between the armature and a pole having a shoe area of 30 × 30 cm² = 0·09 m², with a flux density of 0·9 Wb/m², (i.e. 9,000 gauss) is

$$F = 4{\cdot}06 \times 0{\cdot}9^2 \times 0{\cdot}09 \times 10^4 \text{ kg}$$
$$= 2{,}960 \text{ kg}$$
$$= 6{,}500 \text{ lb}$$

The force of attraction is thus very nearly 3 tons; the flux density actually is not constant owing to the bevel of the pole shoe.

There is a corresponding force of attraction between the pole and the yoke which must, therefore, be stiff enough to withstand these forces in addition to having sufficient cross-sectional area to carry the flux.

Magnetic Hysteresis. When iron is subjected to alternating magnetization, the energy put into the magnetic field when the flux is increasing is not completely given back when the flux dies away, but a certain portion is wasted. This waste is due to the fact that a magnetic substance, once magnetized, tends to retain its magnetism. So that if the magnetization curve for a closed iron circuit be plotted for a complete cycle of alternating magnetization, a "hysteresis loop" is obtained as in Fig. 88, the portion of the curve corresponding to decreasing magnetizing force not coinciding with that for increasing magnetization.

The total amount of energy supplied to the iron during one complete cycle is given by $\int Ni \cdot d\Phi$, this integration being over the complete period. For a non-magnetic substance, $\int Ni \cdot d\Phi$ over one complete period is zero, indicating that the energy put into the magnetic field is completely returned when the field dies away. But for the magnetic substance, the value $\int Ni \cdot d\Phi$, represented by the area of the hysteresis loop, gives at once the value of the energy loss taking place for each complete cycle of magnetization.

For the loop given in Fig. 88, when the magnetizing force is reduced from its maximum value ND to zero, the flux density drops by only a small amount to the "residual density" OC. During this fall in the flux density, a small amount of energy represented, by the section NDC, is restored to the circuit magnetizing the iron. To reduce the magnetic flux density to zero, an actual demagnetizing force is necessary to coerce the magnetism out of the iron.

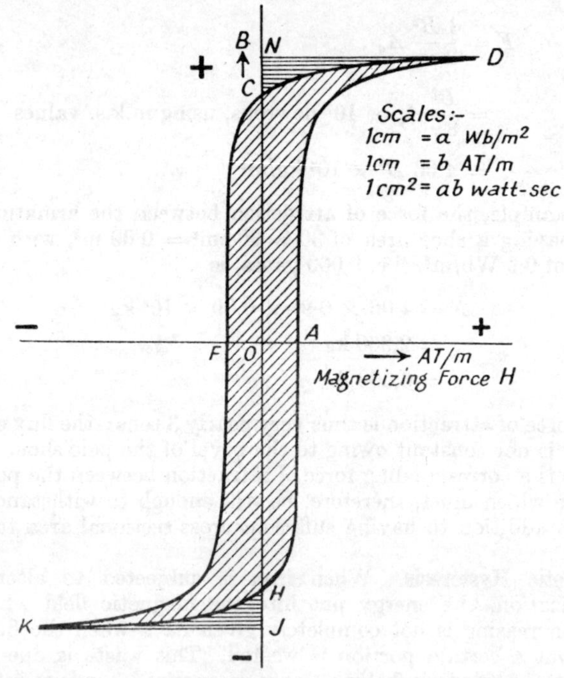

Fig. 88. Hysteresis Loop for Alternating Magnetization

This "coercive force" is represented by OF. For permanent magnets this coercive force is relatively large, and for good dynamo sheet steel it is small. During the time that the flux is reduced from the residual value OC to zero, work is done by the magnetizing circuit in coercing the magnetism out of the iron, the amount of work thus wasted being represented by the area CFO. The iron being demagnetized at F, to remagnetize it in the reverse direction up to the density OJ, energy must be provided by the magnetizing circuit, of an amount represented by the area $FKJO$. Of this energy, only the small portion represented by the area KJH is restored to the circuit when the magnetizing current is reduced to zero. The total energy wasted during the complete cycle of magnetization is thus represented by the area enclosed by the loop.

To determine the value of the energy wasted in an actual specimen it is most convenient to plot the loop in terms of the flux density and the ampere-turns/metre as indicated in Fig. 88. The area of the loop will then, when multiplied by the scales chosen, give the energy wasted per cubic metre of the material in watt-sec.

Finally, if the loop be plotted in terms of the total flux, expressed in webers, and the total ampere-turns, the area of the loop when

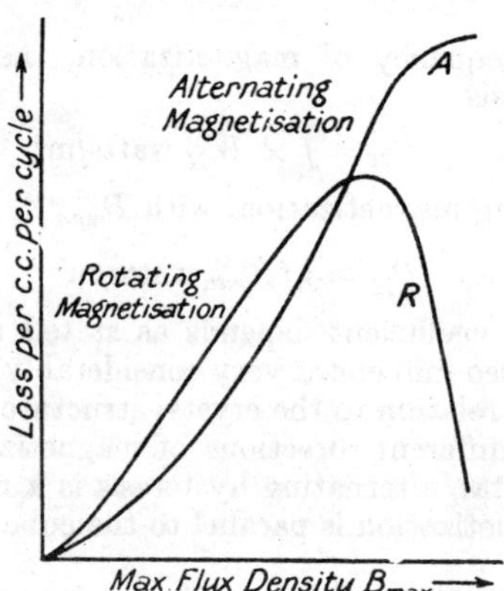

FIG. 89. TYPICAL HYSTERESIS LOSS CURVES FOR ALTERNATING AND ROTATING MAGNETIZATION

multiplied by the scales chosen, will give the energy wasted in the whole specimen, in watt-sec for each cycle. (These points have already been discussed on pp. 116–7.)

Steinmetz found that for iron worked at densities of not more than about 1 Wb/m², the energy wasted in hysteresis due to alternating magnetization could be expressed in the form

$$\text{Energy loss per cycle} = \eta' (B_{max} - B_{min})^x \qquad (105)$$

For normal alternating magnetization $B_{min} = - B_{max}$, and (105) reduces to

$$W_H = \eta B^x_{max} \qquad (106)$$

The index x was found to be remarkably constant for various materials, and to have the value of 1·6, the loss per cycle being

$$W_H = \eta B^{1\cdot6}_{max} \qquad (107)$$

For most grades of steel sheet at present in use, Steinmetz' law holds reasonably well up to $B = 1{\cdot}0$ Wb/m², but at higher flux densities the law no longer holds, the loss per cycle varying in accordance with curve A, Fig. 89. For flux densities between 1·0 and 2·0 Wb/m² the Steinmetz index is of the order 1·8 to 2·0 for ordinary dynamo sheet steel, and in practice the hysteresis loss, for simplicity, is often taken as proportional to B^2 within the above limits of flux density. But at high flux densities, when ferric saturation is reached, the hysteresis loss due to alternating magnetization tends to remain constant, so that in general, Steinmetz' law cannot be applied, as discussed later, page 129.

Power Loss. This is measured by the energy expended per sec, and is thus given by the product of the energy loss per cycle and the frequency.

Thus if f = frequency of magnetization, the power wasted in magnetic hysteresis

$$P_H = f \times W_H \text{ watts/m}^3$$

and for alternating magnetization, with $B_{max} \not> 1{\cdot}0$

$$P_H = \eta f B_{max}^{1\cdot6} \text{ watts/m}^3 \quad . \quad . \quad . \quad (108)$$

The hysteretic coefficient depends as stated upon the magnetic material and is also influenced very considerably by the direction of magnetization in relation to the crystal structure, and so for a given sheet varies for different directions of magnetization in the sheet. For a single crystal, alternating hysteresis is a minimum when the direction of magnetization is parallel to the cube edge of the crystal lattice.

For average conditions, with ordinary dynamo sheet steel η is about 50–55, and for silicon steel may be as low as 27–30, the actual value depending on the percentage of silicon present.*

With cold-rolled grain-oriented silicon steel magnetized parallel to the direction of rolling the hysteresis loss is very much less, but its anisotropy prevents its use in most d.c. machines.

Hysteresis Loss due to Rotating Magnetization. Hysteresis losses also occur in iron subjected to rotating magnetization, i.e. such as would occur when a particle of iron is rotated in a uniform magnetic field. At low values of the flux density the loss of energy due to rotating magnetization exceeds considerably the loss for purely alternating magnetization as shown by curves R and A in Fig. 89. At the knee of the magnetization curve the loss due to rotating magnetization reaches a maximum, after which the loss falls off rapidly, becoming negligible at ferric saturation. Fig. 90 shows the ratio of the hysteresis losses due to the two methods of magnetization, as determined by Brailsford,† to hold very closely indeed for 4 steels with a silicon content ranging from 0·13 to 3·7 per cent. The alternating loss is based on "average" conditions.

Another, and most useful way of stating the hysteresis loss is to give the loss in watts/kg—or per lb—at $B = 1{\cdot}0$ Wb/m² and $f = 50$.

* In c.g.s. units, with B' measured in gauss, and the loss in watts per cm³, the corresponding values of η are $0{\cdot}0012 \times 10^{-7}$ to $0{\cdot}0025 \times 10^{-7}$ for ordinary dynamo sheet steel and $0{\cdot}008 \times 10^{-7}$ for silicon steel.

† cf. *Rotational Hysteresis Loss in Electrical Sheet Steels.* F. Brailsford. *J.I.E.E.* 1939, Vol. 84; pp. 399–407.

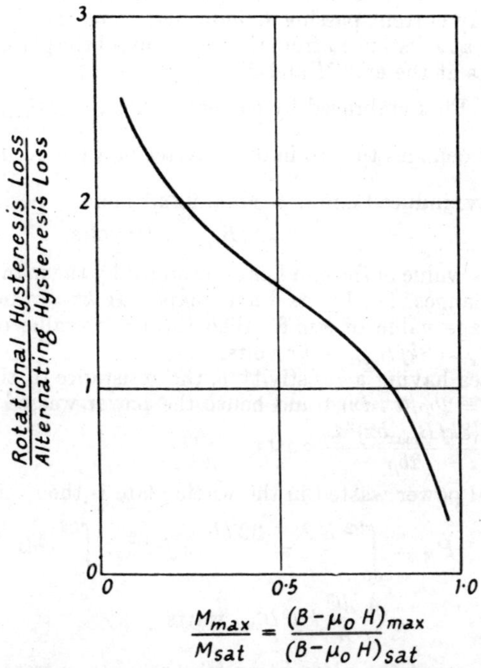

Fig. 90. Ratio of Rotational to Alternating Hysteresis Loss

The power wasted, expressed in this manner, ranges from about 1·8 W/kg for ordinary steel sheet to 1·0 W/kg for silicon steel, or even down to 0·9 with a very high silicon content. These are for "average" conditions, and not for grain oriented steel.

Eddy-current Loss in Iron Subjected to Alternating Magnetization. When a changing magnetic flux permeates any mass of metal, eddy currents are induced. In armature cores, these eddy currents are restricted in amount by building up the core of a number of thin plates, lightly insulated from each other by paper, varnish, etc., and so arranged that the plane of the actual plates coincides with the direction of the magnetic flux. The eddy currents being induced in a direction at right angles to the magnetic field, the paper, etc., offers resistance to their flow and restricts their path to the narrow limits of each lamination.

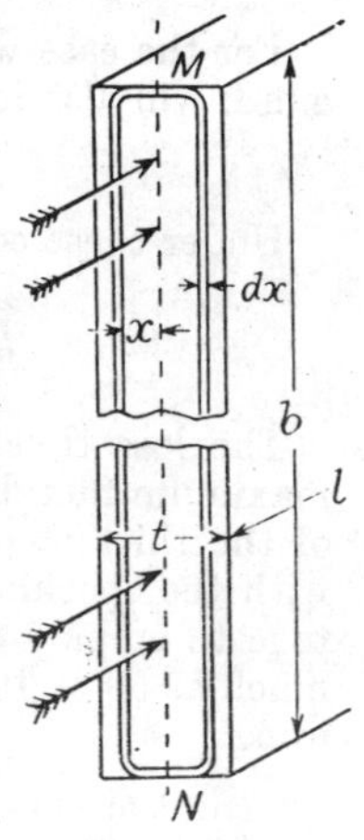

Fig. 91

In Fig. 91 is given an enlarged view of a section through a core plate through which a magnetic flux alternates in the direction indicated. Considering then an eddy current passing as indicated along the elemental path of width dx, at a distance x from the centre line of the plate. Neglecting small effects at the end M and N,

$$\text{Max. Flux embraced by current} = \Phi_{max} = 2B_{max} \times bx.$$

This flux changes to zero in one quarter of a period, i.e. in $\frac{1}{4f}$ sec.

$$\begin{aligned}\text{Hence, av. induced e.m.f.} &= \text{Max. flux/time of } \tfrac{1}{4} \text{ period} \\ &= 8fB_{max} \times bx \text{ volts.}\end{aligned}$$

The r.m.s. value of the e.m.f. is determined by the manner in which the flux changes, i.e. by its wave shape. If k* = r.m.s. value of e.m.f./average value of e.m.f., then the r.m.s. value of the e.m.f. induced, $E_x = 8kfB_{max} \times bx$ volts.

For plates having a resistivity ρ, the resistance of the elemental path is $r_x = 2b\rho/(l \,.\, dx)$,† and hence the power wasted in the path $= \frac{E_x{}^2}{r_x} = \frac{l\{8kfB_{max}bx\}^2dx}{2b\rho}$ watts.

The total power wasted in the whole plate is then

$$P_E = \int_0^{t/2} \frac{E_x{}^2}{r_x} = \frac{32\,bl}{\rho} \,.\, k^2f^2B_{max}^2 \int_0^{t/2} x^2dx$$

$$= \frac{4}{3}\,\frac{blt^3}{\rho}\,k^2f^2B_{max}^2 \text{ watts.}$$

The volume of the plate being blt, the eddy current loss per m³ reduces to

$$P_E = \frac{4}{3\rho}\,k^2f^2t^2B_{max}^2 \text{ watts/m}^3 \quad . \quad . \quad (109)$$

For the case when the flux variations are sinusoidal, the induced e.m.f. will also follow a sinusoidal law, and the form factor is

$$k = \pi/2\sqrt{2} = 1{\cdot}11.$$

Under these conditions the eddy loss is given by

$$P_E = 1{\cdot}64 \times \frac{1}{\rho} \times f^2t^2B_{max}^2 \text{ watts/m}^3 \quad . \quad . \quad (110)$$

The loss thus varies in direct proportion to the square of the maximum flux density, the square of the frequency, and the square of the thickness of the core plates. The loss also varies inversely with the specific resistance of the material. Hence it is an advantage to employ a material like silicon steel for which ρ may be as much as 50×10^{-8} as compared with about 12×10^{-8} for ordinary iron.

* k is defined as the "form factor."

† l is the length of the plate and $l \,.\, dx$, the area of the elemental path.

Since the hysteresis loss is not affected by reducing the thickness, and the iron space factor becomes smaller as the thickness is reduced, nothing is to be gained by reducing the thickness too much. For d.c. machines, plates 0·4 mm to 0·5 mm thick are commonly used for the armature; occasionally a thickness of 0·35 mm is adopted. With Stalloy, advantage is taken of the reduced eddy current losses by working with relatively thicker plates, i.e. 0·5 mm Stalloy being used instead of 0·35 ordinary iron. In practice it is sometimes a convenience to modify the formula to give the loss per kilogramme as in (112). For ordinary iron and steels with a low silicon content the density is 7·8 g/cm^3, for 3·5 per cent silicon steel it is 7·6 g/cm^3.

Total Iron Loss Curves. These curves give the total iron losses, including both hysteresis and eddy currents, plotted as a function of the maximum flux density for various values of the frequency.

Normally the curves are determined experimentally by wattmeter measurements of the actual losses that take place in a suitable sample of the material when subjected to purely sinusoidal alternating magnetization. Since the eddy loss is dependent upon the thickness of the laminations as well as the frequency, the curves hold good only for the particular thickness of material and frequency employed.

The wattmeter method of testing is quite satisfactory for the range of flux densities usual in transformers, but is not suited to the high flux densities of 2 Wb/m^2 or more used in the teeth of d.c. armatures. A much more elaborate method of testing is necessary for high flux densities.* Fig. 92 gives the total loss for sheet steel† 0·019 in. thick, normally used in d.c. machines, as measured by Brailsford and Bradshaw at 50 c/s. (The silicon content is 0·5 per cent.) Ferric saturation of the material is $M_{max} = 2{\cdot}1$ Wb/m^2. Fig. 92 also gives the eddy current loss as calculated by the above authors, using the usual formula, and in addition the hysteresis loss curve as determined by the difference between the measured total loss and the computed eddy current loss. It will be noticed that beyond the point of ferric saturation there is no further increase in the hysteresis loss, the increase in the total loss being entirely due to the increase in the eddy current loss.

It is a convenience to have available total loss curves for various frequencies. With the losses separated out as in Fig. 92, the hysteresis loss and the eddy current loss at any other frequency and for any other thickness can be calculated and hence the total loss determined for a number of values of the flux density. The loss curves in Fig. 93 have been thus based on Fig. 92 for a thickness of 0·5 mm. Such curves represent very closely the actual losses that would obtain in transformers, but the losses in dynamo-electric

* *Iron Losses at High Magnetic flux Densities in Electrical Sheet Steel.* Brailsford and Bradshaw. *J.I.E.E.* 1955, Vol. 102. Part A, pp. 463–71.

† Fig 8 gives the magnetization curve corresponding to this material.

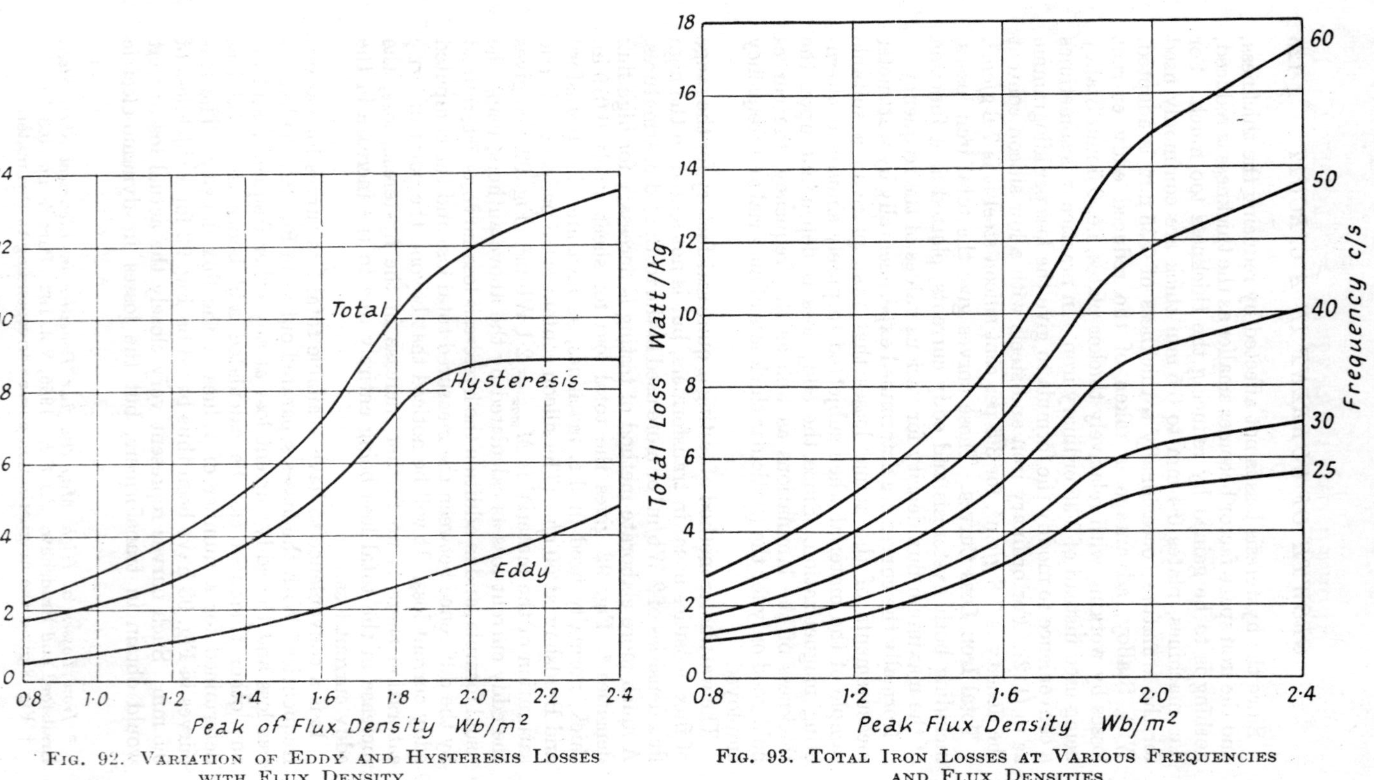

FIG. 92. VARIATION OF EDDY AND HYSTERESIS LOSSES WITH FLUX DENSITY

FIG. 93. TOTAL IRON LOSSES AT VARIOUS FREQUENCIES AND FLUX DENSITIES

machines are invariably higher than those due to simple sinusoidal alternating magnetization as discussed below.

Iron Losses in D.C. Machines. Under normal conditions of steady load, the only portions of the magnetic circuit of a d.c. machine which are subjected to changing magnetization are the armature and the pole shoes.

For both the core and the teeth the frequency with which the magnetization alternates is given by

$$f = np \quad . \quad . \quad . \quad . \quad . \quad (111)$$

A consideration of Fig. 23, p. 38 serves to establish the fact that for any particular particle of the armature core both the intensity of magnetization and the direction of magnetization change periodically as the armature is rotated. Various positions occupied by a particular particle are indicated, and the direction of magnetization shown. The intensity of magnetization is maximum at points in line with the neutral axis, at which points the direction of magnetization is circumferential. At the mid-pole position, the intensity is minimum and the direction radial. The magnetization of the core itself is thus partly of an alternating and partly of a rotating nature, and corresponds to "elliptical rotation."

In the armature teeth, on the other hand, the axis of magnetization is more nearly constant, so that for practical purposes the teeth may be regarded as being subjected to purely alternating magnetization, but by no means sinusoïdal.

This is apparent from Fig. 94*a* which is a reproduction of the field form deduced on page 49. If the armature had an extremely large number of very narrow teeth, the flux in each tooth, *neglecting secondary effects*, would change in a manner corresponding to the field form. The e.m.f. induced in a search coil wound round a tooth would then, as determined by the rate of change of the flux, have the wave shape indicated in Fig. 94*b*, the e.m.f. being very far removed from sinusoidal, being in fact zero for the greater part of each period. For a normal number of slots and when secondary effects are included, the e.m.f. induced in the coil would have the same general wave shape as in Fig. 94*b* but would be a little less extreme. It is then evident that the eddy e.m.f.'s. induced in the teeth of d.c. machines are very far removed from sinusoidal and that a formula based on sinusoidal alternating flux cannot give correct results. Further, for high harmonics, corresponding to a high frequency, the formula will not hold for plates of usual thickness. However it is a convenience to use equation (109) but it becomes imperative to modify the form factor k to the much larger value of from 1·6 to 2, the larger value being necessary when the air gap changes very quickly at the pole tip, as in Fig. 11*b*.

In the past, the discrepancy between the actual eddy current loss in the teeth and that calculated by (110), has been attributed to the effects of filing out the slots, in impairing the insulation between core plates. However, experience with most carefully assembled cores shows that a large discrepancy still remains, and this discrepancy is undoubtedly due to the very rapid changes in the flux in the tooth as it approaches and leaves the pole face.

Pre-determination of Iron Losses. A simple method of estimating the losses is to make use of the ideal total loss curves, as given in

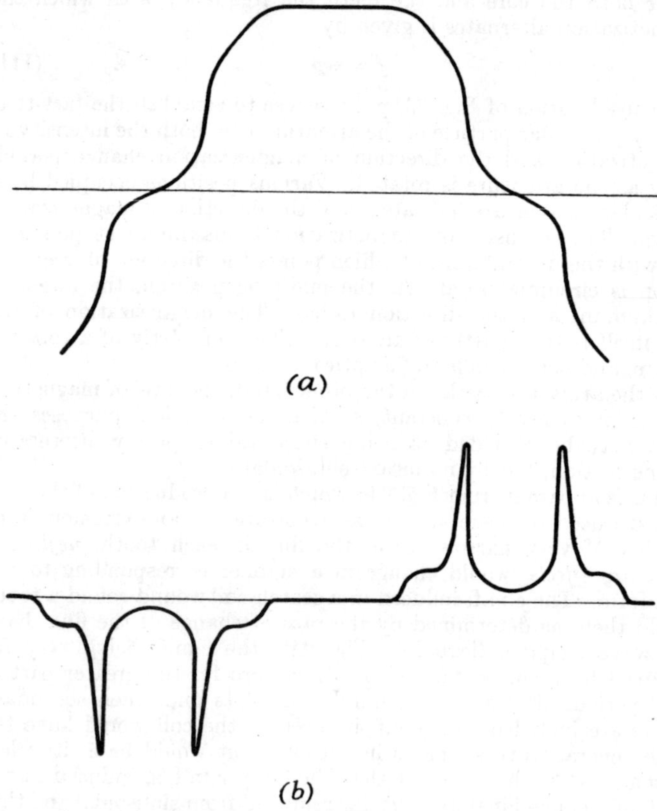

FIG. 94. (a) FIELD FORM; (b) INDUCED E.M.F. IN COIL AROUND ONE TOOTH

Fig. 93, introducing a multiplying factor based on actual test results on finished machines. This method is not scientifically justified, but when applied always to machines of similar proportions may be expected to give reasonably correct results. This method assumes a common value for the multiplying factor for the core and the teeth, and for hysteresis and eddy current losses.

In other instances the factor is given different values for hysteresis and eddy losses, and the constants thus taken assumed to apply equally to the teeth and the core. From the factors thus assumed fresh total loss curves are then calculated and plotted on the lines of Fig. 93.

The author prefers to determine separately the hysteresis and eddy losses. The eddy current loss can rapidly be determined by a

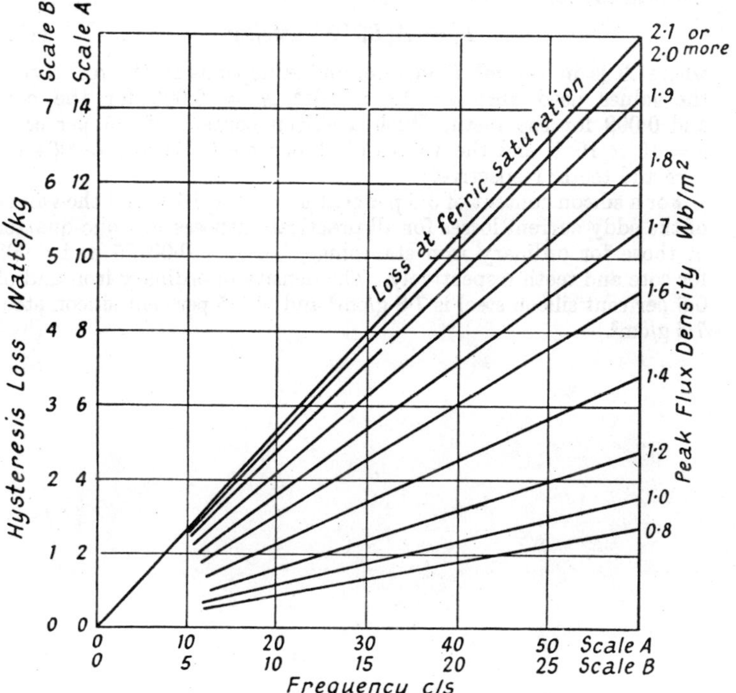

FIG. 95. HYSTERESIS LOSS AT VARIOUS FREQUENCIES AND FLUX DENSITIES

simple slide-rule calculation and there is no need to rely on reference to pre-calculated graphs. For the hysteresis loss it is, however, necessary to have recourse to graphs. With d.c. machines the frequency may have any value up to 75 c/s or so, and the method of plotting the hysteresis loss the author has found most satisfactory, is to plot the loss as a function of the frequency for various given values of the flux density. The graphs thus obtained are simply straight lines passing through the origin, and the same graphs can be used for various ranges of the frequency.

Such a family of graphs is given in Fig. 95. These graphs are based on Fig. 92 and hold for ordinary dynamo sheet steel with sinusoidal alternating magnetization, as determined by Brailsford and Bradshaw, with the exception that a multiplying factor of 1·5 has been introduced. This factor is used for both core and teeth. (It is probably a shade on the low side for the core and too generous for the teeth.)

For the eddy current loss the author uses different and larger multiplying factors for the teeth and the core. Representing the losses in the form—

$$P_e = A(Bft'')^2 \text{ watts/kg} \quad . \quad . \quad . (112)$$

where B is in Wb/m^2 t'' in mm, and A depends on the resistivity, the values used are: $\rho = 12 \times 10^{-8}$, $A = 0{\cdot}005$ for the core and 0·008 for the teeth. With a silicon content of 0·5 per cent, $\rho = 15 \times 10^{-8}$, and the values of A become 0·004 and 0·0064 for core and teeth, respectively.

For a silicon content of 3·5 per cent $\rho = 49 \times 10^{-8}$, and the values of the eddy current losses for all practical purposes are one-quarter of those for ordinary iron stampings, i.e. $A = 0{\cdot}00125$ and 0·002 for core and teeth respectively. The density of ordinary iron and of 0·5 per cent silicon steel is 7·8 g/cm^3 and of 3·5 per cent silicon steel 7·6 g/cm^3.

CHAPTER VI

COMMUTATION

THE e.m.f. induced in each coil on the armature of a d.c. machine alternates with a frequency corresponding to the product of the speed and the number of pairs of poles. With a full-pitched coil, the e.m.f. passes through zero at the instant when the active sides of the coil pass through the magnetic neutral position. To obtain direct currents in the external circuit, it is therefore necessary to reverse periodically the connexions of each armature coil with the external circuit, and the reversal should take place at the instant when the e.m.f. induced in the coil is zero. The commutator and brushes combine to form this reversing switch. Thus, for the ring armature indicated in Fig. 96, the mere passage of coil I from the

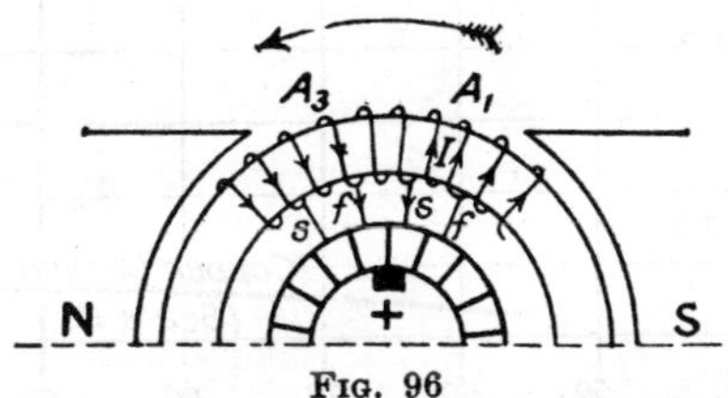

FIG. 96

position shown at A_1 to that at A_3 serves to alter the end nearest the + brush from s to f. For a brief period the commutator segments joined to the coil ends, s and f, are bridged, or short-circuited, by the brush, and during this brief period the coil is transferred from the circuit to the right to that at the left of the brush. It is during this brief interval that the current in the coil has to be reversed or "commutated." The diagram given in Fig. 96 for a ring winding holds also for a simple lap winding; with a wave winding, however, the brush, in bridging two adjacent segments, will short circuit all the coils that are connected in series between those segments, i.e. 2 coils for a 4-pole, 3 coils for a 6-pole, and, in general, p coils for a $2p$ pole armature, as indicated in Fig. 48, page 68.

The general problem of reversing a current in a closed circuit has already been discussed in Chapter V. In dealing with the commutation of the current in a coil of a d.c. armature, the problem is complicated by the fact that the circuit is by no means of a constant resistance, owing to the nature of the brush contact resistance.

Brush Contact Resistance. With d.c. machines, the main resistance in the circuit of the coil undergoing commutation occurs at the brush contacts.

This contact resistance is not constant but depends largely upon the current density at the contact surface, decreasing in a marked manner as the current density is increased. In fact, if a graph be plotted as in Fig. 97, showing the manner in which the voltage drop at the brush contact varies with the current density, a relatively flat curve is obtained. With normal carbon brushes, an increase of some 300 per cent in the current density, i.e. from 20 to 80 A/in.^{2}, only suffices to increase the brush contact drop by about 50 per cent. The graphs given in Fig. 97 cover the normal

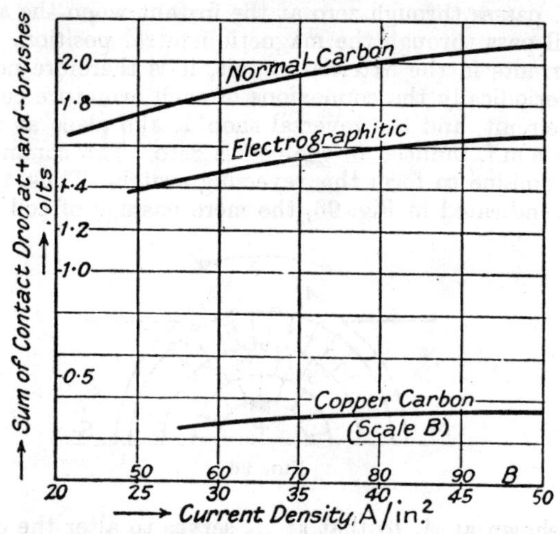

FIG. 97. THE TOTAL CONTACT VOLTAGE DROP AT THE BRUSHES

working current densities, and it will be seen that the variation in the voltage drop is quite small.

The contact drop is influenced slightly by the peripheral speed of the commutator, and depends to a small extent upon the pressure of the brush upon the commutator. Further, the drop has not quite the same value when current is flowing from the commutator to the brush, as when flowing in the reverse direction, so that the voltage drop at the positive brushes is not quite the same as that at the negative brushes. For a generator, at low temperatures the contact drop is less at the positive than at the negative brushes. The converse holds for a motor.

In addition, the brush contact drop depends to a very large extent upon the composition of the brush. This is illustrated by Fig. 97 in which curves are given for various grades of brushes.

For soft, graphitic brushes the normal working value of the current density may be as high as 60 A/in.^{2} or 9 to 9·5 A/cm^2, and the total voltage drop—positive + negative—at this density is about 1·6 volts. These brushes have a low coefficient of friction—

about 0·12—and are suited for high speed commutators. For brushes made of a copper-carbon mixture, the working value of the current density may be about 100 A/in.^{2} or 15 to 16 A/cm^2. The coefficient of friction with these brushes is a little higher than with the graphitic brush, and is about 0·15 to 0·18. Depending upon the relative proportions of graphite and copper, the total voltage drop at the two sets of brushes may range from 0·25 to 0·35. Such brushes are used for very low voltage, heavy current, direct current machines, as used for electro-plating.

With normal carbon brushes, the working value of the current density is from 35 to 40 A/in.^{2} or 5·5 to 6·5A/cm^2. The voltage

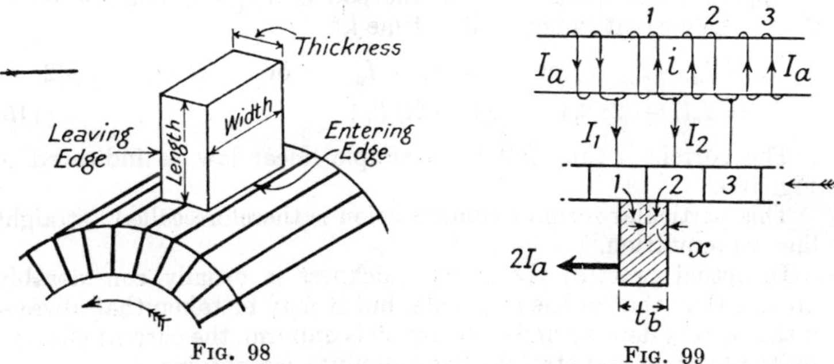

FIG. 98 FIG. 99

drop at the two sets is about 2, and the coefficient of friction from 0·22 to 0·27. Electro-graphitic brushes—normal carbon brushes which have been graphitized by heating in an electric furnace—can be worked at a current density of 55 to 60 A/in.^{2}—8·5 to 9 A/cm^2—and have a total voltage drop of 1·7 to 1·8 for the two brush sets. The coefficient of friction is about 0·22.

TERMS. Much confusion arose in the past owing to various meanings having been given to the same term. The terms in use have been standardized, and are indicated in Fig. 98. This diagram needs no further explanation.

Current Density Under the Brush. At first, it would seem to be desirable to aim at adjusting conditions during commutation so that the current density under the brush is uniform over the whole contact area.

Taking, for the moment, the case when the brush thickness does not exceed that of one commutator segment, and neglecting the thickness of the mica separator between the segments:

t_b = thickness of brush

and x = the breadth of the contact between segment 2 and the entering edge of the brush, as in Fig. 99.

I_1 = the current delivered to the brush by segt. 1.

I_2 = ,, ,, ,, ,, 2.

I_a = the current per circuit in the armature.

Then $I_1 = 2I_a(1 - x/t_b)$ and $I_2 = 2I_a x/t_b$ . . . . (113)

For the coil shown, commutation starts when $x = 0$, and finishes when $x = t_b$. If, then, time is measured from the instant when commutation starts, and the commutator speed is v_C

$$x = v_C t$$

and the time of commutation

$$T_C = t_b/v_C \quad . \quad . \quad . \quad . \quad (114)$$

Applying Kirchhoff's law to the coil undergoing commutation: if i = the current in this coil at time t,*

$$i = I_a - I_2, \quad \text{or} \quad = I_1 - I_a, \quad \text{or} \quad = (I_1 - I_2)/2$$
$$= I_a(1 - 2x/t_b) = I_a(1 - 2t/T_C) \quad . \quad . \quad . \quad . \quad (115)$$

The current i thus follows a simple linear law as indicated in Fig. 100.

This particular form of commutation is therefore called "straight line commutation."

In actual practice the brush thickness is usually considerably greater than that of the segments, but it may be taken that always, if the current density under the brush is uniform, the current changes quite regularly and straight line commutation obtains.

It follows then that to obtain uniform current density under the brush—a condition which used to be considered ideal but which may be somewhat improved on—steps must be taken to compel the current to follow this linear law.

Variable Current Density. If the manner in which the current changes in the coil undergoing commutation is known, there is no particular difficulty in determining the manner in which the current density under the brush varies.

It is evident that if the change in the current is delayed—as compared with straight line commutation—the current I_1 carried by the riser 1, Fig. 99, will not fall so quickly as in the case just considered, and, as a result, the current density at the leaving edge of the brush will be greater, and at the entering edge less than with straight line commutation.

The actual manner in which the current density changes may be deduced from the "short circuit" current graph as follows. In Fig. 101 the curve i represents the manner in which the short circuit current is assumed to vary. The current entering the segment at the leaving edge of the brush being given by $I_1 = I_a + i$, is represented by a curve I_1 precisely similar to i but displaced vertically by an amount corresponding to I_a.

The contact area of the brush on the segment at the leaving edge

* The direction *before* commutation is taken as positive.

varies in a linear law from t_b at $t = 0$ to zero at the end of commutation. This area may then be represented to vary with time according to the ordinates of the straight line AB in the diagram. Then, at time t the current carried by the segment at the leaving edge is

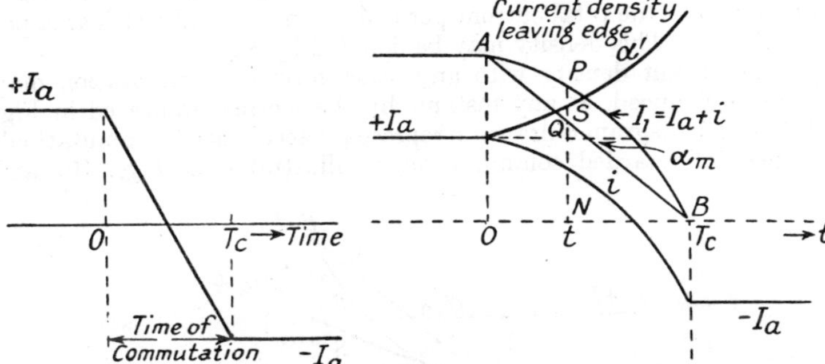

FIG. 100. STRAIGHT-LINE COMMUTATION

FIG. 101. THE CURRENT DENSITY AT THE LEAVING EDGE OF THE BRUSH, WITH RETARDED COMMUTATION

represented by PN, and the contact area by QN. The current density is therefore proportional to PN/QN, and is thus represented by the graph α', shown.

In a similar manner the current density at the other edge of the brush can be determined. For this portion of the brush $I_2 = I_a - i$ so that the graph for I_2 is readily obtained, as in Fig. 102, and the current density α'', at various instants deduced.

The diagrams given indicate the case when the change in the current is retarded or "retarded commutation." It will be noted that under these conditions the leaving edge of the brush carries far more than its fair share of the current, and that, at the instant of this edge finally breaking contact with the coil, the current density reaches* two and a half times its initial value. This initial value is equal to the mean current density over the whole brush, since at $t = 0$ the whole current is carried by the one commutator segment on which the brush rests.

The current density under the entering edge of the brush* reaches its normal value only at the end of commutation, and, for the major part of the commutation period, is less than one half of the normal value as given by dividing the total current by the area of contact of the whole brush, and represented by α_m in Figs. 101 and 102.

In Fig. 101 the straight line AB, in addition to giving the width of brush contact under the leaving portion, also represents the current that would flow in the riser 1 for straight-line commutation.

* For the case illustrated.

Hence the actual current density under the two sides of the brush can very readily be determined by comparing the actual current curve obtaining on short circuit with that for straight line commutation.

With straight-line commutation the current density is constant, and equal to the total current per brush arm ÷ total brush area per brush arm. This density may be denoted by α_m.

The current density with any other form of commutation may then be deduced, at any instant, by the method indicated in Fig. 103, which diagram is drawn to represent "accelerated commutation" instead of retarded commutation, as illustrated in Figs. 101 and

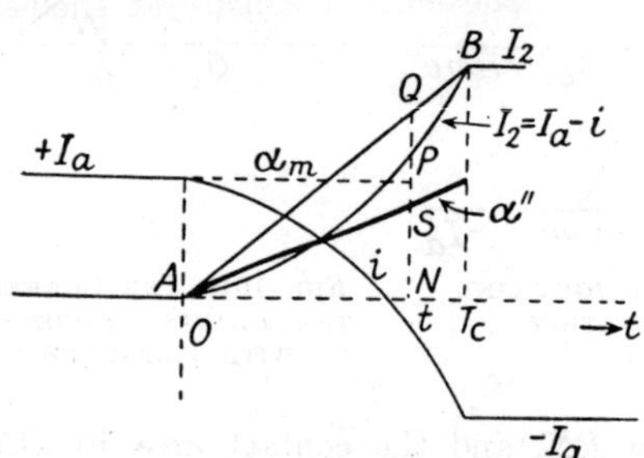

FIG. 102. THE CURRENT DENSITY AT THE ENTERING EDGE OF THE BRUSH WITH RETARDED COMMUTATION

102. The curve i, Fig. 103, indicates the actual short-circuit current, and the straight line AB corresponds to straight-line commutation.

The total current in each armature path is represented by the length OA, hence $AC = RN$, denotes the total current collected by the brush under consideration.

Then $I_1 = I_a + i$ is represented by PN

and $I_2 = I_a - i$ is represented by RP.

The corresponding values of the currents for straight-line commutation are represented respectively by QN and RQ.

Then at the leaving portion of the brush,

$\frac{PN}{QN} =$ the ratio of the actual current to that obtaining with uniform current density under the brush.

Hence the actual current density at the *leaving edge* is,

$$\alpha' = \frac{PN}{QN} \times \alpha_m$$

For the *entering* portion of the brush,

$\frac{RP}{RQ} =$ the ratio of the actual current to that obtaining with uniform current density under the brush.

Hence the actual current density at the entering edge is

$$\alpha'' = \frac{RP}{RQ} \times \alpha_m.$$

At the very end of commutation, the current density at the leaving edge of the brush will be given by—cf. Fig. 103—

$$\alpha' = \alpha_m \times \frac{di}{dt} \div \frac{di'}{dt} = \alpha_m \times \frac{di}{dt} \div \frac{2I_a}{T_C}$$

where i' is the current for straight-line commutation. Hence if when $t = T_C$, $\frac{di}{dt} = 0$, the current *density* at the leaving edge is reduced to zero by such time as the brush breaks contact with the

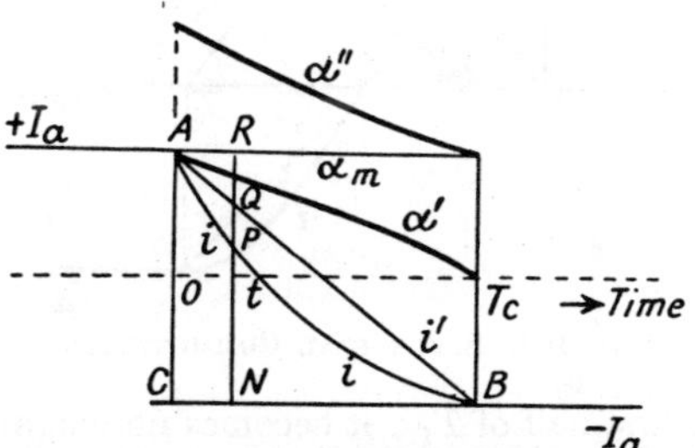

FIG. 103. THE CURRENT DENSITY AT THE TWO EDGES OF THE BRUSH WITH ACCELERATED COMMUTATION

segment, and all possibility of sparking taking place at this instant is removed.

In a similar manner the current density at the entering tip at the beginning of commutation is given by

$$\alpha'' = \alpha_m \times \frac{di}{dt} \div \frac{2I_a}{T_C}$$

A good method of starting commutation is then to arrange that $\frac{di}{dt} = 0$ when $t = 0$, although this is by no means so important as it is to arrange that $\frac{di}{dt} = 0$ at the end of commutation.

The cases illustrated in Figs. 101 and 102 give $\frac{di}{dt} = 0$ when $t = 0$, and for Fig. 103 $\frac{di}{dt} = 0$ when $t = T_C$. Actually, the short-circuit currents for both these have been chosen so that the changes in current density under the *leaving edge* in the one correspond very closely with those obtaining under the *entering edge* in the other, and vice versa.

A consideration of the two forms of short circuit current just discussed would indicate that it is desirable to retard commutation somewhat during the first moments, but that it is also desirable that the later stages should take place more or less in accordance with Fig. 103. In this respect a sinusoidal short circuit current, as indicated in Fig. 104, possesses the advantage that $\frac{di}{dt} = 0$ both at the beginning and end of commutation. Commutation is thus

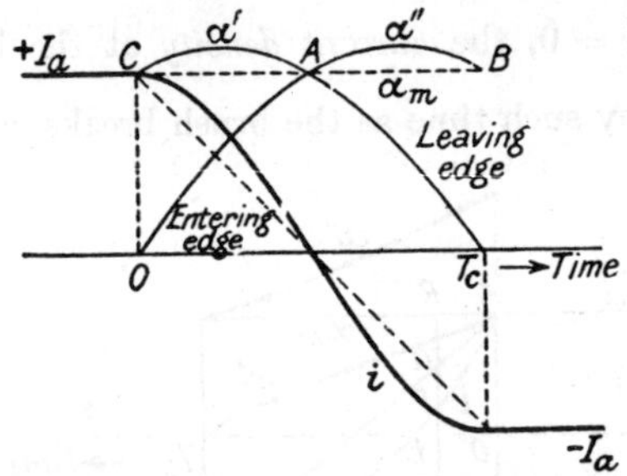

FIG. 104. SINUSOIDAL COMMUTATION

retarded for the first half of T_C; it becomes normal at $t = T_C/2$ and is then accelerated so that during the last instants but little current is carried by the leaving edge of the brush. With sinusoidal commutation the maximum current density under the brush rises only to about 11 per cent greater than the mean density, as indicated in Fig. 104.

Influence of Brush Contact Resistance. With carbon brushes worked at normal current densities the pressure drop at the contact is of the order of 1 volt. Normally this pressure drop is several times as great as that caused by full load current flowing in an armature coil, so that to a close approximation it may be taken that the resistance of the coil itself is negligible in comparison with that of the brush contacts. If, then, no e.m.f. whatever be induced in the coil undergoing commutation, and the resistance of the coil is negligible, the current will divide between the two commutator risers in direct proportion to the areas of the portions of the brush in contact with the two segments, as in Fig. 99. That is, the current density under the brush will be uniform, and straight-line commutation will obtain.

Even if a small e.m.f. be induced in the coil, and the coil resistance is not negligible, it is still evident that if the volt drop at the brushes is the most important term, this resistance at the brush contacts will tend to cause a proper distribution of the current. In fact many machines have been built and are yet in use in which, as discussed later, the brush contact resistance alone has sufficed to ensure satisfactory commutation with fixed brushes under very varying conditions of load.

E.m.f. Induced in Coil Undergoing Commutation. The value of the total e.m.f. induced in the coil undergoing commutation has therefore a very important bearing upon the current distribution under the brush, and hence upon sparking at the brushes. This total e.m.f. is at once given by the total rate of change of the flux linked with the coil, hence it follows that for the induced e.m.f. to be zero the flux linked with the coil undergoing commutation must remain constant. The flux linked with the coil may change due to the motion of the coil-sides through a stationary magnetic field, thereby establishing a "dynamic" e.m.f. in the coil. On the other hand the flux change may be due to the change in current, in the coil itself, or in neighbouring coils, thereby establishing a "static" e.m.f. of self- or mutual induction.

Normally it will be found convenient, instead of dealing with the total e.m.f., to deal separately with the dynamic and static e.m.f.'s. Of these two, the former may or may not be present. With the brushes in the magnetic neutral position, for a non-interpole machine no dynamic e.m.f. is present. If the brushes are, for a generator, behind the magnetic neutral position, a dynamic e.m.f. will be established in a direction tending to maintain the original current in the coil and therefore producing retarded or under-commutation. The extent of this will increase as the lag of the brushes behind the magnetic neutral position is increased. In particular, if the brushes are left at the geometric neutral position, when the dynamo is loaded armature reaction establishes a magnetic field of such a polarity, at the brush axis, as to cause the dynamic e.m.f. to retard commutation.

On the other hand, if with a generator the brushes are rocked beyond the magnetic neutral position, the coil-sides undergoing commutation actively cut through the fringe of the magnetic pole towards which they are moving, and the dynamic e.m.f. is then in such a direction as to tend to accelerate commutation. If excessive lead be given to the brushes, over-commutation will result. For an interpole machine, as discussed later, a dynamic e.m.f. is generated by the motion of the coil sides through the field under the interpole. This field is of such a polarity that the dynamic e.m.f. tends to accelerate commutation.

The static e.m.f. must always be present owing to the inductance of the coil. Further the e.m.f. induced by this means is always such as to retard the change in the current and thus to tend to produce under commutation.

So that the total e.m.f. in the short-circuited coil shall be zero, the sum of the static and dynamic e.m.f.'s must be zero. Since the static e.m.f., i.e. the reactance e.m.f., opposes the change in the current, it follows that to obtain satisfactory commutation a dynamic e.m.f. definitely tending to reverse the current should be generated.

This is in accordance with Chapter V, where it was shown that, to reverse a current in an inductive circuit rapidly, it is *essential to develop a reversing e.m.f. in the circuit,* this reversing e.m.f. being such as to overcome the e.m.f. of self-induction due to the change in the current.

The reversing e.m.f. may be obtained, as stated above by—

1. Giving the brushes of a generator *forward* lead to such an extent that the brush axis corresponds to the fringe of the pole towards which the coils undergoing commutation are moving. This involves moving the brushes to such a position that the m.m.f. developed by the main field coils overpowers that developed by armature reaction. Since both the armature ampere-turns and the e.m.f. of self-induction of the commutated coil increase in direct proportion of the load current, it follows that for perfect conditions the brush lead should be adjusted for each value of the armature current, increasing as the current increases. Good enough commutation can often be obtained by having the brushes fixed at a position corresponding to perfect commutation at about $\frac{3}{4}$ full load; or

2. Using series wound interpoles of the same polarity, for a generator, as the pole towards which the coil is moving. By using interpoles, the brush position is fixed at the neutral axis.

For a motor, with a given direction of rotation and polarity of the magnets, the armature current will be in the opposite direction from that obtaining for a generator. It follows then that the direction of the reversing field for a motor must be opposite to that for a generator. With a non-interpole motor, then, for perfect commutation, the brushes should be given backward lead. For an interpole motor, the interpoles are of opposite polarity to the main poles in advance of them.

Time of Commutation.

Let v_C = the peripheral speed of the commutator, i.e.

$$v_C = \pi D_C \times n$$

t_b = the brush thickness

τ_C = the pitch of the commutator segments, i.e. $= \dfrac{\pi D_C}{C}$

t_m = the thickness of the mica separator between segments.

In Fig. 105*a* the relative positions of brush and commutator are given for the instant at which commutation is just beginning for coil 1, and in Fig. 105*b* are shown the corresponding positions at the instant when commutation ends.

During commutation of a single coil, then, the surface of the commutator moves through a distance equal to the thickness of the brush less that of the mica. This distance being denoted by β_1,

$$\beta_1 = t_b - t_m \quad . \quad . \quad . \quad . \quad (116)$$

Considering now the commutation of two adjacent coils. Commutation of the current in coil 2 begins a little later and ends a

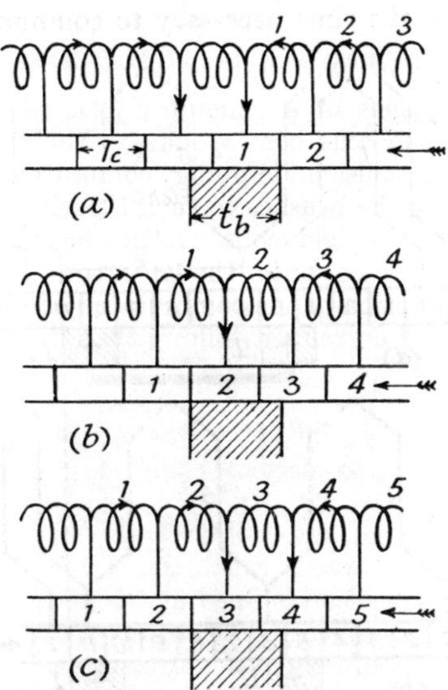

FIG. 105. THE TIME OF COMMUTATION

little later than in coil 1, this time displacement corresponding to the motion of the commutator through a distance of one segment pitch.

During commutation of the current in the two coils, then, the commutator surface moves through a distance,

$$\beta_2 = \tau_C + t_b - t_m \quad . \quad . \quad . \quad . \quad (117)$$

In a similar manner, for x adjacent coils, the total distance moved through is

$$\beta_x = (x - 1)\tau_C + t_b - t_m \quad . \quad . \quad . \quad (118)$$

In dealing with commutation it is a convenience to deal with the commutation of the complete layer of coils in the slot rather than with the individual coil; (118) can be applied at once to the case of x coil-sides per layer in the slot.

The time of commutation for a single coil is therefore

$$T_{C_1} = \frac{t_b - t_m}{v_C} \quad . \quad . \quad . \quad . \quad (119)$$

and for the complete layer of x coils

$$T_{Cx} = \frac{(x-1)\tau_C + t_b - t_m}{v_C} \quad . \quad . \quad . \ (120)$$

For a full pitched winding this time will correspond to the complete reversal of the current in all the conductors in one slot. With a chorded winding, the time necessary to commutate the complete

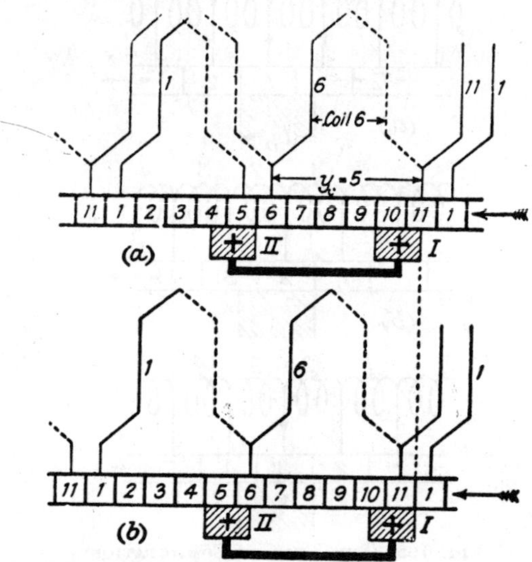

Fig. 106. Illustration of the Duration of Commutation with a 4-Pole Wave Winding having 4 Brush Sets

x coils is not altered, but the upper and lower coil sides in each slot will not simultaneously be commutated. *With a wave winding* provided with only two brush arms, each brush commutates simultaneously as many coils as there are pairs of poles, i.e. 2 coils for a 4-pole and, in general, p coils for a $2p$-pole armature. For this case, independent of the number of poles, we again have

$$\beta_x = (x-1)\tau_C + t_b - t_m$$

and

$$T_{Cx} = \frac{(x-1)\tau_C + t_b - t_m}{v_C}$$

However, wave windings are often worked with the full number of brush spindles, i.e. $2p$, as this permits of the use of a relatively short and cheap commutator. By introducing the extra brush sets, adjacent pairs of positive, and negative, brushes, short circuit the individual coils directly in turn which increases the time of commutation. Fig. 106 illustrates the example of a 4-pole retrogressive wave winding having 11 commutator segments. The positive brushes only are shown, their mutual displacement being 11/2 = 5½ segments. In the upper figure, segments 5 and 6 are on the point of being bridged by brush *II*, and commutation in coil 6 is about to begin, since coils 6 and 11 are connected in series between these two segments. At the same time, the two legs of coil 6 are on the point of being short circuited by the two brushes.

The lower diagram represents conditions occurring after the commutator has moved forward by a distance of one half a segment, and indicates that at this instant segments 11 and 1 are on the point of being bridged by brush *I*. Between these two segments coils 1 and 6 are connected in series. The current in coil 6 will therefore not finally be commutated until segments 11 and 1 are no longer bridged by brush *I*; as already seen, commutation starts in 6 when conditions are as in Fig. 106*a*.* If brush *II* were lifted, the start of commutation would be delayed until the conditions in Fig. 106*b* were realized. The introduction of the second set of brushes therefore increases the time of commutation, the distance moved through by the commutator during the commutation of one coil being increased by one half of the pitch of the commutator segments. For a 4-pole winding, therefore, the effect of introducing the extra brush sets is to increase the total time of commutation in accordance with

$$\beta_1 = \frac{\tau_C}{2} + t_b - t_m; \quad \beta_x = (x - \tfrac{1}{2})\tau_C + t_b - t_m \quad . \ (121)$$

and $$T_{C_1} = \frac{\frac{1}{2}\tau_C + t_b - t_m}{v_C}; \ T_{Cx} = \frac{(x - \frac{1}{2})\tau_C + t_b - t_m}{v_C} \quad . \ (122)$$

In general it may be shown that, for a winding having $2a$ paths in parallel, and $2p$ poles,

$$\dagger T_{C_1} = \frac{\left(1 - \frac{a}{p}\right)\tau_C + t_b - t_m}{v_C}; \ T_{Cx} = \frac{\left(x - \frac{a}{p}\right)\tau_C + t_b - t_m}{v_C} \quad . \ (123)$$

Inductance of Coil Undergoing Commutation. In modern d.c. machines, with slotted armatures, each coil has a considerable leakage inductance.

When the current in an armature coil is commutated, the coil normally lies along the axis of magnetization due to the armature currents. The reversal of the current in the coil does not appreciably affect the value of this total flux due to armature reaction, but as shown below, serves to alter the manner in which this flux links with the coil. Thus, if current were carried by the conductors in one only of the slots, the magnetic field set up would simply encircle those particular conductors, as shown in Fig. 107.

* The effect established in other than the neighbouring teeth under these conditions is of no present interest.

† T is given in sec if either m.k.s. units or c.g.s. units is used throughout.

A difference of magnetic potential is therefore established between the two walls of the slot. When the complete winding carries current, this difference of magnetic potential between the two walls of the slot still obtains, and as a result some of the flux lines leak across the slots as indicated in Fig. 107, the direction in which the lines cross the slot being determined solely by that of the current. Hence, when the current reverses, the direction in which the flux

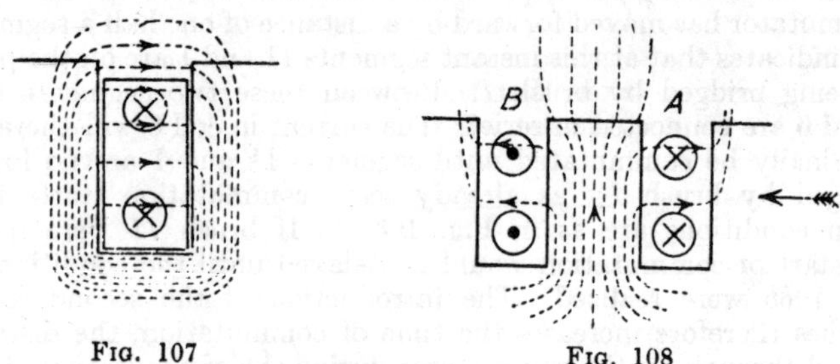

FIG. 107 FIG. 108

traverses the slot also reverses, as illustrated in Fig. 108, slots *A* and *B*.

As a result, during the time that the current in a slot is being commutated, this reversal of the flux leaking across the slot establishes an e.m.f. in the coils undergoing commutation, and this e.m.f. is in a direction tending to retard the change in the current.

The effect is not confined to the portions of the coils embedded in the slots, but also occurs to a reduced extent at the end connexions. The e.m.f. thus established is generally called "the reactance e.m.f." Actually, however, it is not merely the inductance of a coil that has to be considered, but also the mutual inductance of other coils which may be in the process of commutation at the same instant.

To calculate the value of this e.m.f. it is necessary first to determine the magnitude of the flux that crosses the slot, for a given armature current, and hence to obtain also the equivalent inductance of the coil.

Permeance of the Slot. It is a convenience to express the necessary formulae in terms of unit length of core.

For a perfectly uniformly distributed flux crossing a rectangular slot as indicated in Fig. 109 it is at once evident that the slot permeance for each unit length of core is $\mu_0 h_s/b_s$, and the flux established by an m.m.f. of one ampere-conductor is $0{\cdot}4\pi h_s/b_s \times 10^{-6}$ Wb/m of core length.

In actual practice this does not obtain, since the difference of magnetic potential between the two walls of the slot is not constant. Thus, consider Fig. 110 which shows, for convenience, a number of very thin conductors completely filling the slot. The magnetic line leaking across the slot at a height x from the bottom, only links with a fraction of the total conductors in the slot—to be precise it only links with $\frac{x}{h_s}$ of the total conductors. The m.m.f. establishing the leakage line at x is therefore only $\frac{x}{h_s}$ of the total developed by the conductors in the slot. Since the permeability of the iron for the teeth in the interpolar region is very great compared with that of air, it may be taken that the difference of magnetic potential between the walls of the slot will then increase uniformly from zero,

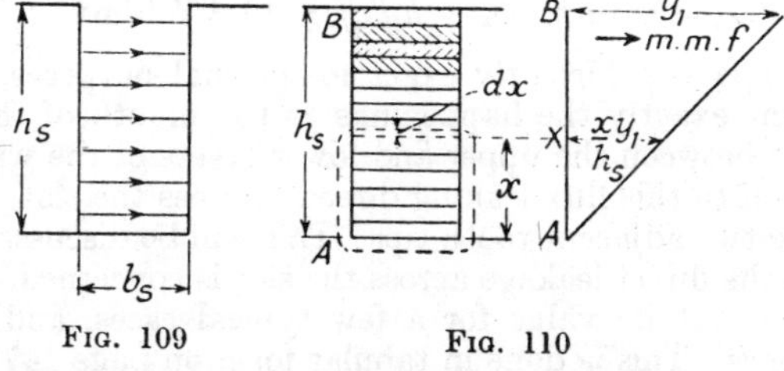

FIG. 109 FIG. 110

at the bottom, to full value at the top of the slot. This is indicated in graphical form in Fig. 110. It is evident that for the same maximum value of the difference of magnetic potential between the walls of the slot, the actual total slot leakage flux for Fig. 110 is only one half of that obtaining for Fig. 109.

So far as the reactance e.m.f. is concerned a further effect must be considered—the e.m.f. depends upon the number of *flux linkages*.

Thus if Z_s be the number of conductors per slot, and F the total m.m.f. per slot, the flux leaking along the elemental path dx (cf. Fig. 110) links with $\frac{x}{h_s} Z_s$ conductors, the corresponding m.m.f. being $\frac{x}{h_s} F$. The effective number of flux linkages produced by the flux is then

$$\phi_x = \frac{\mu_o F Z_s x^2}{h_s^2} \cdot \frac{dx}{b_s} \text{ per unit of core length,}$$

so that the total number of flux linkages established for the whole slot is

$$\phi = \frac{\mu_o F Z_s}{h_s^2 b_s} \int_0^{h_s} x^2 dx$$

$$= \mu_o F Z_s h_s / 3b_s \text{ per unit length of core}$$

This is equivalent to an effective flux $\mu_o F h_s / 3b_s$ linking with all the Z_s conductors in the slot, for unit length of core. So that, with

$F = 1$ ampere-conductor per slot, the equivalent flux per metre length of core is $0{\cdot}4\pi h_s/3b_s \times 10^{-6}$ weber.*

In actual practice the winding does not completely fill the slot, but in fairly large machines is held in place by fibre or bakelite slot wedges. In practice it is thus a blend of those shown in Figs. 109 and 110 and is indicated in Fig. 111.

If then h_c = height of copper in the slot,
and h_t = height above copper, as in Fig. 111

Then from the above it follows that for each ampere-conductor in the slot the effective flux established is approximately

$$\lambda_s = 0{\cdot}4\pi(h_c/3b_s + h_t/b_s) \times 10^{-6}\ \text{Wb/m}\dagger \quad . \quad . \ (124)$$

This formula is sufficiently exact for normal purposes, but does not represent exactly the happenings at the mouth of the slot, or in the space between the upper and lower layers of the winding.

In addition to this flux leaking directly across the slot, flux leaks between the two adjacent tooth tips. This will be discussed shortly.

So far as the direct leakage across the slot is concerned, it may be well to work out its value for a few typical cases, and for some abnormal ones. This is done in tabular form on page 147.

For the normal thickness of the wedge, and making due allowance for the insulation at the top and bottom of the slot, it is accurate enough to take, with large slots $h_t = 5$ and $h_c = h_s - 7$, and hence $h_c/3b_s + h_t/b_s \simeq (h_s + 8)/3b_s$, all dimensions being in mm.

For small machines in which bands of steel binding wire are employed and no wedges used, it suffices to take $h_c/3b_s + h_t/b_s \simeq (h_s + 4)/3b_s$.

In the table for slots of greater depth than 30 mm the expression corresponding to the use of wedges is employed, and for smaller slots that corresponding to the use of wire bands.

A consideration of this table at once indicates the importance of employing suitable slot proportions. With deep narrow slots the effective inductance is very great, and with broad shallow slots it is relatively small. For non-interpole machines it is desirable to employ slots having the ratio slot depth/slot breadth as small as possible. To take full advantage of the increased armature output possible with interpole designs it becomes necessary to use relatively deeper slots. In general it may be taken that for non-interpole designs the ratio slot depth/slot breadth will be from 3 to 4, 3½ being about an average figure. For interpole designs these figures are exceeded, values of 5 being met with on rare occasions, and 4 to 4½ being quite common.

As a rough comparison, three slots having approximately the same "area" may be taken, i.e., 13 × 30, 10 × 40, and 8 × 50.

* I.e. $0{\cdot}4\pi h_s/3b_s$ maxwells/cm of core length.

† I.e. $= 0{\cdot}4\pi(h_c/3b_s + h_t/b_s)$ maxwells/cm. h_t and b_s can be measured in any convenient unit.

THE EQUIVALENT FLUX PER AMPERE-CONDUCTOR, PER UNIT LENGTH OF CORE FOR THE DIRECT SLOT LEAKAGE*

(Tooth tip leakage *not* included)

Slot Breadth (mm)	Total Height of Slot (mm)					
	25	30	35	40	45	50
8	1·52	1·78	2·25	2·51	2·78	3·04
9	1·36	1·58	2·00	2·23	2·46	2·70
10	1·22	1·42	1·80	2·01	2·22	2·42
11	1·11	1·29	1·63	1·82	2·02	2·21
12	1·02	1·18	1·49	1·67	1·85	2·02
13	0·93	1·09	1·38	1·55	1·71	1·87
14	0·87	1·02	1·28	1·43	1·58	1·73
15	0·80	0·95	1·21	1·34	1·48	1·62

To a rough approximation the equivalent slot leakage fluxes per ampere are, respectively, 1, 2, and 3. The advantage, from the point of view of reactance voltage, of employing the broad shallow slot, is then apparent.

Leakage Across Tooth Tips. To a first approximation the paths of this leakage flux may be assumed to be as in Fig. 112, i.e. part

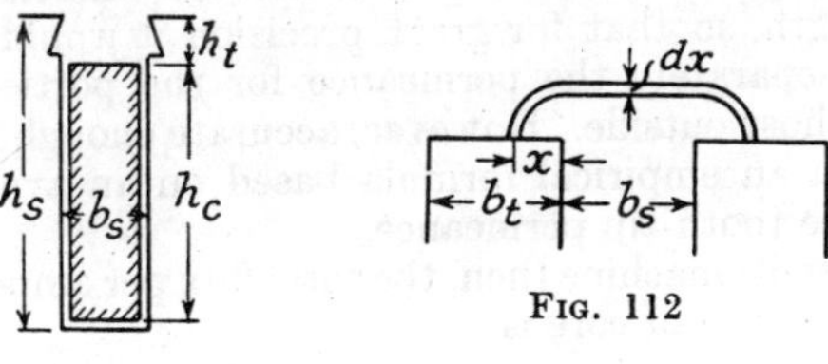

Fig. 111

Fig. 112

circular and part rectilinear. The m.m.f. acting along all the paths is the total m.m.f. developed by the conductors in one slot. The permeance of the elemental path shown is,

$$\text{permeance } \lambda_x = \frac{\mu_0 dx}{b_s + \pi x} \text{ per metre length of core.}$$

The total permeance, extending to one half of the tooth width, is then

$$\lambda_t = \int_0^{\frac{b_t}{2}} \frac{\mu_0 dx}{b_s + \pi x} = \frac{\mu_0}{\pi} \log \varepsilon \left(1 + \frac{\pi}{2}\frac{b_t}{b_s}\right)$$

* In 10^{-8} Wb/m core length or maxwells/cm.

The flux established per ampere-conductor in the slot is then

$$0{\cdot}4\pi \times \frac{2{\cdot}3}{\pi} \log_{10}\left(1 + \frac{\pi}{2}\frac{b_t}{b_s}\right) \times 10^{-6} = 0{\cdot}92 \log_{10}\left(1 + \frac{\pi}{2}\frac{b_t}{b_s}\right) \times 10^{-6} \text{ Wb/m.}$$

The flux will spread much more than is assumed by the distribution given in Fig. 112, and the permeance will be greater than that deduced above. It is well then to take the flux per unit m.m.f. as somewhat larger than the above value, by replacing the constant 0·92 by 1·0.

Hence the tooth-tip leakage for a non-interpole machine, per ampere-conductor, may be taken as—

$$\lambda_t = \log_{10}\left(1 + \frac{\pi}{2}\frac{b_t}{b_s}\right) \times 10^{-6} \text{ Wb/m length of core* } . \quad (125)$$

For an interpole machine, the presence of the shoe of the interpole modifies the alue of the permeance of the path between the tooth tips. An inspection of Fig. 113 will indicate that an approximation to the maximum permeance of the path of the flux crossing from one tooth tip to the next via the interpole shoe—the slot is then centrally under the shoe—is $0{\cdot}5b_{cp}/2l_{g_{cp}}$. The mean value of this permeance will be about 0·75 of the maximum, as fringing effects are relatively large.

The axial length of the interpole is normally much less than the total core length, so that for great precision it would be necessary to calculate separately the permeance for the portions under the shoe and for those outside. However, accurate enough results will be obtained from an empirical formula based on an average value of $b_{cp}/6l_{g_{cp}}$ for the tooth-tip permeance.

For an interpole machine then, the total flux per ampere-conductor for each unit length of core is

$$\lambda_t = 0{\cdot}4\pi \times \frac{b_{cp}}{6l_{g_{cp}}} \times 10^{-6} \text{ Wb/m core length} \quad . \ (126)\dagger$$

The total flux linked with the embedded portion of each conductor in the slot, for unit m.m.f. of one ampere-conductor per slot will then have an effective value of

$$\lambda_i = \lambda_s + \lambda_t$$

$$= 0{\cdot}4\pi \left\{\frac{h_c}{3b_s} + \frac{h_t}{b_s} + \frac{b_{cp}}{6l_{g_{cp}}}\right\} \times 10^{-6} \text{ Wb/m} \quad . \quad . \quad . \ (127)\ddagger$$

for an interpole machine,

* $\lambda'_t = \log_{10}\left(1 + \frac{\pi}{2}\frac{b_t}{b_s}\right)$ maxwells/cm.

† Equation (126) is only an approximation, and will give too small a value of the flux where the gap dimension $l_{g_{cp}}$ is large.

‡ $= 0{\cdot}4\pi \left\{\frac{h_c}{3b_s} + \frac{h_t}{b_s} + \frac{b_{cp}}{6l_{g_{cp}}}\right\}$ maxwells/cm for an interpole machine

$$\text{and} = \left\{0{\cdot}4\pi\left(\frac{h_c}{3b_s}+\frac{h_t}{b_s}\right)+\log_{10}\left(1+\frac{\pi}{2}\frac{b_t}{b_s}\right)\right\}\times 10^{-6}\ \text{Wb/m} \quad . (128)$$

for a non-interpole machine

Inductance of End Connexions, or "Free" Portions of Coil. The exact predetermination of the inductance of the end connexions of the coil (*see* Fig. 114) is not possible. However, the inductance of

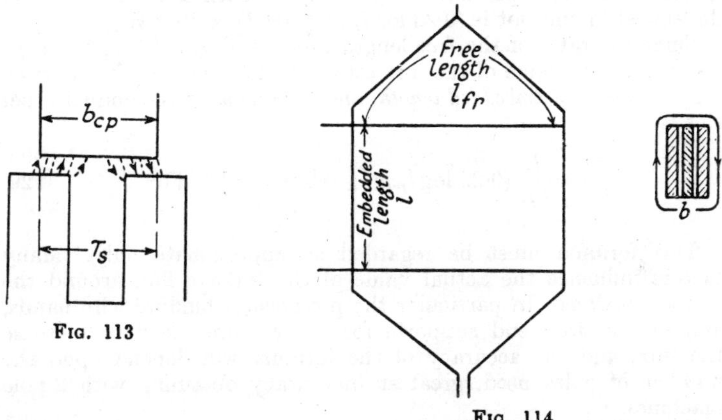

FIG. 113

FIG. 114

these portions is very much less than of the embedded part of the coil, and it is in practice of no moment if the method of calculating the inductance of the end coils is not very precise. For a multipolar machine, to a very rough approximation, the end connexions, or free portions of the armature coil, may be assumed to form one half of a square at each end of the machine, and the inductance may be assumed to correspond to the value holding for a square coil.

The inductance of a square built up of a conductor of rectangular section is* $4a(0{\cdot}46 \log 2a/b + 0{\cdot}14) \times 10^{-6}$ henry where a is the length of each side of the square and b is the periphery of the conductor in metres, and b/a is small. The flux established per metre for a current of 1 A is therefore $(0{\cdot}46 \log 2a/b + 0{\cdot}14) \times 10^{-6}$ Wb.

To a first approximation, then, this formula may be taken to hold for the end connexions of the winding, the length $2a$ being replaced by the total length l_{fr} of the free portion of each armature conductor, as shown in Fig. 114.

The dimension b is simply the periphery of one complete layer of

* "Radio Instruments and Measurements," *Washington Bureau of Standards Circular*, No. 74, page 247.

the winding, and hence of one complete coil group, measured around the bare copper. (As it is a logarithmic term that is involved, great precision in the determination of b is unnecessary.) Thus modified, i.e. with l_{fr} replacing $2a$, the above expression will give the flux per metre for each ampere-conductor *in the coil-group.*

It is a convenience to express this flux in terms of the total *ampere-conductors in the slot.* There are two coil groups in two layers in each slot, so that the ampere-conductors per coil-group is one half of that for the whole slot, and the flux established per metre of the free portion of the coil, for each ampere-conductor developed in one slot is $(0{\cdot}23 \log l_{fr}/b + 0{\cdot}07) \times 10^{-6}$ Wb/m.

Since the ratio of the free length to the embedded length is l_{fr}/L the equivalent value of the flux established around the free portion for *each metre of embedded length,* and for one ampere-conductor per slot is

$$\lambda_{fr} = \frac{l_{fr}}{L}(0{\cdot}23 \log l_{fr}/b + 0{\cdot}07) \times 10^{-6} \text{ Wb}^{*} \quad . \quad . \ (129)$$

This formula must be regarded as approximate only. Many factors influence the actual value of the leakage flux around the end connexions. In particular the presence of binding wire bands, and of cast iron end supports for the winding, tend to increase the flux, and the accuracy of the formula will depend upon the number of poles used, greatest inaccuracy obtaining with 2-pole machines.

Total Inductance. The total equivalent value of the flux per unit core length that is established for each ampere-conductor in one slot, including both the effects of the embedded and the free portions, is then

$$\begin{aligned} \lambda &= \lambda_i + \lambda_{fr} \\ &= (\lambda_s + \lambda_t) + \lambda_{fr} \\ &= \left\{0{\cdot}4\pi\left(\frac{h_c}{3b_s} + \frac{h_t}{b_s} + \frac{b_{cp}}{6l_{gcp}}\right) + \frac{l_{fr}}{L}\left(0{\cdot}23 \log \frac{l_{fr}}{b} + 0{\cdot}07\right)\right\} \\ &\qquad \times 10^{-6} \text{ Wb/m}\dagger \quad . \quad . \ (130) \end{aligned}$$

The actual effective value of the flux thus established for a slot having Z_s conductors each carrying I_a amperes becomes

$$\lambda \times I_a Z_s \text{Wb per unit core length}$$

* $\lambda'_{fr} = \frac{l_{fr}}{L}(0{\cdot}23 \log l_{fr}/b + 0{\cdot}07)$ maxwells/cm.

† $\lambda' = \left\{0{\cdot}4\pi\left(\frac{h_c}{3b_s} + \frac{h_t}{b_s} + \frac{b_{cp}}{6l_{gcp}}\right) + \frac{l_{fr}}{L}\left(0{\cdot}23 \log \frac{l_{fr}}{b} + 0{\cdot}07\right)\right\}$ maxwells/cm.

Since there are two active coil-sides, the total effective flux linked with the whole coil group is simply,

$$2\lambda L \times I_a Z_s^* \quad . \quad . \quad . \quad . \quad (131)$$

Average Reactance e.m.f. of Coils Undergoing Commutation. During commutation, with slots containing a number of coil-sides in each layer, the position of the various coils, with respect to the neutral axis, depends upon their situation in the slot. Further, with the coil-sides adjacent to the walls of the slot, a considerable portion of the commutation period takes place when the current is steady in all the other coil-sides in the same slot, whereas no such conditions obtain for the coils placed in the middle of the layer. It is then desirable to deal with average conditions, and to consider the reversal of the current in the complete layer of coils in one slot, rather than with the individual coil. For a full-pitched winding this is equivalent to considering the commutation of the whole current in one slot, as the upper and lower coil sides are commutated simultaneously.

For such a winding, during the time that the current in the upper layer is commutated, the whole of the current in the slot is reversed, and the effective flux changes from $2\lambda L \times I_a Z_s$ to $-2\lambda L \times I_a Z_s$. The total flux change is then $4\lambda L I_a Z_s$, and, as shown on page 142, this change takes place in the time T_{C_x}.

The *average value* of the static e.m.f. induced in *each turn* of the coils undergoing commutation is then

$$e_{r_{av}} = \frac{4\lambda L I_a Z_s}{T_{C_x}} \text{ volts} \quad . \quad . \quad . \quad (132)$$

This e.m.f. is in such a direction as to oppose the change in flux; it therefore retards the reversal of the current, and tends to produce "under commutation."

If there are N_a turns per coil in the winding, the average reactance e.m.f. per segment, for a simple lap winding, is

$$E_{r_{av}} = 4N_a \lambda L \frac{Z_s I_a}{T_{C_x}} \text{ volts} \quad . \quad . \quad . \quad (133)$$

For a *simple wave* winding having only two brush spindles, and $2p$ poles, there are p coils in series under the brush, and if each of these consist of N_a turns, the mean reactance voltage becomes, for full-pitched coils,

$$E_{r_{av}} = 4pN_a \lambda L \frac{Z_s I_a}{T_{C_x}} \text{ volts} \quad . \quad . \quad . \quad (134)$$

* It will be noticed that the effect of ventilating ducts has been neglected and the embedded length taken to be equal to the total core length. The error thus introduced is negligible in comparison with other errors inevitably introduced, and serves partly to compensate for the omission of the extra inductance due to the use of the bands of steel binding wire used over the end coils.

With normal wave windings, the number of slots is chosen so as to avoid the need of dummy coils, and is therefore not a multiple of the number of poles. This results in the reversal of current in the upper layer of a slot taking place a little earlier or a little later than that in the lower layer, as the winding is not quite full pitched. The reactance e.m.f. is then somewhat reduced, as discussed below for chorded windings.

As already stated, conditions are not the same for all the conductors in the complete layer of the winding, and the average reactance e.m.f. for the coil-sides adjacent to the walls of the slot will be less, and for those in the middle of the layer greater than the values as given in (133) and (134), as mutual inductance is greater for the central coil sides.

If the commutation of a single, centrally placed, coil be considered, the time of commutation is $T_{C_1} = t_b/v_C$,* as shown on page 134. For the central conductors, the average number of coils simultaneously undergoing commutation in the same slot, is given by twice the brush thickness ÷ the segment pitch, i.e. by $2t_b/\tau_C$, as the corresponding upper and lower coils are simultaneously commutated. With straight line commutation in all the coils, the average rate at which the slot m.m.f. is changing is then $2t_b/\tau_C$ times the rate at which it changes for the single coil and is therefore $(2N_aI_a \times 2t_b/\tau_C) \div T_C$. (The maximum rate would correspond to the nearest integer above the actual value $2t_b/\tau_C$.)

In a lap winding, the average reactance e.m.f. for this central coil is then

$$E_{r_{av}} = 2N_a\lambda L \times \frac{2t_b}{\tau_C} \times \frac{2N_aI_a}{T_{C_1}}$$

$$= 8N_a\lambda L \times \frac{t_b}{\tau_C} \times N_aI_a \times \frac{v_C}{t_b}$$

$$= 8N_a^2\lambda LI_av_C/\tau_C \quad . \quad . \quad . \quad . \quad . \quad (135)$$

The peripheral speed, for C commutator segments, is

$$v_C = \tau_C Cn$$

Hence, for the central coil in the slot, the average reactance e.m.f., with straight-line commutation, is

$$E_{r_{av}} = 8N_a^2\lambda LI_aCn \quad . \quad . \quad . \quad . \quad (136)$$

For sinusoidal commutation, the maximum value of the induced e.m.f. is $\frac{\pi}{2}$ times the average value, and would be

$$E_{r\,max} = 4\pi N_a^2\lambda LI_aCn \text{ volts} \quad . \quad . \quad . \quad (137)$$

(The actual inductance of the single coil at the centre of the slot is approximately $2N_a^2\lambda L$ henrys; mutual induction between upper

* Assuming the thickness of mica to be negligible.

and lower layers increases this to $4N_a^2\lambda L$ henrys. The "frequency" corresponding to commutation is proportional to the product of the number of segments and the speed, i.e. nC.)

So far as the reactance e.m.f. for the centrally placed coil-sides is concerned, then, provided the brush thickness exceeds the pitch of the segments, increasing the thickness has no effect upon the average value of the reactance e.m.f. A consideration of the expression (133), for the average e.m.f. for the whole slot, indicates, however, that this latter e.m.f. decreases as the thickness of the brush is increased, provided that the brush does not cover more segments than there are coil-sides per layer.

Again, so far as the maximum e.m.f. is concerned it is desirable that t_b/τ_C should be just a little less than an integer rather than just a little more. If, for example, the brush covers 1·1 segments, for a brief instant 2 coils are simultaneously being commutated, this number is not exceeded if the brush thickness is increased to 1·9 segments.

However, considerations of the reactance e.m.f. are by no means the only ones affecting sparking conditions. If the brush covers too many segments, the conditions for the various coils, when they undergo commutation, are too widely different, especially with a relatively small number of slots per pole.

Reactance e.m.f. with Chorded Windings. In a lap winding, short chorded by one slot pitch, during the time of commutation of one layer—(either upper or lower)—of the conductors in any slot, the current in the other layer is quite steady. The total flux change that is produced for the complete commutation of one coil group is then, so far as the embedded portion is concerned, reduced to approximately one half of that for a full pitch winding. (The one coil-side is at the top of one slot and the other at the bottom of another slot; hence the flux change for the upper side is very much less than for the lower side.)

The "effective" value of λ_s for a winding short chorded by one slot pitch actually reduces to

$$\lambda_s = 0{\cdot}4\pi\left(\frac{5h_c}{24b_s} + \frac{h_t}{2b_s}\right) \times 10^{-6}\ \text{Wb/m}$$

For λ_t the effective permeance reduces to one half of its value for a full pitch winding.

For the free portions of the coil there is no reduction in the flux established, as even with a full pitched winding this flux is due simply to the one layer of the winding, excepting for the short straight portions projecting directly from the slots.

For a winding chorded by one slot pitch the value of λ becomes then

$$\lambda = \left\{0{\cdot}4\pi\left(\frac{5h_c}{24b_s} + \frac{h_t}{2b_s} + \frac{b_{cp}}{12l_{gcp}}\right) + \frac{l_{fr}}{L}\left(0{\cdot}23\log\frac{l_{fr}}{b} + 0{\cdot}07\right)\right\} \times 10^{-6}\ \text{Wb/m} \quad . \quad . \quad . \ (138)$$

and the *total* effective value of the flux linked with the coil group is

$$2\lambda L Z_s I_a$$

To a rough approximation it may then be taken that by chording the winding by one slot pitch the inductance of the embedded portion is reduced to one half of the value that obtains for a full pitched winding.

Occasionally, with lap windings, the number of slots chosen is an odd multiple of the number of pole pairs, so that the winding becomes short chorded by one half a slot pitch. Again, to a sufficient degree of accuracy for practical purposes, it may be taken that the reduction in the inductance of the embedded portion is one half that obtaining for a winding short chorded by one slot pitch.

With wave windings the degree of chording of the coil is usually small, although present. If u represents the fraction of a *slot* pitch by which the winding is under- or over-chorded, it will in general be sufficiently exact to base the inductance of the embedded portion upon the following value for λ_i

$$\lambda_i = 0{\cdot}4\pi\left(1 - \frac{u}{2}\right)\left(\frac{h_c}{3b_s} + \frac{h_t}{b_s} + \frac{b_{cp}}{6l_{gcp}}\right) \times 10^{-6}\ \text{Wb/m} \qquad (139)$$

Where u is greater than unity, the same value will obtain for λ_i as when $u = 1$.

Commutation in Non-interpole Machines. Much mathematics has been written dealing with this subject and based for the most part on constant brush contact resistance. This latter condition holds only for uniform current density under the brush and hence holds only for straight-line commutation. With non-interpole machines with fixed brushes, straight-line commutation cannot possibly obtain except at one particular load, so that an analytical method based upon constant brush contact resistance is not of great value.

In general it is found that satisfactory commutation can be obtained when the total e.m.f. induced in the coil undergoing commutation does not exceed the normal value of the volt drop at the brush contact. This condition is deduced analytically and stated in another form by Arnold, $RT_C/L^* \nless 1$ for satisfactory commutation, where R is the contact resistance of the brush. For straight-line commutation the two methods of expressing the condition are exactly equivalent.

When the brushes are given lead, rather greater values of the reactance voltage may be permitted, but no precise data can be laid down. Actually, non-interpole machines are but rarely used, and then only in cases where commutation conditions are very easy.

* L here denotes inductance not core length.

Interpoles.* Another method of providing the necessary "reversing field" is to fit special commutation poles to the frame of the machine. These poles are placed midway between the main poles, and are excited in such a manner as to produce a commutation field of the required polarity and strength. For a generator, to assist the reversal of the current the polarity of the interpole must obviously be the same as that of the main pole *in advance*, as shown in Fig. 115. For a motor, on the other hand the converse will be true.

Again, since the reactance e.m.f. increases in direct proportion to the current, the reversing field must increase in the same proportion. This involves the interpole windings being connected in series with the armature.

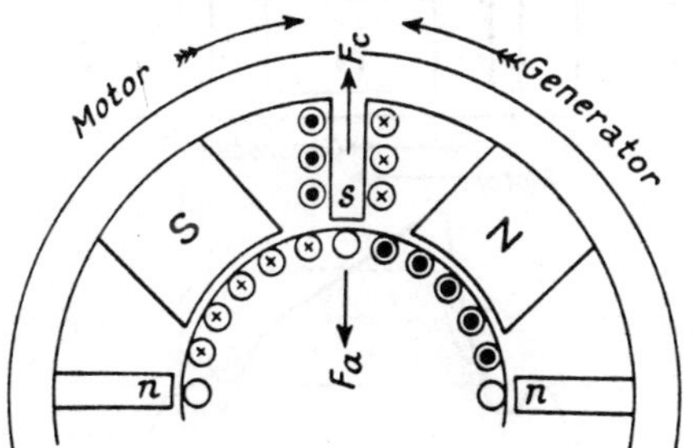

FIG. 115. SHOWING THE ARMATURE AND INTERPOLE M.M.F.'S ACTING IN OPPOSITION

It will be noted that the reversing field required is of opposite polarity to that which armature reaction tends to establish. If, in fact, the interpoles were not excited, commutation would be seriously impaired. By arranging that the m.m.f. developed by the interpole is exactly equal and opposite to that of armature reaction, commutation would take place in the magnetic neutral position, and there would be no reversing field. In short, to establish a reversing field under the interpole shoe, it is essential that the m.m.f. developed by the interpole winding shall overpower that due to armature reaction, the flux established being determined by the *difference* in the magnitude of these two m.m.f.'s, as indicated in Fig. 115.

The interpole winding is local in its action. That is to say, the winding develops an m.m.f. which overcomes armature reaction in the commutating zone only. Thus the distortion of the main field under the pole shoes is not affected by the use of interpoles. This is illustrated by Fig. 116, which gives the resulting field form for a loaded interpole machine and indicates that, in a generator, the interpoles act as outposts of the main poles.

* Interpole and commutation pole are synonymous terms. The latter is usually abbreviated to compole.

By the use of these series-wound interpoles it then becomes possible to obtain satisfactory commutation with brushes in a fixed position—the geometric neutral—under all conditions of load. Further, in motors, satisfactory commutation becomes possible with fixed brushes for both directions of rotation, and over a considerable range of speed obtained by field control. More important still, with interpoles it has been made possible to construct machines, such as relatively high voltage and high speed machines, which could not possibly be built to operate satisfactorily without interpoles. In fact, until interpoles were used, the rated output of d.c.

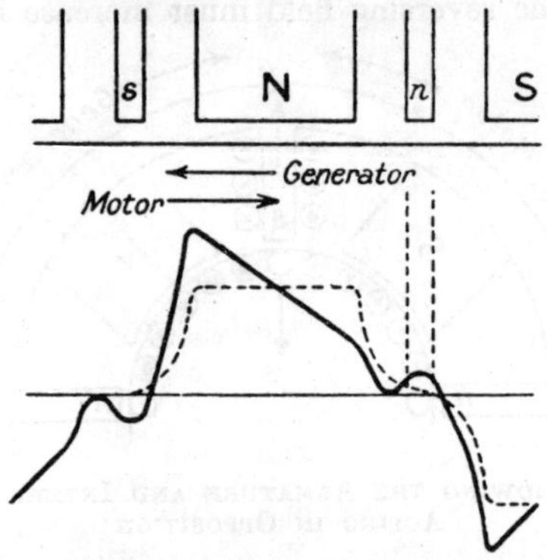

FIG. 116. FLUX DISTRIBUTION CURVE OF A LOADED INTERPOLE MACHINE

machinery was quite often prescribed solely from sparking considerations. With interpoles, temperature rise practically always determines the rated output, so that it has become possible to increase very considerably the rated output of machines of given dimensions, and to take full advantage of the more intensive methods of ventilation now commonly employed. This vastly improved performance and, often, vastly increased rating, is then obtained at the expense of building, assembling, and fitting the interpoles and their relatively powerful coils. The extra expense, moreover, is partly compensated by the fact that it is not necessary with interpole machines to work with such strong field windings on the main poles as in non-interpole machines. In fact, the air gap can be reduced to a figure not much in excess of the minimum necessitated by mechanical reasons. The advantages gained by the use of interpoles are so great that it is standard practice always to employ interpoles except for very small machines.

Fig. 117 shows the field system of a 6-pole machine. The shunt coils on the main poles, and the series-connected coils on the interpoles are clearly shown.

Calculation of Density of Flux Under Interpole Shoe. The mean e.m.f. generated in the coils undergoing commutation, by their rotation in the reversing field under the interpole, must be just equal—and opposite—to the reactance e.m.f., so that the coil is not the seat of any resulting e.m.f.

In terms of the peripheral speed of the armature, v_A, the axial

FIG. 117. FIELD SYSTEM OF 450 kW D.C. GENERATOR, 450 V. 650 REV/MIN
(*Metropolitan-Vickers Electrical Co., Ltd.*)

length of the interpole L_{cp}, and the flux density in the interpole gap B_{cp}, the mean value reversing e.m.f. is

$$E_C = 2B_{cp}L_{cp}v_A \text{ volts per turn} \quad . \quad . \quad (140)$$

For this e.m.f. to balance that due to the coil reactance, etc.,

$$E_C = E_{r_{av}}$$

$$2B_{cp}L_{cp}v_A = 4Z_sI_a \times \frac{\lambda L}{T_{Cx}}$$

i.e.

$$B_{cp} = \frac{L\lambda}{L_{cp}} \times \frac{2Z_sI_a}{\beta_x} \times \frac{v_C}{v_A} \text{ since } T_{Cx} = \frac{\beta_x}{v_C}$$

$$= \frac{L\lambda}{L_{cp}} \times \frac{2Z_sI_a}{\beta_x \times \dfrac{D_A}{D_C}}$$

$$= \frac{L\lambda}{L_{cp}} \times \frac{2Z_sI_a}{\beta_{xA}} \quad . \quad . \quad . \quad (141)$$

where $\beta_{xA} = \beta_x \times \frac{D_A}{D_C}$, and is equal to the distance moved through by the armature circumference during the time of commutation for the complete x coil sides in one layer of the winding.

As shown on page 141, for a simple lap winding, or for a wave winding having only two brush spindles, $\beta_x = (x - 1)\tau_C + t_b - t_m$.

For a 4-pole wave winding having four brush spindles, $\beta_x = (x - 0{\cdot}5)\tau_C + t_b - t_m$.

It may be well to re-state the significance of the various terms in (141).

L/L_{cp} is the ratio of the axial lengths of the armature and of the interpole; λ is the value of the equivalent flux, per metre of embedded length, established around the coil by a magneto-motive force of 1 ampere-conductor in the slot. $2Z_sI_a$ represents the total change in the ampere-conductors per slot for a complete reversal of the current in the whole slot. β_{xA} has been discussed above.

Since the value of the reactance voltage per turn is as a rule small, and only a small percentage of the maximum e.m.f. generated in the coil when under the main pole, it follows that, in general, the value required for B_{cp} will be comparatively low, even if the axial length of the interpole is much less than that of the main pole. It is, however, undesirable to reduce too much the length L_{cp}, partly on account of the increased value of the ampere-turns that would have to be developed by the interpole winding, and partly on account of the fact that there is considerable magnetic leakage, and it is essential that the magnetic circuit of the interpole shall not be saturated, or under-commutation would occur at relatively small overload currents.

In the above calculation, E_C gives the average value of the reversing e.m.f. during the whole period of commutation. If the value of the reversing e.m.f. is at every instant equal and opposite to that of the reactance e.m.f., and the resistance of the coil is negligible in terms of the brush contact resistance, straight line commutation will obtain. With straight line commutation, the reactance e.m.f. is constant under normal conditions, so that to obtain straight line commutation the flux density in the interpole gap must be uniform for one coil-side per layer. To obtain an approximation to sinusoidal commutation, commutation should start and end in a relatively weak field, and it will be necessary for the maximum density in the gap under the interpole to exceed that given above, which, as already stated, represents the average flux density. Obviously with a number of coils per layer it is difficult to arrange that precisely similar conditions shall obtain for each of the coils.

Calculation of Ampere-turns and Number of Turns for Interpole Winding. As the air gap under the interpole is relatively large, and the iron parts of the interpole magnetic circuit are worked well below saturation point, the reluctance of the iron can in general be neglected.

Taking first straight line commutation.

If $l_{g_{cp}}$ = the air gap under the interpole shoe,

$K_{g_{cp}}$ = the Carter gap coefficient,

F_a = the armature reaction ampere-turns per pole

$$= \frac{I_a Z_a}{4p}$$

F_{cp} = compole ampere-turns,

Then $F_{cp} - F_a = 0{\cdot}796 K_{g_{cp}} B_{cp} l_{g_{cp}} \times 10^6$

and $$F_{cp} = F_a + 0{\cdot}796 K_{g_{cp}} B_{cp} l_{g_{cp}} \times 10^6 * \quad . \quad . (142)$$

From the value of the full load current, the number of turns necessary for the interpole winding are then readily obtained.

By slight adjustment of the gap dimension this number of turns can very readily be made into a convenient figure. With large heavy current machines, the number of turns required is very small, and sometimes it may be a convenience to arrange that the interpole windings consist of two circuits in parallel, each carrying one half of the total current.

Depending upon the ratio L/L_{cp}, the dimension $l_{g_{cp}}$, and the proportions of the machine generally, the ampere-turns necessary for the interpole work out to from about 125 per cent to 150 per cent of the armature ampere-turns.

To obtain a commutation curve of the type shown in Fig. 104, i.e. having $\frac{di}{dt} = 0$ at the beginning and end of commutation, it becomes necessary to increase the density of the reversing field during the middle of the time of commutation and to reduce it practically to zero at the beginning and end. This effect can approximately be obtained by working with a relatively narrow shoe on the interpole, and a relatively long air-gap, so that there is an appreciable fringe to the pole. The ratio of the maximum flux density in the gap to the density already calculated for straight-line commutation will depend upon the actual field distribution. Unless the shoe is shaped, the ratio will not approach that obtaining for sinusoidal distribution, i.e., 1·57; normally, the ratio may be taken to vary from about 1·2 to 1·4, according to the actual dimensions chosen for the air gap and pole shoe. For straight-line commutation, then, formula (142) holds. For an approximation to sinusoidal commutation the ampere-turns required are increased owing to the increase in the maximum flux density in the gap, and may be taken to be, for a relatively narrow shoe and long gap,

$$F_{cp} = F_a + 1{\cdot}4(0{\cdot}796 K_{g_{cp}} B_{cp} l_{g_{cp}}) \times 10^6 \dagger \quad . (143)$$

where B_{cp} has the value given in equation (141).

* With c.g.s. units $F_{cp} = F_a + 0{\cdot}796\, K_{g_{cp}} B'_{cp} l'_{g_{cp}}$.

† With c.g.s. units $F_{cp} = F_a + 1{\cdot}4(0{\cdot}796\, K_{g_{cp}} B'_{cp} l'_{g_{cp}})$.

Width of Shoe of Interpole. If straight-line commutation is aimed at, the width of the shoe should be not less than β_{xA}, i.e. the distance moved through by the armature during commutation of the complete layer of conductors in one slot. The ampere-turns will then be calculated in accordance with (142).

For sinusoidal commutation the shoe width may be taken as equal to the slot pitch, and formula (143) used. This latter method

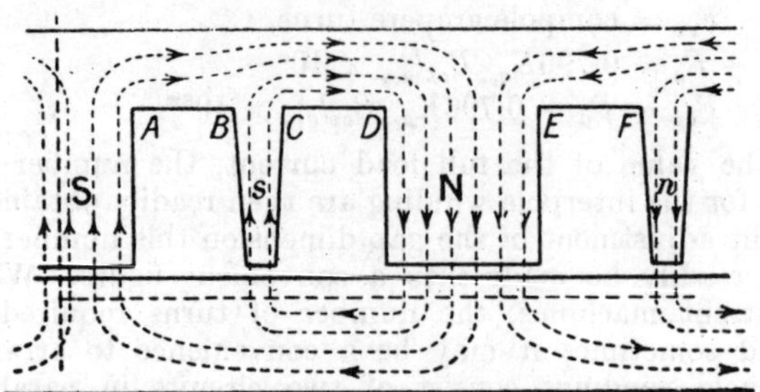

FIG. 118. FLUX DISTRIBUTION IN YOKE AND ARMATURE CORE, WITH AN INTERPOLE MACHINE

gives a stronger reversing field for the central conductors in the layer and is preferred by many designers.

So that the whole of the conductors in the slot shall be commutated in a field of uniform density, the width of the shoe should be not less than the value β_{xA} given above.

For windings short-chorded by one slot pitch, if calculations are based on the value λ_i given in (139) and formula (141) be used, the width of the shoe should be increased to allow of the reversing e.m.f. being induced simultaneously in the two sides of the coil.

Effect of Interpoles upon Flux Distribution in Yoke and Armature Core. The resulting flux distribution in these parts of the machine can be determined by considering separately the main flux and the interpole flux and summing their values for various parts of the magnetic circuit.

Thus considering, as in Fig. 118, the portion of the yoke between the S interpole and the N main pole. Considering their effects separately, both the interpole flux and the main flux will traverse this portion, from C to D, in the same direction, i.e., left to right. For the portion AB, however, the interpole flux is passing from right to left, whereas the main flux passes from left to right. The net result is that between A and B the flux carried is the difference between the fluxes, and between C and D the fluxes add up. Thus the flux carried by the yoke is distributed as indicated in Fig. 118, its values being $\frac{1}{2}(\Phi_p + \Phi_{cp})$ and $\frac{1}{2}(\Phi_p - \Phi_{cp})$ respectively. In the armature corresponding conditions prevail, but the difference is relatively much less, as the armature does not carry the leakage flux. (This leakage flux is not shown in Fig. 118.)

Magnetic Leakage with Interpole Machines. With a machine on full load, the m.m.f. developed by the interpole windings will be of the same order of magnitude as that due to the main poles. Often it will be greater. Between the main pole and its neighbouring interpole of opposite polarity, therefore, very considerable leakage flux passes. Thus, in Fig. 119, the leakage line passing between the tips of the N pole and the S interpole has only to traverse the short distance d_1, and the total m.m.f. acting around the path is the sum of the ampere-turns developed by the main N pole and the S interpole, if we neglect the reluctance of the iron parts. On the other

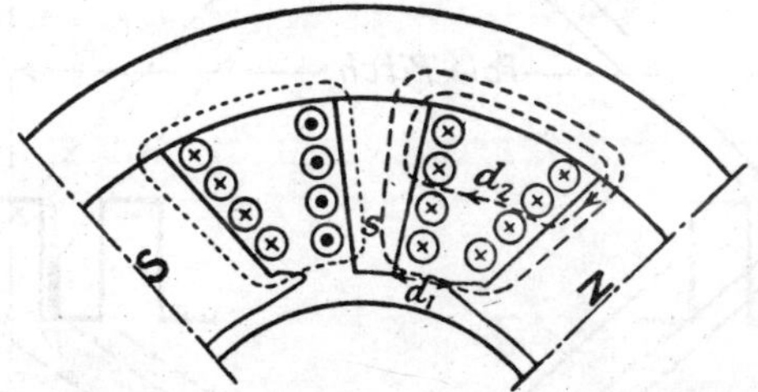

FIG. 119. MAGNETIC LEAKAGE WITH INTERPOLE MACHINES

hand, between main poles and interpoles of like polarity the resulting m.m.f. acting is the difference between the ampere-turns developed by the two windings, and at full load, as these m.m.f.'s are nearly equal, there is but little if any leakage between poles of like polarity.

If the main and interpoles were of the same axial length, to a rough approximation on full load the interpole would not only have to carry its own useful flux but also practically the whole of the leakage flux from the main poles.

The "leakage coefficient" for the interpoles is therefore very large, and the total flux at the root may be as much as three times the useful flux or even more. To prevent the possibility of the interpole becoming saturated at heavy loads, then, it is desirable to employ a larger sectional area at the root than at the pole tip; the interpole with large machines is therefore generally tapered as indicated in Fig. 119.

Commutation with Bar Windings. A winding having simply one coil-side per layer is ideal in that commutation conditions are identical for all the coils, and the phase displacement between the e.m.f.'s in neighbouring coils corresponds exactly to the number of commutator segments. With coil-type windings, such as those indicated in Figs. 40, 261 and 262 for 3 and 5 coil-sides per layer, the e.m.f.'s in the various members of the group are in phase and commutation conditions differ for the various coils.

As a result of this difference in commutation conditions there may be a tendency to blackening of every third segment of the commutator for 3 coil-sides per layer, every fourth segment for 4 coil-sides per layer, and so on.

When 2 coil-sides per layer are used, in conjunction with an odd number of slots per pole pair, the arrangement indicated in Fig. 120 can be used with *bar windings*, i.e. coil No. 1 can be short-chorded and coil 2 long-chorded by one half a tooth pitch. The phase displacement between the e.m.f.'s will then correspond exactly

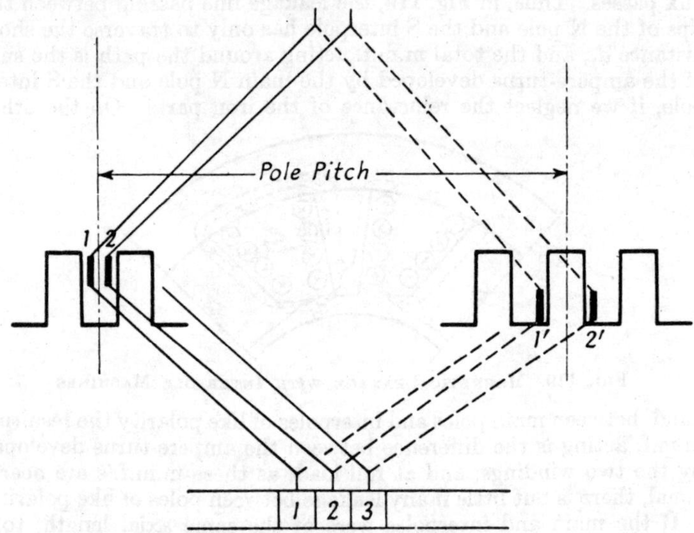

FIG. 120. ARRANGEMENT OF WINDING WITH ALTERNATE COILS UNDER- AND OVER-PITCHED BY HALF A SLOT

to the number of commutator segments, and equal e.m.f.'s are induced in the coils. In short, the arrangement is ideal for commutation.

With three or more coil-sides per layer this exact agreement is impossible. The best that can be done has already been indicated on page 60, i.e. to have one coil spanning one tooth more than the other two, e.g. for, say 39 slots per pole pair, to have a coilspan of 19 teeth for coils 1 and 2, and 20 teeth for coil 3, i.e. slots 1–20 and 1–21.

Interpole Ampere-turns when Compensating Windings are Used. The total m.m.f. acting along the axis of the interpole is due to the joint action of the compensating winding and the interpole winding.

The total number of turns to produce correct commutation as given by (142) or (143) are then divided between the compensating winding and the interpole winding. For the compensating winding the nearest convenient number corresponding to $F_k =$ pole-arc/pole-pitch $\times$ Armature ampere-turns per pole $= \alpha F_a$ is used and the remainder used for the actual interpole coil.

CHAPTER VII

METHODS OF EXCITATION; FIELD WINDINGS

DIRECT-CURRENT machines may be magnetized by currents obtained from a separate source, such as a battery or an auxiliary dynamo called an exciter, or they may generate their own magnetizing current. These two methods of excitation are then called "separate excitation" and "self excitation" respectively.

Separate Excitation. In Fig. 121 the diagram of the essential connexions for separate excitation is given. Two alternative methods

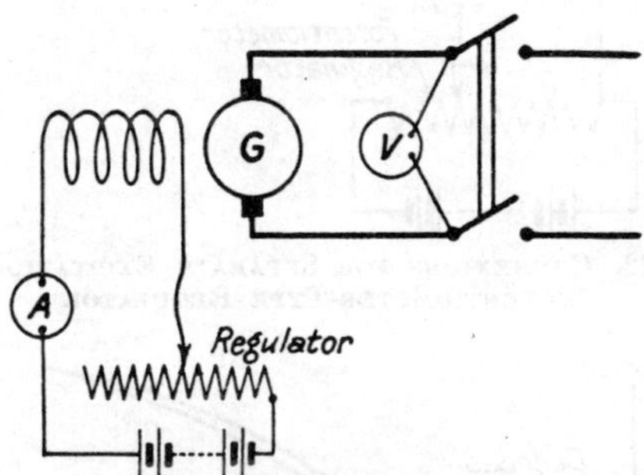

FIG. 121. CONNEXIONS FOR SEPARATE EXCITATION, WITH SERIES REGULATOR

of connecting the regulating resistance are shown. In Fig. 121 the "field regulating resistance" is connected in series with the field coils, and in Fig. 122 the regulating resistance is connected, potentiometer fashion, across the full voltage of the exciter or battery, and the field coils are connected in parallel with a variable portion of the resistance. By adjusting the moving contact, K, the voltage applied to the field coils can be varied from zero to the full voltage of the exciter or battery. The potentiometer method has the advantage that the field circuit of the machine is never actually opened when it is carrying current, and, further, quite low values of the exciting current are readily obtainable without the need for relatively high values of the regulating resistance, as would be required with the plain series regulator.

Separate excitation is largely used in the works testing of d.c. machines for the determination of the open-circuit or magnetization characteristic, to which reference has already been made. This characteristic is obtained by driving the machine at its normal speed, and varying the exciting current in steps from zero to a maximum by means of the field regulating resistance. For each step, readings are taken of the exciting current and terminal voltage, the speed being kept rigidly constant. Since the main circuit of the machine is open there is no current in the armature, so that the value of the terminal voltage is identical with that of the induced e.m.f. Having reached maximum excitation, readings are taken

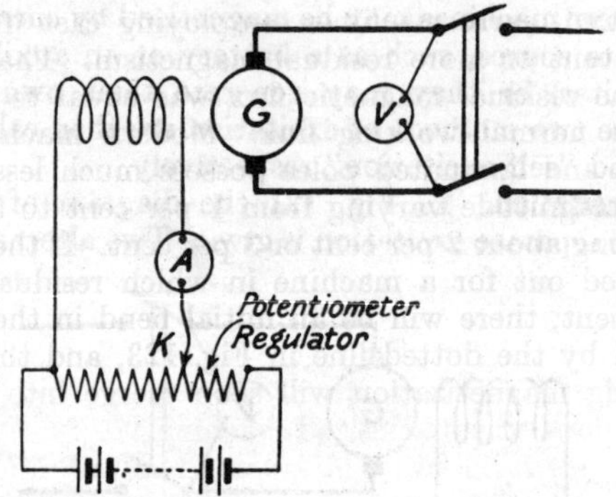

FIG. 122. CONNEXIONS FOR SEPARATE EXCITATION, WITH POTENTIOMETER-TYPE REGULATOR

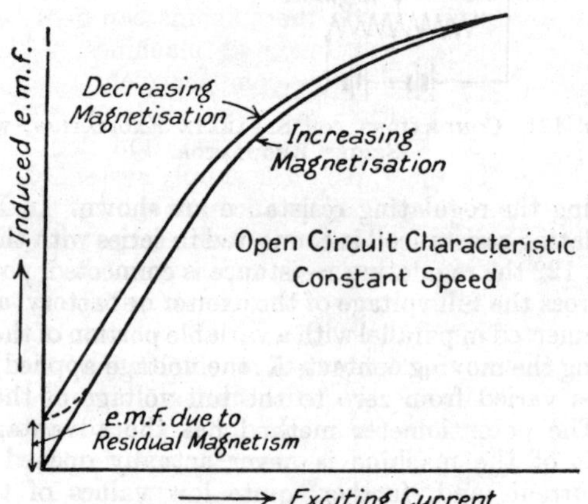

FIG. 123. OPEN-CIRCUIT CHARACTERISTIC OR MAGNETIZATION CURVE

of the voltage and exciting current, as the excitation is reduced, in steps, to zero.

Owing to magnetic hysteresis it is found that the graph representing the happenings with decreasing excitation lies a little above that for increasing excitation as indicated in Fig. 123. (This diagram is drawn for a machine which at first is completely demagnetized.) This fact is chiefly of importance in that it indicates that the magnetic circuit will remain magnetized, although relatively feebly, even when there is no current in the field windings. The magnitude of this residual magnetism is determined largely by the nature of the materials forming the magnetic circuit. If the latter is made up entirely of relatively soft iron stampings, as in certain turbo-generators, there is but very little residual magnetism. On the other hand, early machines, employing cast iron very freely, had 10 per cent or more residual magnetism. That is, with such machines the residual magnetic flux was equal to 10 per cent or more of the normal working flux. Modern machines employing steel frames and laminated poles possess much less residual magnetism, its magnitude varying from 1 per cent to 5 per cent, and normally being about 2 per cent or 3 per cent. If the magnetization test is carried out for a machine in which residual magnetism is already present, there will be an initial bend in the characteristic, as indicated by the dotted line in Fig. 123, and the characteristic for increasing magnetization will soon merge into that obtaining for the completely demagnetized machine.

Separate excitation has the great advantage that the value of the exciting current is entirely independent of that of the load current in the armature, and that it permits a machine to work more or less satisfactorily over a range of voltage extending from zero to the maximum which the machine can generate. For satisfactory operation, with self-excited machines, the range of the working voltage is, generally, comparatively limited, as will be discussed later.

With motors, questions of self excitation do not arise, as the motor is normally excited from the supply system feeding the motor armature, and does not in any circumstances generate its own magnetizing current. Occasionally, with large motors, the exciting current is supplied by a source separate from that feeding the armature. The term "separate excitation," as applied to motors, is then reserved for such cases.

Self Excitation. Self-excited machines may have their field windings connected as a shunt to the main circuit, or in series with the main circuit. The field coils for a shunt-wound machine are then connected as in Fig. 124, and for a series-wound machine as in Fig. 125.

With the *shunt-wound generator*, self excitation takes place in the following manner. Dealing first with the machine when it is run up to full speed before the field circuit is completed, and the main circuit is left open as in Fig. 124. Before completing the field circuit the generator acts as a very feeble magneto dynamo, generating an e.m.f. of 1 per cent to 5 per cent of normal full value, and due entirely to the rotation of the armature in the residual magnetic flux. This residual magnetism is essential to the process of self excitation in all cases.

When the field circuit is completed, the e.m.f. induced by the armature winding, due to its rotation through the residual magnetic flux, sends a current around the circuit consisting of the armature and field winding, together with any regulating resistance that may be included. (Owing to the inductance of the circuit the current builds up slowly, and its value will not be given by dividing the value of this induced e.m.f. by that of the total circuit resistance until steady conditions are obtained.) This current may be in such a direction as to tend to increase the magnetic flux, or such as to

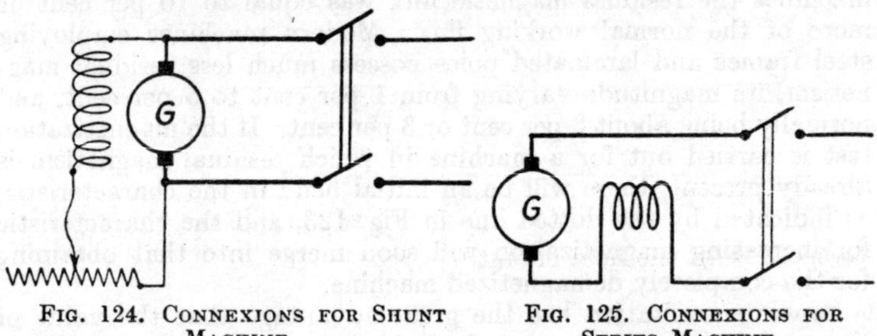

FIG. 124. CONNEXIONS FOR SHUNT MACHINE

FIG. 125. CONNEXIONS FOR SERIES MACHINE

tend to decrease it. With correct connexions the former conditions will obtain, the magnetic field will be gradually strengthened, and as a result the induced e.m.f. will be increased. This, in turn, will increase the exciting current, and so the voltage will build up until steady conditions are obtained. The voltage ultimately attained depends, for a given speed, entirely upon the total resistance of the exciting circuit, and upon the magnetization curve of the machine.

For the magnetic field to be increased, it is necessary that the energy stored in the field shall also be increased. When the machine first begins to build up its field, this necessary energy can be provided by the armature, as the volt-drop in the ohmic resistance of the circuit is less than the armature-induced e.m.f. Ultimately, however, the e.m.f. generated in the armature only just suffices to overcome the ohmic resistance, so that no portion of it is available for supplying energy to increase the value of the magnetic field. The e.m.f. generated will thus not build up beyond this value, unless either the resistance of the circuit is reduced, or the speed increased. Under certain circumstances, the amount of building up may be inappreciable, as may be gathered from the open circuit characteristic given in the inset of Fig. 126, in which the initial portion of the curve is shown plotted to a larger scale. From this curve it is seen that an excitation of 0·05 A gives rise to a generated e.m.f. of 6·0 volts. For the machine to build up to just 6·0 volts, then, the total resistance of the exciting circuit—including armature and brush contacts—must then be 6·0 ÷ 0·05, i.e., 120 ohms. In a similar manner the resistance corresponding to other values of the final steady voltage may be calculated, as indicated in tabular form on page, 168 for the machine having the magnetization curve plotted in Fig. 126. The results of these calculations are further plotted in Fig. 127, and show the voltage to which the machine builds up for

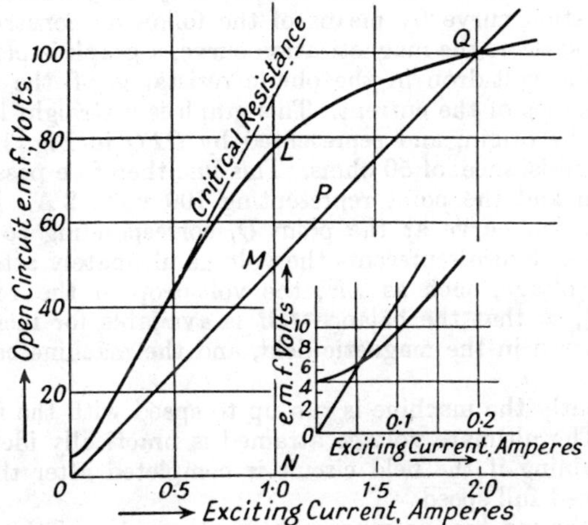

Fig. 126. The Building up of a Shunt Generator

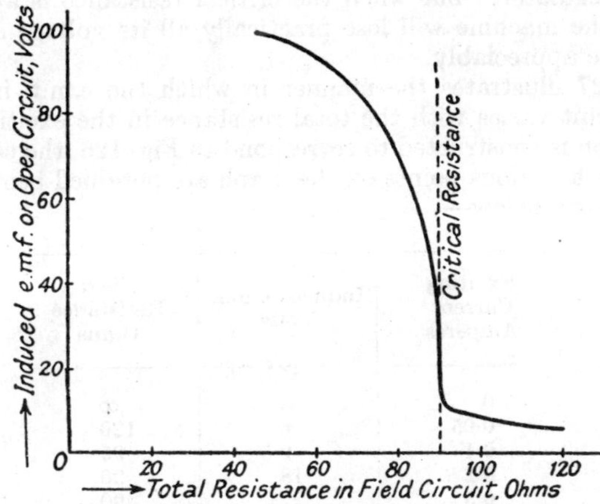

Fig. 127. The Dependence of the Induced e.m.f. upon the Resistance in the Field Circuit

various values of the resistance. This diagram illustrates the fact that until the resistance of the field circuit is reduced to a certain "critical value," the voltage does not build up appreciably. This critical resistance corresponds to the straight-line portion of the magnetization curve, i.e. to the conditions obtaining when the iron parts of the magnetic circuit do not offer any appreciable reluctance to the magnetic flux.

The voltage to which the machine will build up, with any particular resistance in the field circuit, may be obtained from the magnetization curve by means of the following construction; To the same scale as the magnetization curve, a graph is plotted representing the volt-drop in the ohmic resistance of the circuit, for various values of the current. This graph is a straight line passing through the origin, and represented by *OPQ* in Fig. 126, for the case of a resistance of 50 ohms. This line therefore passes through the origin and the point representing 100 volts 2 A. It cuts the magnetization curve at the point *Q*, corresponding to 100 volts, which latter figure represents the voltage ultimately attained. For a lower voltage, such as *LN*, the volt-drop in the resistance is only *MN*, so that the balance *LM* is available for increasing the energy stored in the magnetic field, and the machine continues to build up.

Frequently the machine is run up to speed with the field circuit closed. The ultimate voltage attained is practically identical with that obtaining if the field circuit is completed after the machine has attained full speed.

When the machine is running, then, the value of the voltage on open circuit may be adjusted by varying the resistance in the "shunt regulator," but when the critical resistance is attained, or passed, the machine will lose practically all its voltage and fail to self excite appreciably.

Fig. 127 illustrates the manner in which the e.m.f. induced on open circuit varies with the total resistance in the exciting circuit. This graph is constructed to correspond to Fig. 126, the calculations from which various points on the graph are obtained being given in tabular form below—

Exciting Current Amperes	Induced e.m.f. Volts	Field Resistance Ohms
0	5	∞
0·05	6	120
0·1	9·5	95
0·2	18	90
0·3	27	90
0·4	36	90
0·5	44	88
0·75	62·5	83·5
1·0	77	77
1·25	86·5	69
1·5	92·5	61·5
1·75	97	55 5
2·0	100	50

Time Taken to Self-excite. A precise solution of this problem is not practicable owing to the effect of eddy currents induced in the solid masses of the magnetic circuit. If the effects of these eddy currents, induced by the increasing flux, be neglected, calculations by a step-by-step method can be used. As the inductance varies considerably there is not much point in working out its value, and the following method can be used—

Let $2pN_e$ = total number of turns in series of all the field windings, i.e. no. of coils in series × no. of turns per coil

r_e = total resistance of the field circuit, including armature winding and brush contacts

i, Φ_p, and e = instantaneous values of field current, mean flux in poles, and e.m.f. generated in armature, respectively.

Then the e.m.f. induced in the field winding is $-2pN_e\dfrac{d\Phi_p}{dt}$ and the e.m.f. equation for the complete circuit is

$$e = ir_e + \frac{2pN_e d\Phi_p}{dt}, \text{ i.e., } 2pN_e\frac{d\Phi}{dt} = e - ir_e$$

The time taken for the flux to increase by a small amount, from Φ_{p_1} to Φ_{p_2} is then

$$\delta t \simeq 2pN_e(\Phi_{p_2} - \Phi_{p_1})/\text{average value of } (e - ir_e)^*$$

The building-up-time curve can then be determined step-by-step as given in Tabular form below for the example given in Fig. 126. Calculations are for 750 rev/min, total field resistance $r_e = 50$ ohms, 4-poles, 1,500 turns per field coil.

To illustrate the method:

The time taken to build up the voltage from 18 to 27 is calculated—

Change in pole flux $= (0{\cdot}54 - 0{\cdot}36) \times 10^{-2}$ weber

mean value of $(e - ir_e)$ during this change $= (12 + 8)/2 = 10$ volts

hence $\delta t = 6{,}000 \times 0{\cdot}0018/10 = 1{\cdot}08$ sec

Fig. 128 shows the building-up-time curve as calculated. The calculation being based on negligible eddy current effects may be very different from the actual happenings.

Effect of Speed of Operation upon Self Excitation. For a given magnetic flux, the e.m.f. generated varies in direct proportion to the speed of rotation. If, then, magnetization curves for various speeds

* Owing to the increase in the leakage coefficient with increasing saturation, the e.m.f. generated in the armature e is not exactly proportional to Φ_p.

CALCULATION OF BUILDING-UP-TIME CURVE

750 rev/min; field resis. r_s = 50 ohms; 4 poles, 1,500 turns per coil. $2pN_s$ = 6,000.

Exciting Current	Armature e.m.f. e	Pole Flux Φ_p	Volt drop r_s	$e - ir_s$	Av. value of $e - ir_s$	Change in flux $\delta\Phi$	Change in time δt	Total time t
amperes	volts	weber $\times 10^{-2}$	volts	volts	volts	weber $\times 10^{-2}$	sec	sec
0	5	0·1	0	5				0
					4·25	0·02	0·28	
0·05	6	0·12	2·5	3·5				0·28
					4·0	0·07	1·05	
0·1	9·5	0·19	5	4·5				1·33
					6·25	0·17	1·63	
0·2	18	0·36	10	8				2·96
					10	0·18	1·08	
0·3	27	0·54	15	12				4·04
					14	0·18	0·77	
0·4	36	0·72	20	16				4·81
					17·5	0·18	0·62	
0·5	44	0·90	25	19				5·43
					22	0·44	1·20	
0·75	62·5	1·34	37·5	25				6·63
					26	0·36	0·83	
1·0	77	1·70	50	27				7·46
					25·5	0·25	0·59	
1·25	86·5	1·95	62·5	24				8·05
					20·75	0·15	0·43	
1·5	92·5	2·10	75	17·5				8·48
					13·5	0·15	0·67	
1·75	97	2·25	87·5	9·5				9·15

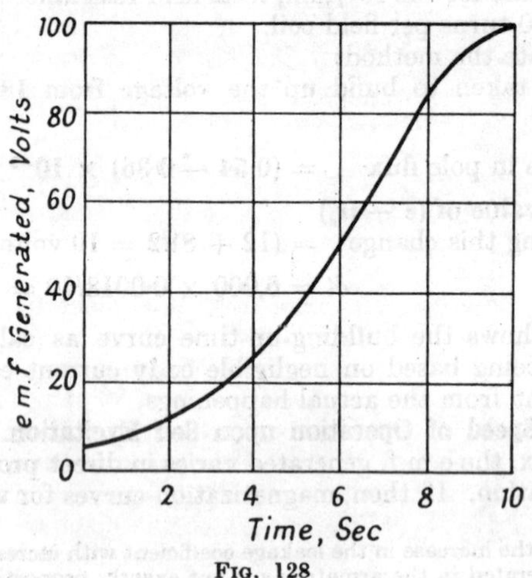

Fig. 128

be plotted on the same base and to the same scale, the ordinates of the various curves for any particular value of the excitation will vary in direct proportion to the speed as in Fig. 129. The "critical resistance" thus varies directly with the speed.

By drawing various resistance lines to represent the total resistance in the circuit of the field winding, as in Fig. 129, it is at once

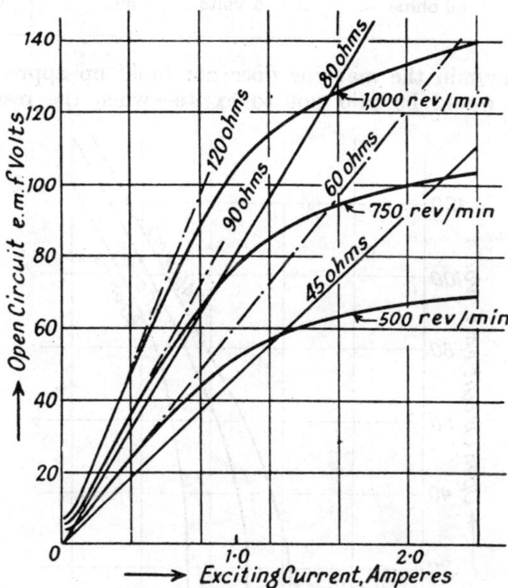

FIG. 129. OPEN CIRCUIT CHARACTERISTICS AT VARIOUS SPEEDS

apparent that the value of the voltage to which the machine builds up, for a given resistance, does not vary in direct proportion to the speed. It is further apparent that the "critical resistance" increases in direct proportion to the speed, so that a generator may not build up appreciably at, say, 500 rev/min, but may excite quite satisfactorily at 750 rev/min.

The graphs in Fig. 129 represent a shunt generator having a normal speed of 750 rev/min and a field winding resistance of 45 ohms. The magnetization curves at 500, 750, and 1,000 rev/min are given, and resistance lines drawn for 45, 60, and 80 ohms, corresponding to 0, 15, and 35 ohms resistance in the regulator, respectively. The diagram also shows the values of the critical resistance for the three speeds, i.e., 60, 90, and 120 ohms.

The final steady voltages to which the machine will build up, at the three given values of the speed, are then as shown in the table on page 172.

Rev/min	500	750	1,000
Total resistance, 45 ohms . .	58 volts	102·5 volts	150 volts
,, ,, 60 ohms . .	18–36 ,, (critical resistance)	94 ,,	138 ,,
,, ,, 80 ohms . .	5 volts	69 ,,	124 ,,

At 500 rev/min the machine does not build up appreciably—in practice it would be said not to excite—when the resistance is

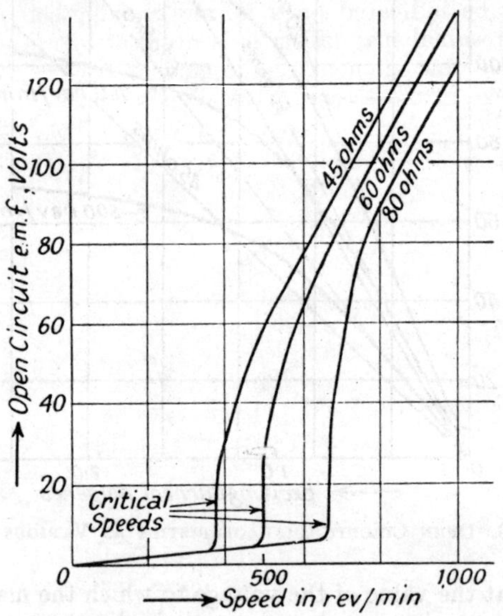

Fig. 130. The Dependence of the Induced e.m.f. upon the Speed of a Shunt Generator

80 ohms; 60 ohms is the critical resistance for this speed. In none of these is the ratio of the voltages even approximately equal to the ratio of the speeds.

To illustrate the manner in which the voltage induced on open circuit varies with the speed, for the case of a shunt generator having a fixed resistance in its field circuit, Fig. 130 has been drawn.

Three graphs are given, corresponding to 45, 60, and 80 ohms, for the total resistance in the field circuit.

Failure to Self Excite. It has been seen that for a shunt generator to build up its field—

1. There must be residual magnetism in the field system.
2. The connexion of the field circuit to the amature must be

such that the direction of the field current established is such as to tend to increase the field already existing.

3. The total resistance of the field winding circuit must not exceed the "critical value," if appreciable building up is to take place, This critical resistance increases in direct proportion to the speed, It follows then that a machine cannot build up if the speed is too low.

Dealing with these three conditions separately.

1. ABSENCE OF RESIDUAL MAGNETISM. This may possibly occur in the case of machines being put into service for the first time. If residual magnetism is present, a small but quite readable voltage will be indicated on a normal moving-coil voltmeter. If there is no residual magnetism it is necessary to separately excite the field for a few moments. The current from a number of dry cells will serve in most cases. Where possible, the best method is to connect the machine to a d.c. supply so that it runs as a motor, its brushes being of the normal polarity, and the direction of rotation also normal. On shutting down, the machine will be left with its residual magnetism of the correct polarity for operation as a generator.

2. FIELD CONNEXIONS REVERSED. The machine will not build up if the connexions to the field are not such that the field current will tend to increase the value of the flux. This state of affairs may arise from an error in connexions, or it may equally arise by running the generator with the wrong direction of rotation, since the direction of the induced e.m.f., and therefore of the field current, will reverse with reversed rotation. Although under these conditions the small current flowing in the field winding will *tend* to demagnetize the machine, owing to magnetic hysteresis it will not reduce to any appreciable extent the strength of the residual magnetic flux in any normal case. If a machine fails to excite due to wrong field connexions, it is therefore simply necessary to change over the connexions from the armature to the field circuit, and maintain the same direction of rotation, or to leave the connexions unaltered and reverse the direction of rotation of the prime mover. Naturally in normal practice the direction of rotation of the machine will not be altered and it then becomes essential to take the former alternative. It will not normally be necessary to remagnetize the machine.

3. FIELD RESISTANCE TOO HIGH. Except during the first installation of a generator the most common cause of failure to excite is too high a resistance in the field circuit. This, of course, includes the case when the field circuit is actually incomplete due to a broken lead in any part of the field circuit, including the shunt regulator and the field winding itself. The resistance of the field circuit also includes that of the armature, brush contacts, etc. In dealing with the theory of self excitation it has been assumed that the resistance of the field circuit is constant. This is not quite exact owing to the fact that brush contact resistance is very much higher for low

values of the current density than for normal values. Failure to excite may on rare occasions be due to too high a brush contact resistance, or to a dirty commutator. Normally the resistance is most likely to be too high owing to either an actual break in the field circuit, or to too large a resistance in the field regulator, for the particular speed of operation. If, of necessity, for some reason the machine is being run at a speed much below normal, it may well happen that the resistance of the field winding itself exceeds the critical resistance. It then is possible to obtain self excitation only

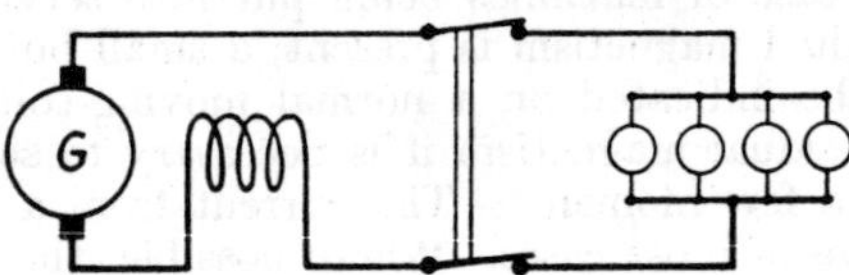

FIG. 131. SERIES GENERATOR LOADED ON LAMP LOAD

by re-connecting the individual field coils, in parallel, or series-parallel, instead of the normal connexion where all the individual coils are connected in series. Thus, for example, a normal 4-pole shunt dynamo may be arranged to self excite at half normal speed by re-connecting the field coils so that there are two parallel paths, the best arrangement being to include the two N pole coils in the one circuit and the two S pole coils in the other, as this arrangement will ensure the same value for the total m.m.f. acting along each complete path, even if the two circuits have not quite the same resistance.

Reversed Polarity. It may happen that a dynamo has the polarity of its residual magnetism reversed. This is more likely to occur with compound than with shunt-wound machines. With normal direction of rotation this will not prevent the machine building up, but it will build up with reversed polarity, i.e. the brush which should be positive will become negative. It then becomes necessary to remagnetize the machine with the correct polarity, from some other source. (A possible laboratory alternative would be to reverse both the connexions to the field coils and the direction of rotation, and avoid the need for remagnetizing.)

Self Excitation of Series Generator. With a shunt-wound generator it is usual, though not essential so far as the generator is concerned, to allow the machine to build up its voltage before the main switch is closed. With a series generator, self excitation is not possible until the load circuit is completed.

So far as the main principles of self excitation are concerned, the action of the series generator differs only in detail from that of the shunt machine. It is now essential that the main switch be closed, as in Fig. 131. Again, the final value of the voltage attained, for any speed of operation, is determined by the resistance of the exciting circuit, which latter is now the complete load circuit of the machine. Further, appreciable building up will not take place until the resistance of this circuit is reduced to a certain critical value determined by the speed of operation, and the magnetization characteristic of the machine. The actual value of the induced voltage can no longer be deduced from the true open circuit magnetization curve of the machine determined by separately exciting the field coils, since the magnetizing action of the armature currents is now not negligible. With the shunt machine, the current taken by the field winding is only of the order of 1 or 2 per cent, or less,

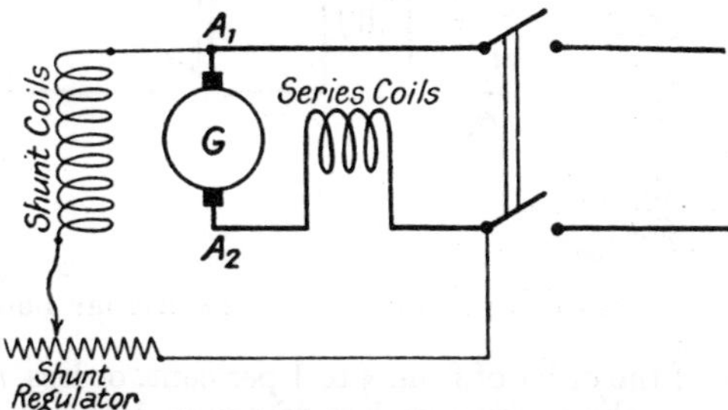

FIG. 132. CONNEXIONS FOR COMPOUND GENERATOR: LONG SHUNT

of the full load armature current. This small current flowing in the armature will have no appreciable effect upon the magnetic field due to the main magnets. With the series machine, no-load conditions can only correspond to practically zero voltage; when the machine is generating normal voltage its armature winding is of necessity as fully loaded as the field winding. The action of the series machine will be discussed more fully in the chapter devoted to the performance of d.c. generators when loaded.

Compound-wound Generator. As in the shunt machine, the compound-wound generator can build up its voltage with the main circuit open, as it possesses both shunt and series field windings, as in Fig. 132. The shunt winding may be joined across the terminals of the machine, or it may be connected across the armature. In the former, Fig. 132, it is said to be a "long shunt," and in the latter, Fig. 133, a "short shunt." It is of no real moment whether short or long shunt connexions be employed. In either case the machine can be used as a plain shunt machine by supplying the external circuit from the two armature terminals, A_1, A_2, as indicated in Figs. 132 and 133. For the short-shunt machine, the arrangement is then obviously simply that of the plain shunt generator. With long-shunt connexions, the field winding in Fig. 132 would consist of the few turns of the heavy winding in series with the normal shunt winding, and the characteristics of the machine would be those of a normal shunt generator.

Field Windings. As already seen, d.c. machines may be shunt, series or compound wound. There is inevitably energy wasted in overcoming the resistance of the coils, this energy being degraded into heat and thus raising the temperature of the coils. Quite apart from any considerations of efficiency, it therefore becomes imperative to keep down the value of the power wasted in the field coils to prevent the temperature rise exceeding the safe limit. With large machines the total value of the power wasted in the field

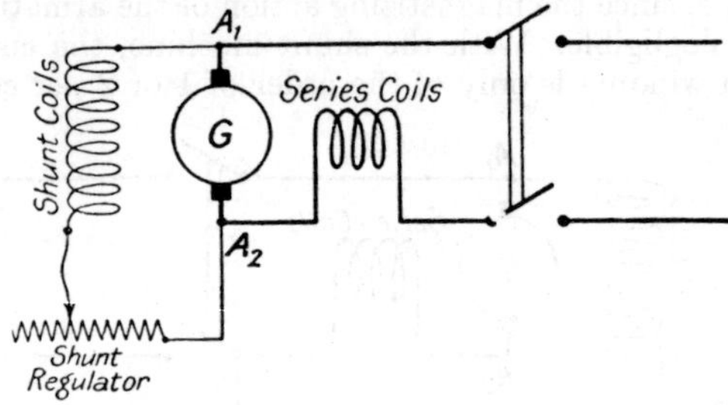

FIG. 133. CONNEXIONS FOR COMPOUND GENERATOR: SHORT SHUNT

coils will be of the order of from ½ to 1 per cent, or less, and for quite small machines the percentage loss of power does not often exceed 3 or 4 per cent of the full load output of the machine.

For shunt windings it follows then that the resistance will be relatively high, the current low, and, in consequence, the number of turns large. Series windings, on the other hand, will be of relatively low resistance; they carry the full load current and thus have relatively few turns.

A 200-kW, 500-V machine may be taken to illustrate this. A reasonable figure for the field winding loss in a machine of this size is about 1 per cent, or 2 kW. For a shunt-wound machine, the field current would therefore be of the order of 2,000/500 = 4 A. For a series machine the current in the field coils would be 400 A and the volt-drop of the order of 5 volts. The resistance of the winding would therefore be 500/4 = 125 ohms for the shunt-wound machine, and 5/400 = 0·0125 ohm for the series. Maybe the coils would have to develop 7,200 ampere-turns per pole, in which case there would be 7,200/4, i.e. 1,800 turns on the shunt coils, and 7,200/400 = 18 turns for the series-wound machine.*

It may be well to consider the manner in which the above figures would be modified if the value of the working voltage were altered. To take a simple example, assume the working voltage to be reduced to one half, i.e. to 250 volts. The values of the load current and the shunt current, for the same percentage loss, would each be doubled. The resistance of the shunt winding would then be 250/8 = 31·25 ohms, and for a series winding 2·5/800 = 0·003125 ohms. Thus, in each type of winding, by halving the working voltage the number of turns is reduced to one half, but the section of the conductor is

* These figures are, of course, only approximate.

doubled, so that the actual ohmic resistance is reduced to one quarter. For the same value of the power wasted, then, the weight of copper required for a coil is independent of the working voltage, and for that matter of the type of coil—whether shunt or series—if the length of the average turn of the coil is constant.

The actual value of the power wasted in the exciting circuit is generally determined entirely from considerations of temperature rise, and a means of predetermining a suitable figure for this loss is discussed later. For the moment it will be assumed that the value of the permissible loss in the exciting circuit is known.

Calculation of Shunt Coils. In calculating the number of turns and sectional area of the conductor for a shunt winding the following data are required—

1. The ampere-turns to be developed by the coil on each pole.
2. The permissible loss in the complete winding.
3. The voltage available for the complete winding.
4. The "mean length of turn" of the coils.

Of these, items 2 and 4 are determined mainly by the dimensions of the poles, although item 2 is dependent upon certain other factors. Item 1 is calculated from the dimensions of the magnetic circuit. With regard to the voltage available for the complete winding, it is necessary in a generator to include a field regulating resistance in circuit with the windings so that the voltage can be varied and so that, if necessary, the voltage can be increased somewhat above normal value. Further, this regulating resistance affords a means of compensating for the increase which takes place in the resistance of the winding as its temperature rises. Normally, depending upon circumstances, the regulating resistance left in circuit under full load conditions will be such as to absorb from 15 per cent to 25 per cent of the total voltage, when the machine is at its final temperature. As an average, the voltage available for the field coils, when hot, may then be taken as being 80 per cent of the terminal voltage of the machine.

The field coils of shunt windings are invariably connected in series, so that there is only one circuit. (With series windings, owing to the very heavy currents handled, it often becomes a convenience to arrange that there are two paths in parallel. For example, the equivalent of $2\frac{1}{2}$ turns carrying the full current is conveniently obtained by arranging the winding with two paths in parallel, each coil having 5 turns carrying one half of the total current.) For a shunt winding, then, let—

P_e = the total permissible loss in the winding, in watts.

V_e = the total available voltage for the whole winding.

F_e = the total ampere-turns to be developed by each coil.

l_e = the mean length of turn of the winding.

N_e = the number of turns per pole.

The remaining particulars of the winding are then readily calculated as follows—

Current, I_e = watts lost ÷ voltage = P_e/V_e . . . (144)

Number of turns per pole, N_e = ampere-turns ÷ current = F_e/I_e . . . . (145)

Resistance = volt-drop ÷ current, $r_e = V_e/I_e$ . . . (146)

Total length of conductor L_e = number of poles in series × $N_e l_e = 2pN_e l_e$ . . . . (147)

Cross-sectional area of conductor $A_e = \rho \frac{L_e}{r_e}$. (148)

The value for ρ, the specific resistance, should be taken to correspond to the final temperature attained. Usually, it may be taken as being 2×10^{-8} ohm-metre.

The above method of calculation is simple and most direct, but it obscures the fact that, to obtain the required ampere-turns, quite independent of the permissible loss and number of turns, one sectional area only for the conductor is possible. This may be shown as follows—

$$F_e = N_e I_e = N_e \frac{V_e}{r_e}$$

$$= \frac{N_e V_e A_e}{2pN_e l_e \rho}$$

Hence
$$A_e = \frac{F_e \times l_e \rho}{V_e/2p} * \qquad (149)$$

The sectional area of the conductor is thus independent of the number of turns employed. It may of course readily be deduced that, by increasing the number of turns, the resistance will increase in the same ratio, and the current will be correspondingly reduced.† The ampere-turns will thus remain constant. The number of turns employed determines the value of the current and hence of the power wasted, the power wasted varying inversely as the number of turns. Since the weight of copper used will vary in direct proportion to the number of turns on each coil, it follows also that the value of the power wasted will vary inversely as the weight of copper used.

Calculation of Series Coils. In calculating shunt windings it was seen that to obtain the required ampere-turns one value only for the sectional area of the conductor was possible. The number of turns could be chosen at will to suit the desired value of the power wasted. With series windings it is exactly the contrary; the number of turns is at once determined from the known values of the current and the ampere-turns. The sectional area of the conductor determines the value of the power wasted.

* The area will be in m² if m.k.s. units are used and in cm² with c.g.s. units.
† Neglecting any change in the length of mean turn.

The data required to calculate completely the series winding differ only from those in the shunt winding in that the value of the current is given instead of the volt drop.

Using the same notation as in the shunt winding, with the exception that the suffix $_{se}$ is used to denote that the symbols apply to the series winding,

$$\begin{aligned} &\text{Number of turns} && N_{se} = F_{se}/I_{se} \\ &\text{The volt-drop} && V_{se} = P_{se}/I_{se} \\ &\text{Resistance} && r_{se} = V_{se}/I_{se} \\ &\text{Cross-sectional area } && A_{se} = \frac{F_{se} \times l_{se}\rho}{V_{se}/2p}\,* \quad . \quad . \quad (150) \end{aligned}$$

* The area will be in m² if m.k.s. units are used and in cm² with c.g.s. units.

CHAPTER VIII

LOAD CHARACTERISTICS OF D.C. GENERATORS

It has already been seen that with a generator of direct currents—whether battery or dynamo is immaterial—the potential difference between the terminals is of necessity less than the induced e.m.f., except when open circuit conditions obtain. The relation between the voltages is

$$V_A = E_A - I_A r_m$$

where r_m is the resistance of the main circuit, i.e. armature + interpoles + series windings + brush contacts.

The manner in which the terminal voltage changes as load is put on the generator is also dependent upon the method of excitation, and is, in addition, influenced by armature reaction.

External Characteristic of Separately-excited Generator. The external characteristic expresses the manner in which the voltage across the load, or external part of the generator circuit, varies with the value of the load-current. It thus is given by a graph showing the manner in which the potential difference between the generator terminals varies with the load current. Normally, generator characteristics are plotted for constant speed. Occasionally, in practice, the drop in speed of the prime mover when load is put on is sufficient to modify the characteristic appreciably.

To determine experimentally this characteristic, it is simply necessary to connect up the generator to a dead load, such as a lamp load, a water resistance, or any other form of load which can be adjusted at will. The generator is driven at normal speed, and the exciting current adjusted so that the e.m.f. on open circuit is the normal working value. The speed and exciting current are maintained at their initial values throughout the test, and the load current increased in steps from zero to 50 per cent overload, readings being taken at each step of the value of the current and of the terminal voltage of the machine. The connexions for the test are indicated in Fig. 134, and typical test results given in the graph, Fig. 135. In carrying out this test it is important that the brushes should be in their normal position, since the effect of armature reaction largely depends upon the brush position. In Fig. 135 is also shown a graph indicating the manner in which the volt-drop in the internal resistance, r_m, of the machine varies with the load. Neglecting slight changes in the brush contact resistance, this graph is a straight line. (The internal resistance includes that of the armature winding, the brushes and brush contacts, and interpole windings if such be used.) By adding the ordinates of the resistance drop graph to those for the terminal voltage, the curve given in dotted lines is obtained, indicating the manner in which the induced e.m.f. falls as the load on the machine is increased. This fall in induced e.m.f. can only be due to a fall in the value of the flux per pole.

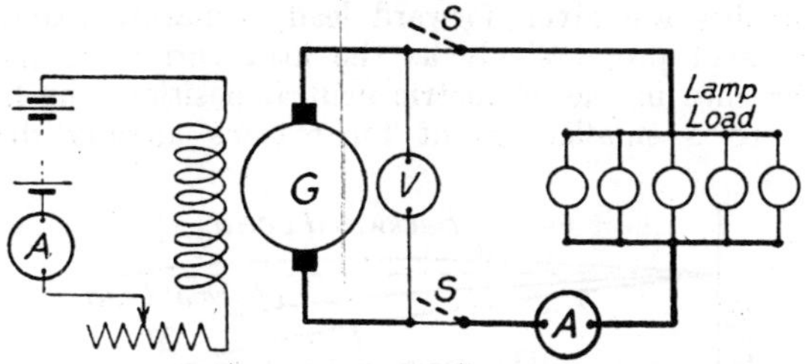

Fig. 134. Connexions for Load Test on Separately-excited Generator

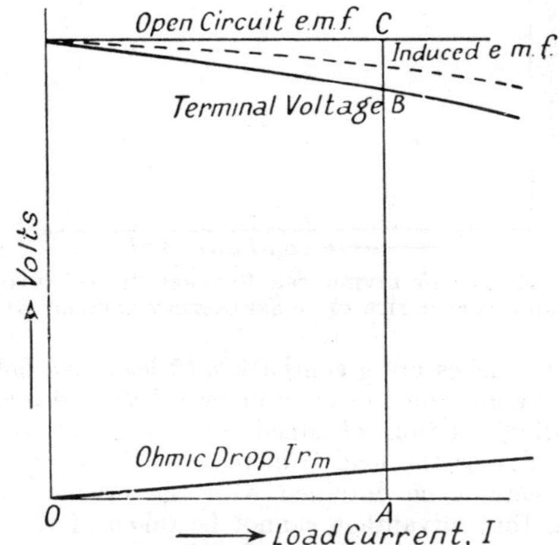

Fig. 135. External Characteristics of Separately-excited Generator

The exciting current being constant, this fall in flux must therefore be due to armature reaction, as already discussed.

This slightly drooping characteristic of the separately-excited generator is well suited to most normal purposes, so far as stability of operation is concerned.

To maintain a constant voltage under all conditions of load it is then necessary, even with a separately-excited generator, to increase the value of the exciting current as extra load is put on. The extra excitation necessary is readily provided by cutting out some of the resistance in the field regulator.

If the external characteristic of a separately-excited generator be determined for excitations much below normal, it is found that much the same type of characteristic is obtained, so that separate excitation is well suited to machines which have to supply a voltage which is fairly constant at all loads, for various values of the voltage.

If the brushes are given forward lead, armature reaction will weaken the field progressively as the load current is increased. With the brushes in the geometric neutral position, the field will be reduced to a smaller extent for reasons already discussed.

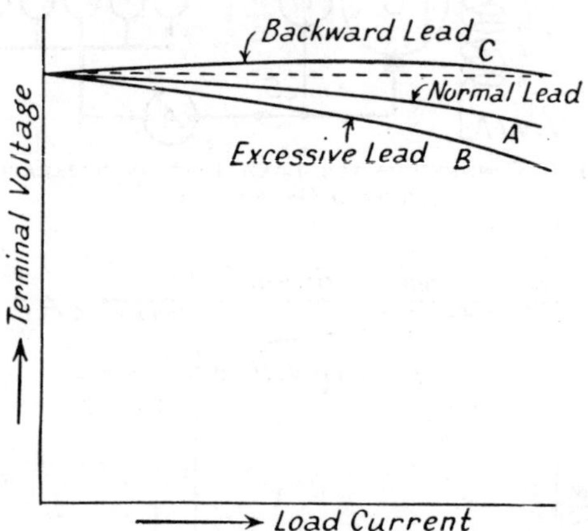

FIG. 136. EFFECT OF GIVING THE BRUSHES "LEAD" UPON THE EXTERNAL CHARACTERISTICS OF A SEPARATELY-EXCITED GENERATOR

Finally, if the brushes are given backward lead, the field strength will tend to be maintained or even increased as load increases, and a "compounding" action obtained, the terminal voltage even increasing slightly as the load is increased. However, satisfactory commutation cannot be obtained with the brushes thus rocked backwards, so that advantage cannot be taken of this method of maintaining the terminal voltage. In Fig. 136 are given typical characteristics for a separately-excited generator, A corresponding to normal lead for a non-interpole machine, B to excessive lead, and C to the case of negative lead.

External Characteristic of Shunt Generator. For a shunt generator, the external characteristic is obtained with constant resistance in the exciting circuit, the connexions necessary for determining the characteristic experimentally being given in Fig. 137. For the normal test, the resistance in the field regulator is adjusted so that normal e.m.f. is generated on open circuit, and the resistance is not varied during the test, which again is carried out at constant speed, and with the machine at its usual working temperature.

With the shunt generator, the fall that takes place in the terminal voltage, when load is put on, is more marked than with the separately-excited machine. The increased drop is due to the fact that a drop in the terminal voltage results in a fall in the value of the exciting current, with a consequent fall in the value of the flux. When load is put on the shunt generator its voltage thus tends to

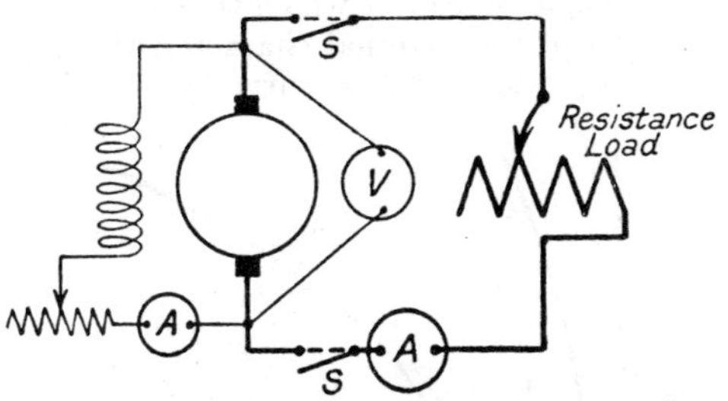

FIG. 137. CONNEXIONS FOR LOAD TEST ON SHUNT GENERATOR

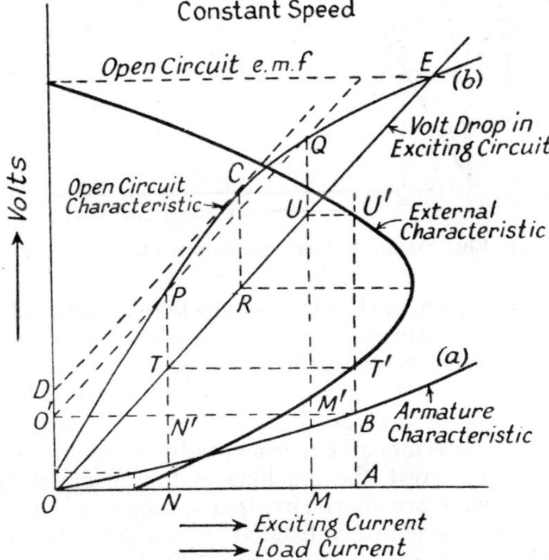

FIG. 138. EXTERNAL CHARACTERISTIC OF A SHUNT GENERATOR

fall by a process of exactly a reverse nature to that of "building up." Unless the load resistance is relatively small, however, steady conditions are soon obtained without too serious a fall in voltage, but when conditions are such that the machine is working on the straight-line portion of its magnetization curve, the drop in voltage is very marked. After the load resistance is reduced to a certain value, any further decrease, instead of increasing the load current, ultimately results in the current decreasing, as indicated by the form of the graph shown in Fig. 138.

When the terminals are short circuited, the shunt winding is inactive, but current is established owing to the presence of residual magnetism.

The actual form of the external characteristic of a shunt generator is largely dependent upon the value of the initial voltage, and is influenced by the angle of lead at which the brushes are set.

The general form of the external characteristic may be deduced from the open circuit magnetization curve by the following method.

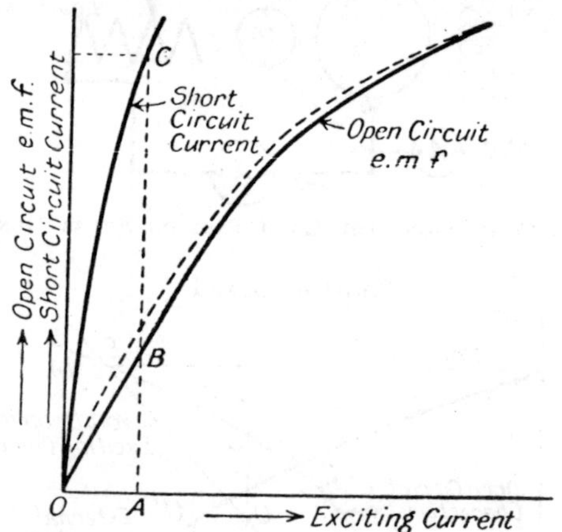

FIG. 139. OPEN-CIRCUIT AND SHORT-CIRCUIT TEST

The method is inexact in that it does not take account of the exact degree of magnetic saturation under all conditions.

In brief, the method replaces the joint effect of armature resistance and armature reaction by an equivalent voltage drop, which voltage drop is assumed to be constant for any particular value of the load current, and is determined experimentally by means of a *short-circuit test.* In this test the machine is first completely demagnetized, the brushes are short circuited through an ammeter, and the field windings separately excited with relatively very feeble currents. The generator is driven at normal speed and readings taken of the short circuit current for various values of the excitation. In addition, an open circuit test is taken, starting with the machine demagnetized, the results of the two tests being plotted as in Fig. 139. Then since, for example, the excitation OA produces an open circuit e.m.f. of AB, or a short-circuit current AC, the equivalent internal voltage drop for the current AC may be taken to be the voltage AB.

From the two curves in Fig. 139 another may be drawn, as in Fig. 140, indicating the manner in which the equivalent voltage drop varies with the load current.

Having determined the equivalent voltage drop by the above method, a *Load Magnetization Curve* can be constructed for any particular value of the load current. This load magnetization curve shows the manner in which the terminal voltage varies with

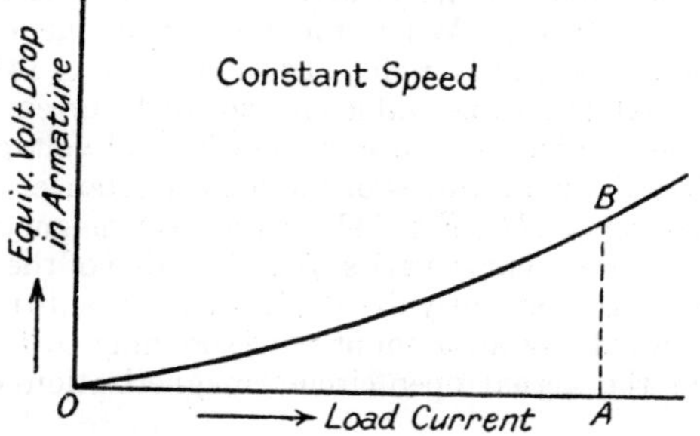

FIG. 140. EQUIVALENT VOLT-DROP IN THE ARMATURE; COMBINING THE EFFECTS OF ARMATURE REACTION AND RESISTANCE

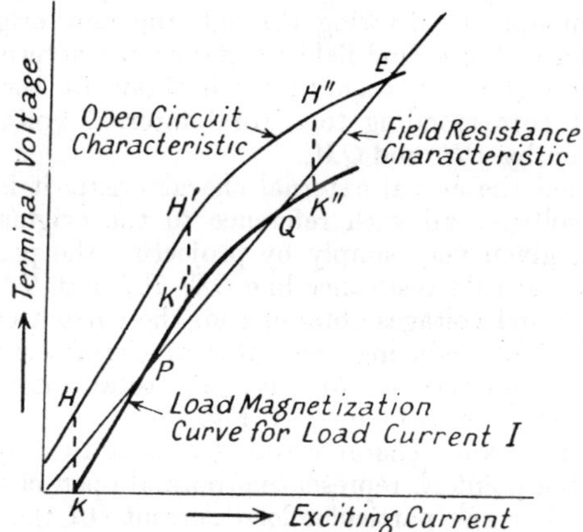

FIG. 141. LOAD MAGNETIZATION CURVE OF A GENERATOR

the exciting current, for one definite value of the load current. The curve is an exact replica of the open-circuit magnetization curve, except that each ordinate is less than the corresponding ordinate of the open-circuit magnetization curve by the constant amount corresponding to the equivalent volt-drop. This is illustrated in Fig. 141, where the load curve is displaced vertically by the constant amount, $HK = H'K'$, etc., below the open circuit curve. (The curve for decreasing magnetization is shown.)

The volt-ampere characteristic for one particular value of the resistance of the field circuit is also shown in Fig. 141. This line will, for reasons already given, cut the open-circuit curve at a point E, giving the voltage to which the machine builds up on open circuit. For similar reasons, it will cut the load magnetization curve at a point—or points—which will give the value of the terminal voltage for the particular load current for which the load magnetization curve is drawn. When the resistance line cuts the load magnetization curve in two points, as shown in Fig. 141 at P and Q, it indicates that the same value of the load current may obtain with two widely different values of the terminal voltage, and hence for two widely different values of the load resistance.

The displacement HK, Fig. 141, of the load magnetization curve from the open-circuit curve varies with the value of the load current. This can most conveniently be dealt with by shifting the origin vertically upwards by an amount corresponding to the value HK, and utilizing the actual open-circuit magnetization curve as the load curve.

This is indicated in Fig. 138, the origin being shifted to O', and the line $O'PQ$ drawn to represent the voltage drop in the field winding, cutting the open-circuit curve at P and Q. This construction therefore amounts to drawing through the new origin O' a line $O'PQ$ parallel to the actual field resistance characteristic OE. The values of the terminal voltage that will obtain for the value of the load current corresponding to a total internal voltage drop OO' are then given by PN' and QM'.

To construct the actual external characteristic it is desirable to plot these voltages all with reference to the original origin, the values being given very simply by projecting the points P and Q on to the actual field resistance line OE, TN and UM then representing the actual voltages obtaining for the current OA.

The method of deducing the external characteristic from the open-circuit characteristic (b), and the total internal drop (a) characteristic is then as follows (cf. Fig. 138).

The field resistance characteristic OE is drawn by joining the origin O to the point E, representing normal open circuit e.m.f., as in Fig. 138. For any particular load current OA the total equivalent voltage drop AB is read off from curve (a) Fig. 138, and through a new origin O', such that $OO' = AB$, $O'PQ$ is drawn parallel to OE, cutting the open-circuit curve at P and Q. These two points projected on to the line OE, as at T and U, give the terminal voltages at which the load current will be OA. Hence, by constructing as indicated, the two corresponding points U' and T' on the load characteristic are obtained.

By taking various values for the load current OA, and displacing the resistance line from OE, by the corresponding amount AB, a number of points on the external characteristic can be obtained.

The maximum current output can readily be deduced by constructing, parallel to OE, a tangent to the open-circuit curve, as indicated by DC in Fig. 138. The vertical displacement, OD, of this tangent from the line *OE*, gives the maximum possible value that can obtain for the equivalent armature drop. The current producing this voltage drop, as obtained from the armature characteristic, gives the maximum current output and the voltage at which it occurs is represented, as indicated, by the point *R*.

A little consideration will establish the fact already stated that the value of the maximum possible current will depend very largely upon the position of the initial working point upon the open circuit

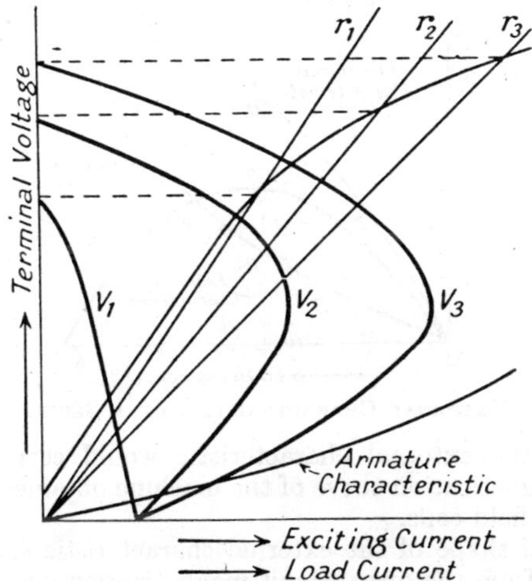

FIG. 142. EXTERNAL CHARACTERISTICS OF SHUNT GENERATOR FOR VARIOUS VALUES OF THE FIELD RESISTANCE

characteristic, and hence upon the resistance of the field circuit. In Fig. 142 are shown resistance lines corresponding to three values for the open-circuit voltage, the first of which is barely beyond the straight-line portion of the open-circuit curve. The corresponding external characteristics are indicated in Fig. 142, and it will be seen that, for the firs case, maximum value of the armature current is attained only wl en the machine is short circuited; this current is simply that due to residual magnetism. To obtain satisfactory voltage regulation, it is then essential that there shall be a certain degree of magnetic saturation present at normal voltage on open circuit, so that the working point on the open-circuit curve shall be well beyond the knee of the curve.

The actual complete external characteristic of the shunt generator may be divided into two parts. For the first part, as the load current increases the terminal voltage decreases, obviously representing *stable working conditions*. For the latter part of the curve, a decreased current corresponds to decreased terminal voltage, so that the action is very unstable.

Normally, shunt generators are worked only on the first, i.e. the stable portion, the working portion of the characteristic being simply a slightly drooping curve just as in the separately-excited machine (cf. Fig. 135), but the droop being rather more pronounced.

External Characteristic of Series Generator. With a series generator, the value of the load current determines that of the flux, and, if armature reaction and armature resistance were negligible, the e.m.f. induced and the terminal voltage would have the same

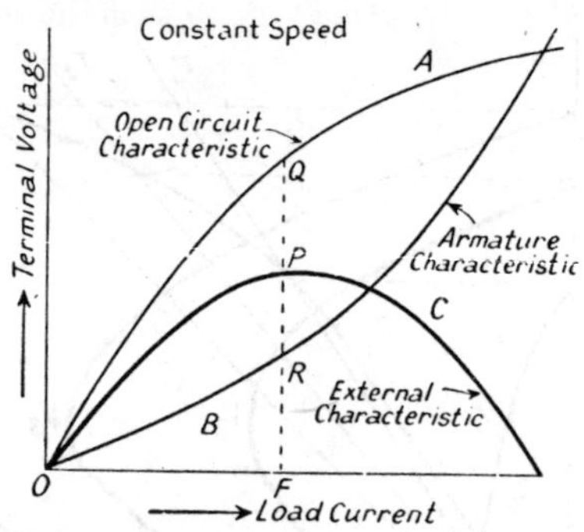

FIG. 143. EXTERNAL CHARACTERISTICS OF A SERIES GENERATOR

value, and the external characteristic would correspond exactly with the magnetization curve of the machine obtained by separately exciting the field coils.

The actual shape of the external characteristic can then readily be deduced from the open-circuit magnetization curve by subtracting, from each ordinate of the curve, the voltage drop equivalent to the joint effect of armature resistance and armature reaction. This voltage drop may again be obtained approximately, as in the shunt generator, by means of a short-circuit test. For the series machine the construction is quite simple, as indicated in Fig. 143. Curve A represents the open-circuit magnetization curve; B represents the equivalent voltage drop, plotted in terms of the current; curve C gives the external characteristic, obtained by subtracting ordinates of B from corresponding ordinates of A. The point at which A and B intersect indicates the load current corresponding to short-circuit conditions. Under these conditions the whole of the induced e.m.f. is absorbed internally.

The load, or external, characteristic of the series generator thus differs widely from that for the shunt machine. With the series machine, the voltage depends entirely upon the value of the load current; with the shunt machine, over the working range the voltage does not vary widely with varying load. The *first part* of the characteristic of the series machine represents *unstable* conditions, as increase in current gives rise to an increased voltage. Beyond the point of maximum voltage, however, the series machine is stable. The series machine will not build up its voltage until the total resistance of the load circuit is reduced to the critical value corresponding to the initial slope of the magnetization curve. Further, a comparatively small change in the value of the load resistance may result in a very marked increase in the value of the current.

As a result, series machines are rarely used as self-excited generators. They are often used as "boosters," in which case the current is controlled by the main generator, so that the series generator can

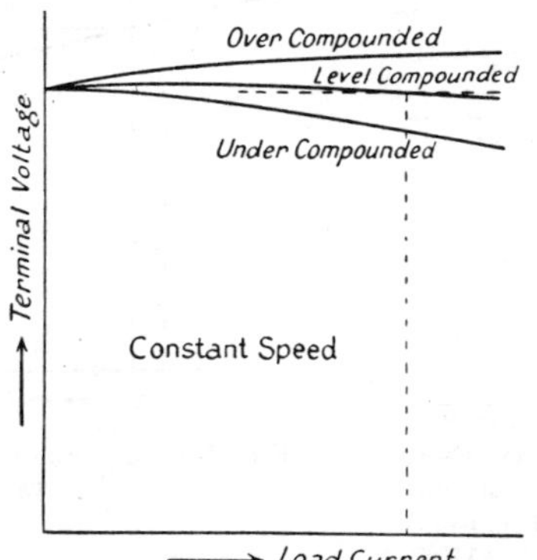

FIG. 144. EXTERNAL CHARACTERISTICS OF COMPOUND-WOUND GENERATORS

then work on all parts of its magnetization characteristic. Series generators have also been used as self-excited machines working on the second, i.e. the stable portion of their load characteristics.

One special use of series machines as self-excited generators occurs in electric traction. For electric braking on tramcars the connexions of the series machines normally used as motors are changed so that they act as self-excited generators, the energy generated being dissipated in the resistances normally used for starting purposes.

External Characteristics of Compound Generators. With d.c. compound generators, the series winding is arranged normally to assist the shunt winding, and the manner in which the terminal voltage varies with increasing load is determined by the relative strengths of the two windings. It is thus also dependent, as in the shunt machine, upon the value of the initial voltage on no-load.

Under suitable conditions, the terminal voltage may remain practically constant from no-load to full load; or by using relatively strong series windings the terminal voltage may increase with increasing load current. In the first case the machine is said to be "level-compounded," and in the other "over-compounded." The load characteristics will then be as in Fig. 144. If a level-compounded machine is run at less than normal voltage it will be found to be over-compounded, and if run at above normal voltage, it will give a drooping external characteristic, as indicated in Fig. 145. (This assumes normal speed of operation.)

The action of a compound generator is not to be explained as merely equivalent to a combination of shunt and series characteristics. The voltage drop with the shunt generator is largely accounted for by the reduction that takes place in the value of the field current. With a level-compound generator, the voltage across the shunt

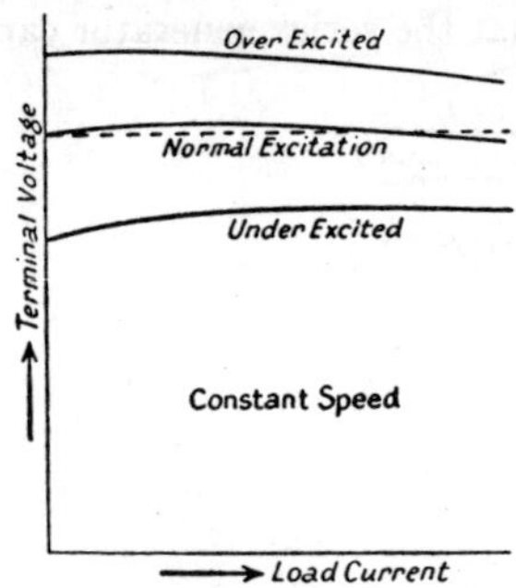

FIG. 145. EXTERNAL CHARACTERISTICS OF A COMPOUND GENERATOR WITH VARIOUS DEGREES OF EXCITATION

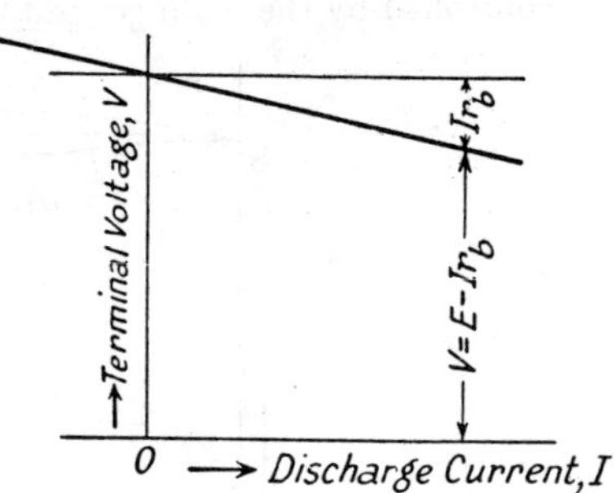

FIG. 146. EXTERNAL CHARACTERISTICS OF A VOLTAIC BATTERY

winding is almost constant, so that the action of this machine is equivalent to that of a separately-excited machine, having in addition a few series turns.

With an over-compounded machine the value of the current in the shunt winding actually increases with increasing load. With the compound machine, then, the series winding is beneficial not merely in so far as it produces a m.m.f. proportional to the value of the load current, but also because it results in the strength of the shunt winding being maintained or even increased as the load increases.

Operation of Generators in Parallel. Consider first the parallel operation of two storage batteries. The manner in which the batteries share the load is determined entirely by their individual external characteristics. With a storage battery, the external characteristic is a straight line, as in Fig. 146, the terminal voltage being given by $V = E - Ir_b$, where r_b is the internal resistance of the battery, and E is the constant e.m.f.

In Fig. 147 are plotted external characteristics for two batteries having e.m.f.'s E_1 and E_2 respectively, and an internal resistance of r_1 and r_2 respectively; the battery having the higher e.m.f. is taken to have the higher internal resistance so that the characteristics cross as indicated.

Since the total current delivered to the load is the sum of the currents in the two batteries, the total current delivered to the external circuit, when the p.d. between the battery terminals has fallen to V, is the sum of the currents given by the external characteristics to correspond with that terminal voltage. Thus, in Fig. 147, the total current delivered by the battery when the terminal voltage is OV, is represented by $VC = VA + VB$. The total external characteristic of the two batteries in parallel is thus

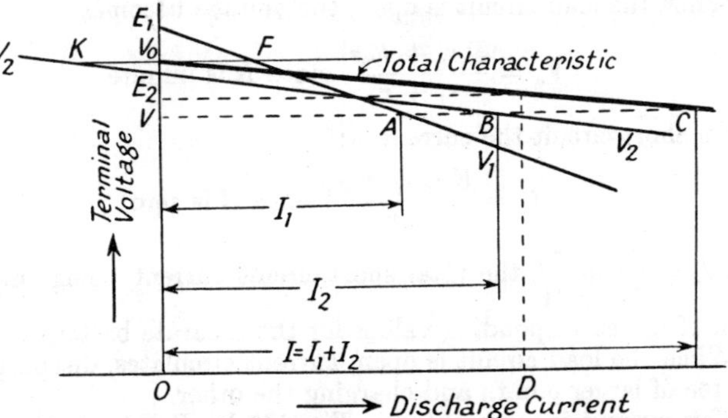

FIG. 147. DIVISION OF THE LOAD BETWEEN BATTERIES CONNECTED IN PARALLEL

obtained by summing the abscissae of the individual external characteristics, as in Fig. 147.

The batteries share the load current equally at the point where their characteristics cross, i.e., when the total load is OD amperes. For smaller current loads than this, the battery having the larger e.m.f. and, for larger current loads, the other battery contributes the greater share of the load current. At the heavy loads, the main factor in determining the distribution of the current is the ratio of the internal resistances and, at the light loads, the ratio of the e.m.f.'s.

At very light loads, and when the complete battery is on open circuit, battery E_2 delivers negative current, i.e. receives a charging current from the other.

Analytically, the distribution of the current between the two sources can be determined as follows—

Let R = the total external resistance of the load.

Then $E_1 = I_1 r_1 + (I_1 + I_2)R$

and $E_2 = I_2 r_2 + (I_1 + I_2)R$

Whence
$$I_1 = \frac{E_1 r_2 + (E_1 - E_2)R}{r_1 r_2 + (r_1 + r_2)R}$$

$$I_2 = \frac{E_2 r_1 - (E_1 - E_2)R}{r_1 r_2 + (r_1 + r_2)R}$$

$$I = \frac{E_1 r_2 + E_2 r_1}{r_1 r_2 + (r_1 + r_2)R}$$

The terminal voltage,

$$V = IR = \frac{E_1 r_2 + E_2 r_1}{\frac{r_1 r_2}{R} + (r_1 + r_2)}$$

When the load circuit is open, the voltage becomes

$$V_o = \frac{E_1 r_2 + E_2 r_1}{r_1 + r_2} \text{ since } R \text{ is infinite.}$$

On short circuit the current is:

$$I_c = \frac{E_1 r_2 + E_2 r_1}{r_1 r_2} \text{ since } R \text{ is zero,}$$

i.e. $I_c = \frac{E_1}{r_1} + \frac{E_2}{r_2}$, the total short-circuit current being simply the sum of the corresponding values for the separate batteries.

When the load circuit is open, current circulates, discharging the source of larger e.m.f., and charging the other.

This current is represented in Fig. 147 by V_oF for battery 2, and V_oK for battery 1. Its magnitude is $I_o = \frac{E_1 - E_2}{r_1 + r_2}$.

Shunt Generators in Parallel. With generators, as with batteries, the subdivision of the load during parallel operation is determined by the nature of their external characteristics. Compared with batteries, however, the subdivision of the load is more easily adjusted, as it is now possible to alter the induced e.m.f. by adjusting the field strength.

With rapid changes in the load, it is not possible by hand regulation to control the field strength for every change in load, and it therefore becomes desirable to investigate the manner in which the machines will automatically share the load, if no adjustment whatever be made to their exciting circuits after they have been put in parallel.

Again, as with batteries, the total external characteristic for a number of machines in parallel is obtained by summing corresponding abscissae of their individual characteristics. Thus Fig. 148 gives the external characteristics of two shunt generators, excited so as to give exactly the same e.m.f. on open circuit, and then put in parallel. For a terminal voltage OV machine I delivers a current of VA, and machine II a current VB; the total current is therefore $VC = VA + VB$. It follows, then, that the machine with the more drooping characteristic takes the smaller share of the load. For perfect operation it is desirable that the machines have similar characteristics, that is to say, the voltage drop at half their normal rated output should be the same for the two machines, similarly the voltage drops at full load and 25 per cent overload, respectively, should be the same for the two machines.

In general it may be taken that shunt generators, having a drooping characteristic, work extremely well in parallel. If, for any reason, one of the prime movers falters and develops less than its normal power, the speed and induced e.m.f. will fall to such an

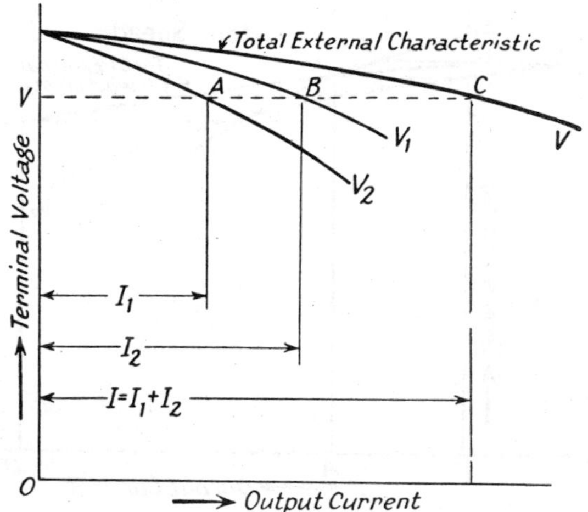

FIG. 148. OPERATION OF SHUNT GENERATORS IN PARALLEL

extent that the necessary balance of power, i.e. prime mover and generator outputs, is obtained. Naturally, other generators working in parallel will be somewhat overloaded, but there will be no tendency whatever for unstable conditions to be established.

Dealing now with machines excited so that their induced e.m.f.'s on open circuit are not equal: This is illustrated in Fig. 149, and corresponds closely to that already given in Fig. 147 for batteries in parallel. The total characteristic is again obtained by adding corresponding abscissae of the individual characteristics. As in batteries of unequal e.m.f., at very light loads the machine having the smaller open-circuit e.m.f. delivers a negative current, i.e. it actually *receives* current at its positive brush and therefore acts as a motor, so that the total load on the other generator is greater than the power delivered to the external circuit. Should the difference in excitation be relatively great, this uneven distribution of load may take place on heavy loads and result in one of the machines being unduly overloaded.

For the case illustrated in Fig. 149, when the main circuit is open the terminal voltage of the generators is OV_o. The one machine generates V_oF amperes and the other receives V_oK amperes. This second machine continues to function as a motor until the load current reaches OA amperes, at which point machine *II* just floats. Beyond this load, both machines share the load, machine *I* taking the greater part until the load *OD* is attained, after which machine *II* carries the greater share.

Series Generators in Parallel. EQUALIZING CONNEXION. Although series machines are not normally used as generators for supply

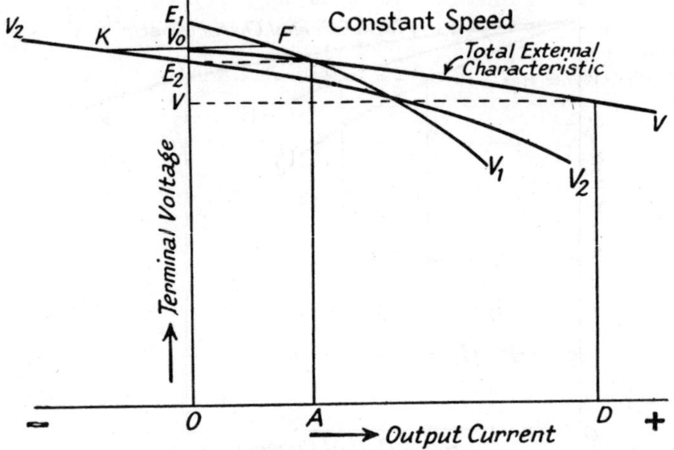

FIG. 149. OPERATION OF SHUNT GENERATORS IN PARALLEL; UNEQUAL EXCITATION

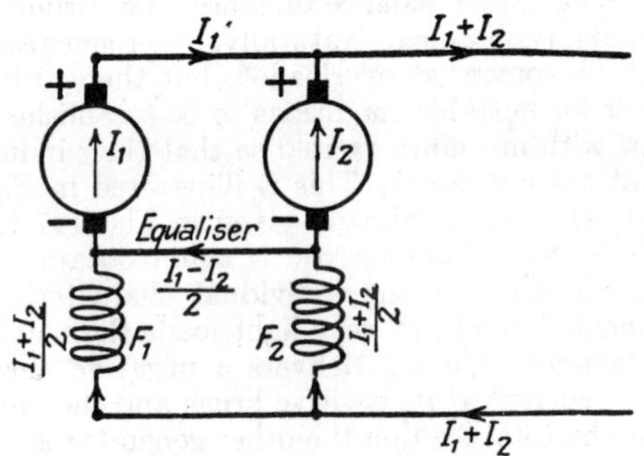

FIG. 150. SERIES GENERATORS IN PARALLEL, SHOWING EQUALIZING CONNEXION

purposes, series motors are frequently arranged to operate as generators during electric braking. In particular, for electric braking, the driving motors on tramcars are utilized as self-excited series generators, operating in parallel.

To ensure satisfactory operation, when series generators are connected in parallel an "equalizing connexion" must be used as in Fig. 150, or the field windings cross connected as in Fig. 151.

When the equalizing connexion is used, the two field coils windings are connected in parallel, so that the flux for either machine is not directly dependent upon its own armature current. Without the equalizing connexion, the value of the flux, and hence of the induced e.m.f., varies when the armature current changes. For machines working under normal conditions, i.e. on the rising portion of the characteristic, Fig. 143, any increase in current in one of the armatures produces an increase in the induced e.m.f. As a result, the machine concerned will deliver a still greater current and relieve

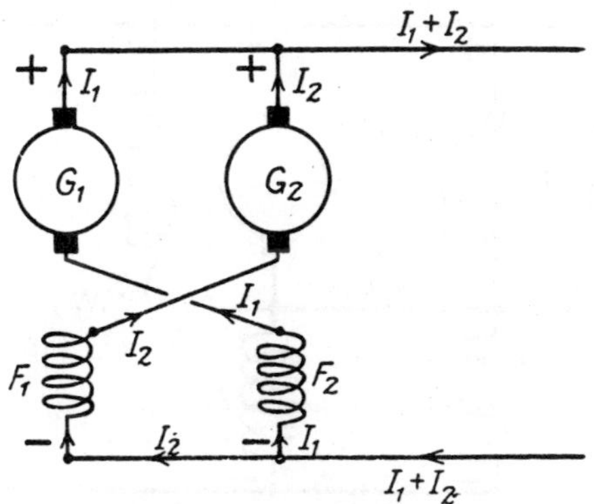

FIG. 151. SERIES GENERATORS IN PARALLEL, WITH FIELD WINDINGS CROSS-CONNECTED

the other machine. The e.m.f. of machine *I* will therefore progressively increase and that of machine *II* decrease until ultimately the current in machine *II* will reverse so that this machine will reverse its voltage. The two e.m.f.'s will then act in series around a circuit of very low resistance and conditions are virtually those of a short circuit on the two machines.

With the field coils connected in parallel, the value of the flux is determined solely by the total load current. If, then, the prime mover driving one of the machines, e.g. machine *II*, momentarily falters, this machine will tend to fall in speed until it takes such a share of the load as is within the capacity of the prime mover, but there is no tendency whatever for the action just discussed to take place, stable conditions rapidly being attained. Even if machine *II* were left in parallel with the other, and steam, etc., shut off from its engine, nothing untoward would happen. Machine *II* would slow down slightly until its induced e.m.f. fell a little below the terminal voltage. Its armature current would then reverse, and the machine would act as a separately excited motor on light load, as the field current would have the same value as in the other machine.

In the case mentioned above, of electric braking on tram motors, both motors *must* run at the same speed, and satisfactory distribution of the load can be obtained by using an equalizing connexion to connect the field coils in parallel. Cross connecting the fields is even more positive in its action since, should the current in one armature fall, the field of the other machine is automatically weakened.

Compound Generators in Parallel. Since these machines possess windings in series with the armature and, normally, having a rising characteristic, it is again necessary to equalize their field currents.

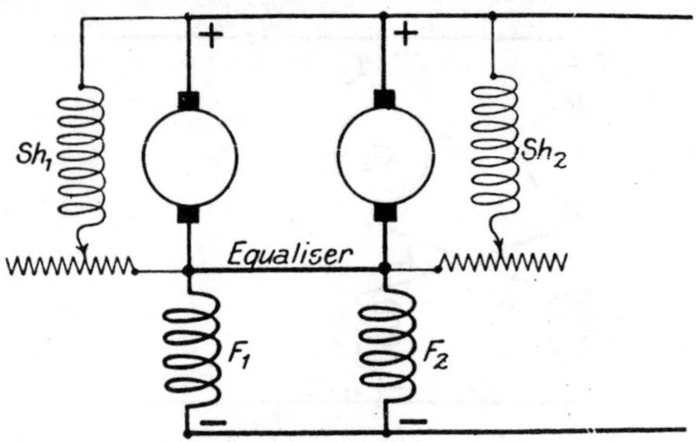

FIG. 152. COMPOUND GENERATORS IN PARALLEL; WITH EQUALIZING CONNEXION

This is usually affected by means of an auxiliary bus bar on the main switchboard, the bus bar being called the "Equalizer bus bar."

This equalizing connexion serves exactly the same purpose as with simple series wound machines.

Essential connexions for the parallel operation of compound generators are then as shown in Fig. 152.

For satisfactory operation, the resistance of the equalizing connexion should be very small in comparison with that of the field coils.

CHAPTER IX

LOAD CHARACTERISTICS OF D.C. MOTORS

Back e.m.f. Induced in Motor Armature. When a current is passed through the armature of a d.c. machine and its field coils excited, torque is established and the armature rotates. The direction of rotation can be reversed by reversing either the armature current or the polarity of the magnets. The actual direction in which the torque acts can readily be deduced from first principles as indicated in Fig. 4, page 7.

Rotation of the armature gives rise to an induced e.m.f. which, according to Lenz's law, will oppose the flow of current. Hence,

if E_A = the numerical value of the induced e.m.f.,

V_A = „ „ „ applied voltage.

The armature current is given by

$$I_A = (V_A - E_A)/r_m$$

hence

$$V_A = E_A + I_A r_m \quad . \quad . \quad . \quad . \quad (151)$$

and the power input

$$V_A I_A = E_A I_A + I_A^2 r_m \quad . \quad . \quad . \quad (152)$$

$I_A^2 r_m$ represents the power wasted in ohmic resistance, and $E_A I_A$ therefore represents the power available for performing mechanical work (in addition to the useful work, some of this power is absorbed in overcoming iron and friction losses).

Electrical energy can only be converted to another form of energy by current thus flowing against a back e.m.f.

The e.m.f. generated by the armature must then have a perfectly definite value for each particular value of the load current, viz.

$$E_A = V_A - I_A r_m.$$

This induced e.m.f. is also determined from ordinary considerations of flux, number of conductors and speed, and is thus

$$E_A = Z_c \times 2p\Phi n$$

From these two expressions for the induced e.m.f., the manner in which the speed of the motor varies with the load current, and the flux, can readily be deduced, and becomes

$$V_A - I_A r_m = Z_c \times 2p\Phi n$$

$$n = \frac{V_A - I_A r_m}{Z_c \times 2p\Phi} \quad . \quad . \quad . \quad (153)$$

Since the value of the power supplied to the armature is $V_A I_A$ watts, and of the power wasted in ohmic resistance $I_A^2 r_m$, the percentage loss due to armature, etc., resistance is $100 \times I_A r_m / V_A$. This percentage is of necessity very low, not merely because the designer aims at producing an efficient machine, but also because it is physically impossible by normal means to keep the temperature rise of the machine within the prescribed limits, if the losses are not relatively small. Normal values for the resistance drop on full load are from 1·5 per cent for large machines to about 5 per cent for small machines. (For very small machines, such as used for table fans, etc., the figure of 5 per cent is much exceeded.)

To a first approximation, then, the speed of a given motor may be taken to vary directly with the applied voltage and inversely with the field strength.

For a motor operating under the normal conditions of constant applied voltage, the speed can therefore readily be varied by varying the value of the flux. Methods of speed control will be discussed later; for the moment attention will be confined to the operation of the more common types of motors at constant voltage under varying conditions of load.

The torque developed by a direct current armature has been shown to be equal to $T = 0{\cdot}1173 \times 2p\Phi I_a Z_a = 0{\cdot}1173 \dfrac{p}{a} \Phi Z_a I_A$ lb ft. For a machine of given dimensions, etc., the torque is thus dependent solely upon the product of the armature current and the flux per pole.

Load Characteristics of Shunt Motor. When a shunt motor has attained its final temperature, the exciting current is constant and quite independent of changes in the value of the load current in the armature. The m.m.f. developed by the field windings is thus not influenced by the value of the load. The flux, however, does not retain its no-load value, owing to the effect of armature reaction. With most modern machines, interpoles are used, and the brushes placed in the neutral position. With many non-interpole motors, rotation in both directions is required, and again the brushes are placed in the geometric neutral position. Under these conditions, armature reaction, just as in the case of a generator, results in a falling off in the flux as the load current increases, on account of the teeth being saturated, as indicated in Fig. 73, page 99.

With motors in which the brushes are given backward lead, as for large non-reversing machines without interpoles, armature reaction, as in the case of generators having forward lead, will also cause the flux to fall in value as the load increases.

With shunt motors, then, the flux automatically becomes slightly weaker as the load increases, and this may partially or entirely compensate the effect of armature resistance, and the speed of the motor may remain sensibly constant at all loads, or it may even increase slightly when the load increases.

Thus, in Fig. 153, the line AB represents the constant supply voltage, and AC represents the actual back e.m.f. of the motor, plotted in terms of the armature current. BC represents therefore the ohmic drop $I_A r_m$. The flux per pole is represented by the graph EF, indicating, approximately, the falling off that takes place due to armature reaction.

Then the speed at any particular value of the load current, such as OG, is proportional to the back e.m.f. GC divided by the flux GF obtaining for that load current.

In Fig. 153 the fall in the flux is relatively less than the drop in the back e.m.f., corresponding to a motor having a relatively strong main

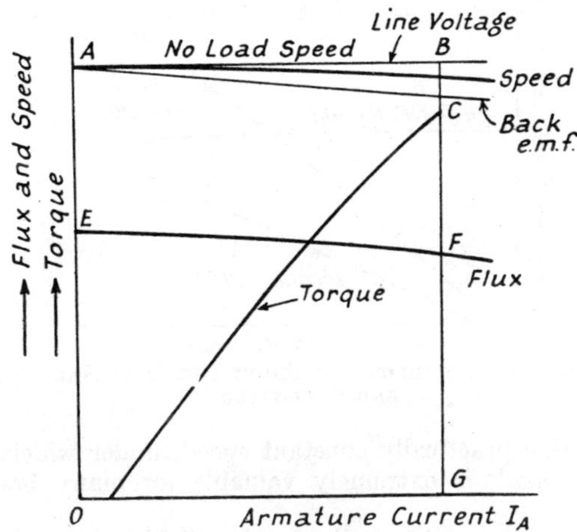

FIG. 153. LOAD CHARACTERISTICS OF THE SHUNT MOTOR

field and devoid of saturation in the armature teeth (i.e. the ratio Field m.m.f./full load armature reaction m.m.f. is at least unity).

The speed-current curve for a shunt motor is thus readily deduced. For this case the curve will be almost a horizontal line drooping slightly at heavy loads.

The current-torque curve would be a straight line passing through the origin if the flux were constant, and iron and friction losses were negligible.

The weakening of the field owing to armature reaction causes the line to droop slightly, and the iron and friction losses cause it to be slightly lower than the line representing the total torque corresponding to $T = 0{\cdot}1173 \times \frac{p}{a} \Phi Z_a I_A$ lb ft, as in Fig. 153.

With a shunt motor operating from a constant voltage supply, there will be a particular set of load characteristics corresponding to each value of the field resistance. For example, with the motor field reduced to one half normal strength, the speed for any given value of armature current will be approximately double, and the torque one half of the value obtaining for the same armature current and full field, as in Fig. 154.

For all practical purposes, then, such a strong field shunt motor may be regarded as a constant speed machine, although the actual speed of operation can readily be varied between fairly wide limits by controlling the exciting current. The property of automatically

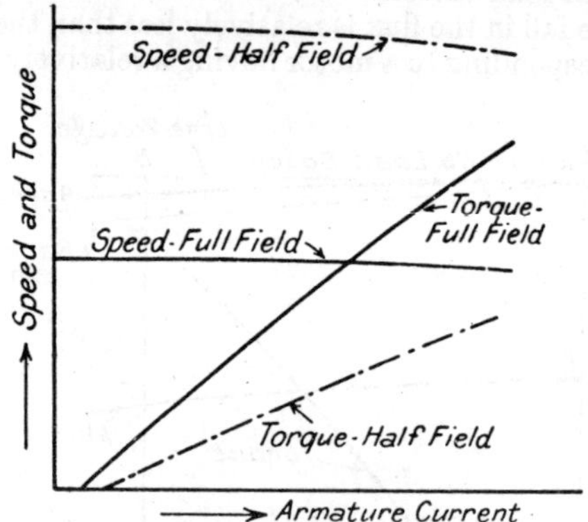

FIG. 154. CHARACTERISTICS OF SHUNT MOTOR AT HALF-FIELD AND FULL-FIELD

maintaining this practically constant speed, under widely varying conditions of load, is extremely valuable for many branches of factory work.

It should, however, be realized that slight—maybe extremely slight—changes in speed are essential to the action of the motor. For example, if the motor is running light, and load is suddenly applied, the immediate effect of the load is to slow down the motor very slightly. This slightly reduces the back e.m.f., and causes a very marked increase in the *difference* between this e.m.f. and the applied voltage, so that the armature current increases appreciably and the armature develops more torque. Since the ohmic drop is only of the order of 2·5 per cent of normal voltage for a large fully-loaded motor, the back e.m.f. has only to fall by about 2·5 per cent to change the armature current from no-load to full-load value.

Series Stability Windings. Most modern machines are designed with the ratio field ampere-turns/armature reaction ampere-turns less than unity and with a fair degree of magnetic saturation in the armature teeth. With such a simple shunt motor the fall in flux with increasing armature current would be greater proportionally than the fall in the back e.m.f. As a result the speed would increase with increasing load-torque as shown at *b* in Fig. 155 instead of falling continuously as shown in *a* for a machine having a relatively strong field m.m.f. Instability would thus result, as the motor would tend to take a further increasing current and might actually "race". Relatively few simple shunt motors, therefore, are now made, and the majority of so-called shunt motors are provided with a relatively weak *series stability winding* connected so as to assist the shunt winding and so tend to increase the field. Such motors are therefore virtually cumulative compound motors as discussed on page 202.

Load Characteristics of Series Motor. With the series motor, the m.m.f. due to the exciting coils increases in direct proportion to the

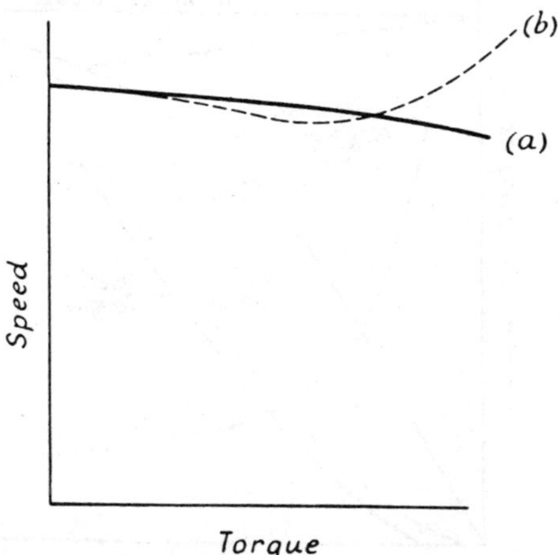

FIG. 155. CHARACTERISTICS OF STABLE AND UNSTABLE SPEED

load current, so that—neglecting the effects of armature reaction—the value of the flux will vary with the load current according to the ordinary magnetization curve. Owing to armature reaction, even with the normal case of brushes in the geometric neutral position, the actual graph representing the total flux will fall below the open-circuit magnetization curve, the fall being quite considerable at heavy values of the current. The graphs given in Fig. 156 may be taken to be typical.

The line representing the manner in which the back e.m.f. varies with the load current is also given. Then since the speed $n = \dfrac{V_A - I_A r_m}{Z_c \times 2p\Phi}$* and the torque $T = 0{\cdot}1173 \dfrac{p}{a} \Phi Z_a I_A$ lb ft, the speed-current, torque-current and speed-torque graphs can readily be deduced from the e.m.f. and flux graphs.

Neglecting the slight effect due to armature resistance, the initial portion of the speed-current graph will be a rectangular hyperbola

* For a series motor r_m = resistance of armature, brushes, and interpole and series coils.

as the initial portion of the magnetization curve is a straight line. The final portion of the graph will merge into a straight line, and the speed is zero when the value of the current is the normal short circuit current of the motor, i.e. equal to the applied voltage divided by the motor resistance. Normally this current is very many times full-load value.

The curve for the total torque will at first be a parabola, until the magnetization curve ceases to be a straight line, but ultimately will merge into a straight line passing through the origin when the

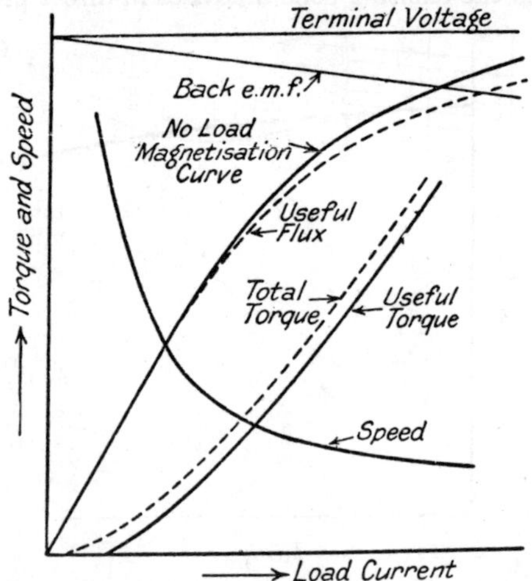

FIG. 156. LOAD CHARACTERISTICS OF THE SERIES MOTOR

magnetic circuit is virtually saturated. The useful torque is, of course, less than the total torque, the relative difference being most marked for small values of the current, owing to the very high speed then attained and in consequence the large friction loss. The speed on no-load will normally be many times the full load speed, and in general would be in excess of the limiting safe speed for the machine. Series motors are then normally not used under conditions where there is any possibility of no-load working being obtained.

In Fig. 156 the current-torque and the current-speed graphs are given, corresponding to the other graphs of back e.m.f. and flux.

Again, as with the shunt motor, it is possible to arrange that a series motor shall work to various load characteristics all of the type of those in Fig. 156. This may be accomplished by "diverting" various amounts of the field current or by "tap field" control as instanced later.

Load Characteristics of Compound-wound Motors. The series field coils of compound motors may be connected so that their m.m.f. assists or opposes that due to the shunt winding, as in Fig. 157. In the former, the motor is said to be "cumulative-compound," and in the latter, "differential-compound."

Differential-compound motors are rarely used. Specially-designed motors, with weak armature reaction, large air gaps and little magnetic saturation, have been used for constant speed work. Such differential arrangement of the windings is unsafe unless the total weakening of the field due to all causes merely corresponds to the fall in the back e.m.f. For such purposes only a relatively weak

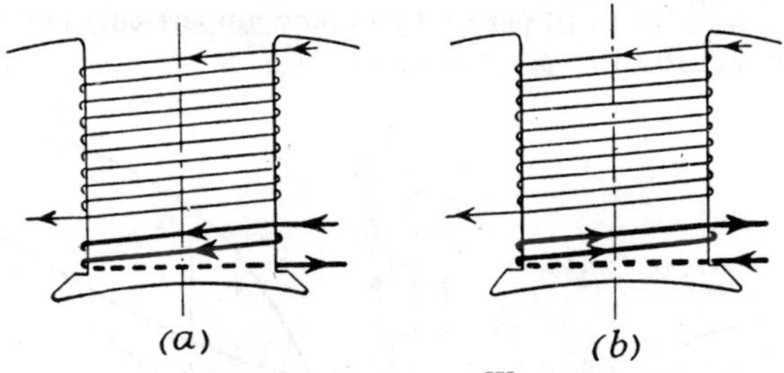

FIG. 157. COMPOUND WINDINGS
a. Cumulative-compound; *b.* Differential-compound

series winding is necessary. In this connexion, if a level-compound generator be used as a motor, the connexions remaining unaltered, differential action between the shunt and series coils will result. For the same polarity of the brushes, the direction of the shunt current will be the same for generator and motor action, but the main current in the armature and series windings will be reversed when the change is made from generator to motor operation.

The differential arrangement of the windings is a disadvantage in very heavy overloads, as the field may be weakened too much and the motor tend to race, taking a very large current. This may ultimately result in the series winding overpowering the shunt, causing the motor to reverse. At starting, too, if the starter handle be moved too quickly so as to permit too heavy a current to flow in the series coils, the series coils will overpower the shunt and the motor start up with wrong rotation.

With a motor intended for constant speed work, however, the series winding is normally so weak that these troubles are not likely to occur.

Cumulative-compound motors are very widely used for individual drives of certain machines, lifts, winches, etc., where a speed characteristic of the type possessed by the series motor is required, with the exception that the no-load speed must be limited to a safe value. Depending upon the relative strengths of the two windings any desired type of characteristic may be obtained, as indicated in Fig. 158. The cumulative arrangement is particularly suited to driving machines, such as presses, in which the load is intermittent but heavy. By fitting a flywheel to the machine, the load on the supply system can to a large extent be equalized. At the instant when the heavy duty is required of the press, etc., the motor slows down and the flywheel is thereby enabled to give up some of its stored energy. For the remainder of the load cycle the motor speeds up and imparts energy of motion to the flywheel, in readiness for the next operation.

As already stated, the ordinary industrial motor—frequently called a shunt motor—is most often a cumulatively-compound motor, being provided with a relatively weak series winding developing from 10 to 15 per cent as many ampere-turns as armature reaction, i.e. 10 to 15 per cent of $0{\cdot}5I_aZ_a/2_p$.

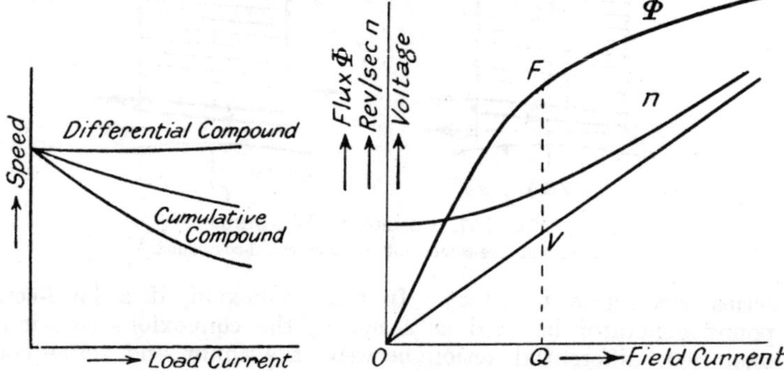

FIG. 158. LOAD CHARACTERISTIC OF COMPOUND MOTORS

FIG. 159. VARIATION OF SPEED OF SHUNT MOTOR WITH APPLIED VOLTAGE

Speed-voltage Characteristics of Motors. Neglecting the effect of armature reaction and armature resistance, the speed of a motor varies directly as the voltage applied to the armature, and inversely as the field strength.

For a motor having constant excitation, therefore, the speed will vary in direct proportion to the armature applied voltage. This principle is utilized in the Ward Leonard method of speed control as discussed later (*see* page 213). For a *series* motor, the field strength is determined entirely by the value of the load current, so that for a given load current, neglecting resistance effects, the motor speed will vary in direct proportion to the applied voltage.

For a shunt motor, however, variations in the voltage applied to the armature affect equally the value of the field current. Neglecting resistance effects, for the straight-line portion of the magnetization curve, the field strength will vary in direct proportion to the applied voltage, and the motor speed will be unaffected by changes in the applied voltage. As the magnetic circuit approaches saturation, however, the speed will tend to vary directly as the applied voltage. This is illustrated by Fig. 159, which shows the magnetization curve of a shunt motor and, in addition, gives the voltage necessary to establish the various values of the field current. For the field current OQ the voltage applied is QV, and the flux established is QF. The speed is then proportional to VQ/FQ, and the speed graph is therefore of the form indicated.

Parallel Operation of Motors. Occasionally it is necessary to couple two motors mechanically so that they drive the same load. It then becomes necessary to consider the manner in which the motors will share the load, and, in particular, will share momentary

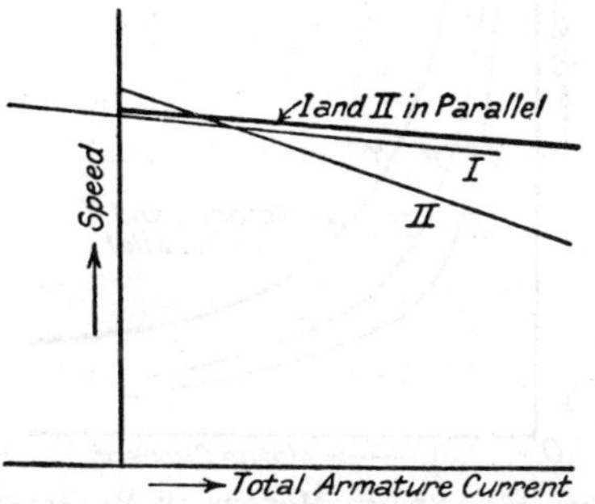

FIG. 160. SHUNT MOTORS IN PARALLEL

overloads. The problem is akin to that concerning the parallel operation of generators, the manner in which the machines share the load being determined by their load characteristics. With motors, as distinct from generators, it must however be remembered that there will be no switchboard attendant to make adjustments to field regulators, should occasion arise, to restore the balance of load between two machines.

Fig. 160 illustrates the load characteristics of two shunt motors of slightly different no-load speeds, and one curve being "flatter" than the other owing, maybe, to slight differences in the brush settings. With these motors connected electrically in parallel and coupled to the same load, the resulting load characteristic of the combination is obtained by summing corresponding abscissae of the two graphs, as indicated in the diagram, which shows speed plotted against armature current, this latter being roughly proportional to the torque developed. For the case indicated, the heavy overloads are taken almost entirely by the machine having the flat characteristic, but at light loads this particular machine functions as a generator and a heavy load is put on the other motor, as this latter tends to run at a higher speed. The coupling of shunt motors in parallel should then be carried out only when field regulators are provided and the motor characteristics can be made practically identical.

With series motors, the speed characteristics are of such a type that parallel operation is quite satisfactory even if the motor characteristics differ appreciably. This is illustrated by Fig. 161, which gives speed current curves for two supposedly identical motors which have been constructed with slightly different values for the air gap. The total characteristic is again obtained by summing the corresponding abscissae—thus $PT = PQ + PR$—and although the machine with the larger air gap takes rather more of the load than its companion, the disparity is never very serious, and

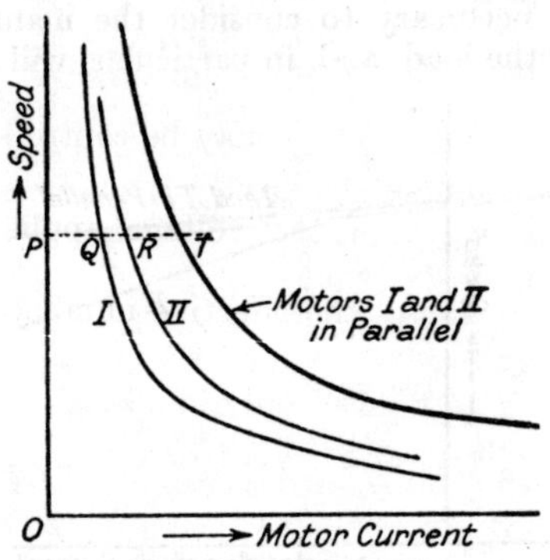

Fig. 161. Series Motors in Parallel

there is no possibility at light loads of current circulating between the machines due to one being driven as a generator.

Actually, series motors connected in pairs in parallel are used to a very great extent for traction, with perfectly satisfactory results.

Series Operation of Motors. In order that motors may work satisfactorily when connected in series, and mechanically coupled to the same load, practically the sole necessity is that their armatures shall be designed for the same current. It is not essential that the motors shall have similar characteristics, and comparatively large differences in their relative excitation are of no moment.

For the one machine, the induced e.m.f. is

$$E_1 = Z_{c1} \times 2p_1\Phi_1 n$$

and for the other

$$E_2 = Z_{c2} \times 2p_2\Phi_2 n$$

Neglecting the ohmic drop in the main circuits, the total applied voltage is

$$V = E_1 + E_2$$

The load is thus shared in proportion to the values of the induced e.m.f.'s, as both motors carry the same current.

For the particular case of similar machines connected in series, the load is shared in the ratio of the field strengths of the two machines—to a very close approximation.

CHAPTER X

CONTROL OF D.C. MOTORS

It has been shown on page 197 that when steady conditions of speed are obtained,

$$n = \frac{V_A - I_A r_m}{Z_c \times 2p\Phi}$$

Hence the speed of a d.c. motor may be controlled by

1. Varying the value of the flux;
2. Varying the value of the voltage applied to the motor armature; or
3. Varying the value of the effective number of conductors in series.

This last method can be applied only in special cases, and involves rocking the brushes through an appreciable angle. Normally, the speed can then only be controlled by varying the field, or by varying the applied voltage.

Most motors work off constant voltage supply systems. The only simple means of varying the voltage applied to the armature is by including resistance in the main armature circuit. This involves considerable waste of energy, so that the method is used only where very low speeds are occasionally required, and other considerations outweigh those of efficiency.

The most common method of speed control is by varying the field strength of the motor.

Constant Voltage, Variable Field. In a shunt motor the field strength can readily be adjusted by means of a field regulating resistance as indicated in Fig. 162.

If a motor is running at a steady speed and the field strength weakened, the immediate effect is that, owing to the momentary reduction in the back e.m.f., the current which at first flows in the armature is increased out of all proportion to the reduction in the flux. As a result, in spite of the weakened field, the torque is momentarily tremendously increased, and will exceed considerably the value corresponding to the load. The surplus torque, thus available, causes the motor to accelerate, and the back e.m.f. to rise. Steady conditions of speed will ultimately obtain again when this back e.m.f. has risen to such a value that the current flowing in the armature develops, with the slightly weakened flux, only just sufficient torque to drive the load.

Consider as an example a 100-V motor working with an armature current of 50 A, the flux per pole being 0·03 Wb, the speed 800 rev/min, and the resistance of the armature, etc., 0·1 ohm.

The back e.m.f. is then $E = 100 - 50 \times 0{\cdot}1$

$= 95$ V

If now the flux is suddenly reduced to 0·025 Wb, the back e.m.f. will fall momentarily to $95 \times \frac{0{\cdot}025}{0{\cdot}030}$, i.e. to 79·2 V. The current flowing in the armature will then rush up to $I = (100 - 79{\cdot}2) \div 0{\cdot}1$, i.e. to 208 A, if inductance effects can be neglected.

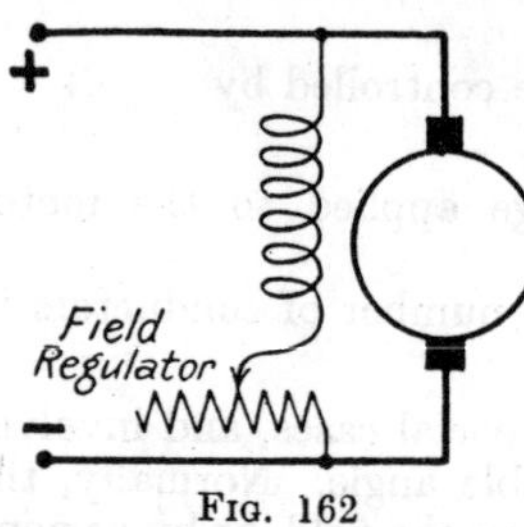

Fig. 162

The torque developed before weakening the field is therefore temporarily much exceeded, and, the torque being proportional to the product of flux and armature current, is increased in the ratio of $(208 \times 2{\cdot}5)/(50 \times 3)$, i.e. 3·47 to 1. This additional torque rapidly accelerates the motor, and if the increase in speed has no effect upon the total value of the load torque to be developed, the speed will increase until the current falls to a value I_2 such that $I_2 \times 2{\cdot}5 = 50 \times 3$, that is, $I_2 = 60$ A.

This ultimate value of the speed then becomes

$$N_2 = N_1 \times \frac{V - I_2 r_m}{V - I_1 r_m} \times \frac{\Phi_1}{\Phi_2}$$

$$= 800 \times \frac{100 - 6}{100 - 5} \times \frac{3}{2{\cdot}5} = 950 \text{ rev/min}$$

If dangerous current rushes are to be avoided, it follows that the field strength must be reduced slowly.

The manner in which the motor slows down when the field strength is increased is also of importance. When the field is increased, the immediate effect is to increase the value of the e.m.f. generated in the machine, i.e. the back e.m.f. Quite a small increase in the flux will normally serve to make this e.m.f. exceed the applied voltage, and, in consequence, will not merely reduce the value of the current flowing in the armature, but will cause it to reverse direction. This can readily be observed on an ammeter connected in the supply circuit. The machine will therefore act temporarily as a generator, and feed energy to the supply system, this energy being obtained at the expense of the kinetic energy associated with the machine and its load, with the result that a rapid reduction in speed takes place. This reduction in speed will cease when the back e.m.f. has fallen to a value such that the motor takes just sufficient current to develop—with the increased value of the flux—the torque the load requires. By thus increasing the field strength the motor speed is reduced by a form of "regenerative braking."

Consider the effect of increasing suddenly the flux from 0·03 to 0·033 Wb in the example just discussed.

The back e.m.f. rises at once to $95 \times 3{\cdot}3/3 = 104{\cdot}5$ volts.

So that the current the machine takes *from the supply* is

$$\frac{V - E}{r_m} = \frac{100 - 104{\cdot}5}{0{\cdot}1} = -\,45 \text{ A}$$

The machine thus acts as a generator.

If the torque remains constant, the final value for the current is given by

$$I_2 = 50 \times \frac{3}{3{\cdot}3} = 45{\cdot}5 \text{ A}$$

and hence the speed ultimately reached

$$N_2 = 800 \times \frac{100 - 4{\cdot}55}{100 - 5} \times \frac{3}{3{\cdot}3}$$
$$= 731 \text{ rev/min}$$

With a shunt motor, then, the speed can be controlled by a field regulating resistance; from the magnetization curve of the machine, the value of the exciting current corresponding to any particular flux is at once obtained, and hence the value of the field regulating resistance necessary to obtain any given speed can be deduced.

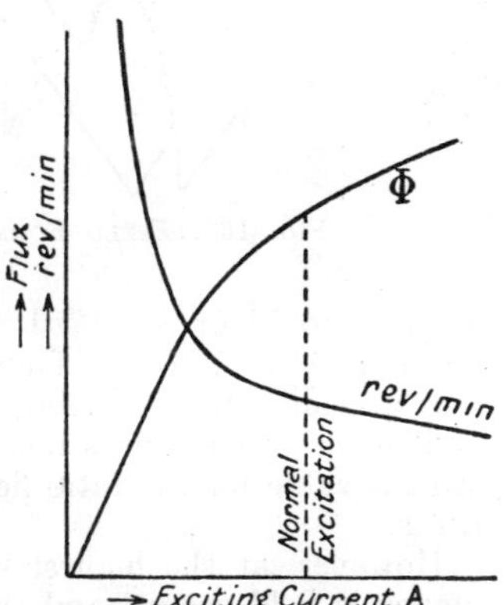

FIG. 163. SPEED CONTROL BY FIELD VARIATION

If the volt-drop in the armature circuit be neglected, the speed for constant armature voltage is inversely proportional to the flux. A graph showing the manner in which the speed varies for different values of the field current is then readily obtained, as in Fig. 163. The ordinates of the speed curve are simply proportional to the reciprocals of those for the flux curve.

Since, at full excitation, modern machines work with an appreciable degree of saturation in the armature teeth and other parts of the magnetic circuit, the exciting current necessary to produce one half of full field is well less than one half of the current for full field. Further, armature reaction modifies the value of the flux and so influences the speed.

A drawback to this method of speed control arises out of the fact that the top speeds are only obtained at a reduced torque owing to the very weak field, so that advantage cannot be taken of the high speeds to increase the power output of the motor. Further, the operation at high speeds with weak field renders commutation difficult unless interpoles are used. For normal machines without interpoles, if a wide range of field control is required, it is necessary to design the machine with a relatively long air gap, so that with full field the ratio: Field ampere-turns ÷ Armature ampere-turns is much greater than usual. Otherwise it is not possible to obtain a satisfactory commutating field even when the brushes are rocked backwards from the geometric neutral. If the field windings are not made relatively strong in this manner, the maximum speed range that can well be obtained from a non-interpole machine is

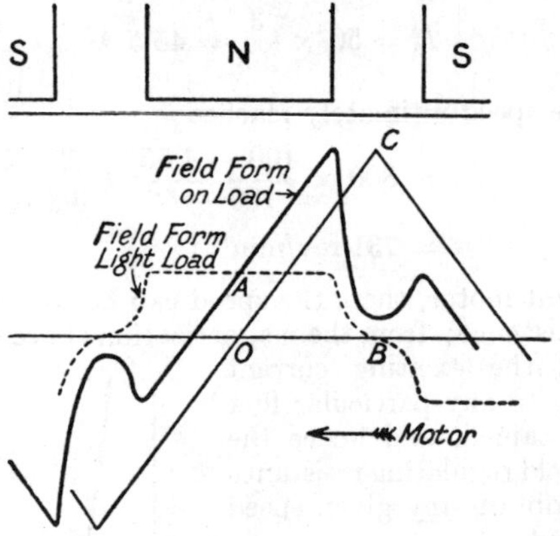

FIG. 164. FIELD FORM FOR MOTOR WORKING WITH A WEAK FIELD

about 1 to 1½ or so. With interpole machines the interpoles can be designed so that satisfactory commutation will take place over a very wide range of speed, speed ratios of 1 to 4 being common, and much larger ranges not abnormal and this without using too large a value for the ratio field ampere-turns to armature ampere-turns.

However, at the highest speed, most marked distortion of the magnetic field occurs, and the direction of the flux in the air gap at one tip of a pole may actually be reversed. This is indicated in Fig. 164, which is constructed for a machine in which at full field and full load current the ratio Field m.m.f. ÷ Armature m.m.f is 1·5, and for 3 times normal speed the excitation has to be reduced to one-fifth of full value.*

For shunt motors working with very weak fields, owing to the relatively powerful effect of armature reaction, instability may result. To avoid this, such machines are often fitted with a relatively weak series field winding, connected so as to assist the shunt. No danger then arises from momentary heavy loads, as the heavy current would strengthen the field and tend to reduce the speed.

* *See also* Fig. 69, p. 96.

The speed of *series motors* may similarly be controlled by varying the exciting current, but in this the regulating resistance must be connected *in parallel* with the field winding, as shown in Fig. 165. By this means a varying amount of the total current can be diverted, and the field winding can carry any desired fraction of the main current.

A difficulty, however, arises with this arrangement if the field winding is diverted by means of a non-inductive resistance. Any

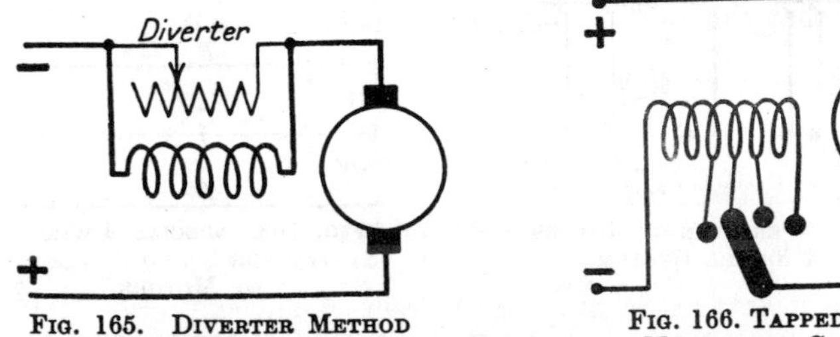

FIG. 165. DIVERTER METHOD OF FIELD CONTROL

FIG. 166. TAPPED-FIELD METHOD OF CONTROL

sudden change in the load current will not at once affect the field winding owing to the high inductance of the field circuit impeding any violent changes in its current. For satisfactory operation under violently changing loads, the diverting resistance should also be highly inductive.

An alternative method of changing the flux, in series motors, is to vary the number of field turns that are utilized, that is, to employ "tapped-field" control as indicated in Fig. 166. For full-field operation the whole of the series turns is used; for higher speeds many of the turns are left out of circuit.

This latter method is used to a considerable extent in electric traction.

When the speed of a normal d.c. motor is controlled by varying the field strength, the voltage being maintained constant the main motor characteristics are similar for all settings of the field resistance. Thus, whether working with full field or a reduced field, the shunt motor with series stability windings maintains a steady speed at all reasonable loads, the characteristics being as indicated in Fig. 154.* Similarly, for a series motor operating with tap-field control, the characteristics would be as indicated in Fig. 156.

Variation of Voltage Applied to Armature. The drawback to varying the speed by field control has been stated to lie in that the high speeds are only possible at relatively low torques. To utilize fully the active material it is therefore desirable to work the machine with full flux, and to control the speed by varying the voltage applied

* With a plain shunt machine these considerations will hold only if the field ampere-turns are relatively great. Otherwise when the field current is reduced appreciably below normal, instability may result and *racing* develop.

to the armature. For normal motors, this voltage cannot economically be controlled if the motor is fed from a constant voltage system.

When the motor is fed from a 3-wire system, two working pressures are available, as the armature may be connected either across the outers for full speed, or between one outer and neutral for half-

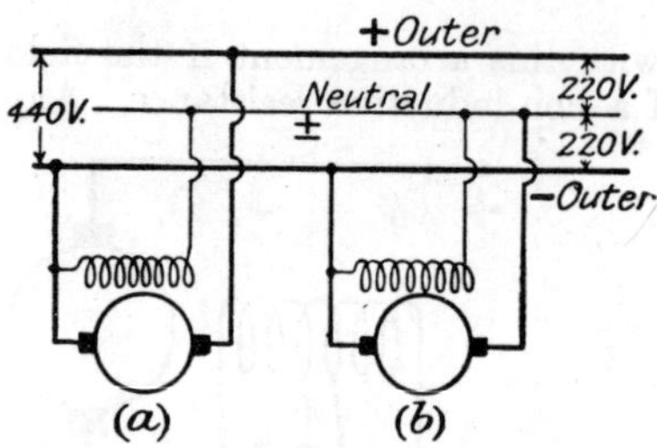

FIG. 167. OPERATION OF MOTORS OFF A 3-WIRE SYSTEM

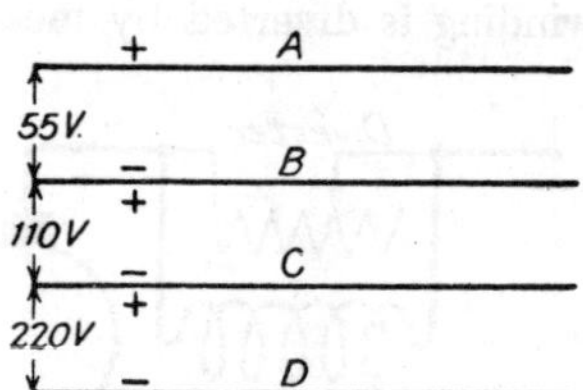

FIG. 168. SPECIAL 4-WIRE SYSTEM FOR SPEED CONTROL OF MOTORS

speed working, as indicated in Figs. 167(*a*) and 167(*b*). The field connexions are the same for the two cases.

In large factories, special 3- or 4-wire systems are sometimes used, giving a number of graded values for the armature voltage. With a 4-wire system as indicated in Fig. 168, the armature may be

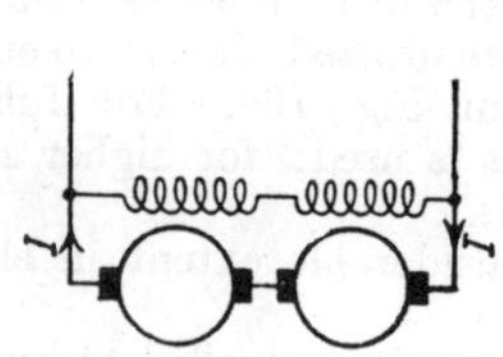

FIG. 169. MOTORS IN SERIES

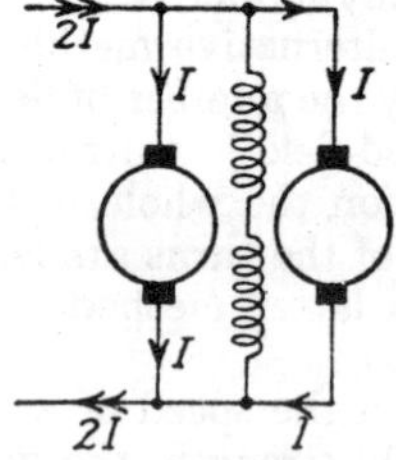

FIG. 170. MOTORS IN PARALLEL

supplied at no less than 6 different values of the voltages, i.e. for the case illustrated, 55 V between *A* and *B*; 110 V between *B* and *C*; 165 between *A* and *C*; 220 between *C* and *D*; 330 between *B* and *D*; and 385 between *A* and *D*.

Series Parallel Control. By using two similar motors mechanically coupled to the same loads, and connecting them in series as in Fig. 169 each motor armature will receive one-half the normal voltage obtained when the motors are connected in parallel, as in Fig. 170, so that two operating characteristics are possible. These characteristics are shown plotted in terms of the current in each motor for both series and shunt motors,* in Figs. 171 and 172. For

* With series stability windings.

the case of shunt motors, the voltage applied to the field windings will be the same for both series and parallel working. It is a matter of convenience whether the field coils are connected in parallel across the full voltage, or in series, as shown.

To a first approximation, for a given armature current, the torque developed by each motor has the same value for both series and parallel connexions, and the speed at which this torque is developed varies in direct proportions to the voltage applied to the armature.

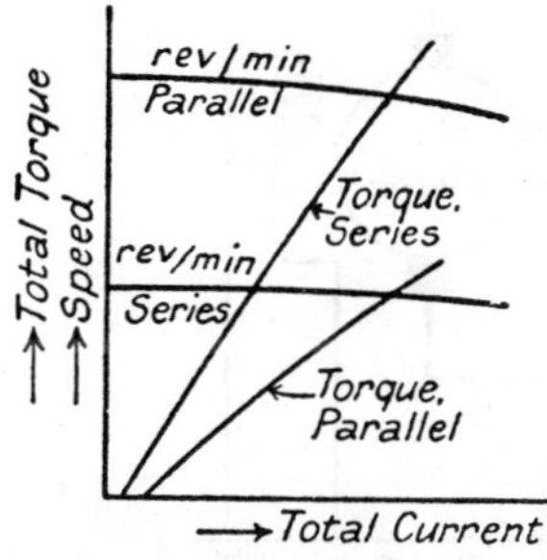

FIG. 171. TOTAL CHARACTERISTICS OF A PAIR OF SHUNT MOTORS WITH SERIES PARALLEL CONTROL

FIG. 172. TOTAL CHARACTERISTICS OF A PAIR OF SERIES MOTORS WITH SERIES PARALLEL CONTROL

The combination when working in parallel and developing a given torque thus takes approximately twice the current that is taken with series connexions for the same torque—but the speed attained with parallel connexions is double that obtained for series working, as shown by the torque current and torque speed curves given in Figs. 171 and 172.

Variable Voltage Control by Ward-Leonard Method. This method of control not only gives a wide range of operating speeds, but reduces to the very minimum the waste of energy that may take place at starting and stopping.

In the Ward-Leonard method the motor is supplied by its own special generator, and the whole of the control is effected by varying the excitation and hence the voltage of this generator. The motor armature is connected directly to the variable voltage generator, as in Fig. 173. Both the generator and the variable speed motor are separately excited. For normal working, the motor excitation is constant, although adjustments may be made by means of a field regulator.

The main control is performed by varying the generator field current—a convenient method being to employ a potentiometer regulator as indicated. By connecting one end of the field winding to the mid point of the regulator, and the other to the floating contact, not only can the exciting current be varied from maximum to zero, but its direction can be reversed. Zero field current will obtain when K is at the mid point. Moving K to the right makes K at a lower potential than C, and to the left at a higher potential.

The variable voltage generator is driven by a d.c. or an a.c. motor, the whole thus forming a motor generator set, which for certain duties is provided with a heavy flywheel—as first proposed by Ilgner. The motor generator set runs continuously, but the generator is excited only during the actual moments when the variable voltage

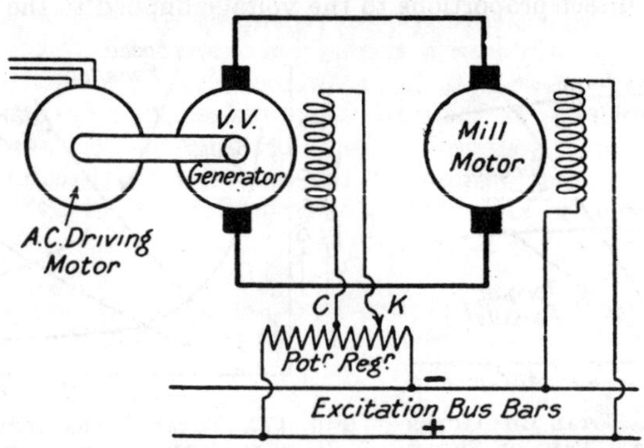

FIG. 173. WARD-LEONARD METHOD OF SPEED CONTROL

motor is working. The field current of the variable voltage motor is unaltered during any given working period, the field being thus energized when the motor is at a standstill, unless the stop is of long duration. To start up this latter motor, the generator field is *very* feebly excited, by moving K a small distance in the direction corresponding to the required rotation. To produce full load current, only a very small percentage of maximum working voltage is necessary. As the motor speeds up, the generator field current—hence its voltage—is steadily increased, so that the current circulating between the two armatures remains sensibly constant, and rapid acceleration takes place even when the motor is heavily loaded. Any position of the potentiometer contact, K, gives a definite steady speed, and the waste of energy in resistances is limited to that in the potentiometer regulator, and is thus relatively small.

This method of control is much used for large reversing rolling mill motors, and for colliery winding, in which much of the energy of motion of the motor and its load can be recuperated, when bringing the speed to zero, by causing the variable voltage motor to act as a generator and feed back energy to the motor generator set. Thus, to stop the motor and its load, the exciting current of the *generator* is steadily reduced, so that the e.m.f. induced by the generator falls, and remains below that developed in the motor armature—the latter still revolving in a fully excited field. The motor current reverses, and flows in the direction of its induced e.m.f. The mill motor therefore functions as a generator, converting into electrical energy its energy of motion and that of the load, and supplying this energy to the generator of the converter set. This latter machine will, in turn, act as a motor and the speed of the converter set be increased, the action of this set being completely reversed so that for a brief period it will actually restore energy to the main supply system. This holds both for a.c. and d.c. supply systems, where no flywheel is provided. When a flywheel is used it is arranged that when heavy load is taken from the variable voltage generator, the converter set slows down appreciably. As a result the flywheel supplies energy to the generator, thus reducing the demand on the main supply system. During "braking" of the winding motor, etc., the energy restored to the converter set is mainly stored by increasing the kinetic energy of the flywheel. Further, when the winding motor is at rest between "winds," the converter set speeds up, and still more approaches its steady "no-load" speed, and so energy is again stored in readiness for the next load.

The flywheel thus acts as a load equalizer and, in spite of the violent changes in the actual input to the winding motor, the demand on the supply system fluctuates but little above and below the average demand. Very real savings in the cost and losses of the main transmission system may thus result.

For reversing rolling mills, the mill motor, when controlled by this method, is reversed by simply moving contact K of the potentiometer regulator steadily from the one side to the other of the fixed contact C.

Variable Voltage Booster. This method differs from the Ward-Leonard method in that the auxiliary variable-voltage dynamo is connected in series with the d.c. supply mains. The whole of the control may then be effected by varying the field current of this auxiliary booster. At starting, the booster is excited in such a direction that its e.m.f. opposes that of the main supply, the difference just sufficing to permit full starting current to flow in the motor. As this latter speeds up, the field current of the booster is *steadily diminished* until, at about half speed, its value is reduced to zero. Beyond this point, the booster field is reversed so that its e.m.f. acts in conjunction with the main supply, and the motor receives a voltage greater than that of the latter supply, full speed being attained at approximately double the line voltage.

The booster is normally motor driven as indicated in Fig. 174. The booster functions as a generator so long as its e.m.f. acts in conjunction with that of the mains, i.e. from half to full speed.

When its e.m.f. *opposes* that of the mains, the booster operates as a motor and drives its fellow machine which in turn restores energy to the supply system. With this method, the motor booster set need be rated at only about one half the full load input to the mill motor, since the booster generates only approximately one-half the voltage applied to the mill motor under full load conditions.

As with Ward-Leonard connexions, every position of the booster field regulator gives economical working. The main motor field may be adjusted to suit best any particular load, but normally the whole of the control is effected on the field regulator of the booster.

Operation of Motors on Rectifier Supplies. D.c. motors are frequently operated on current supplied by mercury-arc rectifiers. This arises both when the motor is one of a number which has to be operated where only a.c. is available and a rectifier unit is installed

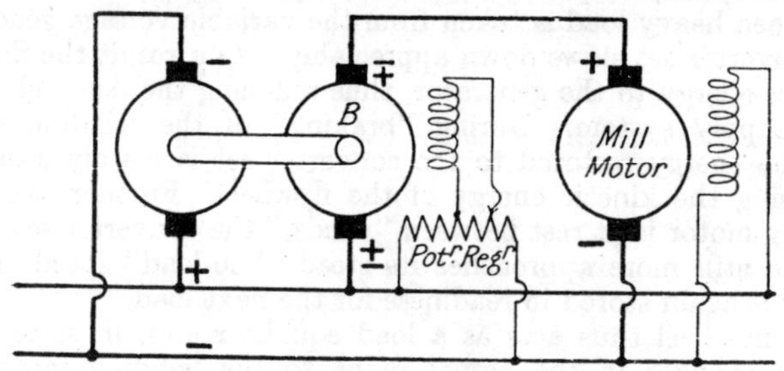

FIG. 174. SPEED CONTROL BY MEANS OF AN AUXILIARY BOOSTER

to provide a nominally constant voltage d.c. supply; also when, by reason of the wide speed range required, a single motor is supplied by a grid-controlled rectifier as an alternative to a Ward-Leonard motor-generator set. A feature of the output of mercury-arc rectifiers is a ripple of much greater magnitude than that produced by rotating machines. A motor supplied by a rectifier is thus supplied not only with d.c. but also with a.c. of frequency and amplitude depending on the number of phases of the rectifier, the inductance of the motor circuit, and, where grid control is used, the amount by which the voltage is reduced below the maximum. The alternating component of the current produces additional losses in both the iron and copper of the motor, hence, it increases the motor heating on any given load. The additional iron loss is greater with series than with shunt motors. Commutation is also made more difficult by the alternating component of current. These effects are slight when the rectifier has six or more phases and is not grid-controlled to low voltages. They may be important with three-phase rectifiers and with those where a low motor speed and high torque are obtained by large angles of delay in grid-controlled rectifiers.

When the speed is very low, especially if the circuit has a low inductance, the ripple of the rectifier output has a greater amplitude than the d.c. component. The current then becomes discontinuous, i.e. it consists as separate "blocks" of current instead of merely varying in magnitude with time. This transition produces a change in the effective output voltage of the rectifier and, since it occurs with a reduction of current at any given position of the regulator the speed of the motor can vary considerably with change of load.

With such Ward-Leonard rectifiers the motor field is supplied from a separate source.

There is of course no energy storage in a rectifier system comparable with that of a flywheel, hence, although Ward-Leonard sets may be replaced by grid-controlled rectifiers, there is no alternative to the Ward-Leonard-Ilgner set.

Starting Torque and Acceleration. The torque developed at starting by the armature of a d.c. motor is determined by the product of the total flux and the total ampere-conductors on the armature, as shown in Chapter I. This torque is utilized partly in overcoming friction, partly in accelerating the armature itself, and partly in accelerating the load. The value of the starting torque required from a motor will then depend very largely upon its duty. To start up against a torque equal to that on full load, the motor must develop rather more torque than on full load, to provide the extra torque necessary to accelerate the armature and the load. If the motor is a constant flux machine, as with a shunt motor, the armature current at starting is a direct measure of the starting torque. The starting current necessary, depending upon the duty, will then vary from about 50 per cent to, in special cases, 200 per cent or more, of normal full load current, and in the particular case of a motor starting up against full load will be about 125 per cent of full load current.

At standstill there is no back e.m.f. of rotation, so that to circulate full load current in the armature but a very small voltage is required. Neglecting the effects of self-induction, the voltage that must be applied to the armature at starting must then represent solely the resistance drop of the main motor circuit. (For an interpole machine the interpole coils are assumed to form part of the armature circuit, and for compound wound or series motors the resistance of the series coils also should be included.)

Where the armature voltage can be increased steadily from zero, as with the Ward-Leonard system or grid-controlled rectifiers, no extra starting equipment is required but with constant voltage d.c. supply systems, the only method of supplying the armature circuit at starting with this reduced voltage is to connect in series with the armature a suitable starting resistance, and to waste in this resistance the surplus voltage. Where a generator is used solely to supply current to one particular motor, it is more economical to effect the control by varying the value of the supply voltage, as discussed on page 213.

In starting up a motor off constant voltage supply mains, resistance is therefore included in the armature circuit. The value of this resistance is arranged so that on closing the circuit the required current is sufficient to develop the necessary starting torque. As the motor speeds up, back e.m.f. is generated and the current falls, and with it the value of the torque developed by the armature. If the resistance were left in circuit, stable conditions of speed would soon obtain when the current had fallen to a value such that the torque developed merely sufficed to overcome motor friction, etc., and to provide the load torque, and no further torque was available for acceleration. With the large resistance necessary to limit the current at starting, this steady speed would be very low, and in addition there would be considerable waste of power in the resistance. It therefore becomes necessary to cut out the whole of the starting resistance, step by step, until "free running" is obtained,

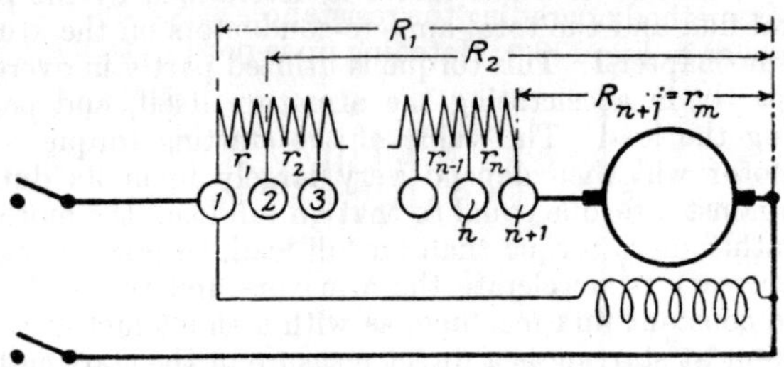

FIG. 175. DIAGRAM OF SHUNT MOTOR STARTER

and the motor is connected directly across the supply system and all external resistance cut out.

For satisfactory operation, it is essential that the steps in the starting resistance shall be graded so that "notching up" can take place without abnormal rushes of current.

Grading of Starting Resistance. Dealing first with a *constant flux motor*, i.e. one in which changes in the armature current do not influence the value of the useful flux :

Number the studs, 1, 2, 3, etc., from and including the starting stud, denoting the resistance elements by r_1, r_2, r_3, etc., and the values of the total resistance in circuit—including the main motor circuit—R_1, R_2, etc., all as in Fig. 175.

Then for n resistance elements, i.e. for $n + 1$ steps and $n + 1$ studs, *excluding* the "off" stud,

$$R_1 = (r_1 + r_2 + \ldots + r_n) + r_m,$$

r_m being the resistance of the main motor circuit. If then on the qth stud the current has dropped to I_2 and notching up takes place, the resistance in circuit is reduced from R_q to R_{q+1}. The motor speed being unaltered for a brief instant, the back e.m.f. E obtaining before notching up, retains its value for the first instant after notching.

Hence $$I_2 = \frac{V - E}{R_q}$$

and $$I_1 = \frac{V - E}{R_{q+1}}$$

where I_1 = the value of the current at the instant following notching up.

It follows then that

$$\frac{I_1}{I_2} = \frac{R_q}{R_{q+1}} \quad . \quad . \quad . \quad . \quad . (154)$$

The ratio of the currents is thus determined by that of the total resistances in circuit on the two studs.

A common method of grading the resistances is based on a current-time diagram as in Fig. 176.* Notching up is presumed to take place

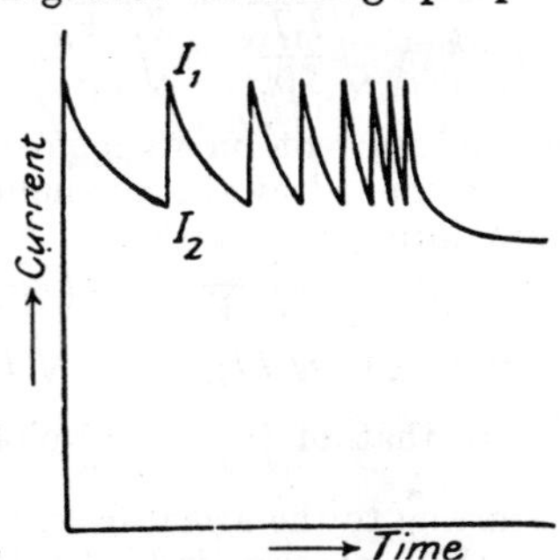

FIG. 176. VARIATION OF STARTING CURRENT WITH TIME

always when the current has fallen to a certain minimum value I_2, and the maximum current on each notch is taken to be I_1. For the starter of a constant flux motor—in particular, for a shunt motor—with the resistance graded on these lines

$$\frac{I_1}{I_2} = \frac{R_1}{R_2} = \frac{R_2}{R_3} = \ldots = \frac{R_{n-1}}{R_n} = \frac{R_n}{R_{n+1}} \quad . \quad . (155)$$

The resistance values thus form a geometrical progression, the common ratio of which is simply the ratio of the currents. Hence if

$$k = I_1/I_2,$$

$$k^n = \frac{R_1}{R_2} \times \frac{R_2}{R_3} \times \ldots \times \frac{R_{n-1}}{R_n} \times \frac{R_n}{R_{n+1}} = \frac{R_1}{R_{n+1}} = \frac{R_1}{r_m} \quad (156)$$

n being the number of resistance elements, as in Fig. 173, and $R_{n+1} = r_m$ = the resistance of the main circuit of the motor, i.e. armature + interpole windings.

Hence
$$k = \sqrt[n]{\frac{R_1}{r_m}} \quad . \quad . \quad . \quad . (157)$$

This gives at once the common ratio k if the number of elements and the resistances R_1 and r_m are known.

* These calculations omit inductive effects which reduce the peak values.

If it is desired to keep the ratio k to a fixed value, the number of steps necessary is given by

$$n \log k = \log \frac{R_1}{r_m}$$

i.e.
$$n = \log \frac{R_1}{r_m} \div \log k \qquad . \qquad . \qquad . \ (158)$$

Where the maximum current on each notch is specified (this is the normal case),

$$k = \sqrt[n]{\frac{I_1 R_1}{I_1 r_m}} = \sqrt[n]{\frac{V}{I_1 r_m}} \qquad . \qquad . \qquad . \ (159)$$

and when the minimum current on each notch is specified (this minimum current is determined by the maximum value of the load torque at starting and during acceleration),

$$k = \sqrt[n]{\frac{V}{I_1 r_m}} = \sqrt[n]{\frac{V}{k I_2 r_m}} = \sqrt[n+1]{\frac{V}{I_2 r_m}} \qquad . \qquad . \ (160)$$

the ratio $\frac{V}{I_1 r_m}$ is simply that of the total voltage, to the voltage actually applied at starting to the armature circuit of the motor.

A consideration of (158) at once indicates that the number of resistance elements necessary for a starter is determined partly by the ratio $\frac{R_1}{r_m}$ and partly by the permissible fluctuation in the current.

As an example, the maximum current at starting is not to exceed 125 A for a 500-V motor. The resistance of the armature + interpole windings is 0·25 ohm. Calculate the values of the resistance elements for a 12-element starter.

$$R_1 = 500/125 = 4 \text{ ohms.}$$

$$R_1/r_m = 4/0{\cdot}25 = 16. \quad n = 12$$

$$k = (R_1/r_m)^{\frac{1}{n}}$$

$$\log k = 1/12 \times \log 16 = 1{\cdot}204/12 = 0{\cdot}1003$$

whence $k = 1{\cdot}26.$

The resistance values are then—

$R_1 = 4 = 4{\cdot}00$

$\therefore r_1 = R_1 - R_2 = 0{\cdot}825$ ohm

$R_2 = 4/1{\cdot}26 = 3{\cdot}175$

$\therefore r_2 = R_2 - R_3 = 0{\cdot}655$ „

$R_3 = 3{\cdot}175/1{\cdot}26 = 2{\cdot}52$

$\therefore r_3 = R_3 - R_4 = 0{\cdot}520$ „

$R_4 = 2{\cdot}52/1{\cdot}26 = 2{\cdot}00$

$\therefore r_4 = R_4 - R_5 = 0{\cdot}413$ „

$R_5 = 2{\cdot}00/1{\cdot}26 = 1{\cdot}587$

$\therefore r_5 = R_5 - R_6 = 0{\cdot}327$ ohm

$R_6 =$ etc. $= 1{\cdot}26$

$\therefore r_6 = R_6 - R_7 = 0{\cdot}260$ „

$R_7 =$ „ $= 1{\cdot}00$

$\therefore r_7 = R_7 - R_8 = 0{\cdot}206$ „

$R_8 =$ „ $= 0{\cdot}794$

$\therefore r_8 = R_8 - R_9 = 0{\cdot}164$ „

$R_9 =$ „ $= 0{\cdot}63$

$\therefore r_9 = R_9 - R_{10} = 0{\cdot}130$ „

$R_{10} =$ „ $= 0{\cdot}50$

$\therefore r_{10} = R_{10} - R_{11} = 0{\cdot}103$ „

$R_{11} =$ „ $= 0{\cdot}397$

$\therefore r_{11} = R_{11} - R_{12} = 0{\cdot}082$ „

$R_{12} =$ „ $= 0{\cdot}315$

$\therefore r_{12} = R_{12} - R_{13} = 0{\cdot}065$ „

$R_{13} = r_m$ „ $= 0{\cdot}25$

With an ordinary slide rule the above calculations may very rapidly be carried out, one setting of the slide sufficing for all the values in turn to be read off.

Since $$k = \frac{R_1}{R_2} = \frac{R_2}{R_3}, \text{ etc.}$$

it follows that each of these ratios is also equal to

$$k = \frac{R_1 - R_2}{R_2 - R_3} = \frac{R_2 - R_3}{R_3 - R_4}, \text{ etc.}$$

Hence the values of the individual resistance elements, r_1, r_2, etc., also from a geometrical progression having a common ratio of $k = \frac{I_1}{I_2}$.

Further $$r_1 = R_1 - R_2 = R_1\left(\frac{k-1}{k}\right) \quad . \quad . \quad . \ (161)$$

Hence there is no need to calculate out the value of R_2, R_3, etc., but the values of the individual elements can be deduced directly.

An example will now be taken when the ratio $\frac{I_1}{I_2} = k$ is not to exceed 1·5 for the motor data given in the previous example.

The number of elements, n, must first be determined*

$$n = \log \frac{R_1}{r_m} \div \log k$$

$$= \log 16 \div \log 1{\cdot}5 = \frac{1{\cdot}204}{0{\cdot}1761}$$

$$= 6{\cdot}83$$

* Determinations of the nth root can rapidly be effected with sufficient accuracy on a slide rule having a log-log scale.

Taking the nearest integer greater than this gives $n = 7$. For this number of elements the ratio of the currents will be a shade less than the maximum allowed in the example, and becomes

$$k = \sqrt[7]{\frac{4}{\cdot 25}} = 1{\cdot}486$$

The total resistance as before is $R_1 = \frac{500}{125} = 4$ ohms.

$$\text{Hence } r_1 = R_1\left(\frac{k-1}{k}\right) = \frac{4(1{\cdot}486 - 1)}{1{\cdot}486} = 1{\cdot}309 \text{ ohm}$$

$$r_2 = \frac{1{\cdot}309}{1{\cdot}486} = 0{\cdot}880 \text{ ohm}$$

$$r_3 = \frac{0{\cdot}880}{1{\cdot}486} = 0{\cdot}592 \text{ ,,}$$

$$r_4 = \frac{0{\cdot}592}{0{\cdot}486} = 0{\cdot}398 \text{ ,,}$$

$$r_5 = \frac{0{\cdot}398}{1{\cdot}486} = 0{\cdot}268 \text{ ,,}$$

$$r_6 = \frac{0{\cdot}268}{1{\cdot}486} = 0{\cdot}181 \text{ ,,}$$

$$r_7 = \frac{0{\cdot}181}{1{\cdot}486} = 0{\cdot}122 \text{ ,,}$$

$$\text{The total is then } \Sigma(r_1 \text{ to } r_7) = 3{\cdot}750 \text{ ohms}$$

This 3·75 ohms added to the motor resistance of 0·25 ohm gives 4 ohms for the total resistance in circuit at starting.

If the resistance were not graded, but divided into equal steps, the current rush would be very marked on the last few steps, and is often disastrous on the last notch.

Taking the above example, and having 7 elements, each of 3·75/7 = 0·54 ohm, approx. The ratio of the current after notching on to the last notch to that before notching up is (0·54 + 0·25)/0·25, i.e. 3·16, since the motor resistance is 0·25 ohm.*

For the preceding notch, the ratio is (0·79 + 0·54)/0·79 = 1·68, and so on, as illustrated by Fig. 177.*

Graphical methods of determining the values of the resistance elements of shunt motor starters are in use, but the whole calculation can be carried out much more quickly on a slide rule. A graphical construction is given on page 228.

Motor Starters. Motor starters, for industrial work, are provided with overload and no-volt releases. Fig. 178 gives a diagram of

* These calculations omit inductive effects.

connexions of a typical "faceplate" starter for a shunt motor. The starter handle is fitted with a spring—not shown in Fig. 178—tending to restore the handle to the off position. This handle carries a soft iron armature *S* which, on the final stud, is attracted by the

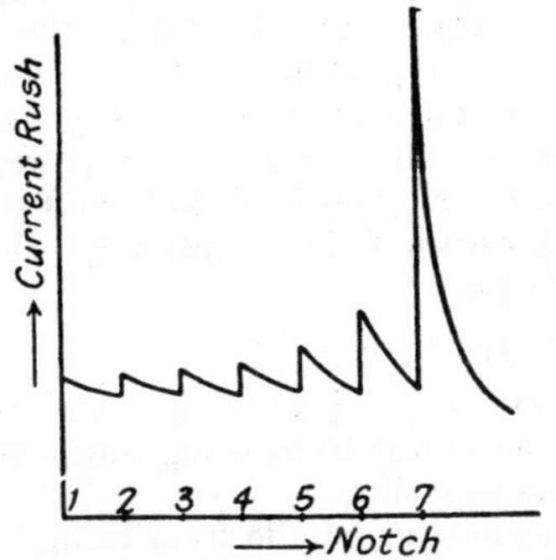

FIG. 177. VARIATION OF STARTING CURRENT WITH TIME FOR UNIFORMLY-GRADED STARTING RESISTOR

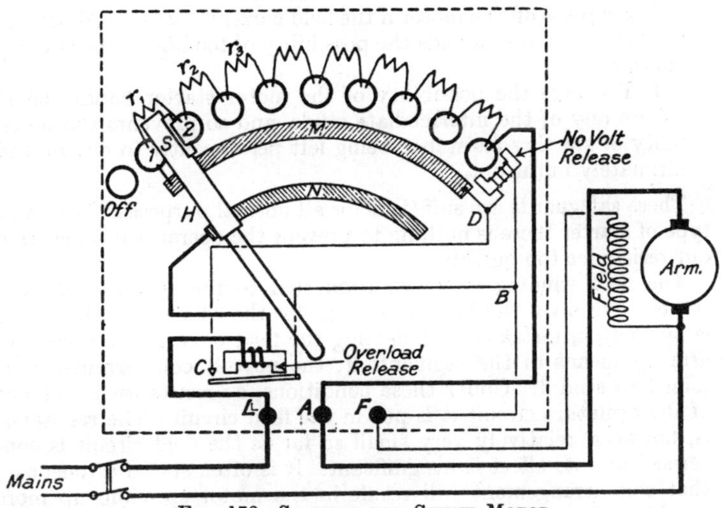

FIG. 178. STARTER FOR SHUNT MOTOR

electro-magnet marked "no-volt release coil," and thus held in position against the action of the above spring.

The main current passes through a small electromagnet designated "overload release" in Fig. 178. This overload magnet attracts a pivoted iron armature as shown, and, when the current reaches a certain amount, causes the armature to make contact with *C* and thus to short circuit the terminals *DB* of the no-volt coil. This results in the spring restoring the handle to the off position. The maximum current that the motor can carry may be regulated by adjusting the control of the armature of the overload release.

The no-volt coil is normally connected in circuit with the shunt winding, except in motors intended for a large range of speed control by field resistance. (For this latter, the no-volt coil is connected to the full supply pressure.) If, with the connexion shown in Fig. 178, the field current is advertently or inadvertently too much reduced, the no-volt magnet is overpowered by the control spring and the starter arm flies back to the off position.

If the supply pressure fails temporarily, the handle is again taken to the off position.

A starter of this type then—

1. Prevents too heavy a load being taken from the motor, and hence prevents the motor from being burnt out due to its being loaded beyond its capacity.
2. Prevents the starter handle from being left in the "full on" position when the motor is shut down or the supply voltage fails, and so prevents the motor from being switched on without resistance in circuit with its armature.
3. Switches off the motor if the field current is weakened beyond a safe limit, and so avoids the possibility of too high a speed being obtained.
4. Prevents the possibility of the motor starter handle being left on any of the intermediate studs, and so prevents the possibility of starting resistance being left permanently in circuit and ultimately burnt out.

These safeguards are sufficient for all normal purposes. With this type of starter there is nothing to prevent the operator from cutting out resistance too quickly.

The field winding is shown connected to the segment M, and therefore receives the full line voltage for the whole time the motor is on. Often, instead of connecting the field circuit to the contact arm by means of the segment M, the no-volt coil terminal D is joined to stud 1. Under these conditions, as resistance is cut out of the armature circuit it is put in the field circuit. The resistance is, however, relatively very small so far as the field circuit is concerned, and its effect is insignificant. It is often stated erroneously that this arrangement will result in the motor speeding up more quickly. Actually the converse is true since the reduction in the field current entails a reduction in the torque, and hence reduces the acceleration of the motor. For both methods of connecting the field circuit it will be noted that, when the starter arms fly back to the "off" position, the field circuit is left closed upon the armature and starting resistance. This avoids any sudden breaking of the field current, which therefore dies away without giving rise to any dangerously high e.m.f.

Grading Starting Resistance for Series or Compound Motor. Dealing now with a variable flux machine, i.e. one in which a change in the armature current gives rise to a change in the main flux, and, in particular, that of a series motor—

Let Fig. 179 represent the manner in which the flux varies with the armature current, and let Φ_1 and Φ_2 respectively be the flux

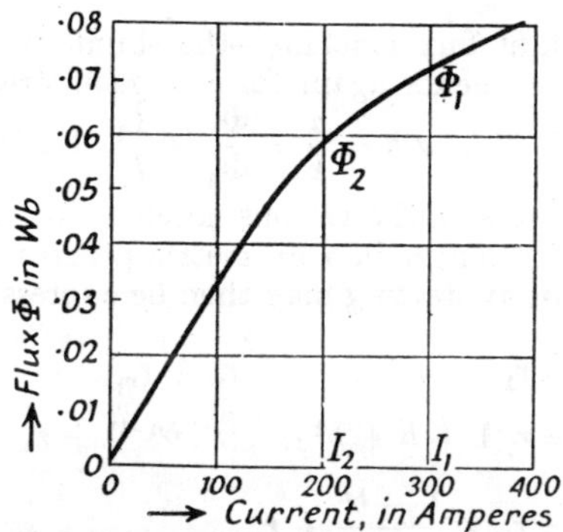

FIG. 179. MAGNETIZATION CURVE OF SERIES MOTOR

established when the current is I_1 and I_2. Further let $\Phi_1/\Phi_2 = a$. Then, dealing with the qth and $(q + 1)$th studs, let E = the back e.m.f. generated at the instant when the current is I_2 on the qth stud. Then, since R_q is the total resistance in circuit,

$$E = V - I_2 R_q$$

If notching up then takes place, and the current at once rises to the value I_1, the back e.m.f. will, in virtue of the increase in current and hence of the flux, at once attain the value

$$E' = \frac{\Phi_1}{\Phi_2} E = aE$$

To obtain the current I_1, the total resistance in circuit on the $(q + 1)$th notch must therefore be

$$R_{q+1} = \frac{V - E'}{I_1} = \frac{V - a(V - I_2 R_q)}{I_1}$$

$$= R_1(1 - a) + \frac{a}{k} R_q \quad . \quad . \quad . \quad . \quad (162)$$

for the case when the current ratios are the same on all notches, and $k = I_1/I_2$, as before.

This expression is perfectly general and will hold for all values of q.

Hence by substituting $q - 1$ for q,

$$R_q = R_1(1 - a) + \frac{a}{k} R_{q-1} \quad . \quad . \quad . \quad (163)$$

The value of the resistance element between the qth and $(q + 1)$th studs is then

$$r_q = R_q - R_{q+1} = \frac{a}{k}(R_{q-1} - R_q)$$

$$= \frac{a}{k} \cdot r_{q-1} \quad . \quad . \quad . \quad . \quad . \quad . \quad (164)$$

As in the constant flux machine—the shunt motor—the values of the resistance elements again form a geometrical progression. The common ratio is now $b = \frac{a}{k} = \frac{\Phi_1}{\Phi_2} \times \frac{I_2}{I_1}$. So far as the values of the total resistances on the various notches, R_1, R_2, etc., are concerned, these are no longer in a geometric progression. The total resistance in circuit at starting may then be expressed

$$R_1 = r_1 + r_2 + \ldots + r_n + r_m$$

$$= r_1(1 + b + b^2 + \ldots b^{n-1}) + r_m$$

$$= r_1 \times \frac{1 - b^n}{1 - b} + r_m \quad . \quad . \quad . \quad . \quad . \quad (165)$$

The value of r_1 is obtained

$$r_1 = R_1 - R_2 = R_1 - \left\{R_1(1 - a) + \frac{a}{k} R_1\right\}$$

$$= aR_1 \left(\frac{k - 1}{k}\right) \quad . \quad . \quad . \quad . \quad . \quad . \quad (166)$$

$$= b(k - 1)R_1$$

Since both b and a depend upon the ratio $k = I_1/I_2$, in the general case the equation can be solved only graphically.

One simple case occurs, however, and that is for a machine working solely upon the straight-line portion of the magnetization curve, so that $a = k = \frac{\Phi_1}{\Phi_2} = \frac{I_1}{I_2}$ and $b = 1$. Under such conditions it is obvious that all the resistance elements have exactly the same value.

Further, since then $r_1 = bR_1(a - 1) = (a - 1)R_1$

and also $$r_1 = \frac{R_1 - r_m}{n} \quad . \quad . \quad . \quad . \quad . \quad (167)$$

the ratio of the currents obtained by equating these two expressions for r_1 becomes

$$a = 1 + \frac{1}{n}\left(1 - \frac{r_m}{R_1}\right) \quad . \quad . \quad . \quad (168)$$

For the general case it becomes necessary, as stated above, to obtain a graphical solution based upon the actual magnetization curve as indicated in an example later. It is however evident that with series motors the number of steps necessary is far less than for a shunt motor having the same value for the ratio I_1/I_2. Further, it must be remembered that the inductance of the field winding will help considerably in preventing the current rush from reaching the figure on which the above calculations are based.

Taking now the general case, equation (165) may be written

$$R_1 = bR_1(k-1) \times \frac{1-b^n}{1-b} + r_m$$

$$1 - \frac{r_m}{R_1} = b(k-1) \times \frac{1-b^n}{1-b} \quad . \quad . \quad . \ (169)$$

In general, the values of r_m and $R_1 = \frac{V}{I_1}$ will be known, and n, the number of resistance sections, given. It remains then to grade the resistance, which in turn involves the determination of the minimum current on each notch. Since the values of a, k, and b all depend upon the minimum current I_2, it becomes necessary to evaluate the expression $b(k-1) \times \frac{1-b^n}{1-b}$ for various assumed values of I_2. The correct value to take for I_2 can then be obtained graphically as instanced for the following example—

Calculate the values of the resistance elements for the starter of a series motor working off a 300-*V supply system, the motor resistance being* 0·12 *ohm, the maximum current at starting* 300 *amperes, and the magnetization curve as indicated in Fig.* 179.

There are to be 4 resistance elements, i.e. 5 steps,

$$R_1 = \frac{300}{300} = 1 \text{ ohm}; \ \frac{r_m}{R_1} = 0{\cdot}12,$$

hence I_2 must be chosen so that $b(k-1) \times \frac{1-b^n}{1-b} = 0{\cdot}88.$

Denoting $b(k-1) \times \frac{1-b^n}{1-b}$ by X, the values of X are calculated for $I_2 = 250$, 230, 220, and 200 A respectively, the calculations being shown in tabular form on page 228.

The values of X are then plotted in terms of I_2, as in Fig. 180, and from the graph so obtained the value of I_2 giving $X = 0{\cdot}88$ is read off, and found to be 226 A.

At 226 A, $\Phi = 0{\cdot}0645$ Wb, whence $a = 7{\cdot}25/6{\cdot}45 = 1{\cdot}124$ and $k = 300/226 = 1{\cdot}326$, and $b = 1{\cdot}124/1{\cdot}326 = 0{\cdot}847$.

Then $r_1 = bR_1 (k - 1) = 0{\cdot}847 \times 1{\cdot}0 \times 0{\cdot}326 = 0{\cdot}276$ ohm
$r_2 = br_1 = 0{\cdot}847 \times 0{\cdot}276 = 0{\cdot}234$ „
$r_3 = br_2 = 0{\cdot}847 \times 0{\cdot}234 = 0{\cdot}198$ „
$r_4 = br_3 = 0{\cdot}847 \times 0{\cdot}198 = 0{\cdot}168$ „

The sum is then: $r_1 + r_2 + r_3 + r_4 = 0{\cdot}876$ „

I_2	300	250	230	220	200
Φ_2 in Wb	0·0725	0·0675	0·065	0·0635	0·06
$a = \frac{\Phi_1}{\Phi_2}$		1·075	1·115	1·142	1·208
$k = \frac{I_1}{I_2}$		1·200	1·305	1·363	1·500
$b = \frac{a}{k}$		0·896	0·854	0·838	0·805
$b^n = b^4$		0·644	0·532	0·495	0·420
$1 - b$		0·104	0·146	0·162	0·195
$1 - b^4$		0·356	0·468	0·505	0·580
X		0·613	0·835	0·950	1·200

This value accords sufficiently well with the figure required, i.e. $1{\cdot}0 - 0{\cdot}12 = 0{\cdot}88$ ohm.

It will be noticed that for this example, in which there are only 4 resistance elements, the ratio of largest and smallest elements is $r_1/r_4 = 0{\cdot}276/0{\cdot}168 = 1{\cdot}64$. This should be contrasted with the example given on page 222 for a shunt motor starter having 7 elements, for which the ratio $r_1/r_7 = 1{\cdot}309/0{\cdot}122 = 10{\cdot}7$.

It is then evident that with a variable flux machine of this series type, departures from the exact calculated values for the resistance elements are not likely to be of so much moment as with constant flux motors, such as a shunt motor.

Graphical Constructions for Determining Grading of Starting Resistances. Shunt Motor Starting Resistance. To determine the values of the resistance steps, the construction shown in Fig. 181 can be used if the maximum and minimum currents on each notch are given.

BE represents, to scale, the total resistance in the main circuit on the first notch, i.e. R_1. To any convenient scale BO is drawn to represent the maximum current I_1 and then OC marked off to represent I_2. The rectangle $BOAE$ is completed and the construction indicated in Fig. 181 performed. Then $HD = CD \times \frac{I_2}{I_1}$, and therefore represents R_2.

The value of the first element is then $r_1 = BJ$. In a similar manner the values of the other elements are obtained as shown in

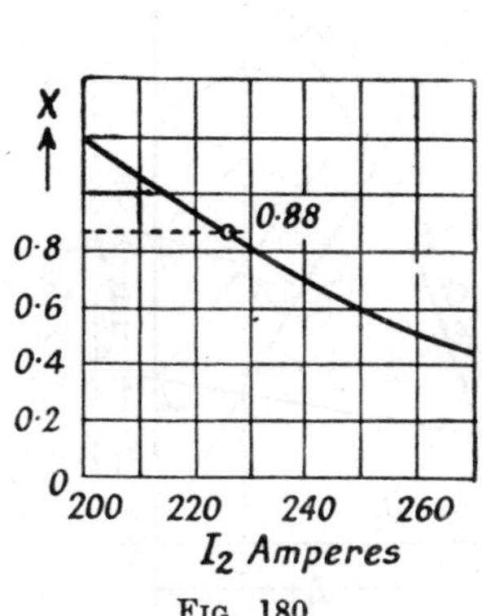

FIG. 180

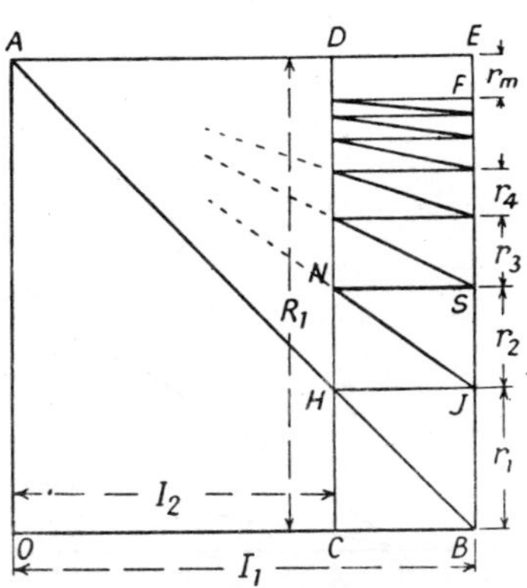

FIG. 181. GRAPHICAL METHOD FOR DETERMINING THE RESISTANCE STEPS FOR A SHUNT MOTOR

the diagram. FE is made equal to the actual resistance of the main motor circuit, r_m.

SERIES MOTOR STARTING RESISTANCE. If the ratio I_1/I_2 and Φ_1/Φ_2 be assumed known, construction can be made as indicated in Fig. 182.

The magnetization curve is shown as ODE. OB represents the maximum current I_1, and, to the same scale, OC represents the current I_2. The fluxes (or e.m.f.'s at some constant speed) at these two currents are represented by BE and CD respectively. BE also represents the supply voltage (to a different voltage scale) and the total resistance R_1.

On starting, the back e.m.f. builds up, and by the time the current has fallen to the value OC, the back e.m.f. has attained the value CH. ($BE = I_1R_1$ and $HT = I_2R_1$.) On notching up, this back e.m.f. at once increases in the ratio Φ_1/Φ_2, i.e. from CH to BJ. The resistance R_2 is then represented by JE, and hence BJ represents the elemental resistance r_1. In a similar manner, the values of the other resistances, r_2, r_3, etc., are determined.

In the diagram, CH, CN, etc., represent the values of the back e.m.f. on notches 1, 2, etc., at the instant before notching up takes place; BJ, BS, etc., represent their values on notches 2, 3, etc., the instant after notching up has taken place.

EF represents the resistance of the motor circuit, r_m. The complete magnetization curve need not be drawn; only the points E and D are required.

With either of the graphical methods just given there is no means of determining at once the correct value to take for I_2 for any

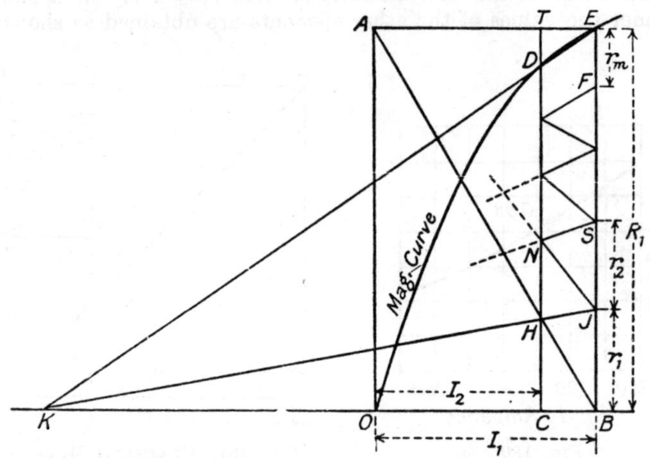

FIG. 182. GRAPHICAL CONSTRUCTION FOR DETERMINING THE VALUES OF THE RESISTANCE STEPS FOR A SERIES MOTOR

particular number of notches. This value with these methods must be found by trial and error.

Fig. 181 for the shunt motor is lettered to correspond with Fig. 182 for the series machine. Fig. 181 is simply the limiting case of Fig. 182, Φ_1/Φ_2 now being unity, the point K will be at infinity for Fig. 181 and the lines drawn to that point from H, N, etc., will be parallel to OB.

Waste of Energy in Plain Rheostatic Starter. During the accelerating period, the armature current, neglecting the effect of self inductance, is

$$I_A = \frac{V - E_A}{(R_x + r_m)}$$

R_x being the external resistance in the armature circuit

Hence $$V = E_A + I_A R_x + I_A r_m,$$

and the power supplied

$$VI_A = E_A I_A + I_A^2 R_x + I_A^2 r_m \tag{170}$$

Of the three terms to the right of the equation.

1. $E_A I_A$ represents the power utilized in performing useful work—together with that absorbed in iron and friction losses.

2. $I_A^2 R_x$ represents the power wasted in the starting resistance.

3. $I_A^2 r_m$ represents the power wasted in the motor resistance r_m.

Hence, neglecting the iron and friction losses, the efficiency at any instant during the starting period is

$$\eta = \frac{E_A I_A}{V I_A} = 1 - \frac{I_A^2 R}{V I_A} \text{ where } R = R_x + r_m$$

$$= 1 - \frac{I_A R}{V}$$

If the current is maintained sensibly constant, the acceleration may be taken to be uniform, and for this case, the induced e.m.f. E will increase in direct proportion to the time. Hence if T is the total time of the acceleration period, and t the actual time from the start as indicated in Fig. 183,

$$E_A = kt \text{ where } k \text{ is a const.}$$

At the end of the acceleration period,

$$E_A = kT = V - I_A r_m$$

whence $$k = (V - I_A r_m)/T$$

and $$E_A = (V - I_A r_m)t/T$$

During the accelerating period:

Total energy input $= V I_A T$

Energy usefully employed $= \int_0^T E_A I_A dt = \int_0^T (V - I_A r_m) I_A t/T \, dt$

$$= 0{\cdot}5(V I_A - I_A^2 r_m)T$$

and the efficiency η $$= 0{\cdot}5(V I_A - I_A^2 r_m)T/V I_A T$$

$$= 0{\cdot}5(1 - I_A r_m/V) \quad . \quad . \quad . \quad (171)$$

Hence, approximately 50 per cent of the total energy supplied to the motor during the starting period is wasted in the external resistance. Actually, neglecting hysteresis and frictional losses, the waste of energy in the external resistance is just equal to the energy usefully employed in speeding up the motor and its load.

This may be illustrated by Fig. 183, in which OP represents the constant applied voltage, and LN indicates the back e.m.f. at any time t from switching in. Since the current is taken to be constant, the lines PK and OR may also be taken to represent the graphs of *power* input and useful power respectively.

The total *energy* taken from the supply during the starting period is therefore represented by the rectangle *PKQO*, and that wasted in the external resistance by the triangle *OSR*. The waste in the motor resistance is represented by the rectangle *PKRS*.

The waste of energy in starting resistances may then be of great importance for motors working to a very short time-cycle—i.e. for motors which are repeatedly stopping and starting heavy loads.

An examination of Fig. 183 will establish that it is during the first half of the starting period that the major part of the energy is wasted. Actually, the energy wasted in attaining half speed is about three times as great as that in accelerating from half speed to full speed, and the energy wasted in attaining half speed is no less than three times the amount usefully employed. In accelerating from one half to full speed, on the other hand, the energy wasted is only one-third of the amount usefully employed. The large waste of energy during the first half of the starting period can be considerably reduced if two motors are operated in series-parallel—half speed corresponding to series working.

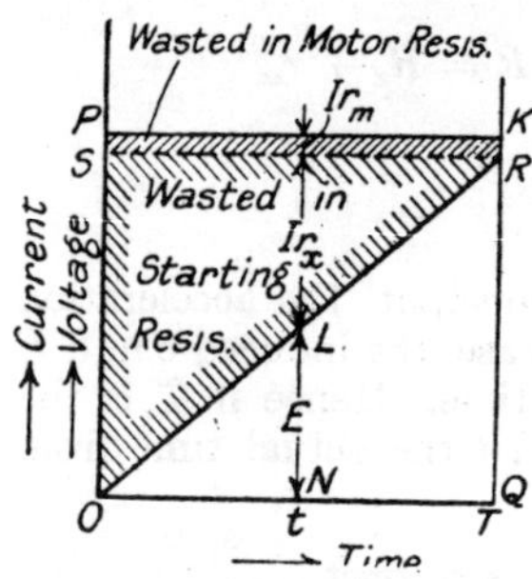

FIG. 183. THE WASTE OF ENERGY IN SIMPLE RHEOSTATIC START

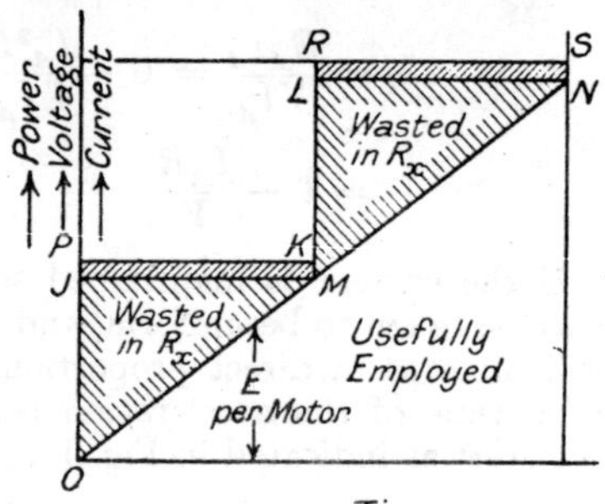

FIG. 184. THE WASTE OF ENERGY IN SERIES PARALLEL STARTING

Economy Gained by Series Parallel Starting. The *motor* current is again assumed to remain constant, to correspond to constant acceleration. The *total* current from the supply will be equal to the motor current when the motors are in series, and to twice the motor current when they are grouped in parallel. The current-time diagram will therefore be of the stepped form, *OPKRS*, shown in Fig. 184. As the supply voltage is constant, this diagram may also be taken to represent the manner in which the power input varies with time.

With the motors in series, the total back e.m.f. will increase according to the line *OL* (that per motor according to *OM*), so that the useful power will increase according to the line *OM*.

For parallel working, the useful power continues to increase along the line *OMN* as the back e.m.f. of the two motors in parallel increases in accordance with MN, and the total current is now twice that in each motor.

The waste of energy in attaining half speed is thus reduced to the amount corresponding to the area of the triangle OJM, and the total waste of energy in the starting resistance, whilst obtaining full speed, as represented by the shaded areas, in Fig. 184, *is reduced to one half* of the waste that occurs with a plain rheostatic start of

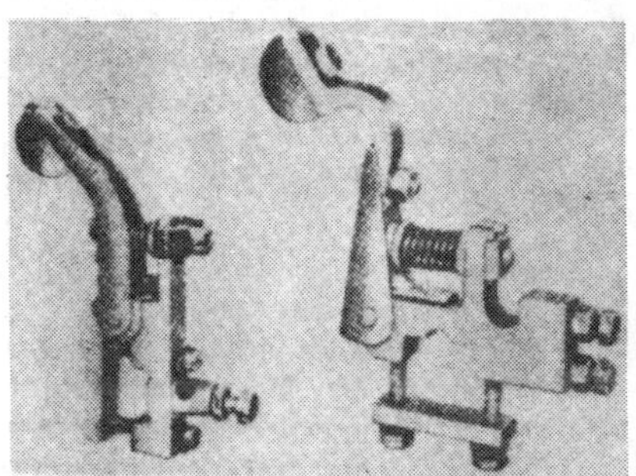

Fig. 185. Contact-finger Assemblies for Drum Controllers
(*Metropolitan-Vickers Electrical Co., Ltd.*)

a single motor, the waste of energy being approximately one half as much as that usefully employed.

The actual figures for the efficiency of the starting period—neglecting motor losses—are then 50 per cent for the plain rheostatic start, and 66·7 per cent for series-parallel starting.

These figures, as stated elsewhere, do not convey a correct impression of the relative merits of the two methods. The consideration of practical importance is the cost of the energy wasted—with the plain rheostatic start this cost is twice that obtaining with series-parallel connexions.

Starters. Drum Controllers. The "faceplate" type of starter, shown diagrammatically in Fig. 178, page 223, in which the operator moves an arm across the face of the starter to cut out the sections of the starting resistor, is not suitable for all purposes. In particular, it is not suitable for large or high-voltage motors, for motors which have to be stopped and started frequently, or for series motors such as hoist motors. The drum type of starter is frequently used for such purposes with medium-size motors. In this, the handle rotates a drum carrying insulated copper segments which make the required connexions by touching fixed "fingers" (Fig. 185) in sequence as the drum rotates. Magnetic blow-outs are fitted where necessary, consisting of electromagnets connected in series with the contacts, and which provide a strong flux at right angles to the arc path. The arc is thus compelled to move in a direction at right angles to both its own length and to the flux, i.e. the arc becomes elongated and thus the current is interrupted more quickly. To prevent the operator moving through two steps in quick succession, a ratchet or racking device may be fitted, so that the handle can be moved only one step at a time. Alternatively, where it is appropriate, the drum may be motor driven, so that the rate of acceleration is taken out of the control of the operator. Fig. 186 shows a manually-operated drum-type controller, with the arc-shield swung clear of the contacts for inspection and maintenance.

Fig. 186. Drum Controller with Cover Removed, showing Arc-shield Swung Clear of Contacts

(*Metropolitan-Vickers Electrical Co., Ltd.*)

Contactor Controllers. With large currents or high voltages, it is not advisable to permit drum finger contacts to carry and break the motor current directly. Drum controllers may then be used in conjunction with "contactors." Contactors are switches or circuit breakers operated by electromagnets, or some other method of remote control. They are naturally quicker and more positive in their action than manually-operated contacts. Contactors breaking large currents are fitted with blow-out coils. With contactors making and breaking the motor circuits, the drum-controller contacts carry only the small currents needed to energize the operating coils of the contactors. It is now easy to arrange that, irrespective of how quickly the operator moves his controller handle, the contactors cannot close too soon and thus cause excessive peaks of current and torque.

In addition to their use with drum controllers, contactors are used for starters for motors of all sizes where acceleration is to be automatic, the operator merely initiating the sequence by pressing a push-button. This includes motors which run continuously for long periods, and those which execute a sequence of reversals, as with planing-machine drives in which the reversals are caused by the operation of limit switches at the ends of the required travel, and in which the speed in the two directions may not be the same.

Automatic Acceleration. When acceleration is automatic, the short-circuiting of the resistance steps of the starter takes place at a rate determined by the starter and not by the operator. Two fundamentally different methods are available, and may be used separately or in combination. In the first method, the short-circuiting of a section of the starting resistor is delayed until the current has fallen to the required value, i.e. to I_2 of equation (154). The second method uses simple timing mechanisms so that the intervals between the short-circuiting of the resistor sections is predetermined and is independent of the load on the motor. Since the time-controlled starter has to be adjusted with the time intervals long enough to accelerate the maximum load, the current-controlled starter will be more rapid in operation with all loads less than the maximum. If, however, an attempt should be made to start with a load greater than the maximum for which the starter is adjusted, almost maximum current will flow through the starter and motor, and the resistor will not be cut out, thus overheating not only the starter but also the motor, since the resulting low speed will provide little ventilation. With a time-controlled starter, however, the motor would start as long as the load was within its capabilities, although excessive currents would be taken. The resistors would thus not greatly overheat and the motor cooling would be better than with the current-controlled starter. The choice has to be made according to the conditions of the particular application. It is possible, with some forms of control, to obtain a greater time delay with heavier motor loads, thus having the best of both forms.

Current-controlled Starters. There are four common ways of arranging for the current itself to control the closing of the contactors—

1. Series relay method.
2. Lock-out contactor method.
3. Volt-drop method.
4. Back e.m.f. method.

1. Series Relay Starters. The series relay method uses both relays and contactors. The motor current, flowing through the operating coil of the relay, holds its contacts open. When the current has fallen to the required value, the relay closes, thereby closing the contactor and energizing the next relay. Fig. 187*a* is a diagram of a simple series relay starter. Pressing the start button energizes the operating coil of Contactor 1 which, closing, connects the motor and resistor in series across the supply. Auxiliary contacts maintain the circuit through the operating coil when the button is released and also energize the contact of current relay 1. The latter however has opened, since the motor current passes through its operating coil. When the current has fallen to I_2, the relay recloses and thus energizes the coil of Contactor 2, which, in closing, short-circuits the first section of the resistor. Contactor 2 has auxiliary contacts similar to those of Contactor 1 and thus the cycle is repeated until the closing of the last contactor puts the motor armature straight across the supply.

2. Lock-out Starters. Lock-out contactors have two magnetic circuits. The flux in one, which saturates, tends to close the contactor. The flux in the other, which does not saturate but has a longer air-path, tends to hold the contactor open. The two magnetic circuits may be excited by different coils or the same coil. The coils are connected in series with the motor so that when the current is

high the unsaturated lock-out magnetic circuit exerts the greater pull and the contactor remains open. When the current has fallen to the required value, the flux in the lock-out coil will have fallen in proportion, whereas that in the saturated closing magnetic circuit will have decreased very little. The pull of the closing circuit now predominates and the contactor closes. Fig. 188 is a diagram for a

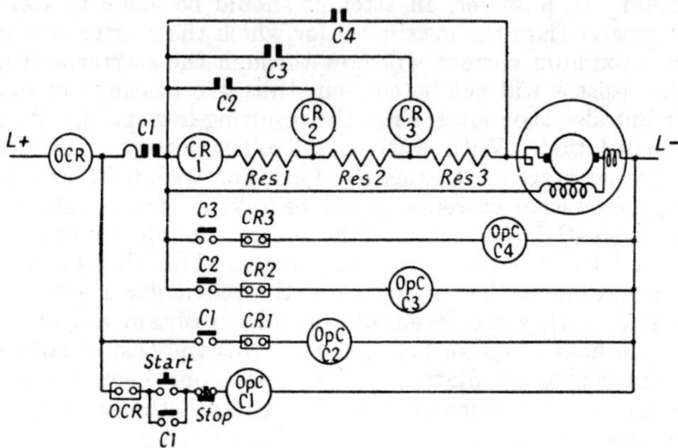

FIG. 187 (*a*). SERIES RELAY STARTER

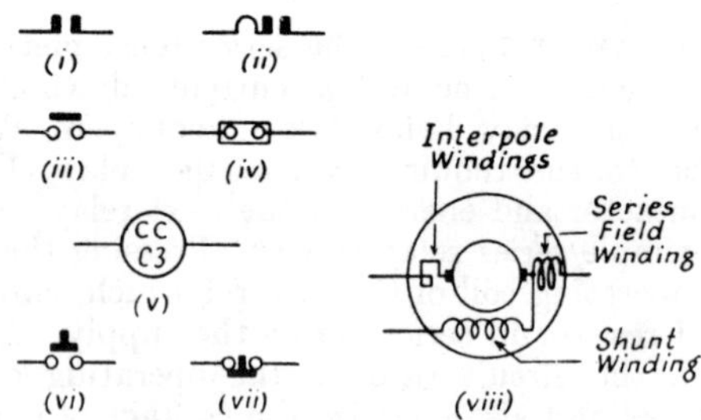

FIG. 187 (*b*). GRAPHICAL STARTER SYMBOLS

Note on Starter Diagrams. Main contactors are shown as in Fig. 187*b* (i) or, if fitted with a blow-out coil, as in (ii). Auxiliary contacts of the main contactors and relay contacts, if open in the off position, are shown as in (iii), but as in (iv) if normally closed. The operating coils of contactors and relays are shown as in (v). Push buttons for starting and stopping are as shown as in (vi) and (vii) respectively. No attempt is made to show operating coils and contacts of either contactors or relays in correct relative positions, identification of coils and contacts is possible only by the letters and numbers. The significance of the letters is as follows—

C	Contactor	*OCR*	Over-current relay
CC	Closing coil	*OpC*	Operating coil
CR	Current relay	*R*	Relay
HOC	Hold-on coil	*Res*	Resistor
LOC	Lock-out coil		

Thus *C*2, denotes contactor No. 2, $\frac{OpC}{C2}$ is the operating coil of contactor No. 2 etc.

Fig. 187*b* (viii) shows the method of representing the d.c. motor in control diagrams.

The diagrams are merely illustrative and should not be taken as the only, or necessarily the best, method of connexion in all circumstances.

series lock-out starter for a series motor. Operation of the start button closes Contactor 1, which has an auxiliary contact to maintain the current in its closing coil when the button is released. The motor then starts with all resistance in series with the armature. The current is flowing through the coil of Contactor 2 and when the

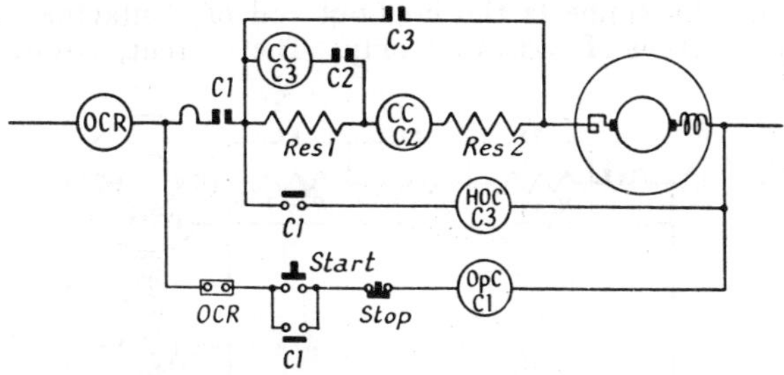

FIG. 188. LOCK-OUT CONTACTOR STARTER*

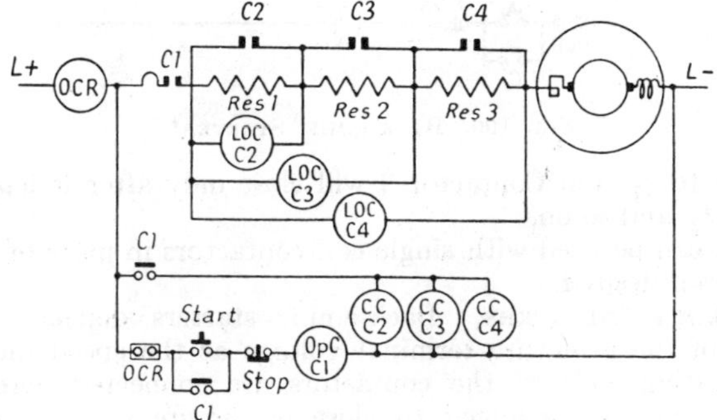

FIG. 189. VOLT-DROP STARTER*

current has fallen to I_2 this contactor closes, thereby short-circuiting the first section of the resistor and energizing the coils of Contactor 3, and so on.

VOLT-DROP STARTERS. In principle the volt-drop starter is similar to the lock-out starter. The essential difference is that the lock-out coil, instead of being connected in series with the motor, is connected in parallel with the section of the resistor to be short-circuited. The voltage across this section, and hence the current in the hold-out coil, decreases as the motor current falls. The closing coil is excited from the line and will, therefore, close the contactor when the motor current has fallen sufficiently. The diagram of a volt-drop starter is shown in Fig. 189. When the start button is pressed, Contactor 1 closes the motor circuit through the starting resistance and, by means of two auxiliary contacts, makes its own

* *See* footnote on page 236.

retaining circuit and energizes the closing coils of the other contactors. The hold-out coil of Contactor 2 is connected across only the first section of the resistor, while those of the remaining contactors are connected across more than one section. Hence when the current has fallen to I_2, Contactor 2 will close but no others. When Contactor 2 closes and short-circuits the first section of the resistor it automatically connects the hold-out coil of Contactor 3 across the second section of resistance only. The current, however, has

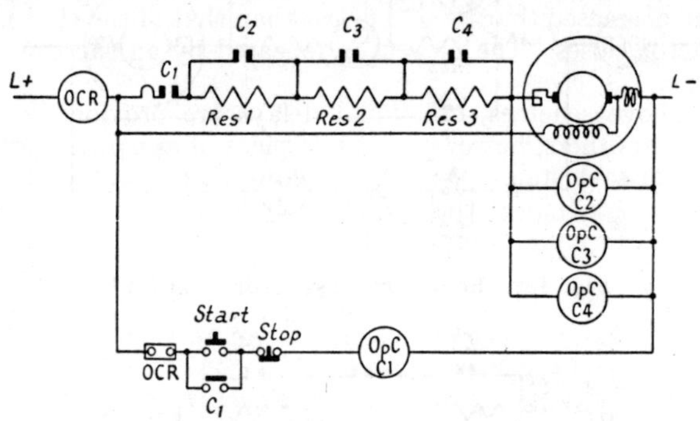

FIG. 190. BACK-E.M.F. STARTER*

increased to I_1, and Contactor 3 will close only after it has fallen again to I_2, and so on.

Relays can be used with single-coil contactors in place of special contactors if desired.

BACK-E.M.F. STARTERS. Back-e.m.f. starters depend on the increase of the armature terminal voltage as the speed increases. The operating coils of the contactors are connected across the armature and are arranged to close on progressively increasing voltages. A compound motor starter diagram is shown in Fig. 190. Pressing the start button closes Contactor 1, thus completing the motor circuits and also its own retaining circuit. When the armature voltage has risen to, say ½ of rated value, Contactor 2 closes. At, say 0·8 of rated voltage, Contactor 3 closes, and finally, at say 0·9 of rated voltage, Contactor 4. Since the contactors close on rising voltage, they do so relatively slowly, and the contacts tend to weld in if this method is used for motors exceeding about 3 h.p.

Time-controlled Starters. The required time delays may be provided by a variety of methods. The inherent operating time of the contactors provides a short delay which may be adequate if acceleration is required to be rapid. If longer delay is required, dash-pots may be fitted to all contactors other than the last. The dash-pot pistons are connected to the contactor by springs and carry auxiliary contacts. Thus the auxiliary contacts close at a given time

* *See* footnote on page 236.

after the main contacts, and, in so doing, energize the operating coil of the succeeding contactor.

Inductance can also be used to provide a time delay. In addition to the closing coil, each contactor has a coil tending to keep it open. The latter is connected across the section of the starter resistor which is shorted by the preceding contactor, so that when the latter closes, the flux which is holding the contactor open begins to decrease slowly, dissipating its energy in the shorted coil. When the flux has so decreased that its pull is less than that of the closing coil, the contactor closes. The process is repeated by all the contactors in sequence.

In the above schemes, individual delays are provided for each contactor. In others, however, a single piece of apparatus provides the delays for each step. One such employs a motor-driven drum to control the energizing of the contactor coils, and the speed of the drum thus determines the delays. If the current taken by the motor to be started is not too large a single multi-fingered contactor may be used, with the fingers arranged to close in sequence. The shaft may be motor driven or solenoid driven with a delay provided by a dash-pot, escapement mechanism, fan or eddy-current brake.

Eddy-current brakes can easily be arranged to provide increased time delays with heavier loads by exciting the brake coil with the motor current. Fig. 191 shows a solenoid-operated starter of this type.

The main contactors at the bottom of the panel are fitted with magnetic blow-outs, the pole-pieces of which can be raised to give access to the contacts. The large solenoid in the centre of the top of the panel moves a copper bar, on the right, upwards and to the right to make contact in sequence with six carbon brushes, thus shorting out successive sections of the resistor which is behind the panel. The last carbon contact is itself shorted out by the copper contact above it, thus avoiding the volt-drop and heating associated with carbon contacts. Since the copper contact neither makes nor breaks the circuit, its contacts do not arc or weld in. The movement of the solenoid plunger also turns the eddy-current disc situated inside the circular box on the left of the panel, thus providing the time delay. The brake magnet is excited by the coil seen below the disc. The braking flux, because the coil is connected in series with the motor, is stronger when the motor current is greater, i.e. on heavy loads, and thus reduces the rate of acceleration. In the middle of the panel is the over-current relay which opens the circuit of the operating coil of the main contactor if the load applied to the motor is such that it takes an excessive current.

All contactor starters have the usual over-current trips irrespective of their method of control. No-volt release is automatically provided, however, without any additional device because the main contactor falls out when the supply fails, there being no current to energize its operating coil. The main contactor is usually fitted with

FIG. 191. SOLENOID-OPERATED STARTER
(Watford Electric and Manufacturing Co., Ltd.)

a blow-out coil in series with its contacts, since it is the only one actually breaking current.

Electric Braking. As stated on page 12 there is no fundamental difference between a d.c. motor and a d.c. generator, the difference being merely the magnitude of the e.m.f. relative to the terminal voltage. The criterion whether a particular machine is functioning at any given time as generator or motor is merely whether its e.m.f. is greater or less than the terminal voltage. Thus, when a machine is connected to a mechanical load which may drive it by virtue of kinetic or gravitational energy, it may motor under some conditions and generate under others. The characteristics shown on pages 181 to 206 for generators and motors are thus not limited to a single quadrant but may be extended into two or more quadrants in order to get a complete picture of the behaviour.*

This reversibility of the d.c. machine provides a convenient method of braking d.c. motors. If the field current be increased so that the e.m.f. is greater than the supply voltage, the armature current of a shunt motor will reverse and so also will the torque, i.e. the motor "regenerates." This will provide a braking torque down to the speed at which the e.m.f. equals the supply voltage. This provides only a limited range of speeds for which braking is possible, since with falling speed the e.m.f. cannot long be maintained greater than the supply voltage. If braking below the speed corresponding to full field is required, it is necessary to disconnect the armature from the supply and to load it as a generator with a resistance load. Some torque is then available for braking at all speeds down to standstill, when the torque becomes zero. This method is called "rheostatic" or "dynamic" braking in contrast to the "regenerative" braking described above. The series generator being a somewhat unstable machine, it is not possible to brake a series motor so easily. For regeneration it is necessary to excite it separately as described in books on electric traction. Rheostatic braking is still possible, however, although with machines operating in parallel it is necessary to treat them as described on page 194.

An alternative method of braking is "reverse current" braking or "plugging." In this the supply to the armature is reversed thus producing a large torque right down to standstill. The motor will of course then reverse if the supply is not disconnected at standstill. This method is most effective, but is wasteful of power and will require a series resistance of about twice the value of the starting resistance in order not to exceed the maximum permissible current.

* As in Fig. 149, p. 194 and Fig. 160, p. 205.

CHAPTER XI

EFFICIENCY: TEMPERATURE RISE: VENTILATION

In converting from electrical to mechanical energy, or vice versa, waste of energy is inevitable, some of the energy being degraded into heat. In dynamo-electric machinery, loss of energy occurs in the electric circuits, and in those portions of the magnetic circuits that are subjected to varying magnetization. Further losses occur due to mechanical friction.

These losses are of importance not merely because they constitute a source of inefficiency, but also because most of them result in the temperature of the windings rising appreciably, and hence determine the rated output of the machine.

The losses may conveniently be divided into—

1. Losses due to changing magnetization—or Iron losses.
2. Electrical losses—mainly Copper losses.
3. Mechanical losses—Friction losses.

The nature of the iron losses has already been discussed in Chapter V.

Effect of Load upon the Iron Losses. With the normal d.c. machine, the effect of load currents in the armature is to distort the flux distribution curve, and to cause the flux to crowd towards the leaving side of the shoe for generators and towards the entering side for motors, as indicated in Fig. 69, page 96.

As a result, the maximum flux density occurring in the teeth is increased appreciably when the load is increased. The iron losses in the teeth therefore increase when the load is increased, even if the value of the useful flux per pole is kept constant at its no-load amount. For the portions of the armature core immediately below the roots of the teeth the losses increase on load, for the same reason.

Not only are the tooth losses considerably increased but, owing to the high degree of magnetic saturation set up in the teeth, an appreciable portion of the flux may be forced out into the core end plates, giving rise to eddy current losses therein.

The pulsation losses in the pole shoes will also increase somewhat when the field form becomes distorted. The net result, therefore, is that the value of the iron losses for a d.c. machine working on full load is considerably greater than on no-load, 50 per cent greater being fairly normal. The manner in which these "extra" iron losses vary with the load depends upon the degree of magnetic saturation in the teeth, but for approximate purposes the extra loss may be assumed to vary in proportion to the square of the armature current.

Electrical Losses. The main ohmic losses are—

1. In the Main Series Circuit of the Armature. They include the loss in—

(*a*)	The armature winding	$I_A{}^2r_a$
(*b*)	The interpole winding	$I_A{}^2r_{cp}$
(*c*)	The series field winding	$I_A{}^2r_{se}$
(*d*)	The compensating winding	$I_A{}^2r_k$
(*e*)	The brush contacts	I_Av_b

In addition there are certain small losses in the commutator segments.

Of the above, (*d*) is normally not present, as armature reaction is only compensated in rare cases; (*c*) is absent for shunt wound machines, and (*b*) for non-interpole designs.

All the above losses, with the exception of those at the brush contact, vary in proportion to the square of the load current in the armature.

For normal approximate purposes the voltage drop v_b at the brush contacts may be assumed to be constant at all values of the armature current, so that the power wasted at the contacts may be taken to vary in direct proportion to the armature current.

2. In the Shunt Exciting Circuit. These losses include the power wasted in the winding itself, and, in addition, that wasted in the regulating resistance. For a level-compounded dynamo, or for a shunt motor working at constant voltage, the loss in the shunt circuit is independent of changes in the load current.

When a shunt dynamo is loaded, to maintain constant terminal voltage the induced e.m.f. must be increased as the load increases, owing to the voltage drop that takes place in the main circuit of the machine. As a result, the flux must be increased slightly as the load is increased; in practice, owing to the drop in speed of the prime mover which invariably accompanies an increase of load, further increase in the flux is necessary. Moreover, armature reaction tends to weaken the flux. The net effect is that, to maintain constant voltage, the exciting current must be increased as the load is increased, so that the loss in the shunt circuit will increase as the load increases.

With an over-compounded dynamo, too, the shunt exciting current will increase with the load. To a first approximation the increase that takes place in the shunt loss of an over-compounded dynamo may be taken to be proportional to the load current, provided the total over compounding is relatively small.

Mechanical Losses. These losses include—

1. Brush friction. 2. Bearing friction. 3. Windage.

Brush Friction. The power wasted in overcoming brush

friction can readily be calculated if the coefficient of friction be known. Thus,

Let μ = coefficient of friction.
p' = brush pressure on commutator in g/cm².
A'_b = total contact area of all the brushes in cm².
v_C = peripheral speed of commutator in m/s.

$$\text{Then power wasted} = \mu \times p'A'_b v_C \times 981 \times 10^{-5} \text{ watts} \quad . \quad (172)$$

The coefficient of friction is found to depend somewhat upon the surface speed v_C, decreasing at high speeds. It is also largely influenced by the type of brush, being relatively low for brushes consisting largely of pure graphite. For such soft brushes the coefficient is about 0·12. For harder graphitic brushes the figure rises to about 0·15, and for the type of brush normally used for 500-V motors the coefficient ranges from 0·2 to 0·3. As an average value 0·25 may be taken.

The brush pressure is usually of the order of 1¼ to 1¾ lb/in.²—roughly 100 to 150 g/cm².

Bearing Friction. For large machines, having sleeve bearings, the loss, being of the nature of fluid friction, may be calculated by a formula of the type

$$\text{Bearing loss} = \text{Const.} \times D_B L_B \times (v_B)^{1\cdot5}$$

where D_B and L_B are the diameter and length of the bearing, respectively, and v_B is the peripheral speed of the shaft in the bearing. If the dimensions D_B and L_B are in cm and v_B in m/sec the loss, for good bearings, may be calculated

$$\text{Bearing loss} = 0\cdot3 D'_B L'_B v_B^{1\cdot5} \text{ watts} \quad . \quad . \quad (173)$$

For motors having ball or roller bearings, the bearing losses are so small in comparison with windage losses that as a rule they are not calculated separately, but included with the windage loss.

Windage. Efficient ventilation is essential in order to obtain the maximum possible rated output. With large d.c. machines it is not common practice to employ large ventilating fans, on the lines of those which are now so widely fitted to machines of relatively small output—200 h.p. or less.

The value of the windage loss will depend upon too many factors for a general formula to be of much use. For a large machine, with no fan, it generally suffices to assume that the windage loss varies in proportion to the square of the speed and is roughly proportional to the barrel surface of the armature. At high speeds the loss increases more rapidly, may be as the cube of the speed.

For surface speeds up to about 50 m/s a rough approximation to the loss is

$$\text{Windage loss} = 0\cdot08 D L_A v_A^2 \text{ watts} \quad . \quad . \quad (174)$$

where D is the armature diameter, L_A is the overall axial length in metres of the completed winding, and v_A is the peripheral speed of the armature in m/sec.

For smaller machines, in which a fan is used, the windage loss is dependent upon the design of the fan and ventilating system. No general formula can therefore be given. The actual loss due to the fan increases in proportion to a high power of the speed, the index ranging from 2·5 to 3.

Efficiency. The efficiency of a dynamo-electric machine is defined as the ratio: Power output/Power input.

Expressed as a percentage, the efficiency becomes

$$\text{Percentage efficiency} = 100 \times \frac{\text{Output}}{\text{Input}}$$

Denoting the efficiency by η, its value may also be expressed

$$\eta = \frac{\text{Output}}{\text{Input}} \quad . \quad . \quad . \quad . \quad . \quad (175)$$

$$= \frac{\text{Input} - \text{Losses}}{\text{Input}}$$

$$= 1 - \frac{\text{Losses}}{\text{Input}} \quad . \quad . \quad . \quad . \quad (176)$$

This latter is very often a more convenient manner of expressing the value of the efficiency.

The term Losses/Input may be defined as the *deficiency*, and denoted by δ.

As a criterion of the performance of the machine, this latter value —the deficiency—is perhaps to be preferred to the one more commonly used, i.e. the efficiency. Take as an example two machines each of 1,000 kW output, the one having an efficiency of 93 per cent and the other of 95 per cent. The difference between these two figures does not at first seem to be serious—a mere difference of 2 in 93. The corresponding deficiency figures of 7 per cent and 5 per cent, however, throw a truer light on the relative merits of the machines and indicate that the one machine is about 40 per cent more wasteful than the other.

For the 1,000 kW machine the input at 93 per cent efficiency is 1,000/0·93 = 1,075·3 kW, and at 95 per cent efficiency the input is 1,000/0·95 = 1,052·6 kW. For the less efficient dynamo the losses are therefore 22·7 kW more than for the other machine, i.e. about 43 per cent more. At one penny per unit, the use of the more efficient instead of the other machine would therefore result in a saving of about two shillings an hour—in all about £400 per year for a 4,000-hour year. It would, therefore, be well worth while to buy the more efficient machine even if it cost several hundred pounds more than the other.

Efficiency Curve. For most machines, and in particular for constant speed and constant voltage machines, the losses may be divided into three main classes—

(*a*) Constant losses, i.e. those which do not vary if the load current varies.

(*b*) Losses which vary in direct proportion to the load current.

(*c*) Losses which vary in direct proportion to the square of the load current

Thus, for example, in a shunt motor, to a first approximation the losses include—

(*a*) *Constant Losses*: bearing friction and windage, brush friction; shunt excitation losses; "no-load iron losses."

(*b*) *Loss Proportional to the Armature Current*: Brush contact loss.

(*c*) *Losses Proportional to the Square of the Armature Current*: ohmic losses in armature and interpole windings; extra iron losses.

The losses may then, to a first approximation, be expressed in the form

$$P_{\mathrm{L}} = a + bI + cI^2 \qquad . \quad . \quad . \quad (177)$$

where I denotes the total current taken by the motor armature. The efficiency of the motor then is represented by an expression of the form, if the total current be assumed to be approximately equal to I.

$$\eta = \frac{VI - P_{\mathrm{L}}}{VI} = 1 - \frac{a + bI + cI^2}{VI}$$

$$= 1 - \left(\frac{A}{I} + B + CI\right) \qquad . \quad . \quad . \quad . \quad (178)$$

where $A = \frac{a}{V}$, $B = \frac{b}{V}$, and $C = \frac{c}{V}$

The deficiency is then expressed by

$$\delta = \frac{A}{I} + B + CI \qquad . \quad . \quad . \quad (179)$$

Of the three terms contained in this expression, the first, representing the effect of the constant losses, becomes zero when I is indefinitely large. Its maximum practical value obtains at the minimum value of I, i.e. when $VI = a$; the value then is unity. The graph representing $1 - \frac{A}{I}$ plotted in terms of the current, is then the hyperbola shown in Fig. 192 by curve *a*. This would represent the actual efficiency curve for the machine if all the other losses were negligible.

For the losses which vary in direct proportion to the current, it is evident from (179) that their effect is to reduce the efficiency by a constant amount. If these losses *only* were present, the efficiency curve would then be the horizontal straight line shown in Fig. 192 by the line *b*.

Finally, the efficiency curve obtaining if all losses were negligible excepting those varying according to the square of the current, would be represented by $1 - CI$, i.e. by the straight line *c* in Fig.

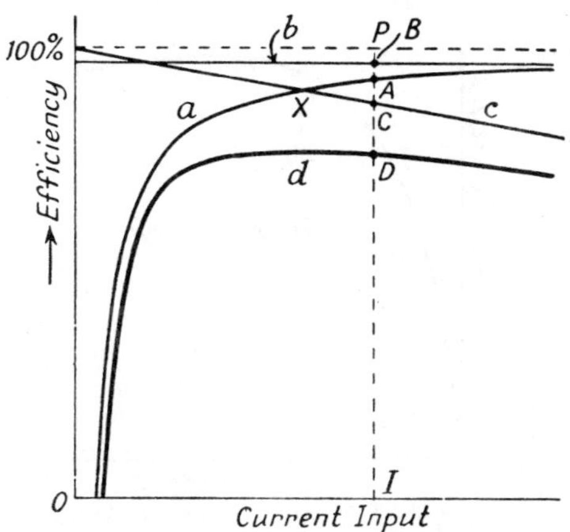

FIG. 192. DERIVATION OF EFFICIENCY CURVE OF SHUNT MOTOR

192, indicating an efficiency which, starting from unity at zero current, falls quite regularly as the current increases.

For a load current of I amperes, as in Fig. 192, the deficiency due to the constant losses is then represented by PA, that due to the losses in class *b* by PB, and that due to losses in the series circuits etc., by PC. The total deficiency is then the sum of PA, PB, and PC, and is therefore represented by PD. D is thus a point on the actual efficiency curve *d* of the machine. The complete efficiency curve can thus be obtained and the effect of the various types of losses upon its shape appreciated. It will be apparent that at light loads the actual efficiency curve merges into curve *a*, and that the constant losses predominate. At heavy loads, on the other hand, the actual efficiency curve merges into curve *c*, and the important losses are those varying with the square of the current.

Maximum Efficiency. Fig. 192 indicates that at some definite load maximum efficiency occurs. This maximum efficiency will obtain when the current is such that $\frac{d\delta}{dI} = 0$.

Hence for maximum efficiency $\frac{d}{dI}\left(\frac{A}{I} + B + CI\right) = 0$

i.e.
$$A = CI^2 \quad . \quad . \quad . \quad . \quad (180)$$

The efficiency of a machine of the type considered will then have its maximum value when the current is such that the losses varying in proportion to the square of the current are equal in magnitude to the constant losses.

Referring to Fig. 192, maximum efficiency will occur at the load for which the graphs *a* and *c* cross, as shown at X. If then P_F = the constant losses, and r_m = the resistance of the main motor

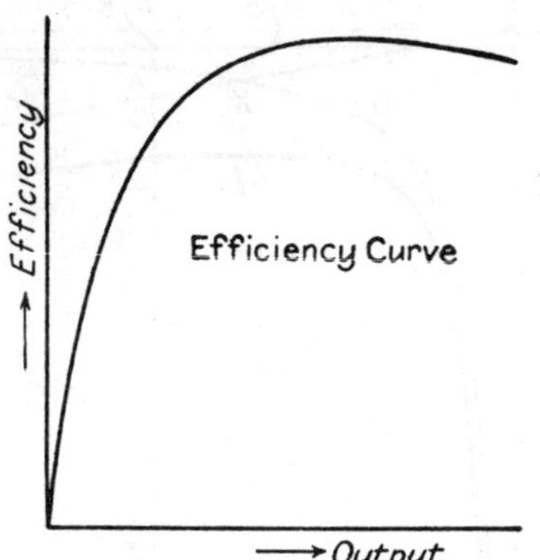

FIG. 193. TYPICAL EFFICIENCY CURVE

circuit, if the extra "load iron losses" be neglected, maximum efficiency will occur when the current is

$$I = \sqrt{P_F/r_m} \quad . \quad . \quad . \quad . \quad (181)$$

The load at which this maximum efficiency occurs, i.e. whether at full load, half load, etc., is determined very largely by the design proportions of the machine, and the speed of operation, etc. The introduction of large ventilating fans on the shafts has permitted of the armature loading being considerably increased, so that maximum efficiency with such machines occurs between half load and full load, in spite of the additional constant losses due to the fan.

From the point of view of efficiency, the best machine for a given duty is the one which, during a working year, wastes the smallest amount of energy. For average purposes a machine designed to give maximum efficiency at about $\frac{3}{4}$ load is well suited.

The efficiency curve in Fig. 192 is shown plotted in terms of the motor input current. From this curve, the motor output can be at once deduced by multiplying the power input by the efficiency. From the data so deduced the efficiency curve as normally used can be plotted, as in Fig. 193, showing the manner in which the efficiency is dependent upon the output. The shape of this latter efficiency curve differs but little from that given in Fig. 192 over the normal working range of the motor, except in that zero efficiency now evidently obtains at the zero point on the output scale. At very heavy overloads, however, the actual efficiency curve—plotted in terms of the output—is of an entirely different form from that shown in Fig. 192, as discussed later.

Maximum Output. For a motor of the type being considered, maximum torque is developed when full voltage is applied to the armature at standstill and the field is fully excited. Normally

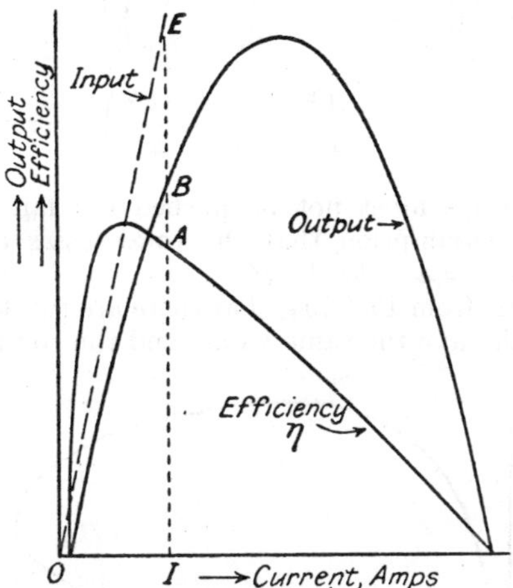

FIG. 194. COMPLETE EFFICIENCY CURVE SHOWING MAXIMUM OUTPUT

the current thus obtained is well beyond the capacity of the armature conductors. At this torque and current the whole of the power supplied is wasted in the motor resistance and the output is zero. Between the two limiting conditions at which the output is zero, i.e. the normal no-load condition and the standstill condition, the output, starting from no-load current, will increase steadily as the load current increases, until at a certain current maximum output is obtained. Beyond which point, although the current increases, the actual output falls off, owing to the marked reduction in the efficiency. This is illustrated by Fig. 194, in which the complete efficiency curve, deduced exactly as in Fig. 192, is given from no-load to standstill conditions. The motor input is represented by the straight line OE, and the output IB at any particular current OI is then proportional to $IE \times IA$. The output current curve thus takes the form shown.

Maximum output, with a machine of normal proportions, occurs at a load current which is practically equal to one half of the current obtaining at standstill, i.e. at a load for which

$I = \frac{1}{2}V/r_m$. The actual load at which the output is maximum is obtained—

Maximum output obtains when $VI - (a + bI + r_mI^2)$ is max.,

i.e. when $\frac{d}{dI}(VI - \{a + bI + r_mI^2\}) = 0$,

i.e. when $I = \frac{V - b}{2r_m} = \frac{V}{2r_m}$ approx., since b is relatively small

hence
$$P_{max} = \left[\frac{V^2}{2r_m} - \left(a + \frac{bV}{2r_m} + \frac{V^2}{4r_m}\right)\right]$$
$$= \frac{V^2}{4r_m} - \left(a + \frac{bV}{2r_m}\right) \text{ approx.} \quad . \quad . \quad . \quad (182)$$

(These conclusions must not be pushed too far. They are all based upon the assumption that the total losses can always be represented by $P_L = a + bI + cI^2$.)

It then appears, from Fig. 194, that there are two loads for which the efficiency will have the same value, and the complete efficiency

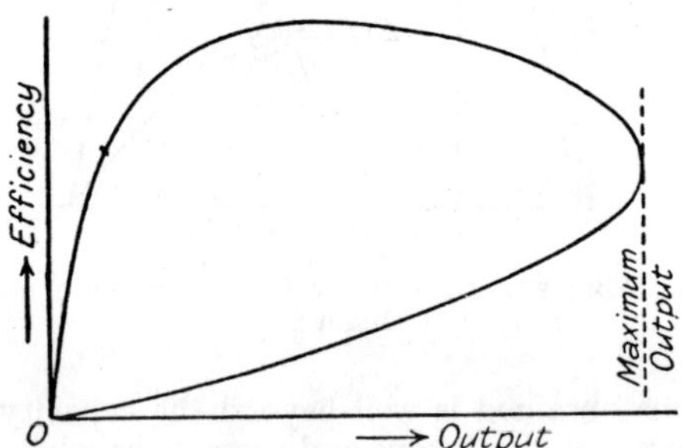

FIG. 195. COMPLETE EFFICIENCY CURVE OF D.C. MOTOR

curve turns back as indicated in Fig. 195, which curve has been drawn to correspond exactly to Fig. 194.

In general, electric motors are not worked at a load which is within range of their maximum possible output, but in fractional horse-power machines the load may be not far removed from this limiting output.

The efficiency curve for a shunt or compound-wound generator, working at constant terminal voltage, can be deduced in a similar manner. Thus, for a level-compounded generator, the total losses can again be expressed in the form

$$P_L = a + bI + cI^2$$

where I is now the output current, and the output VI.

The power input being $VI + P_L$, the efficiency is

$$\eta = \frac{VI}{VI + a + bI + cI^2}$$

The efficiency is maximum when $\frac{1}{\eta}$ is minimum, i.e. when

$$\frac{d}{dI}\left(1 + \frac{a + bI + cI^2}{VI}\right) = 0$$

and thus again occurs when the constant losses are equal to those which vary in proportion to the square of the armature current. The efficiency curve will then be of much the same form as that given in Fig. 192, but zero efficiency will obtain at zero value of the output current.

For series motors, the form of the efficiency curve is of the same general type as that given in Fig. 193 and in Fig. 195. The losses however do not vary according to the same simple law as in the shunt machine, since the speed of the motor is not constant, and, as a result, the friction and windage losses depend upon the value of the load current. Further, the flux is not constant but decreases as I decreases. This decrease is counteracted to a certain extent by the fact that the frequency of the flux reversals increases with decreasing current.

Temperature Rise. The various losses all being degraded into heat, when a generator or motor is working its temperature will tend to rise above that of its surroundings until, ultimately, a steady temperature is reached at which the machine dissipates heat at the same rate as it is produced.

This final steady temperature must not exceed 105°C for machines having Class A insulation, for if such insulation is persistently worked at a temperature in excess of this value it loses its mechanical properties, becomes friable, and a little vibration will result in a breakdown of the insulation.

Temperature Curve. Starting from "cold," when a machine is run under steady full load conditions its temperature will at first increase at a rate which is determined by the power wasted per unit volume of active material, as discussed later. As the temperature rises, the active parts dissipate heat partly by conduction, partly by radiation, and, in most cases, largely by means of air cooling. The more the temperature rises, the greater is the effectiveness of each of these methods of cooling. Hence, as the temperature rises, its rate of increase falls off, and the temperature curve takes the form shown in Fig. 196.

For field coils the temperature curve may readily be obtained experimentally. For the armature, obvious difficulties arise.

Equation of the Temperature Curve. Consider, for simplicity, a homogeneous body in which heat is produced uniformly at a steady rate. The initial rate of increase of the temperature is obtained at once by dividing the quantity of heat produced per second by the thermal capacity of the body.

At any other temperature, the rate of temperature increase is

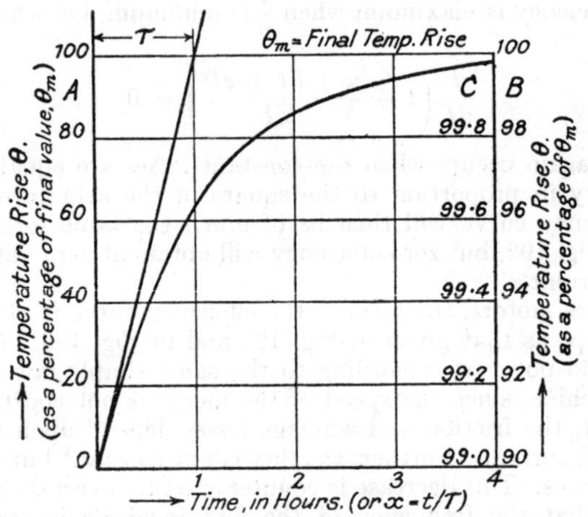

FIG. 196. TEMPERATURE-RISE CURVE

given by dividing the thermal capacity into the difference between the quantities of heat generated and of heat dissipated per second.

Hence if

H = the thermal capacity in calories/°C (i.e. the product of the mass and the "specific heat").

C = the rate at which the body dissipates heat for each 1°C rise in temperature, measured in watts.

θ = the temperature rise at any time t.

P = the power wasted in watts.

Then the final temperature rise is

$$\theta_{max} = \frac{P}{C}$$

The rate at which heat is produced in excess of that dissipated is given by $P - C\theta$, hence the temperature increases at the rate—

$$\frac{d\theta}{dt} = \frac{1}{4{\cdot}2H}(P - C\theta)^* \qquad (183)$$

i.e.
$$\frac{dt}{4{\cdot}2H} = \frac{d\theta}{P - C\theta}$$

* Since 4·2 watt sec = 1 calorie.

This equation is of exactly the same type as that already given in Chapter V, page 107, for the rise of current in an inductive circuit. Its solution is obtained—

Integrating: $$\int_0^t \frac{dt}{4{\cdot}2H} = \int_0^\theta \frac{d\theta}{P - C\theta}$$

whence $$\frac{t}{4{\cdot}2H} = -\frac{1}{C}\log_\varepsilon \frac{P - C\theta}{P} = -\frac{1}{C}\log_\varepsilon \left(1 - \frac{C}{P}\theta\right)$$

and $$\varepsilon^{-\frac{Ct}{4\cdot 2H}} = 1 - \frac{C}{P}\theta$$

$$\theta = \frac{P}{C}(1 - \varepsilon^{-\frac{C}{4\cdot 2H}t}) \quad . \quad . \quad . \quad . \quad (184)$$

ε being the base of the system of natural logarithms.

The above equation may be simplified into the form

$$\theta = \theta_{max}(1 - \varepsilon^{-\frac{t}{T}}) \quad . \quad . \quad . \quad (185)$$

where T is a constant defined as the *temperature time constant*, and is equal to $\frac{4{\cdot}2H}{C}$.

The temperature thus rises according to an exponential law.

The time constant is given numerically by dividing the thermal capacity, expressed in watt-sec for 1°C rise in temperature, by the power that can be dissipated in watts for each 1°C rise.

Alternatively, it is given by dividing the total quantity of heat stored at the maximum temperature, measured in watt-sec, by the power wasted in watts.

The final steady temperature rise, $\theta_{max} = \frac{P}{C}$, i.e. is equal to the power wasted divided by the rate at which heat is dissipated for each 1°C rise in temperature.

To keep this temperature rise within the prescribed limits, the power wasted must be kept small, or the machine must be well ventilated and the heat removed by circulating a large quantity of cooling air.

The initial rate of temperature rise being $\frac{d\theta}{dt} = \frac{P}{4{\cdot}2H}$, if this were maintained for a time $T = \frac{4{\cdot}2H}{C}$, the rise would then be $\frac{4{\cdot}2H}{C} \times \frac{P}{4{\cdot}2H} = \frac{P}{C}$, i.e. the actual final steady temperature θ_{max}. Hence, as shown in Fig. 196, the time constant also represents the time that would be taken for the body to obtain its actual maximum temperature θ_{max}, if there were no heat dissipation whatever, and the initial rate of temperature increase were maintained.

The temperature rise at any time t from the start may be expressed in the form

$$\theta = \theta_{max}(1 - \varepsilon^{-x}) \quad . \quad . \quad . \quad (186)$$

where $x = t/T$ denotes the time measured in terms of the actual time constant. By adopting this method, one temperature curve becomes at once applicable to all cases. On page 110 values are given for ε^{-x} from which the temperature curve $\theta = \theta_{max}(1 - \varepsilon^{-x})$ can at once be plotted.

When $x = 3{\cdot}0$, the value of e^{-x} becomes 0·04979, so that, for the temperature rise to be within 5 per cent of its final value, it is necessary to run a machine for a time equal to 3 times its temperature time constant. When $x = 4{\cdot}0$, the temperature rise is within 2 per cent of its ultimate value.

As indicated when dealing with the rise of current in inductive circuits, the exponential curve given in Fig. 196 can be utilized with various scales for its ordinates. Thus, by using scale A shown against the left ordinate, the curve gives the complete temperature curve starting from cold. By using scale B, to the right, the curve gives the portion of the temperature curve starting from 95 per cent of maximum temperature rise, or by using scale C, from 99 per cent of the maximum.

The temperature rise when $t = T$ becomes

$$\theta = \theta_{max}(1 - \varepsilon^{-1}) = 0{\cdot}632\theta_{max}$$

The Temperature Time Constant depends directly upon the thermal capacity and inversely upon the capacity of the machine for dissipating heat. Thus a small highly ventilated machine will have a relatively small value, and a large totally-enclosed machine a large value for the time constant. In a d.c. machine, it is normally the field system which is the last to attain final temperature, and this determines the time constant that must be used; the field winding cannot normally dissipate its heat so effectively as the armature. As examples, may be quoted a highly ventilated 20 h.p. motor running at 1,500 rev/min, a 5 h.p. motor running at 1,500 rev/min and having no fan, and a 60 h.p. totally enclosed motor running at 250 rev/min. The approximate values for the time constant were 20, 80, and 240 minutes, respectively.

The actual temperature rise of a generator or motor differs from that just considered in that the material is by no means homogeneous, the total rate of heat production is not constant but changes as the temperature rises, and the heat is produced locally and not uniformly throughout the whole mass of the machine. This results in a difference in temperature between various parts of the machine and, owing to the different cooling conditions, the time constant has different values for the armature and field windings.

The temperature curve for any particular point on the machine, however, does not deviate in any marked manner from the exponential curve just discussed.

Rating of Electrical Machinery.* In common with other types of electrical machinery, d.c. machines are rated upon a temperature basis. Before the introduction of interpoles, the rated output of a machine was often prescribed from considerations of commutation. The use of interpoles has largely removed the "sparking limit," so that normally a d.c. machine can be rated solely upon this temperature basis, although occasionally other factors arise. It has then become possible to increase very considerably the rated output of machines of given dimensions by the use of fans of comparatively large diameter.

The permissible temperature at which insulating material may be worked without unduly rapid deterioration depends upon the type of insulation. For Class A insulation consisting of cotton, paper and similar organic materials treated with varnish, 105°C is considered the extreme limit of temperature for continuous working. For insulation based on mica, asbestos and woven glass with varnish treatment, the limit is dependent upon the maintenance of the mechanical strength of the binding materials, and may be taken as about 130°C for Class B insulation in which ordinary organic binding materials and varnishes are used, and much higher (180°C) with Class H insulation in which only inorganic materials and silicone varnishes are used. The maximum temperature of the ambient air is assumed to be 40°C for normal purposes, thus it might be expected that a temperature rise of 65°C would be permitted with Class A insulation. The difficulties of accurate measurement of temperature rise, however, necessitate a margin which varies according to the method of measurement. For example, a thermometer on the surface of the insulation of a coil will read something appreciably lower than the temperature of the insulation adjacent to the copper even though its bulb is covered by a small pad of felt, and 55°C rise is accepted as the limit for this method for the armature windings of large machines. For shunt field coils the limit is 55°C when measured by thermometer as above, but 60°C when measured by the increase of the resistance of the windings, the method used being dependent on the size of the machine. Increase of winding resistance gives the mean temperature rise of the copper. This again is not ideal since what is really important is the temperature rise of the hottest part of the insulation.

For machines up to 50 h.p. per 1,000 rev/min the permissible temperature rise at full load is somewhat less (40°C) since these

* The various specifications issued by the British Standards Institution include B.S. 2613 which covers industrial machines and main generating equipment, B.S. 170 which covers fractional horse-power machines, B.S. 173 which covers motors used for electric traction, B.S. 638 which covers arc-welding plant, and B.S. 741 which covers mining equipment.

machines are expected to be capable of 25 per cent overload for two-hourly periods without sustaining any damage.

For motors working on an intermittent load, "short-time ratings" are used. Such machines may be rated on a one-hour or half-hour basis, depending on the duty cycle. For such motors the temperature rise at the end of the specified period must not exceed the given values.

If insulation is operated at higher temperatures its working life is reduced but may still be reasonable under certain conditions. For special purposes therefore different temperature rises may be permitted.

Pre-determination of Temperature Rise. Cooling conditions differ for various parts of the same winding and depend very greatly upon details of construction. It is, then, not to be expected that precise formulae of universal application shall be available for pre-determining temperature rise. In practice, calculations are based upon direct experiments giving the actual power that can be wasted for normal temperature rise, at various angular speeds. This applies in particular to standard lines of motors, where the introduction of fans of large diameter results in formulae previously accepted being generally inadmissible, and the actual temperature rise is so largely dependent upon both the fan details and the details of the ventilating system generally.

For large machines, fans are not generally favoured and "natural" ventilation is still employed. In these again the power that can be dissipated from a given armature or field winding is dependent upon the speed of rotation, and can be determined experimentally for various values of the speed. With such machines having well ventilated cores and all built on the same standard lines, the power that can be dissipated may be expressed as a function of the cooling surface and the peripheral speed v_A of the armature. In general it is found sufficiently exact to assume that the permissible loss, for standard temperature rise, may be expressed in the form—

$$\text{Permissible loss} = (\text{Cooling surface}) \times (A + Bv_A{}^{\gamma})$$

where A, B and γ are constants.

It must be realized that over a wide range of speed this simple law will not be followed, and formulae of this type are dangerous to apply to abnormal cases.

If applied over a relatively limited range of surface speed, sufficiently close agreement to actual test results can be obtained by taking the index, γ, to be unity, that is, the actual loss curve over the limited range of speed can be replaced by a mean straight line, which will at no point deviate by more than a small percentage from the actual curve. It then becomes possible, with sufficient accuracy, to express the permissible loss in the form

$$\text{Permissible loss} = (\text{Cooling surface}) \times (A + Bv_A) \quad . \ (187)$$

This method, due to Kapp, must be used with discretion, and not applied indiscriminately over a wide range of speed. Further complications are introduced owing to the fact that all portions of the cooling surface are not equally effective.

Temperature Rise of Armature. For precise calculations, it is necessary here to take into consideration the number and width of the ventilating ducts, the external and internal surfaces of the armature core, and the surfaces of the free portions of the winding. Provided machines of normal proportions are employed, accurate enough results can be obtained by basing the calculations upon the external or barrel surface only of the armature winding. Under these conditions the following formula has been found by the author to give fairly good results—

$$\text{Temp. rise in °C} = \frac{250 \text{ to } 300 \times p'}{1 + 0{\cdot}09 v_A^{1\cdot3}} \quad . \quad . \quad (188)$$

The index of v_A, it will be noted, is 1·3, a value recommended by Hawkins p' in the dissipation in W/cm².

The cooling surface is given by multiplying the overall axial length of the complete winding by the external circumference of the armature. The power wasted includes the armature copper loss, the no-load iron losses, and the "extra iron losses," that is, the total estimated armature losses under full load conditions. (Some formulae given do not include these stray losses.) As a rough figure, with a well-ventilated armature of large diameter, 4 ft upwards, the total loss for normal temperature rise and for each square inch of the armature barrel surface may be taken to be about 5 watts at a surface speed of 20 m/sec, i.e. 4,000 ft/min. Naturally, the constant will depend partly upon the ratio of the core length to the overall length.

As an alternative method of estimating the temperature rise, Kapp's formula may be used for speeds, v_A, ranging from 15 to 25 m/sec in the form

$$\text{Temp. rise in °C} = \frac{150 \text{ to } 200 \times p'}{1 + 0{\cdot}1 v_A} \quad . \quad . \quad . \quad (189)$$

the constants given being those suggested by Parker Smith.

Temperature Rise of Field Windings. For these windings, the manner in which the permissible loss varies with the surface speed of the armature is dependent very largely upon the type of field coil employed. With a sectionalized or ventilated coil it is possible to employ a formula of the type already given for the armature. When much of the cooling is effected by conduction through the poles, the index γ of v_A^γ in equation will have a different value from that for the armature.

For general purposes the temperature rise may again be expressed in the form

$$\text{Temp. rise} = p'/(A + Bv_A^\gamma)$$

For normal self-ventilated generators the effect of changes in peripheral speed are much less marked upon the cooling of the field windings than in the case of the armature. This will not apply in fan-ventilated machines, where again the only safe plan is to base calculations upon test results of the actual line of machines.

Since field coils vary appreciably in form, it is well to base calculations upon the total surface of the coil instead of the actual external barrel surface. For a normal field coil—not of the ventilated type—it may then be taken that for the temperature rise by thermometer not to exceed 40°C it is necessary to provide from 16 to 12 cm² of cooling surface for each watt lost, the low value holding for high-speed machines. The actual method of insulating the field coils, however, profoundly affects the temperature rise that obtains in practice. For coils having a winding depth of not more than 4 or 5 cm on large machines, the temperature rise agrees fairly well with the formula—

Temp. rise in °C

$$= 1{,}400 \text{ to } 1{,}600 \times p'/(1 + 0{\cdot}07v_A) \quad . \quad . \quad . \quad (190)$$

and the necessary cooling surface for 40°C rise in temperature, by thermometer,

$$\text{Cooling surface} = \frac{35 \text{ to } 40}{1 + 0{\cdot}07\, v_A} \text{ cm}^2/\text{W} \quad . \quad . \quad (191)$$

These figures will not hold for very low or very high values of the surface speed.

Temperature Rise of Commutator. The temperature rise of the commutator is very largely influenced by the fan action of the risers connecting the bars to the armature coils. With long commutators there is a very appreciable temperature gradient along the bars from the outer edge of the commutator to the risers, and again empirical formula will only hold for machines in which the details of construction are similar. A guide as to the probable temperature rise of a commutator in which no especial steps are taken to promote cooling but for which the risers exercise a good fan action is given by the formula—

$$\text{Temp. rise in °C} = 120 \times p'/(1 + 0{\cdot}1\, v_C) \quad . \quad . \quad (192)$$

These constants are in accordance with those given by Kapp, and are based upon the cylindrical surface of the commutator, neglecting the risers, and upon the surface speed of the commutator in *metres per second*. For long commutators, the maximum temperature will exceed the value thus calculated. The losses include both brush contact and friction.

For adequately ventilating long commutators projections are sometimes left on certain of the commutator segments so as to act as fan blades.

Ventilation of d.c. Machines. With normal machines, the heat developed in the active materials is dissipated partly by radiation and partly by conduction, but the major portion is removed by means of a regular supply of cool air circulating through the machine. With large machines the natural ventilation produced by the fan action of the spider is relied on to ventilate the machine. In special cases more definite ventilating arrangements are used. With small machines, that is, for motors up to 100 or 200 h.p., the rated

FIG. 197. ARMATURE OF MOTOR SHOWING THE VENTILATING FAN
(English Electric Co., Ltd.)

output was considerably improved by the adoption of definite ventilation systems, the volume of air passing per second through the machine being much increased, compared with earlier designs, by fitting a fan of large diameter on the shaft, as in Fig. 197.

The use of such fans naturally increases the no-load loss, but permits of much heavier current loadings on the armature than otherwise would be possible, in addition to improving the cooling conditions for the field windings.

The fan is made of a diameter as large as is convenient for the end bracket, and is fitted on the end remote from the commutator. In most cases a single fan only is used, but in some the fan is divided into two distinct parts, one part serving to draw air under the commutator through axial ventilation ducts in the core, and the other drawing air over the commutator, over the armature surface, and between the field coils.

Other examples of fan ventilated machines are given in Figs. 271, 275 and 276.

Sometimes pipe-ventilated motors are used; in particular such motors are necessary in dusty atmospheres. In these the motor is totally enclosed except for an air inlet pipe and an air outlet pipe, these pipes leading to a purer atmosphere.

With such pipe-ventilated machines the full-load temperature rise is approximately 10°C higher than with the same frame size using normal "protected-type" of construction.

Totally-enclosed Machines. Totally-enclosed machines lose their heat almost entirely by simple radiation. Normally the natural fan action of the armature is sufficient to circulate the internal air

FIG. 198. TOTALLY-ENCLOSED GEARED MOTOR UNIT
(*Lancashire Dynamo and Crypto, Ltd.*)

effectively and equalize the temperature of the various parts. In such, nothing whatever is to be gained by fitting a fan, in fact the fan losses are likely to be detrimental. To improve the radiating capacity of the machine, ribs are sometimes provided in the end brackets to increase the surface. The external surface required for each watt lost is unfortunately very large, and for a frame of normal construction is approximately 15 to 16 cm^2 for each watt lost and for 50°C temperature rise of the windings as measured by thermometer.

The heat that can be dissipated from a given frame is therefore only a small fraction of the amount that can be removed with efficient ventilation. It becomes necessary, therefore, to use relatively large and costly motors for totally-enclosed machines. For a given frame it may be taken that at a given speed the output possible for 50°C rise, totally-enclosed, is only from 25 per cent to 60 per cent of that possible for a protected type machine of the same dimensions and 40°C rating. The smaller limit will obtain in high speed machines, and the larger limit in small, very low speed machines.

Since the temperature rise of a totally-enclosed machine is determined mainly by the total electrical losses, it is evident that with such machines the design should be such as to give the minimum possible losses. This involves the use of entirely different loading constants from those obtaining with protected type machines, and further the constants should be chosen to fit each particular case. With such machines low-loss core plates should be used, such as Stalloy.

Fig. 198 shows a totally-enclosed motor, with ribbed end covers, mounted on the same bedplate with reduction gearing.

To increase the heat dissipation from totally-enclosed machines, fan cooling of the external surface is much employed in "totally-enclosed fan-cooled machines," a fan being fitted to the shaft outside the end-bracket. Illustrations of such totally-enclosed fan-cooled motors and of a pipe-ventilated machine are in Chapter XV.

CHAPTER XII

TESTING

The more important performance tests upon d.c. machines include—

1. The magnetization or open circuit test;
2. The load characteristic;
3. The determination of the efficiency curve;
4. The temperature test.

In addition, when a machine is first put on test at the constructors', it must be given an insulation test, and tests must be made to ensure that commutation is satisfactory. The method of carrying out the open circuit and load characteristic tests has already been discussed in Chapters VII and VIII. To measure directly the efficiency of a generator it is necessary to measure the power input and the power output. The former involves driving the generator through a transmission dynamometer, the drive taking place through a spiral spring. The angle through which the spring is twisted is measured by suitable means and permits of the value of the torque being calculated. The method is but rarely used, and then mainly for laboratory work. The power output of the generator can be measured with any desired degree of accuracy by determining the terminal voltage and the output current.

For a motor, the power input can readily be determined. If the machine is loaded on a brake, as discussed later, the torque developed can be determined, and, the speed being also measured, the power output deduced.

For laboratory testing, or in special cases, it is possible to couple the machine under test to a second special machine, normally d.c., in which the yoke or stator (i.e. the outer part which does not normally rotate) is supported in bearings so that a limited rotation is possible. In accordance with Newton's third law, the torque on the armature of this special machine is balanced by an equal and opposite reaction torque on its yoke which can be measured directly by weights or weighing-machine. Since the torque on the armature of the machine under test is equal to that on the armature of the second machine, its torque can be measured directly except for a small bearing friction torque. The special machine is called a "dynamometer" and may be operated as a generator or a motor according to whether the machine to be tested is a motor or generator. In some cases the power generated can be returned to the supply.

Since in a brake test the whole of the power output is wasted, and a dynamometer test requires a special machine of appropriate rating and speed, the direct method of determining the efficiency is therefore little used except for motors of small output. For larger machines, intended for operation either as motors or generators, the efficiency is determined by the "indirect method," the losses being determined separately, as discussed below.

For the temperature test, with large machines or with machines having very poor ventilation such as totally-enclosed motors, the temperature time constant is fairly large and it may be necessary to run the machine under full load conditions for six hours. For small, well-ventilated machines the final temperature is attained much more quickly and a six hours' test is not necessary. Since it is undesirable to waste all the energy output, dead load methods are used in practice, even for comparatively small machines, only if it is inconvenient to arrange for recuperating some of the energy.

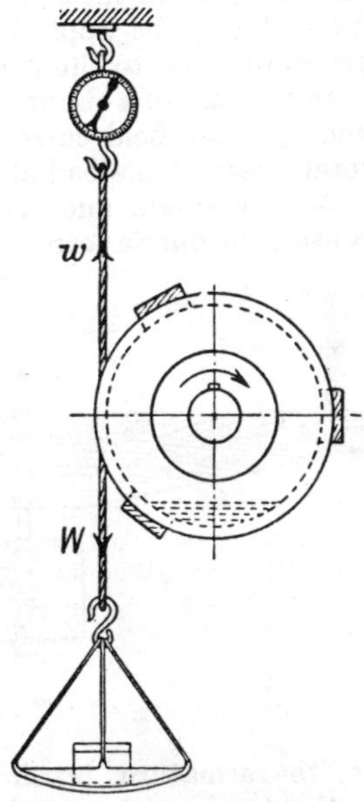

FIG. 199. ROPE BRAKE

Brake Tests. When it is required to determine directly the efficiency of comparatively small motors, the motor is loaded directly by means of a mechanical brake, or by means of an eddy current brake, or a calibrated air fan.

A common type of mechanical brake employed in testing is the rope brake indicated in Fig. 199. The load is regulated by the weight carried by the scale pan, the actual load torque being given by

$$T = (W - w)R^{*} \text{ lb ft} \quad . \quad . \quad . (193)$$

where R is the effective radius in feet, i.e. $\frac{1}{2}$(diam. of pulley + diam. of rope). Where the motor output exceeds about 2 h.p., or where the test is of long duration, it is necessary to use a water-cooled pulley.

The power developed is given by

$$P = \frac{2\pi Tn}{550} \text{ h.p.} \quad . \quad . \quad . \quad . (194)$$

Another form of brake is the Pröny brake, illustrated in Fig. 200.

For temperature tests it is more usual to employ the motor on a useful load by belting it up to a generator and adjusting the voltage of the latter so that it can operate in parallel with the supply system.

Swinburne's Method of Computing Efficiency of Shunt Machine. In this method the iron and friction losses are determined by

* W and w are expressed in lb wt.

measuring the input to the machine on no-load, the machine being run as a motor at normal voltage and speed. The copper losses are calculated from measured values of the various resistances. The method may be applied to the case of shunt or level-compound generators and to shunt motors.

In the case of a shunt motor, let I_o = the total current on no-load and I_s = the field current, V being the supply voltage. Then the total losses on no-load are VI_o.

A very small, and normally negligible, portion of these losses consists of ohmic losses in the main series circuit of the machine,

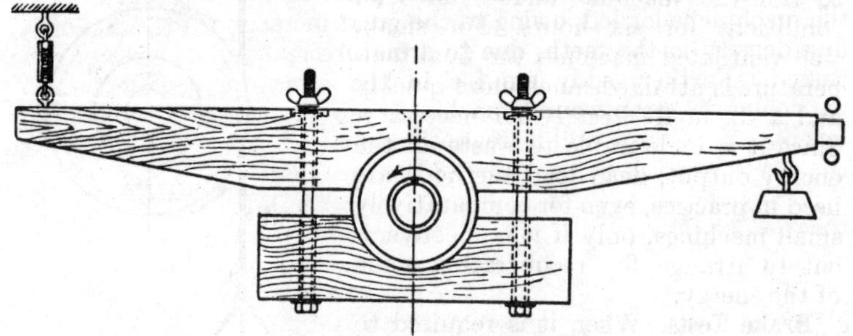

FIG. 200. PRÖNY BRAKE

i.e. the armature, brush contacts and interpole windings. If r_m = this total resistance of the main motor path, and

$$I_{A_o} = I_o - I_s = \text{the armature current}$$

the total iron + friction losses,

$$P_o = VI_{A_o} - I_{A_o}^2 r_m \quad . \quad . \quad . \quad (195)$$

the total "fixed" or "constant" losses,

$$P_F = VI_s + P_o = VI_o - I_{A_o}^2 r_m \quad . \quad . \quad (196)$$

and for an input current I, the armature current $I_A = I - I_s$ and the losses in the main circuit, $P_c = I_A^2 r_m$.

Hence the efficiency,

$$\eta = 1 - \frac{\text{Losses}}{\text{Input}}$$

$$= 1 - \frac{P_F + P_c}{VI} \quad . \quad . \quad . \quad . \quad (197)$$

this value holding for an output,

$$P = VI - (P_F + P_c) \quad . \quad . \quad . \quad (198)$$

The motor resistance is determined by measuring the volt-drop produced in the main series circuit when normal current is passed through at standstill. The exciting current I_e should be the normal value, and the losses P_c should be based upon the value of r_m when final steady temperature is obtained, say on a value corresponding to a temperature rise of 60°C by resistance. To a close approximation the resistance to be taken should be 125 per cent of that measured "cold."

The Swinburne method is useful in so far as it gives a figure of merit for the machine, but the value deduced by the method invariably exceeds the actual value realized on load. This is mainly due to the large increase that takes place in the iron losses when the machine is loaded, owing to the great increase in the maximum flux density in the teeth, due to armature reaction as already discussed. Further, eddy e.m.f.'s in the armature conductors, in particular in strap-wound machines, prevent the current density from being uniform, and increase the armature copper losses beyond the value based upon the measured value of r_m.

Against this, the method as here given includes all the losses in the field circuit—shunt regulator as well as winding, so that losses are included which do not actually occur in the machine. This very slightly counteracts the error due to the omission of the extra iron losses. Further, to be precise, owing to the variation of the brush contact resistance with varying current, the resistance r_m is not strictly constant.

The errors introduced due to these latter two inaccuracies are quite insignificant compared with that due to the assumption that the iron losses retain their no-load value at all loads.

As an alternative to the above the iron and friction losses of a d.c. machine can be measured, as shown later (page 281), by means of an auxiliary driving motor.

The British Standard 269; 1927, *Methods of Declaring Efficiency of Electrical Machinery* is fundamentally based on the above, but a specified allowance is made for stray losses; the losses in field regulator resistances *are included* as part of the total losses; and *total* temperatures of 75°C are assumed for all windings.

Differential Test on Two Similar Shunt Machines. In the original Hopkinson test two similar shunt machines were mechanically coupled and connected electrically in opposition, as indicated in Fig. 201, and driven by belt. With the field currents adjusted so that equal e.m.f.'s were induced in the two armatures, no current whatever circulated between the armatures, but weakening the field of the one machine destroyed the balance of e.m.f. and current circulated in the main armature circuit, just as current circulates between two batteries of differing e.m.f. when connected in parallel. The current circulating was varied by the simple expedient of progressively weakening the field of one machine. The machine with its field so weakened functioned as a motor and the other as a generator. The losses in the two machines were, in the original

Hopkinson test, supplied mechanically via the belt, and their magnitude determined by a Hefner-Alteneck dynamometer.*

In practice a modification of the original Hopkinson test is employed in which the losses are supplied electrically. The modification originated from Kapp, but the test is usually called the Hopkinson test.

Fig. 202 gives the essential connexions for the Hopkinson test—as modified by Kapp—for two similar d.c. shunt machines. To

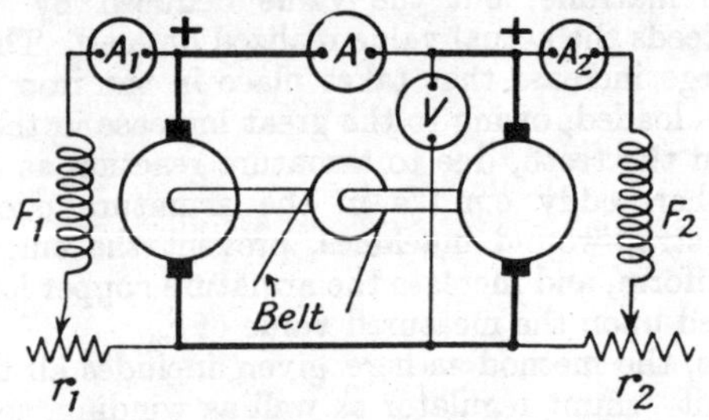

FIG. 201. HOPKINSON TEST ON TWO SHUNT MACHINES
Losses supplied mechanically

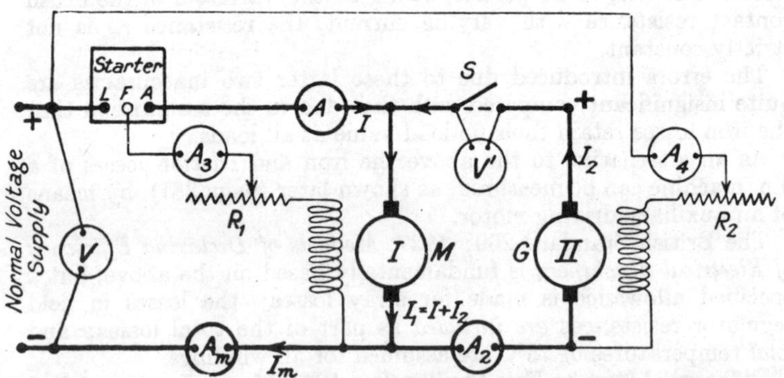

FIG. 202. CONNEXIONS FOR HOPKINSON TEST ON A PAIR OF SHUNT MACHINES

perform this test one of the machines is run up to speed in the normal manner as a shunt motor; the other machine, mechanically coupled to the motor, is then excited to such a degree that it generates an e.m.f. exactly equal to the voltage of the supply feeding the motor. If the generator were loaded on a dead load, such as a water resistance, the set would operate as a plain motor generator, but a considerable waste of energy would result. Instead, the generator is first put in parallel with the supply system, the excitation being

* Cf. *Original Papers on Dynamo Machinery and Allied Subjects*, by John Hopkinson.

adjusted until V' reads zero, and the switch S then closed. Under the conditions stated, the generator will then "float," neither taking nor giving current to the supply. Any desired load can then be put on the generator by increasing the value of its induced e.m.f.; this, in turn, can be accomplished either by increasing the generator exciting current or by decreasing the exciting current of the motor, or both adjustments may be made. The current generated can thus be utilized to help drive the machine operating as a motor, and the total current taken from the supply system is therefore simply that necessary to supply all the losses, i.e. the difference between the motor input and the generator output.

In actual practice it is not essential to start up the set in the manner indicated, but the two machines may be connected in parallel at starting and run up to speed through a single starter. In this case the machines will not start up unless they both develop torque in the same direction. Failure to start is rectified by reversing the field current in one of the machines. To prevent overheating and possible burning out of the no-volt release of the motor starter it is desirable to arrange that the current for one of the field windings is not carried by the starter. Finally, to avoid uneven distribution of the current during running up it is necessary to arrange that the exciting current has much the same value for the two machines.

By adjusting the two field rheostats R_1 and R_2, it can be arranged that each machine acts as a motor and takes exactly the same armature current, and the set runs at the correct speed when supplied at normal working voltage. Under these conditions the reading of the ammeter A will be double that of A_2. To put load on the machines the field current of either machine can be weakened. As a result, this particular machine will speed up the set slightly, and thus will act as a motor and the other machine as a generator. To maintain normal working speed it is necessary that the load shall be obtained by both weakening the field of one machine and strengthening that of the other. With machines of fairly large output, the total difference in the exciting currents necessary for the machine acting as a generator to be fully loaded, is comparatively small, as the difference in the values of the induced e.m.f.'s of the generating and motoring machines will only be about 5 per cent. With comparatively small machines having a relatively large ohmic drop in their armature circuits, the difference in the exciting currents under full load conditions is more important.

With these connexions, data for the complete efficiency curve from no-load up to, say 50 per cent overload, can be obtained in a very short time, the whole control being effected very simply by means of the field rheostats, and the total power taken from the supply system is very small, being simply the losses for the two machines—say 100 kW to 120 kW when testing two 1,000 kW sets at full load. With these connexions, also, the full-load temperature test can be carried out with the minimum waste of energy. To determine the efficiency, readings should be taken of—

The total current I supplied from the mains to the two armatures, as read by A.
The generator armature current I_2 as read by A_2.
The motor shunt current i_1 as read by A_3.
The generator field current i_2 as read by A_4.

During the test the supply voltage should be kept at its normal value, and as stated the test should be carried out at normal speed. At the end of the test, when the windings have reached their final steady temperature, the resistance r_1 and r_2 of the armature circuits —armature windings + brush contacts + interpole windings— should be measured by passing full load current through them at standstill and measuring the voltage drop with the fields not excited. The total copper losses can then be calculated for any particular set of readings. The difference between these copper losses and the total input from the supply system, will then represent the iron and friction losses for the two machines; and one half this amount may, with quite reasonable accuracy, be allotted to each machine. The actual figure thus obtained for the sum of the iron and friction losses invariably and considerably exceeds the corresponding value obtained on no-load. This increase is in the main due to the increased iron losses consequent upon the distortion of the field form.

The efficiency is then deduced—

Motor armature circuit copper losses $= I_1^2 r_1$

Generator do. do. do. $= I_2^2 r_2$

Total iron + friction losses for the two machines $= VI - (I_2^2 r_2 + I_1^2 r_1)$

Total iron + friction losses per machine $P_o = \frac{1}{2}\{VI - (I_2^2 r_2 + I_1^2 r_1)\}$ (199)

Motor input power $P_{I1} = VI_1 + Vi_1$ . . (200)

Motor losses $P_{L1} = P_o + I_1^2 r_1 + Vi_1$. (201)

Motor output power $P_{O1} = P_{I1} - P_{L1}$ . . (202)

Motor efficiency $\eta_M = 1 - \dfrac{P_{L1}}{P_{I1}}$ . . . (203)

Generator output* power $P_{O2} = VI_2$

Generator losses $P_{L2} = P_o + I_2^2 r_2 + Vi_2$

Generator input power $P_{I2} = P_{O2} + P_{L2}$

Generator efficiency $\eta_G = 1 - \dfrac{P_{L2}}{P_{I2}}$ . . . (204)

* If the machine is providing its own field current the output is $V(I_2 - i_2)$.

As determined by this method, the efficiency will differ but little from the actual value, with all except machines of comparatively small output. At first the difference between the values of the useful flux in the two machines might seem to indicate that the iron losses would have widely different values. It must, however, be remembered that the machine with the weak flux acts as the motor and its armature current is greater than in the other machine. As a result, armature reaction will be more pronounced in the motor

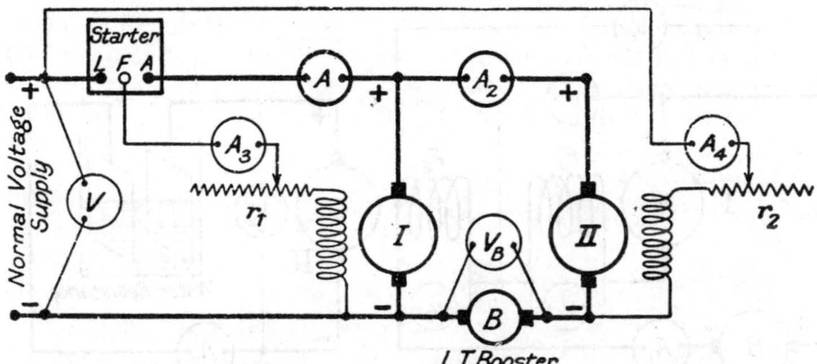

FIG. 203. CONNEXIONS FOR DIFFERENTIAL TEST ON A PAIR OF SHUNT MACHINES

Circulating current controlled by l.t. booster

than in the generator, and with the high degree of magnetic saturation normally reached in the teeth, the maximum flux density in the teeth will not differ appreciably for the two machines.

There are a number of approximate formulae which may be used for determining an efficiency figure, but the above method is in the author's experience more dependable. It allocates to each machine as accurately as is possible its correct quota of the total losses, and obtains a precise figure for the efficiency of each machine and for the output at which this efficiency figure obtains.

Another modification of the Hopkinson test is one in which the need for working the two machines at different field strengths is avoided by including a low tension booster in the armature circuits of the two machines as in Fig. 203. With this arrangement, the field strengths of the two machines are first adjusted so that with the booster unexcited each machine takes the same current, i.e. its normal no-load motor current. From this point the field regulators of the two machines on test are not touched, the value of the circulating current being regulated entirely by altering the excitation of the low tension booster—which is separately excited.

To a close approximation, the low tension booster will provide all the copper losses in the main series or circulating circuit, and the main supply system will provide the iron friction and excitation losses, although, to be precise, a certain small portion of the armature copper losses is supplied from the normal voltage system. For exact calculations the efficiency may be deduced by a method precisely similar to that already given. The total armature losses now are given by the sum of the input, VI, from the main supply, and $v_B I_2$, the booster output.

Methods of Testing a Pair of Similar Series Motors. D.c. traction motors being invariably arranged for series-parallel control, such series machines are usually available in pairs.* A simple motor

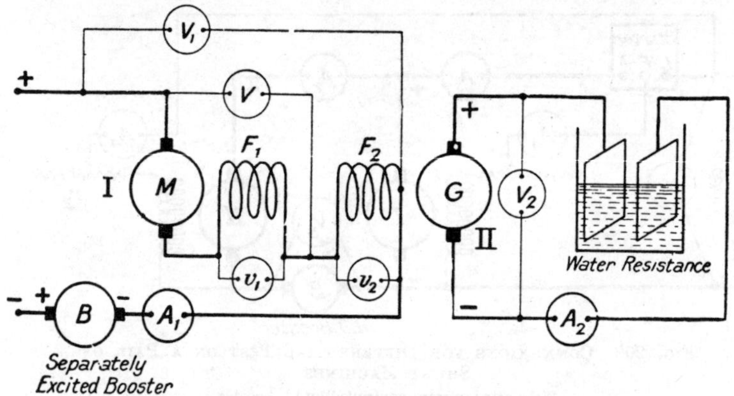

FIG. 204. CONNEXIONS FOR FIELD'S EFFICIENCY TEST ON A PAIR OF SERIES MOTORS

generator test cannot be carried out over a wide range of output unless the machine used as a generator is separately excited, since the series machine will not build up appreciably until the load resistance is reduced to a value not much above that corresponding to full load. Further, with series motors it is desirable to take steps to safeguard against the load being too greatly reduced, or dangerous speeds will result.

M. B. Field's arrangement, as shown in Fig. 204, overcomes the difficulty of obtaining readings at comparatively light loads by utilizing the motor current to excite the generator, the generator armature being connected directly to the load resistance. This latter resistance should be of such a nature that it is impossible to break its circuit completely, so that the whole load on the motor cannot be removed inadvertently. To obtain normal voltage between motor terminals, the total supply voltage must exceed this latter value by an amount corresponding to the resistance drop in the generator field winding; the total voltage required increases as the load is increased. If the test is carried out from a constant

* There are no constant losses so that a load test involving two similar machines is preferable to a summation of losses method. This is in agreement with B.S. 173.

voltage supply system, it is then necessary to include a low voltage booster in the supply circuit, as indicated in the actual diagram.

Since the two machines have the same exciting current it may be taken to a close approximation that the iron losses have the same value for the two machines.

If V and V_2 denote respectively the voltage for the motoring and generating machines as indicated, and I_1 and I_2 the corresponding currents, then for an armature resistance r_a and a field winding resistance r_f,

Total iron + friction losses for the two machines,

$$2P_o = VI_1 - V_2I_2 - I_1^2(r_a + r_f) - I_2^2r_a \quad . \quad (205)$$

and the iron + friction loss for each machine,

$$P_o = \tfrac{1}{2}\{VI_1 - V_2I_2 - I_1^2(r_a + r_f) - I_2^2r_a\} \quad (206)$$

The motor input $P_{I1} = VI_1$

„ losses $P_{L1} = I_1^2(r_a + r_f) + P_o$

„ output $P_{O1} = P_{I1} - P_{L1}$ . . . . . (207)

and „ efficiency $= 1 - \dfrac{P_{L1}}{P_{I1}}$ . . . . . (208)

A relatively small inaccuracy in the measurement of the motor input or generator output may result in a relatively large error in their difference, especially in the case of large efficient motors, such as those for railway work. It is therefore desirable to measure directly this difference and so obtain a more exact determination of the losses. Further, in Field's test, the output of the generator is entirely wasted on a dead load. Traction motors are given a one-hour rating, so that the waste of energy in a full load temperature test is not serious in the case of comparatively small motors. For railway motors of maybe 500 to 600 h.p. the waste of energy is undesirable, and loading back tests—on the lines of the Hopkinson test—are used. Various arrangements are possible, the two given below being used in practice and simple to carry out.

Parallel Hopkinson Test. In this test the main connexions are the same as for Field's test. Instead of wasting the output of the generator, however, a low-tension booster is used to add to the generators such voltage as will cause the combination to feed back into the motor. In Field's test the voltage between the generator brushes is less than the line voltage by an amount corresponding very closely to the sum of the ohmic drops in all the windings. Before the generator can be made to feed back into the line it is therefore necessary to make up this voltage by connecting the booster in circuit as shown in Fig. 205. (The connexion of the booster is thus similar to that used in Fig. 203 for shunt machines.) The booster is usually motor driven and separately excited. By altering its excitation, the circulating current and, hence, the load on the machines under test, can be adjusted at will. Since a series motor runs at a dangerous speed on very light loads, either the generator, etc., must be parallelled with the motor, at a very reduced voltage, or an auxiliary load must be provided for the generator as indicated in Fig. 205. In this the set is started up as in Field's test with the generator load circuit completed; the parallelling switch S is open. The low tension booster B is then excited to such a value that there

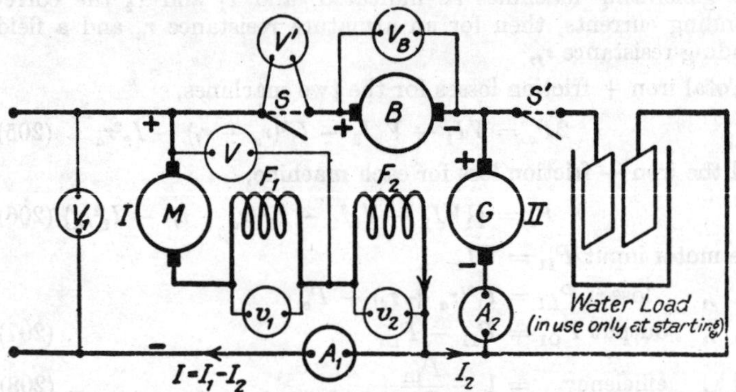

FIG. 205. CONNEXIONS FOR DIFFERENTIAL TEST ON A PAIR OF SERIES MOTORS

is no difference of potential across the parallelling switch, the volt meter V' being connected across the switch. (This involves the sum of the terminal voltages of booster and generator being equal and in opposition to the line voltage.) The parallelling switch is then closed. By increasing the resistance of the dead load the generator, aided by the booster, will begin to feed back into the local circuit between the two machines. The dead load can then be removed by opening the switch S' and the circulating current and, hence, the load on the two machines is then controlled entirely by regulating the excitation of the booster.

To maintain constant voltage at the motor terminals, the total supply voltage, as in Field's test, should increase as the load current increases, to allow for the volt-drop in the generator field winding. A small booster is used where the necessary slight control of the main voltage is not possible. This booster is connected as in Fig. 204, but is not shown in Fig. 205.

To a rough approximation, the copper losses are mainly supplied by the booster B, and the iron and friction losses by the main supply. The current taken from the main supply is thus relatively small, and in carrying out the test it is usually found that the value of this current varies but little until high speeds are reached, and the frictional losses become considerable.

The efficiency is deduced from the observed figures in a manner similar to that already discussed with other forms of Hopkinson tests, and with Field's test. The total losses are obtained by summing the input from the main supply system and that from the booster. Thus, if V_1 = total supply voltage

$$\text{Total losses} = V_1 I + v_B I_2$$

Hence the iron and friction losses per machine,

$$P_o = \tfrac{1}{2}\{(V_1 I + v_B I_2) - (I + I_2)^2(r_a + 2r_f) - I_2^2 r_a\} \quad . \ (209)$$

When the supply voltage can be completely controlled, the artificial load can be entirely dispensed with. The parallelling switch is omitted, and the circuit through the load booster and armature permanently completed. To start up, a small voltage is applied to the main circuit and the load booster is unexcited. The current will divide, part passing through one armature, and part through the other. The machine which is to function as generator will carry rather more current than the other, and where the motors are direct coupled and not geared on to the same countershaft the set may rotate very slowly. On exciting the booster with the correct polarity, the set will start up, the current in the "generator" will reverse, and the current taken from the main supply fall off in value. With this method of starting up, not only is the need for the artificial load done away with but, in addition, the absence of the parallelling switch does away with the possibility of this switch being inadvertently opened during the test, and of the load being entirely taken off the motor.

As an alternative method of starting, and avoiding the need for the artificial load, the booster may first be excited so as to circulate about 25 per cent overload current. Again, with correct connexions for the test, the set should not rotate. The main circuit may then be completed through the starting resistance and the set will start up, the current through the booster falling off at once. The booster excitation can then be increased and the starting resistance cut out step by step, so that the motor can be run up on load.

Series Hopkinson Test. Fig. 206 gives a diagram of connexions for this test. Compared with Field's test, and the parallel Hopkinson test just discussed, the method differs in that now the two armatures carry the same current and the field currents have different values, whereas in the other tests the field currents have the same value and the armature currents are not equal. In the present test the main circuit comprises simply one complete motor and the armature of the other machine. The field of this second machine is separately excited, and in such a direction that, when the set is running, this second machine generates an e.m.f. acting in the same direction as the main supply, whereas the other machine, functioning as a motor, generates e.m.f. opposing the supply. As a result, the voltage applied to the motor terminals is the sum of the supply voltage and the voltage between terminals of the second machine, which operates as a generator. The actual voltage necessary from the supply system is thus much less than normal working voltage—for large railway motors amounting only to about 10 to 15 per cent of normal voltage, under full load conditions and less at lighter loads.

The load current and motor voltage are controlled by varying the excitation of the generator, and the supply voltage respectively.

The total iron and friction losses are at once obtained by subtracting from the main input the copper losses in the main circuit,

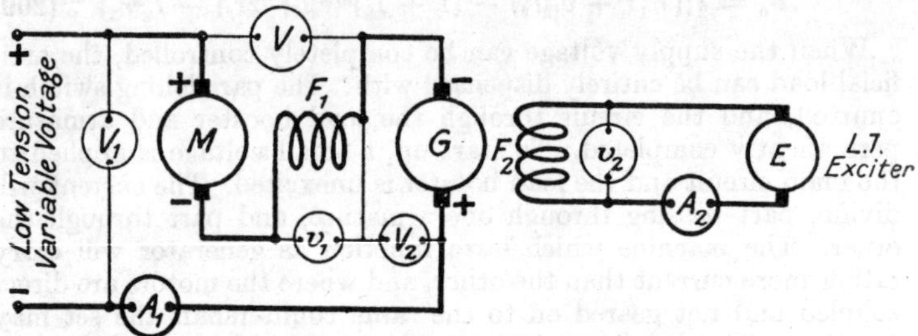

FIG. 206. DIAGRAM OF CONNEXIONS FOR SERIES HOPKINSON TEST ON A PAIR OF D.C. SERIES MOTORS

so that the iron and friction losses for each machine may, approximately, be taken to be

$$P_o = \tfrac{1}{2}\{V_1I_1 - I_1^2(2r_a + r_f)\} \quad . \quad . \quad . \quad (210)$$

For this test the motor and generator armatures carry the same current, but the motor excitation is greater than that of the generator. The iron losses will therefore be somewhat greater in the motor than in the generator.

Compared with the parallel connexions for the loading back test, the series method has the great advantage that the possibility of attaining dangerous speeds is remote—if the excitation of the generator fails, the voltage supplied to the motor at once falls to a value corresponding to quite a low speed. On the other hand, it is somewhat difficult to maintain the voltage applied to the motor at its exact value, and with careless handling the voltage applied to the motor may easily rise to a value much in excess of normal.

The set is started up with the generator field not excited, and with a much reduced voltage applied to the main circuit—about 10 per cent of normal. The machines will then rotate slowly. The generator field is then excited; if the direction of excitation is correct the set will speed up, and the voltage across the motor terminals rise. If, on the other hand, the direction of excitation is incorrect for the test, the set will slow down slightly, and the voltage across the motor terminals fall, since both machines will then act as motors.

Temperature Tests. The normal duration of the temperature test on large machines is six hours. Where two similar machines are available a differential test will invariably be employed.

Failing this, steps will be taken to employ usefully the output of the machine wherever possible. Thus, for the temperature test on a motor, the motor will be belted or direct coupled to one of the test-house generators, and the output of this generator usefully employed. Normally, the generator would be put in parallel with the supply system feeding the motor, so that similar arrangements to those with the Hopkinson test will prevail.

On the other hand, to carry out a temperature test on a single generator, the generator would be motor driven and put in parallel

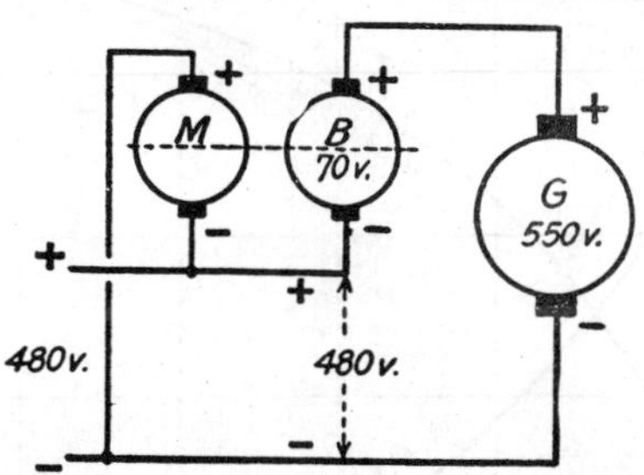

FIG. 207. LOADING A 550-V DYNAMO ON TO A 480-V SYSTEM

with the works supply system. If, as it will be generally, the generator is arranged for a different voltage from that of the works supply, an auxiliary booster machine is connected in the circuit so as to add or subtract the necessary amount from the generator voltage, as indicated in Fig. 207, which shows the arrangement for a generator voltage of 550 and a supply voltage of 480. The booster is motor driven, and the output of the generator is adjusted by adjusting the booster excitation. The excitation of the generator is made to correspond to the required terminal voltage. When, as illustrated, the generator voltage is greater than that of the supply system, the "booster" itself functions as a motor, the other machine working then as a generator.

Short-time Temperature Tests. To avoid the waste of energy and time necessary for continuing a temperature test until the temperature rise approximates to the ultimate steady value, short-time tests are sometimes made and the ultimate temperature rise predicted from readings taken during the short-time test.

A simple method of computing the ultimate temperature rise is as follows.

In the temperature rise equation $\theta_t = \theta_{max}(1 - \varepsilon^{-t/T})$,* by replacing $\varepsilon^{-\frac{1}{T}}$ by k, the equation takes the much simpler form

$$\theta_t = \theta_{max}(1 - k^t) \quad . \quad . \quad . \quad . \quad (211)$$

Thus, when $t = 1 \ldots \theta_1 = \theta_{max}(1 - k)$

$t = 2 \ldots \theta_2 = \theta_{max}(1 - k^2)$ and so on.

* *See* (185) page 253.

Hence $\theta_2/\theta_1 = 1 + k$

and $k = (\theta_2 - \theta_1)/\theta_1$ . . . (212)

and $\theta_{max} = \theta_1/(1 - k)$ . . . (213)

For simplicity the above relates to when the initial temperature

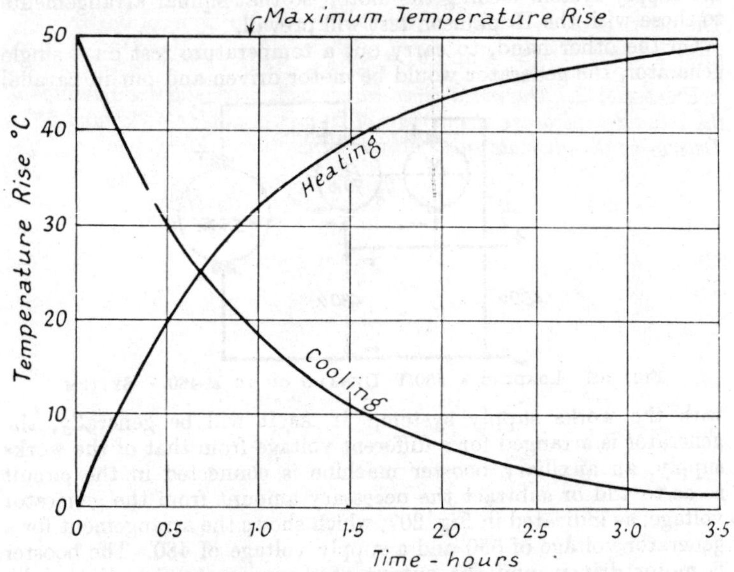

FIG. 208. HEATING AND COOLING CURVES

"rise" is zero. But it follows at once, cf. Fig. 208, that the ultimate temperature rise above the initial value θ_o is always given by

$$(\theta_1 - \theta_o)/(1 - k)$$

and the ultimate temperature rise above the ambient temperature is

$$\theta_{max} = \theta_o + (\theta_1 - \theta_o)/(1 - k) = (\theta_1 - k\theta_o)/(1 - k) \quad . (214)$$

Similarly, it follows that k is always given by the ratio of the actual increases in temperature during two successive equal intervals of time, i.e.

$$k = (\theta_2 - \theta_1)/(\theta_1 - \theta_o) = (\theta_3 - \theta_2)/(\theta_2 - \theta_1), \text{ etc} \quad . (215)$$

If then any three readings on the temperature rise curve at equal time intervals be known—t being taken in any convenient unit, e.g. 15 min, or 30 min—the ultimate temperature rise can be calculated most simply, as in the following examples—

EXAMPLE 1. *The temperature rise of a motor, with constant losses, is 20°C after 30 min and 32°C after 60 min. Determine the ultimate temperature rise, the equation of the temperature rise curve, and the temperature rise after 2 hours.*

Taking unit time as 30 min—

$$k = (32 - 20)/20 = 0{\cdot}6$$

Ultimate rise $\theta_{max} = \theta_1/(1 - k) = 20/(1 - 0{\cdot}6) = 50°C$

The curve is $\theta_t = 50(1 - 0{\cdot}6^t)$ where t is in units of 0·5 hr.

after 2 hours, $t = 4$ and $\theta_4 = 50(1 - 0{\cdot}6^4) = 43{\cdot}5°C$.

EXAMPLE 2. *During a temperature test, 3 consecutive readings of the actual temperature at intervals of 20 min are 36°C, 48°C, and 54°C. Determine the ultimate value of the temperature.*

$$\theta_2 - \theta_1 = 6;\ \theta_1 - \theta_o = 12$$

hence $k = 6/12 = 0{\cdot}5$

and (cf. Fig. 208) $\theta_{max} = \theta_o + (\theta_1 - \theta_o)/(1 - 0{\cdot}5) = 36 + 12/0{\cdot}5$

The ultimate temperature is then $36 + 24 = 60°C$.

Cooling Curve. The cooling curve is shown in Fig. 208 together with the heating curve, the one being the mirror image of the other, displaced by the amount θ_{max}.

The equation for the cooling curve can readily be deduced exactly on the same lines as for the kindred problem on page 112 and is

$$\theta_t = \theta_{max}\varepsilon^{-\frac{t}{T}} = \theta_{max}k^t \quad . \quad . \quad . \ (216)$$

The constants will have the same values as for the heating curve provided that cooling conditions are identical for both.

Graphical Methods. It is a convenience to have a quick graphical method of predicting the ultimate temperature rise from readings taken on a short-time test. The following method avoids the need for any calculation.

In Fig. 209 A and B are points on the temperature rise curve for θ_1 and θ_2 at $t = 1$, and 2 respectively, t being any convenient time interval.

Constructed as shown by dotted lines, AB and NC meet at Q such that $QL = AN/(1 - k) = \theta_{max}$,

For $QL/AN = QN/CN = AN/(AN - BC) = 1/(1 - BC/AN)$
$= 1/(1 - k)$.

This very simple construction can also be applied to any three consecutive readings at equal time intervals, as shown in Fig. 209,* and need not be based on the initial reading.

Representation of the Temperature Rise Curve by a Straight-line Graph. To do this we can either—

(*a*) use a uniformly-divided time scale, with a suitably expanded temperature scale.

* Drawn to illustrate Example 1 (page 276).

or (*b*) use a uniformly-divided temperature scale in conjunction with a suitably contracted time scale.

(*a*) UNIFORM TIME SCALE. From the equation $\theta_t = \theta_{max}k^t$ for the cooling curve it is apparent that ordinary logarithmic division is required for the temperature scale. A convenient method of plotting is then as in Fig. 210 showing θ/θ_{max} in terms of *t*, the logarithmically-divided scale being used "inverted."

The one linear graph serves for both heating and cooling with the scales arranged as in the diagram.

Although multiple scales can be used in the manner indicated in

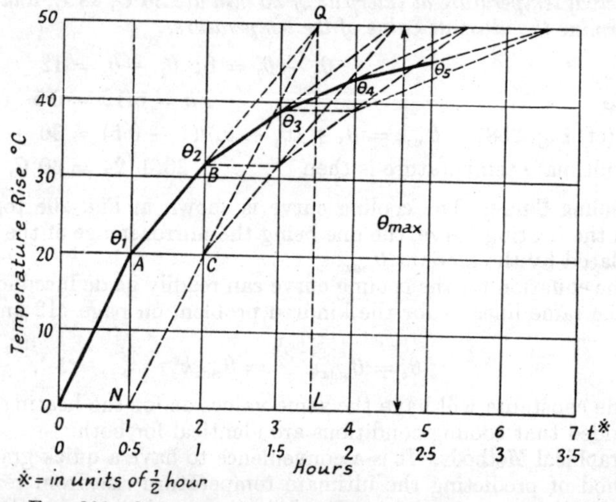

FIG. 209. GRAPHICAL CONSTRUCTION OF HEATING CURVE

Fig. 210, the diagrams based on uniform temperature scale are perhaps of more value.

(*b*) UNIFORM TEMPERATURE SCALE. In Fig. 211, graph (*i*) is a linear graph *OA* for the complete temperature-rise curve from zero to θ_{max}.

The necessary degree of contraction of the time scale follows at once from the previous section. Thus

$$N_1N_2/ON_1 = N_2N_3/N_1N_2, \text{ etc.} = k$$

and, in terms of the total length *OB* of the time scale

$$ON_1 = (1 - k)OB$$

Or the sub-division of *OB* can be made by the simple graphical method indicated by dotted lines.* The units of time in which *t* is

* There are other applications of the above constructions which can be applied to most problems involving the exponential growth and decay curves.

measured can be chosen to suit the particular case—10 min, 15 min, 20 min, etc.—and for reasonable accuracy should be such that θ_1/θ_{max} does not exceed 0·25, i.e. $k = 0{\cdot}75$. If this is done no error

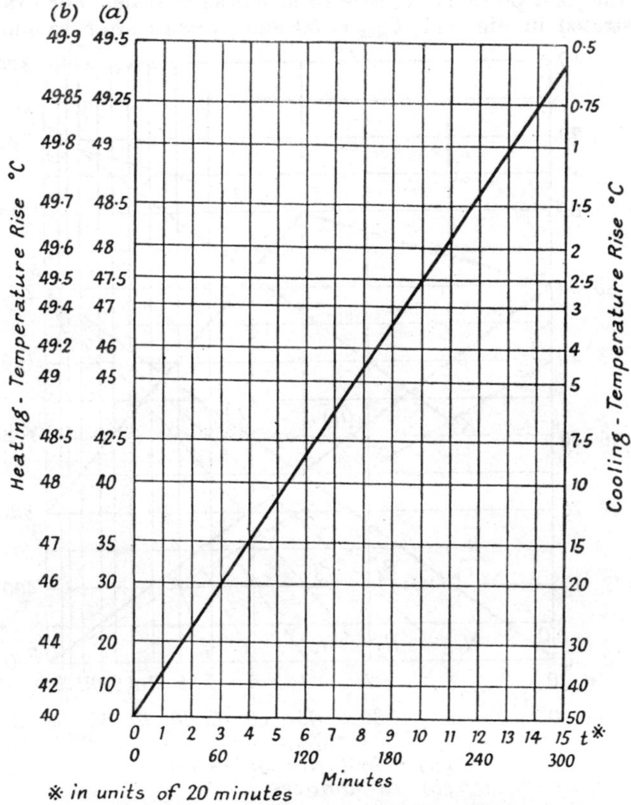

FIG. 210. UNIFORM TIME-SCALE STRAIGHT-LINE HEATING AND COOLING CURVE

of any practical importance will result by assuming any necessary subdivision to be uniform over each unit time section.

The disadvantage inherent in the contracting time-scale is completely outweighed by the fact that *the same set of ordinates can be used for every exponential temperature rise or fall problem, based on the same value of k.*

Further, by choosing as unit time, as in Fig. 211, the time taken to attain 20 per cent of the ultimate value, i.e. $k = 0{\cdot}8$, the same division of the time scale can be used for all cases. (For extreme accuracy the sub-division could be based on $k = 0{\cdot}9$, unit time now

being that necessary to attain one-tenth of the ultimate value.) *CB*, **or graph (ii), gives the normal cooling graph. Graphs (i) and (ii) not only hold for the complete heating and cooling curves, but also for the *final* portions of these to any desired scale. For example, illustrated in Fig. 211, $\theta_{max} = 50$ and $\theta_1 = 10$ after 15 min., i.e.**

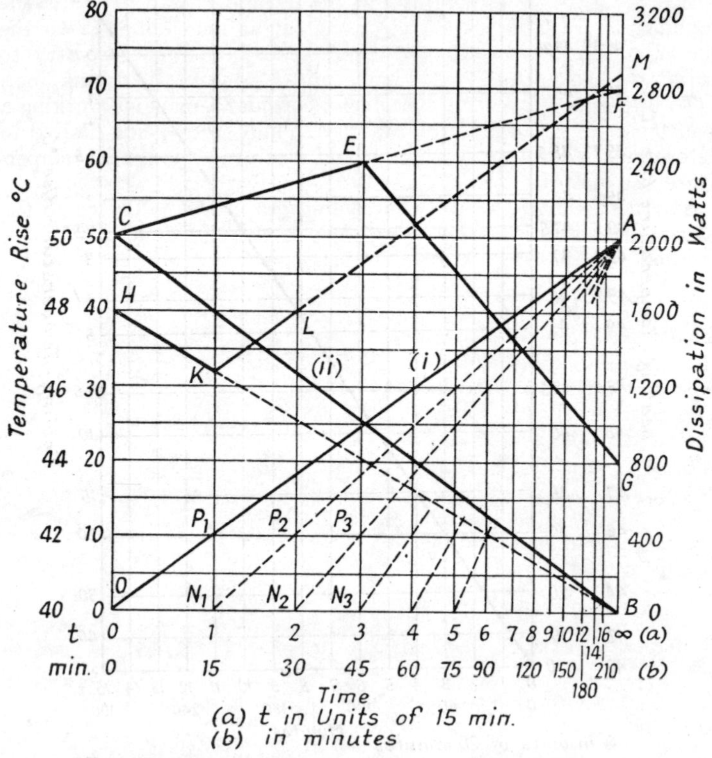

FIG. 211. UNIFORM TEMPERATURE-SCALE STRAIGHT-LINE HEATING AND COOLING CURVE

$k = 0{\cdot}8$. Graph (i) gives also the final portion of the heating curve from 40° to 50° using the supplementary scale shown. For cooling, (ii) can be used for any portion, ending at 0°.

Application to Load-factor and Kindred Problems. For a given machine θ_{max} is proportional to the total power losses, and it is a convenience to indicate the loss corresponding to various values of θ_{max} on the appropriate scale in Fig. 211, where the example corresponds to a loss of 2,000 W for $\theta_{max} = 50$°C. Not only does this permit of the complete heating and cooling curves being speedily drawn for various load conditions, but it is especially helpful when load conditions change.

For example, if the machine under review is already at 50°C when a steady overload is applied such that the losses increase from 2,000 W to 2,800 W its temperature curve will be given by the line CF, F corresponding to 2,800 W loss. After 45 min θ has reached 60°C, at the point E on CF, and if then there follows a period of very light load such that the total losses fall to 800 W the cooling graph EG will be simply a straight line joining E to the 800 W point G. (This graph could be expanded if necessary to obtain greater accuracy by transferring E back to the datum line.)

To illustrate a load factor problem, consider a machine working a load factor of 15 min on, followed by 15 min off, it being desired to determine the power loss which would result in a maximum temperature rise of 40°C.

Starting from H, at 40°, the cooling graph HK is drawn to the 15 min ordinate. The heating graph is then obviously KL. By producing KL to M the corresponding power loss is given as 2,880 W. This method can of course be used only when the cooling conditions are identical during heating and cooling parts of the load cycle. Should these conditions be different, separate heating and cooling charts can readily be made or a single chart could be used with separate time-scales for heating and cooling with little difficulty. The chart can of course be used for much more complicated load cycles than the simple one discussed.

Separation of the Iron and Friction Losses. One method of determining the value of the friction losses and of the iron losses in a d.c. motor or generator is to couple the machine directly to an auxiliary driving motor. The power put into the armature of this auxiliary motor is then measured, both when running light—uncoupled from the machine under test—and when driving this latter machine unexcited, and with various values of the exciting current. Preferably, the machine should be separately excited. If the load on the auxiliary motor is always relatively light, the total losses in its armature and in friction and windage will be practically constant. The copper losses in the series circuit of the motor can, as a precaution, be calculated and allowed for, if necessary, as shown in Fig. 212.

Generally, this test can be carried out simultaneously with the magnetization test. The total input to the armature of the auxiliary motor may be plotted in terms of the exciting current of the machine under test, the test being carried out at constant speed; the open circuit e.m.f. of the machine may be plotted on the same graph. The total losses in the armature circuit of the auxiliary motor can readily be calculated and these, added to the measured constant armature and friction losses of this machine, represent the total that must be subtracted from the motor input to give the losses in the machine under test. Thus, in Fig. 212, curve A represents the input current to the auxiliary motor armature and also the total power input. Curve B represents the total losses in the armature, etc., of the auxiliary motor. These losses are normally nearly constant. The friction losses in the machine under test are therefore represented by *PQ*, and the total losses at any given excitation *OF* are represented by *HK*. Graph *C* in Fig. 212 is drawn to represent the total losses, and by subtracting the friction losses from the total, the iron losses are obtained, as indicated in the figure.

By performing this test at various values of the speed, a number of "armature core loss + friction" curves can be obtained. (It should

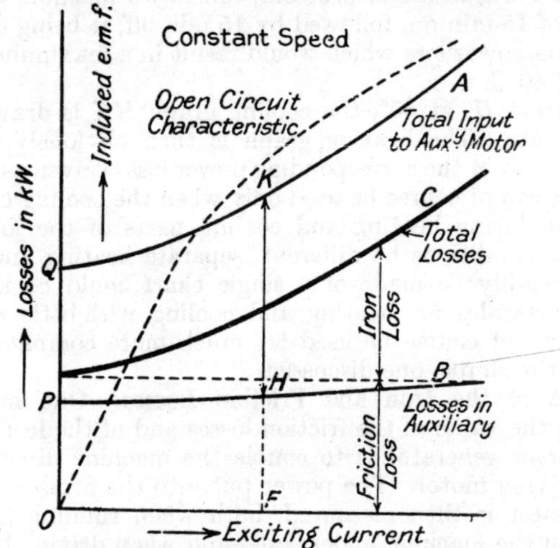

FIG. 212. DETERMINATION OF IRON AND FRICTION LOSSES BY MEANS OF AN AUXILIARY DRIVING MOTOR

be mentioned that to obtain satisfactory readings it is essential that the brush gear and commutator shall be in excellent condition, otherwise steady readings of the current will not occur.) These curves are of value to the designer, as from them he is able to separate the hysteresis from the eddy losses, in addition to the information they provide as to the total iron loss and the total friction loss under all working conditions. The method of separating these losses is given later.

Where it is not practicable to employ an auxiliary driving motor, the losses may be separated from measurements made with the machine operating on light load under various conditions of excitation and applied voltage.

The total "core loss + friction loss" curve for various exciting currents and constant speed is obtained by varying the voltage applied to the motor and adjusting the exciting current so as to maintain constant speed. Since the light load current of fairly small motors is not very large, the voltage applied to the armature may very conveniently be varied by means of a potentiometer-type regulator.

For large machines it is desirable to feed the armature and field circuits from separate supply systems. Failing this, the armature voltage may be regulated by means of a series resistance.

If I_A is the armature current, and V_A the voltage applied to the armature, the ohmic losses in the armature circuit $= I_A{}^2 r_m$, so that the total friction + iron loss $P_o = V_A I_A - I_A{}^2 r_m$. This total loss

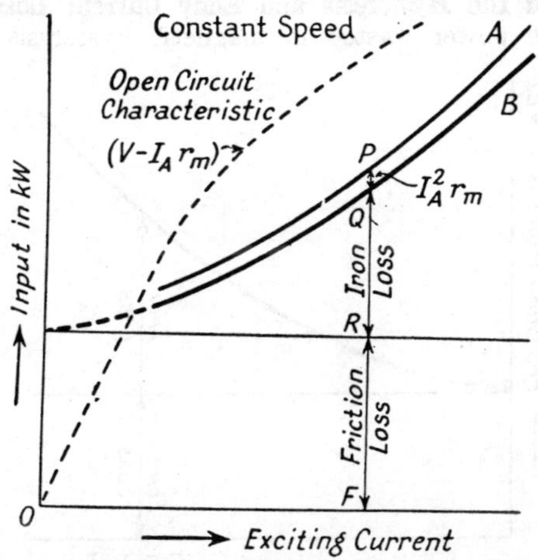

FIG. 213. SEPARATION OUT OF IRON AND FRICTION LOSSES FOR A SHUNT MOTOR

can be plotted in terms of the exciting current for a given speed, as in Fig. 213. From the same readings the magnetization curve can be deduced since the induced e.m.f. $E = V - I_A r_m$. Naturally, readings, at constant speed, cannot be continued right down to zero excitation, so that the final portion of the graph, shown by the dotted line, must be extrapolated—preferably by assuming the graph to be parabolic. Since at zero excitation the iron losses must be practically zero, the intercept of the graph on the vertical axis gives the value of the friction loss, so that the total iron loss at any particular excitation—and voltage—for the particular speed can at once be obtained as indicated in Fig. 213. In this diagram, curve A represents the total input to the armature; from the ordinates of A an amount corresponding to $I^2{}_A r_m$ is subtracted to give B, representing the total "iron + friction" loss. The open circuit magnetization curve is obtained by plotting the values of $E = V - I_A r_m$.

In performing this test, as the applied voltage is reduced the armature current—necessary to maintain constant speed with the reduced field strength—at first falls a little. Ultimately, however, the current will actually increase, and become quite important when low values are used for the applied volts. The friction and windage torque being constant, the current necessitated by these losses is inversely proportional to the back e.m.f., and hence the flux.

For analytical purposes it is desirable to plot the value of the iron losses as a function of the flux, or of E/n which is proportional to Φ.

Separation of the Hysteresis and Eddy Current Losses. For a given flux, the power wasted in magnetic hysteresis is directly

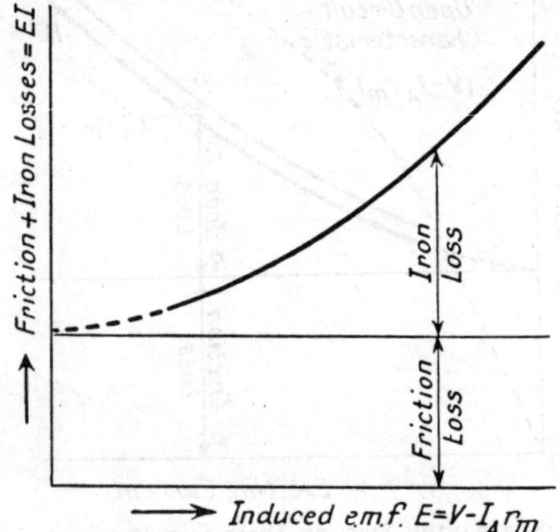

FIG. 214. SEPARATION OF IRON AND FRICTION LOSSES IN SHUNT MOTOR

proportional to the frequency of the flux reversals, and hence to the speed. The power wasted in eddy currents, on the other hand, increases in proportion to the square of the frequency and, hence, also of the speed. If, then, the total iron loss be known for a given flux at two different values of the speed, it is a matter of simple algebra to separate the hysteresis from the eddy current loss.

A simple method of separating out the losses, devised by the author, is to determine the total iron loss curve at normal speed by the method just described, and to repeat the test at either one half or twice normal speed, whichever is the more convenient. The results are then plotted in terms of E/n as indicated in Fig. 215. Taking the case when the second test is at half normal speed—

If P_H and P_E represent respectively the hysteresis and eddy current losses at normal speed for any given flux—

Core loss at normal speed $\qquad P_{i1} = P_H + P_E$

" $\quad$ " $\quad \frac{1}{2}$ speed, for same flux, $P_{i2} = 0{\cdot}5P_H + 0{\cdot}25P_E$

Hence $P_{i1} - 2P_{i2} = 0{\cdot}5P_E$, and $\quad P_E = 2(P_{i1} - 2P_{i2})$.

Thus if, as in Fig. 215, the total iron loss curve be represented by A, and the curve B represents twice the total loss at half speed, PL being the total iron loss at normal speed and PN twice the loss at half speed for the same flux, then LN is one half of the eddy current loss at full speed. Hence, on constructing a third graph C by subtracting from each ordinate of B an amount $NM = LN$, the

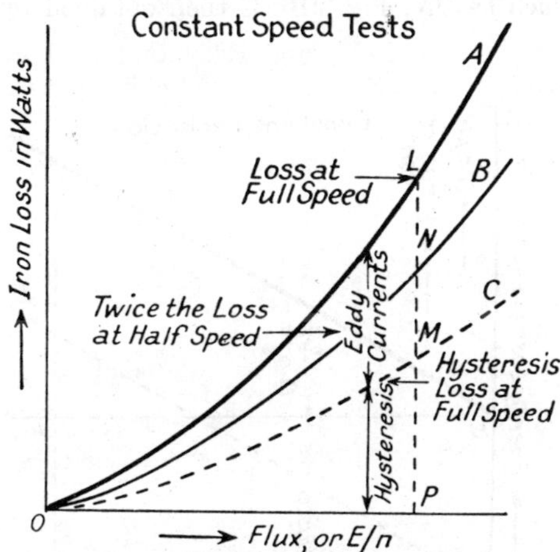

FIG. 215. SEPARATION OF HYSTERESIS AND EDDY-CURRENT LOSSES

ordinates PM of this third graph will give the hysteresis loss at full speed, and the value of ML will give the eddy current loss.

Having thus separated out the losses for one particular speed over the whole range of working flux, the losses at any other speed are at once obtained by taking the hysteresis loss to be proportional to the speed, and the eddy loss proportional to the square of the speed.

If the second test be carried out at twice normal speed the method of separating out the losses is equally simple.

Torque Wasted in Iron Losses. The power wasted in hysteresis varies in direct proportion to the frequency, whereas that wasted in eddy currents varies in proportion to the square of the frequency. For a given maximum value of the flux density, then, the torque necessary to overcome hysteresis is constant at all speeds, whereas that to overcome eddy currents increases in direct proportion to the speed. Put in another manner, the energy wasted in magnetic hysteresis in one revolution of the armature is constant, whereas that wasted in eddy currents increases directly with the speed. If then a graph be plotted, as in Fig. 216, showing the manner in which the energy wasted in the armature iron during one complete revolution varies with the speed, a straight line will be obtained as shown by *HT*, Fig. 216. The intercept *OH* made by this line on the torque axis gives the constant amount of energy wasted in magnetic hysteresis during each revolution, and since this energy is equal to the product of the torque wasted and the constant 2π, the intercept *OH* also represents, to scale, the torque wasted in hysteresis.

The energy wasted per revolution in eddy currents at any particular speed, such as *ON*, Fig. 216, is then obtained by subtracting

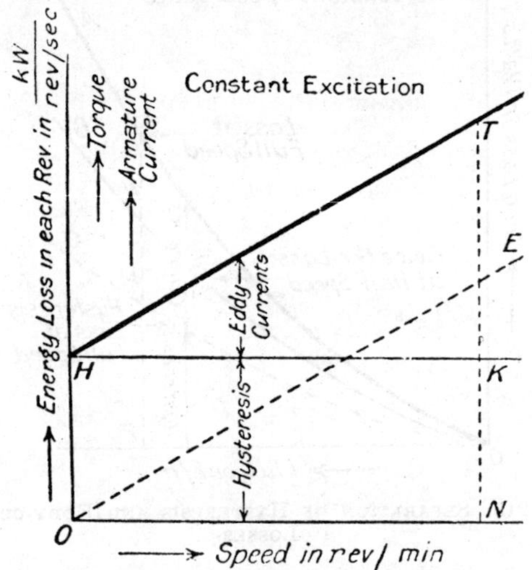

FIG. 216. SEPARATION OF HYSTERESIS AND EDDY-CURRENT LOSSES

from the total loss *NT* the amount *NK* wasted in hysteresis. Alternatively, by drawing through the origin *O* a line *OE* parallel to the total loss line, the ordinates of *OE* will give a waste of energy per revolution in eddy currents. It then becomes possible to separate the hysteresis from the eddy current losses experimentally by determining the value of the iron loss at various speeds for constant excitation, and plotting results as indicated in Fig. 216. This method is very commonly used, a family of graphs being deduced, each holding for one particular value of the working flux.

Retardation Test. If a motor be brought up to speed and then switched off, it will slow down as represented by the running-down curve (Fig. 217).

The angular retardation is at any instant directly proportional to the retarding torque at that instant, and also varies inversely as the moment of inertia of the rotating parts.

If, then, at a speed represented by n the power wasted is P, the retarding torque T is given by

$$2\pi Tn = P, \text{ whence } T = P/2\pi n$$

Hence the angular retardation $\frac{d^2\theta}{dt^2}$ is

$$\frac{d^2\theta}{dt^2} = \frac{T}{J} = \frac{P}{2\pi Jn} \quad . \quad . \quad . \quad (217)$$

where J is the moment of inertia.

The power wasted is thus

$$P = kn\frac{d^2\theta}{dt^2} \quad . \quad . \quad . \quad . \quad (218)$$

where k is the constant $2\pi J$. The angular retardation $\frac{d^2\theta}{dt^2}$ is directly proportional to the slope of the retardation curve. Hence, at a

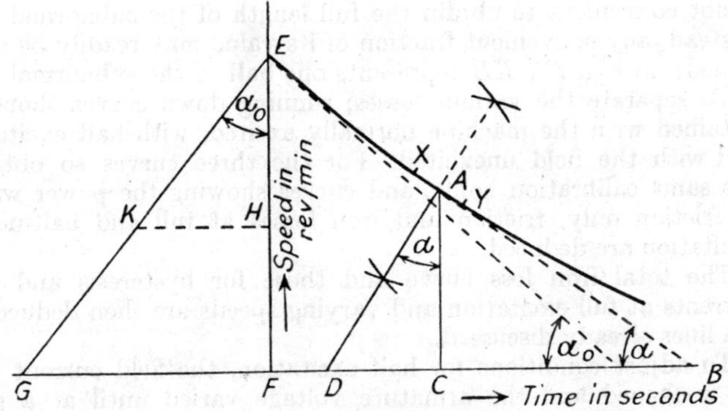

FIG. 217. RETARDATION TEST

speed represented by the point A on the retardation curve, the retarding torque will be proportional to the slope at A, i.e. to $\tan ABC$, and the retarding power will be proportional to

$$n \times \tan ABC, \text{ i.e. to } AC \times \tan \alpha,$$

then if AD be drawn perpendicular to AB, as shown,

$$\frac{DC}{AC} = \tan DAC = \tan \alpha$$

$$DC = AC \tan \alpha \propto Tn \propto P$$

Hence the subnormal DC is directly proportional to the power loss at the point A. It is then necessary simply to determine the value of the subnormal DC for various values of the speed, i.e. for various points on the curve, and a graph showing the manner in which the losses vary with the speed is at once obtained.

The determination of the actual loss at any given speed serves to calibrate the curve.

Thus, to determine the iron and friction losses for a shunt motor, the motor would be run up to normal speed with normal excitation, and careful readings of the input to the armature taken. The total iron and friction loss is obtained by subtracting the armature copper loss from this input. The retardation curve with full excitation is then obtained by opening only the armature circuit and taking readings of the speed at regular time intervals. The subtangent FG at the starting point E then corresponds to the observed iron and friction losses, so that the power loss for a subtangent of unit length is obtained.

To draw the normals to the curve the construction indicated may be adopted: the points X, Y being taken as close together as is practicable, and the normal obtained as in the usual method for bisecting the chord XY. Owing to limitations of space it usually is not convenient to obtain the full length of the subnormal GF.* Instead, any convenient fraction of its value may readily be determined; in Fig. 217 KH represents one half of the subnormal at E.

To separate the various losses, running down curves should be obtained with the machine normally excited, with half excitation, and with the field unexcited. For the three curves so obtained the same calibration holds, and curves showing the power wasted in friction only, friction and iron losses at full and half-normal excitation are deduced.

The total iron loss curve and those for hysteresis and eddy currents at full excitation and varying speeds are then deduced on the lines already discussed.

To adjust conditions for half excitation, the field current must be weakened and the armature voltage varied until at a given speed the motor generates exactly one half of the back e.m.f. obtaining at the same speed with full field.

Commutation Tests. Before a machine, in particular a large machine, is put into service it is necessary to test that the strength of the interpole windings is correct. Owing to the contact drop at the brushes, apparently sparkless commutation can be obtained if the interpole windings are a little too weak or a little too strong.

Correct commutating conditions obtain when the dynamic e.m.f. due to the motion of the coil through the interpole field just neutralizes the e.m.f. of self-induction due to the changing current in the coil. The difference between these two is now called—perhaps not very happily—the *differential reactance voltage*, and it is this voltage which determines the tendency to sparking. The limiting permissible value of the differential reactance voltage is determined by the nature of the brush and is practically independent of the peripheral speed of the armature. On the other hand, the commutating e.m.f.

* In any case the normal at E cannot directly be constructed accurately, and its value should be found by extrapolation, values of GF being plotted as a function of the speed AC.

due to the interpole flux—cf. eqn. (140), page 157—is directly proportional to the speed. As a result, the permissible tolerance in the interpole strength, i.e. in both B_{cp} and the interpole winding m.m.f., above and below the ideal value is inversely proportional to the peripheral speed. For large high-speed machines the margin may be as low as ± 0·5 per cent whereas for low surface speeds it may exceed ± 5 per cent.

One method of obtaining the correct strength of the interpole coils is to run the machine on full load with its interpoles separately

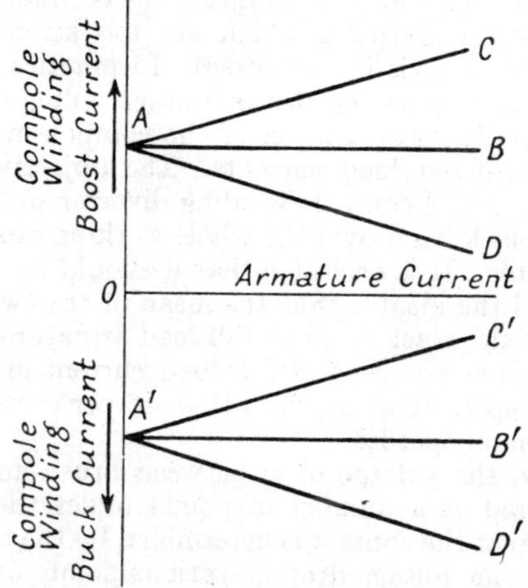

FIG. 218. ELEMENTARY "BLACK-BAND" DIAGRAM

excited from a suitable heavy-current booster. The interpoles are then under-excited to such an extent that sparking is just visible, and then over-excited until sparking again just appears. The correct excitation is taken to be the mean of the two values thus determined.

Completely separate excitation of the interpole windings is undesirable, mainly because the booster has to handle the total current, and it is better to leave unaltered the main circuits of the test machine and to connect the booster armature in parallel with the interpole windings, a control resistor being included in the circuit. The booster then has to carry only the boost or "buck" current. By adjusting the separate excitation of the booster, its e.m.f. can be more or less than the normal p.d. between the terminals of the compole winding thus giving any desired degree of boost or buck to the current and enabling the maximum values of boost and buck currents for dark (i.e. sparkless) commutation to be determined. Frequently the test is made for the complete range of armature current and these limiting values plotted as in Fig. 218, giving the "black band" within which commutation is sparkless. In practice it is often convenient to plot the buck and boost currents as percentages of normal full-load current. The ideal black-band graphs are shown by AB and $A'B'$, i.e. two horizontal straight lines equidistant from the datum line, so that the working condition, represented by the datum line, is midway between the limits and represents perfect commutation. The equality of OA and OA' indicates that the brushes are in the correct neutral position; inequality would indicate that the brushes were not in the neutral position. Incorrect compole strength results in the graphs being inclined so that AC and $A'C'$ indicate weak interpoles and AD and $A'D'$ indicate interpoles which are too strong, the brushes being on the neutral axis in both cases. In practice the graphs are not linear, due partly to the brush contact voltage characteristic, and more especially to saturation of the compole magnetic circuit which occurs with overload currents. The objective is to adjust the brush position and compole winding diverter so that the datum line lies in the black band over the whole working range of armature current if possible. It is desirable that it should be well within the black band, and the ideal is thus the mean of the two graphs.

The width of the black band at full-load armature current varies from as little as one per cent of full load current in machines with high peripheral speed to as much as 10 or 15 per cent in machines of very low peripheral speeds.

Alternatively, the voltage drop between brush and commutator may be measured at a number of points under the brush. Since the voltage drop at the contact is determined by the current density, a knowledge of the voltage drop at various points over the contact arc suffices to give the value of the current density at those points.

To measure the voltage drop, one of the brush blocks may be replaced by a wood or fibre block of the same dimensions, and this block pierced at the centre and close to the two edges to permit a hard pencil lead to be inserted. The p.d. between this pencil lead and the brush box is determined by a potentiometer method, as indicated in Fig. 219, to avoid troubles due to contacts with the auxiliary brushes.

As already discussed in Chapter VI, for straight-line commutation the current density under the brush is constant. Under these conditions, the volt-drop at the brush contact should be constant. For accelerated commutation in which an attempt is made to obtain $di/dt = 0$ at the end of commutation, the current density under the leaving edge of the brush and also the voltage drop is reduced considerably below the average value. For retarded commutation they would be increased beyond the average value, and for pronounced under-commutation may reach very high values.

If, then, the brush drop results be plotted as indicated in Fig. 220—

1. For straight-line commutation the graph will be a horizontal line;

2. For accelerated commutation the graph would droop towards the leaving edge of the brush;

3. For under-commutation the graph would rise towards the leaving edge of the brush;

4. For pronounced over-commutation, the current may actually be reversed under the leaving edge, and the brush drop line would cross the datum line.

As an effective and simple method of adjusting the interpole strength for straight-line commutation, the interpoles are adjusted so that there is no difference of potential between the two tips of the brush.

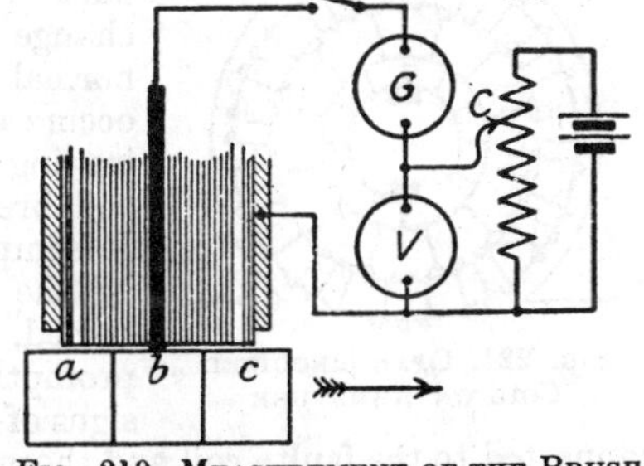

FIG. 219. MEASUREMENT OF THE BRUSH CONTACT DROP

The necessary adjustment is made by altering the length of the air gap under the interpole. If the interpoles as first assembled are too strong, one of the sheet-iron liners is removed at the root of the pole, and if too weak, another is inserted.

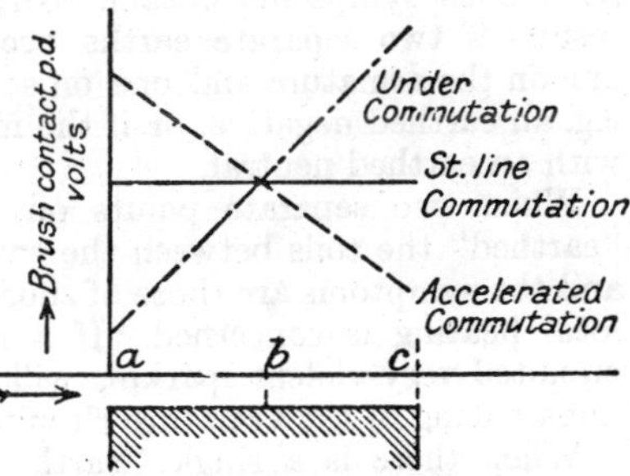

FIG. 220. VARIATION OF BRUSH CONTACT DROP UNDER THE BRUSH

Armature Faults. The main faults that may develop in the armature winding are—

1. A short circuit of a coil or coils;
2. An open circuit in one of the coils;
3. An "earth" on one or more of the coils.

1. SHORT-CIRCUITED COIL. If this fault develops when the machine is running, intense local heating of the faulty coil will occur, quite independent of the load on the machine, owing to the large local current that will circulate in the faulty coil. The local closed circuit is the seat of the alternating e.m.f. normally induced in the coil, so that the current is limited solely by the impedance of the short-circuited coil. The heavy local current very often results in the burn-out of the coil, and the fault then becomes an open circuit. The short circuit may take place between the risers, or it may take place between adjacent turns of a multi-turn coil. The heavy current is purely local, so that it does not overload the brushes.

2. OPEN-CIRCUITED COIL. The symptoms of this fault are most spectacular when the machine is loaded, but are inappreciable on

light load. Fig. 221 indicates a bipolar armature with an open-circuited coil *C*. The armature circuit containing the faulty coil, i.e. that to the left of the brush axes, is idle; the other circuit therefore carries double its normal current. The brushes are not overloaded, however, and trouble does not occur until the faulty coil passes under the influence of the brush, and in so doing is transferred from the one armature circuit to the other, thereby causing the current in the *whole* of the sound coils to be changed from double normal value to zero, or *vice versa*. The reactance voltage associated with this change is therefore many times the normal value, and a most vicious spark occurs at both sets of brushes when the leaving segment connected to the faulty coil breaks contact with the brush It is imperative to shut down promptly or the commutator will rapidly be ruined. If the machine is shut down promptly, the commutator will show signs of serious pitting at the segments connected to the faulty coil and those one-pole-pitch distant from it.

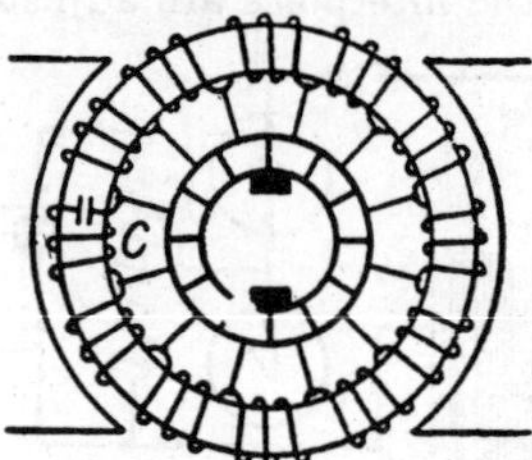

Fig. 221. Open-circuited Coil on Armature

3. Earthed Coil. If only a single earth occurs on the machine no serious symptoms obtain. Serious trouble, however, at once results if two separate earths occur, either on the armature, or one on the armature and one on some other part of the machine—e.g. an earthed negative, or if the machine is a three-wire generator with an earthed neutral.

When two separate points on the armature winding become "earthed" the coils between the two faults become short circuited, and the symptoms are those of short-circuited coils so far as intense local heating is concerned. If a number of coils are thus short circuited very violent sparking will take place owing to the marked unbalancing of the armature circuits.

When there is a single "earth" on the armature and another "earth" on the machine or system it supplies, serious sparking is certain as the machine will periodically be short circuited by the two "earths." Thus a generator may work with an earthed negative as in traction systems, and have one armature coil earthed. When the faulty coil is under the negative brush, conditions are normal; but as the coil moves from that position an increasing portion of the winding is short circuited—in this the negative brush, but not the positive, will carry this heavy current. Ultimately, when the faulty coil is under the positive brush, the whole winding is short circuited. Vicious sparking, particularly at the negative brush, and general blackening of the commutator would therefore occur.

Testing for Faults with Armature Stationary. A simple method of armature testing is the "drop test." This test can be carried out with the armature either in position or removed from its bearings.

If in position, current is passed through the armature, any two convenient brush sets being used and the others lifted. If removed from its bearings, temporary leads can be clipped on two risers or other suitable arrangements made. The exact placing of the temporary "brushes" is of no moment, but most conveniently they may be placed a pole-pitch apart. The current is obtained from any suitable low voltage source. The potential difference between pairs of adjacent segments is then determined by means of a suitable low reading moving-coil voltmeter, as in Fig. 222. If the winding be perfect, the difference of potential will be the same for all pairs of segments between the brushes in the direction *ACB*, and for segments in the direction *ADB*, the p.d. will have a constant value, which value will however only be equal to that for the other segments when the brushes divide the winding into two equal paths in parallel.

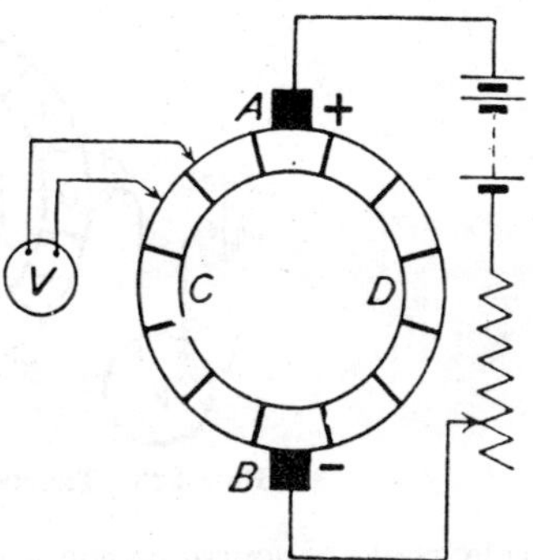

FIG. 222 CONNEXIONS FOR "DROP TEST" ON ARMATURE

A *Short-circuited Coil* is at once indicated by the fact that, owing to its reduced resistance, the voltdrop between the two segments to which it is joined will be zero for a dead short circuit, and much less than normal for a partial short circuit.

An *Open-circuited Coil* prevents the passage of current in its circuit. The voltmeter will then give no reading at all when testing the particular circuit containing the fault, until it bridges across the segments joined to the faulty coil when the reading will correspond to the full voltage applied to the brushes.* For a lap winding, the faulty coil is thus at once located. For a wave winding the fault will be in one of the coils which are connected in series between the adjacent commutator segments, i.e. 2 for a 4-pole machine, 3 for a 6-pole machine, and so on. For brushes placed one pole-pitch apart with a multipolar wave-wound armature, the p.d. between segments will change sign at the points distant by one or more pole-pitches from the + brush. When testing for an open circuit with such a machine, $2p$ pairs of segments will be found having no circuit between them. The faulty coil will in every case be in the circuit between these segments and so can be isolated.

Always, when carrying out the drop test, it is desirable to perform the test twice, moving the commutator round a little between the

* With multipolar lap-wound armatures, having equalizing connexions, the presence of the equalizers will limit the voltage to that obtaining between adjacent equalizers. The readings of the voltmeter will be normal, except for the section of the winding—between two adjacent equalizers—containing the faulty coil.

two tests so that different coils become short circuited by the brushes.

Earthed Coil. The insulation resistance of the winding may first be measured by a "megger." Failing that, a d.c. supply system or battery may have its positive pole connected to the positive terminal of a voltmeter reading full supply voltage, and the negative pole connected to the shaft. (This must not be done on a system without permission. For a system having one side "earthed," the earthed

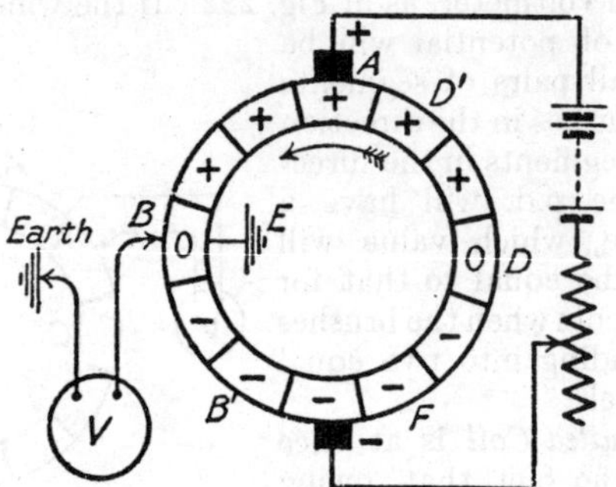

Fig. 223. Testing for "Earthed" Coil

side would of course be connected to the shaft.) The other terminal of the voltmeter is connected to any commutator segment. For a dead "earth" the voltmeter will give full reading of the supply pressure. In general, if the voltmeter reading be v volts, the supply voltage be V, and the resistance of the voltmeter r, the insulation resistance of the winding is

$$R = r \times \left(\frac{V}{v} - 1\right) \quad . \quad . \quad (219)$$

If the winding is faulty, the faulty coil may be located by connecting as indicated in Fig. 223, the voltmeter being joined permanently to "earth," i.e. the frame at one terminal, and the other terminal joined to the wandering contact. When the wandering contact passes the segments connected to the faulty coil its readings reverse. The reading of the voltmeter thus progressively falls, starting from the positive brush A, the fault being at the point where the voltmeter reading reverses. With this test a slight "snag" occurs, as there will be a segment in the other part, ADB, which will have the same potential as the faulty segment. Thus in Fig. 223 there would on testing appear to be two "earths," at B and D respectively. The second of these, D, is called a "phantom earth." The real fault can readily be obtained by repeating the test after moving the commutator round slightly. The faulty segment will always indicate zero voltage, whereas the actual segment, which is a phantom, depends entirely upon the position of the fault relative to the positive brush. Thus the "phantom" for B' corresponding to moving the commutator through two segments is F, a segment displaced by 4 from the first phantom, which is now at D'. When the fault is really bad, current may be led in at one brush and out by means of the shaft. By connecting the voltmeter as before, the real fault is at once obtained; the phantom earth does not then exist.

The reader will no doubt be able to realize in what respect the behaviour of a wave winding would differ from that of a lap winding when tested for "earths." When testing multiplex windings, too, the results obtained in all these drop tests may appear surprising unless the nature of the winding is understood.

CHAPTER XIII

SPECIAL MACHINES

Shunt, or Separately-excited Booster. It is often necessary to supply a motor, or other apparatus, with a voltage differing from that of the main supply system. For example, the main supply system may have a voltage V_1, and the required voltage may be $V_2 = V_1 + v$ the current being I_2.

An obvious method is to employ a motor-generator, with the motor operating off the main supply system at V_1 volts, driving a generator of output $V_2 I_2$.

Then if

$$\eta_1 = \text{efficiency of the generator}$$

$$\eta_2 = \quad ,, \quad ,, \quad \text{motor}$$

$$\text{Motor input} = V_2 I_2/(\eta_1 \eta_2)$$

$$\text{Motor current } I_1 = I_2 \times V_2/V_1 \times 1/\eta_1\eta_2$$

$$\text{and the total losses } P_L = V_1 I_1 - V_2 I_2$$

$$= V_2 I_2 (1/(\eta_1\eta_2) - 1) \quad . \quad . \quad . (220)$$

This method is unnecessarily wasteful; a more economical method is to generate the required additional voltage v, and to add it to the voltage of the supply system, as indicated in Fig. 224. For this purpose, the generator used is called a booster. It is driven by a shunt motor operating from the main supply system.

$$\text{The booster output} = vI_2$$

$$\text{The motor input} = vI_2/(\eta_1'\eta_2') \text{ where } \eta_1' \text{ and } \eta_2' \text{ are the booster, and motor efficiencies}$$

$$\text{The motor current } I = I_2 \times v/V_1 \times 1/(\eta_1'\eta_2')$$

$$\text{and the total losses } P_L = V_1 I - vI_2$$

$$= vI_2(1/(\eta_1'\eta_2') - 1) \quad . \quad . \quad . (221)$$

Compared with the motor generator set, the booster set has then the advantage that its output is really only $vI_2 = (V_2 - V_1)I_2$ instead of $V_2 I_2$, i.e. its relative output is only $vI_2/V_2 I_2 = (V_2 - V_1)/V_2$. A considerable saving in first cost is then possible when the additional voltage v is relatively small.

In addition, the losses are reduced to $vI_2(1/(\eta_1'\eta_2') - 1)$, instead of $V_2 I_2 (1/(\eta_1\eta_2) - 1)$. The product $\eta_1'\eta_2'$ for the booster set will be rather less than for the motor generator, owing to the smaller size of the units. The losses will then be very considerably reduced, but not quite in the ratio of $(V_2 - V_1)/V_2$.

Fig. 224 illustrates stepping up the voltage from 110 to 220, and the figures for the currents are based on unity efficiency.

A booster set, as indicated in Fig. 224, can be employed equally for stepping up or stepping down the voltage. To step down the voltage from 220 to 110 the connexions would be exactly as indicated in Fig. 224, but the directions of all the currents would be reversed, the booster B then acting as a motor driving M as a generator. To convert the 10 A at 220 V into 20 A at 110 V, the booster set has to provide the additional 10 A at the expense of a reduction of 110 V in the pressure. This surplus 110 V is absorbed in overcoming the back e.m.f. of machine B, which, as stated, functions as a motor.

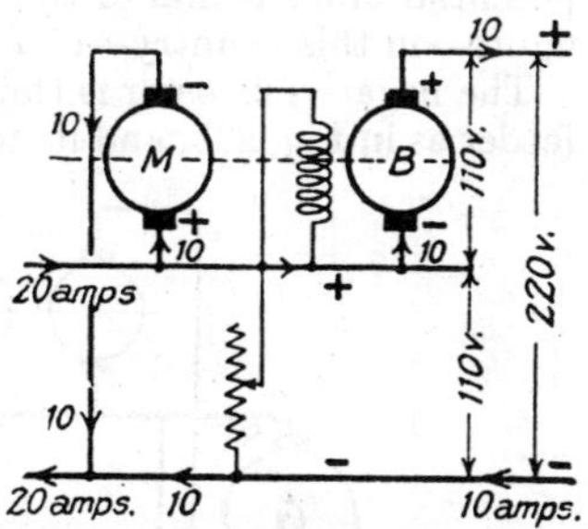

FIG. 224. SEPARATELY-EXCITED BOOSTER

Series Boosters for Traction Work. The voltage drop in a conductor is given by $v = \frac{I}{A}\rho l$. It follows then that for long feeders, as used on low-tension d.c. traction systems, the sectional area of the conductor becomes very heavy if the voltage drop is to be kept within narrow limits, and the conductor has to be worked at a current density much below its normal rating. A great economy in the cost of the feeders, and at the same time a much better voltage regulation at the receiving end of the feeder, can be obtained by connecting a small series generator in the feeder circuit as shown in Fig. 225. This machine—commonly called a positive series booster—is driven at constant speed by a shunt motor, and is designed so that when full load current flows in the feeder, the booster generates, and adds to the main voltage, an amount exactly equal to the volt-drop in the positive feeder. The potential of the distant end of the feeder is thus kept the same as that of the positive bus bar at the generating station. By designing the booster to work entirely on the initial straight-line portion of its magnetization curve, the voltage generated will be proportional to the line current, so that the volt-drop in the feeder will automatically be compensated at all values of the current.

It is necessary only to connect boosters in circuit with long, heavily-loaded feeders. The voltage of the main part of the system can be maintained sufficiently closely by using over-compounded generators. The voltage of these latter machines is determined by the total load on the whole system and is applied indiscriminately to all the feeders, whether long or short, heavily or lightly loaded. The action of the series booster is, on the contrary, entirely controlled by the load on, and affects solely the feeder to which it is connected.

Not only are series boosters used in circuit with the positive feeders, but similar machines also are connected in circuit with the negative or track feeders, so that the distant ends of these feeders can be maintained at approximately earth potential, and hence the potential of all points of the rail system kept within the prescribed figure—in this country—of 7 volts above "earth."

The negative booster is thus connected in series with the negative feeder as in Fig. 225, and its terminal *A* becomes at a lower potential

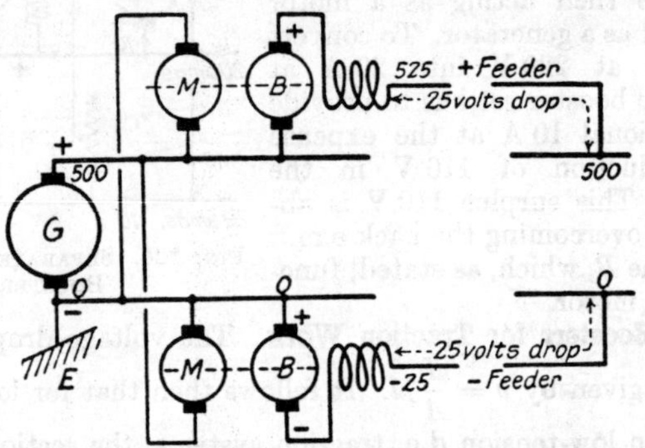

FIG. 225. SERIES BOOSTER FOR TRACTION WORK

than "earth." As a result the booster sucks most of the current out of the rails, and causes it to return by means of the negative feeder. Fig. 225 shows both positive and negative boosters, and the approximate potential of various points on the system is indicated. For large line-currents it is sometimes convenient to construct the booster field-winding for a reduced current, the field coils being shunted by a suitable diverter. Or the booster excitation may be obtained from a small exciter mounted on the same shaft.

Battery and Generator in Parallel. Buffer Batteries. Storage batteries are occasionally used in parallel with d.c. generators, their purpose being to act as load equalizers; during heavy overloads the battery has to assist the generators, and during light loads the battery has to be charged.

The simplest method is to connect the battery directly in parallel with the station bus bars, as in Fig. 5. The load characteristic for the battery is a straight line, inclined but very slightly to the horizontal, as indicated by *BDA* in Fig. 226. If such a battery were connected in parallel with a shunt generator having, as is normal, a drooping characteristic, and the generator e.m.f. on open circuit were adjusted to be equal to that of the battery, the battery would carry the greater portion of the load under all conditions. By increasing the excitation of the generator so that its open circuit e.m.f. exceeds that of the battery, it can, however, be arranged that when the load is relatively light, the generator actually charges the battery, and during heavy loads the battery discharges, and assists the generator. In particular, violent momentary overloads can thus be supplied mainly by the battery. Thus, in Fig. 226, *SRH* represents the load characteristic of a shunt generator working in parallel with a battery, the open circuit e.m.f., *OS*, being somewhat greater than that of *OB* the battery. The total load characteristic

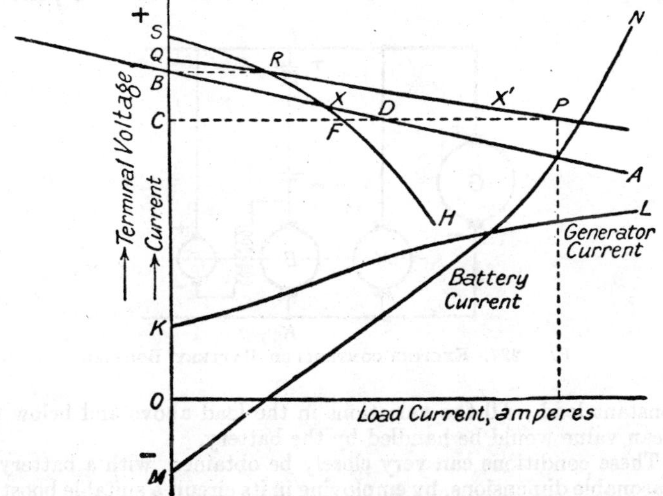

FIG. 226. GENERATOR AND BATTERY IN PARALLEL

of the combination is obtained as already discussed on page 192, and is represented by *QRP*, the point *P* being such that $CP = CF + CD$. When the load current corresponds to *BR*, the battery floats, and the whole of the load current is supplied by the generator. For lighter load currents, the battery receives a charging current, and for heavier loads the battery discharges. The two characteristics cross at *X*, the corresponding point on the total characteristic being *X'*. At this point the load is shared uniformly; for heavier loads, the battery supplies the greater part of the current. The effect of the battery is thus to equalize the load on the generator. The graphs *KL* and *MN* indicate the manner in which the generator and battery currents vary in terms of the total current. These graphs are obtained respectively by plotting *CF* and *CD* in terms of *CP*.

In actual practice, the action of a buffer battery is not quite in accordance with Fig. 226, the difference being due partly to slight polarization effects, and partly to hysteresis effects resulting in the external characteristic of the generator departing from the simple graph *QRP*.

The actual e.m.f. of the battery varies according to its state of charge. Since the whole distribution of the load will be seriously affected by such changes in the battery e.m.f., frequent hand regulation of the number of end-cells in circuit is necessary.

If the battery were extremely large, so that its internal resistance were negligible, its external characteristic *BDA* would be horizontal. Under these conditions the bus bar voltage would be maintained rigidly constant, and so the generator would work on a perfectly

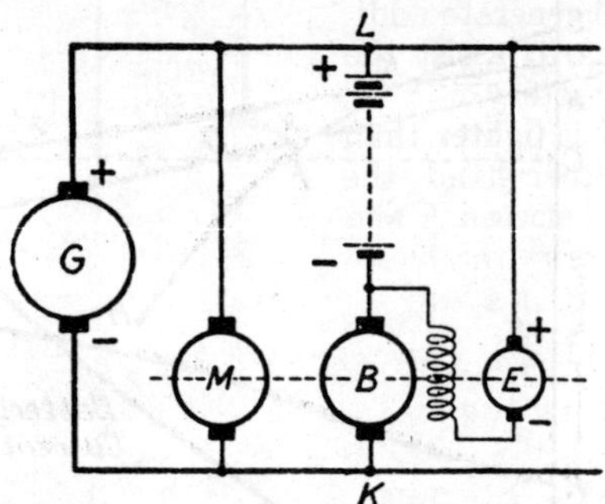

FIG. 227. EXCITER-CONTROLLED BATTERY BOOSTER

constant load. All the variations in the load above and below the mean value would be handled by the battery.

These conditions can very closely be obtained, with a battery of reasonable dimensions, by employing in its circuit a suitable boosting generator—or booster; further, this booster can also be arranged to counter-balance any variations that may take place in the battery e.m.f.

A simple arrangement of such a battery booster is shown in Fig. 227.* The booster and battery are connected in series across the main bus bars. The booster *B* is motor driven at constant speed, and separately excited, the voltage applied to its field circuit being the difference between the voltage of the level-compounded exciter *E* and the voltage between battery terminals. The exciter generates a constant voltage of the same value as the normal voltage between bus bars and the booster is designed so that it generates a voltage just equal to the voltage applied to its field coils. The booster thus automatically provides just the correct voltage to counterbalance any change in the p.d. between battery terminals, and the voltage between *L* and *K* is maintained constant, quite independent of the state of charge of the battery or of the current passing therein.

Further, with this arrangement the main generators may be disconnected during emergency or during periods of very light

* Many other types of battery booster have been used in practice but need not be discussed here.

load; the battery and booster will maintain the voltage between bus bars at its normal value.

The action of the set is then as follows. At normal average load on the station, the battery and booster just float on the bus bars, neither taking nor delivering current. If the load increases, the p.d. between the terminals L and K tends to fall and the battery discharges. As a result, the voltage between the battery terminals falls; the field current in the booster therefore increases, causing the booster to generate additional e.m.f. so as to assist the discharge of the battery.

When the load is lighter than normal, on the other hand, the p.d. tends to rise between L and K, so that the battery receives a charging current. The booster field current falls due to the increase in the terminal voltage of the battery. This reduced excitation reduces the booster e.m.f. and further assists the battery to be charged; in addition, the voltage between L and K is prevented from rising appreciably.

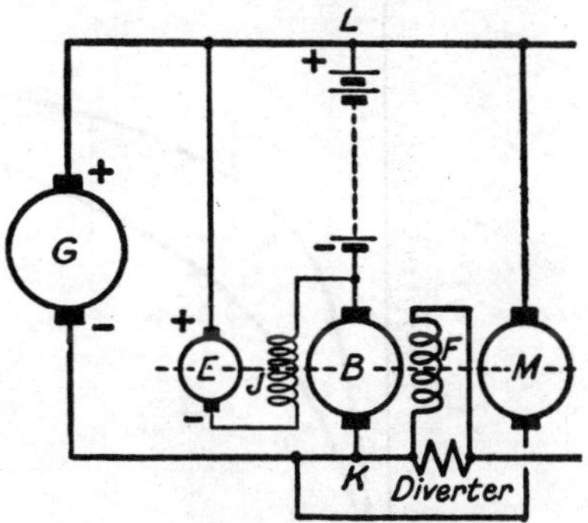

Fig. 228. Highfield Booster

The whole action of the booster is thus dependent upon the auxiliary generator E, called an exciter; the use of exciter-controlled boosters of this type is due to Highfield.

In another form of the Highfield booster, the discharge and charge of the battery is further assisted by providing an additional heavy current field winding F on the booster, this winding being connected so as to carry the main load current, or a definite fraction of this current, as in Fig. 228, the coil being connected in parallel with a diverter. (This avoids the need of arranging for the field coils of the comparatively small booster to carry very heavy currents.) The coil F tends to establish a booster e.m.f. assisting the battery to discharge. On installation, the exciter voltage is adjusted so that, when normal current flows in the main circuit, the battery and booster just float on the bus bars.

Any increase in the load current increases the m.m.f. due to the diverted winding F, and thereby increases the booster e.m.f. by an amount which makes the battery discharge at a rate closely equal to the amount the load current has risen above normal. In a similar manner, if the load current be less than normal the booster e.m.f. falls, and the battery receives a charging current corresponding very closely to the fall in the load current. Any change in the value of the battery e.m.f. due to an unusually long discharge period, is entirely compensated by the action of the exciter and fine wire coil J, the connexions of which are exactly as in Fig. 227.

With a battery thus connected in circuit with a booster, it is necessary periodically to charge the battery fully, to gassing point.

For the booster to respond immediately to changes in the load, it is essential that the major part of its magnetic circuit should be laminated. This condition is sufficiently met by the use of machines having laminated poles. In addition, the booster must be designed so that its magnetic circuit is unsaturated, and the induced e.m.f. strictly proportional to the excitation.

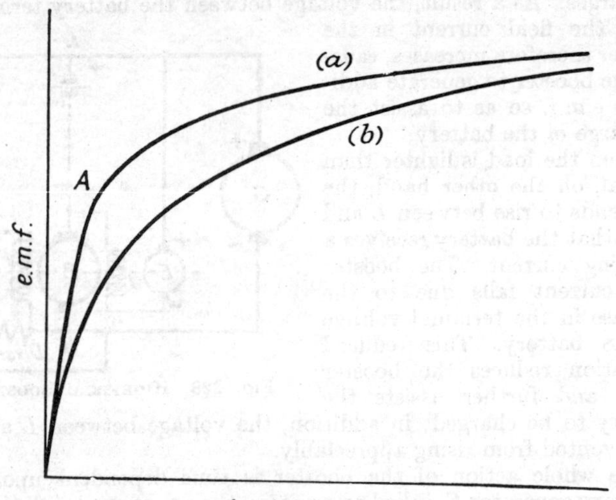

FIG. 229. MAGNETIZATION CURVE OF (*a*) NORMAL SHUNT GENERATOR, (*b*) SHUNT-WOUND EXCITER WITH SATURATION PLATES

Shunt-wound Exciter. A simple special machine is the shunt-wound exciter used to supply the fields of both d.c. and a.c. machines which do not justify the use of a double-exciter unit consisting of a separately-excited exciter whose field is supplied from a constant-voltage pilot- or sub-exciter. The range of voltage required to supply the field of a main a.c. generator so that the voltage may be adequately controlled from no load with a cold field winding to full load with a hot field winding, may be considerable. A shunt generator of normal design, however, can only be controlled from maximum voltage down to the upper end (*A* on curve *a* of Fig. 229) of the straight part of the magnetization characteristic, i.e. until the field circuit resistance has reached the critical value. To extend this range it is necessary for the bend to start as near the origin as possible, i.e. for saturation to begin at low flux densities. One way of achieving this is to separate the poles from the yoke with thin steel spacers of restricted area which saturate when the pole flux is quite low, but which, being thin, add only a small effective air-gap through which the flux has to pass at higher densities. The characteristic is thus modified as in curve *b* Fig. 229. Even so, for large generators connected to big power systems, the range of voltage obtained is inadequate and a pilot exciter has to be used in addition to the main exciter.

The Dynamotor. Instead of using two entirely separate machines to form a motor generator set, or a simple booster set, or a balancer

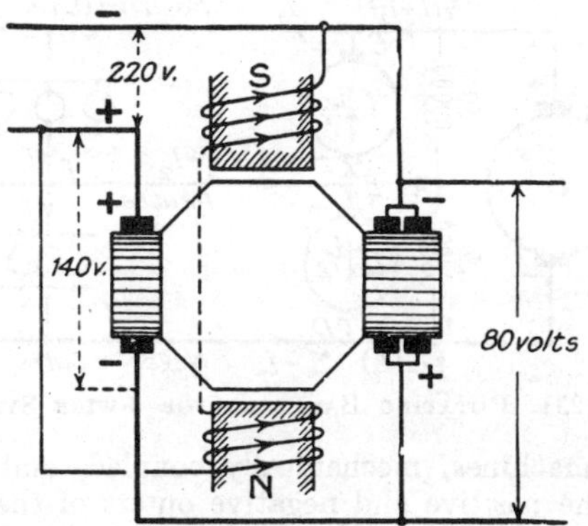

FIG. 230. DYNAMOTOR USED AS A "CRUSHER" SET

set, a doubly-wound armature may be used, working in a single field system. Such an arrangement is called a dynamotor.

Each armature winding is provided with its own commutator, these being arranged one at each end of the machine. The field winding is excited from the supply system.

The arrangement is indicated in Fig. 230, the connexions there shown being for a booster, or as it is sometimes called, a *crusher* set for reducing from 220 V to 80 V.

The two windings on the armature are entirely separate. With this arrangement, due to the combined action of the two windings, the net armature reaction would be zero if the machine were of unity efficiency. In addition, the shaft has to transmit but little torque. When interpoles are fitted to such a machine, the windings would then be relatively feeble under normal conditions, as armature reaction has not to be neutralized. To prevent sparking at sudden momentary overloads on the L.T. side, it is well to have powerful windings on the interpoles, and the reluctance of the interpole magnetic circuit may with advantage be increased to permit of this, by inserting non-magnetic distance pieces between the interpole and the yoke.

As compared with the standard arrangement of two separate machines, the dynamotor has a greater efficiency owing to the reduction in the number of bearings and the use of a single-field winding.

As compared with a motor generator set, the performance differs in that the secondary voltage is not affected by changes in speed of the set, but, neglecting the resistance drop in the two windings, is strictly proportional to the primary voltage.

Balancers. For maintaining the neutral wire of a three-wire system at its correct potential a rotating balancer may be used, as indicated in Fig. 231. The balancer set consists of two identical

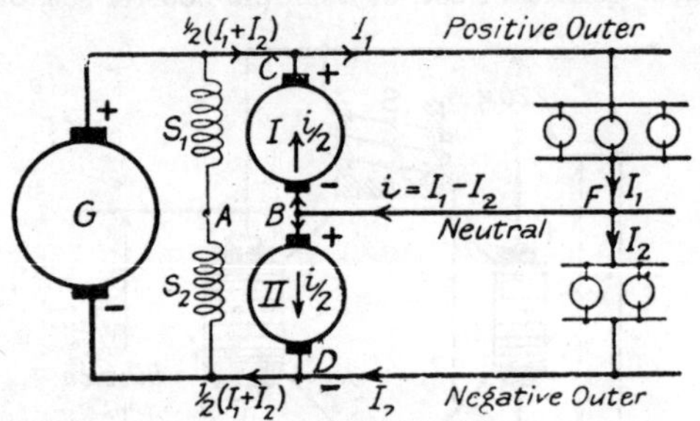

FIG. 231. ROTATING BALANCER FOR 3-WIRE SYSTEM

shunt-wound machines, mechanically coupled, and connected in series across the positive and negative outers of the system. The set can be installed either at the main generating station or at any convenient substation. When there is no-load on the system, the set operates as two motors in series on light load, and the motors are adjusted so that under these conditions they generate exactly equal back e.m.f.'s. The neutral point B will then have a potential just midway between that of the outers C and D. When the system is loaded, the current flowing in the neutral is the difference between the currents in the two outers. If the positive side is the more heavily loaded, current flows back to the balancer terminal B, and if the load on the negative side is the heavier, current flows from B into the neutral wire.

Neglecting the losses in the balancer set, the current in the neutral divides at B, half passes upwards, for the case indicated, through machine I, and the other half downwards through machine II. In the former machine, the current flows IN the direction of its induced e.m.f., so that this machine acts as a generator; in machine II the current flows AGAINST the direction of the induced e.m.f., so that this machine works as a motor. The total load on the two sides of the system is thus maintained constant or *balanced*; that on the side heavily loaded with lamps, etc., is reduced by the fact that machine I acts as a generator and takes part of the load; on the other side, the total load is increased owing to machine II acting as a motor.

The arrangement shown in Fig. 231 indicates that the field windings of the two machines are connected in series across the outers and hence carry exactly equal currents. The induced e.m.f.'s

E, in the two armatures will then have the same value. For the machine working as a generator, the terminal voltage is lower than the induced e.m.f., whereas for the other, functioning as a motor, the terminal voltage is greater. With the arrangement shown, therefore, the potential of the point B will rise when the positive side is heavily loaded. Since, under these conditions, the load point F is of necessity at a higher potential than B, the potential of F will rise still more, and hence the voltage between positive outer and neutral at the load centre will progressively fall, and that between neutral and outer progressively rise as the out-of-balance load increases on the positive side.

If r_o is the resistance of the neutral wire from F to B,

r_m is the resistance of the main circuit of each machine, i.e. armature + interpole windings + brush drop.

V = voltage between outers; i = out-of-balance current flowing from F to B.

V_G = terminal voltage of machine which is generating,

V_M = " " " " motoring.

$V_G = E - r_m i/2;\ V_M = E + r_m i/2.$

$V = V_G + V_M = 2E.\ \therefore E = V/2$ under all conditions.

The voltage on the positive side at F

$$V_1 = V_G - ir_o$$

i.e.

$$V_1 = V/2 - i(r_m/2 + r_o) \quad . \quad . \quad . \quad (222)$$

and the voltage on the negative side at F

$$V_2 = V/2 + i(r_m/2 + r_o) \quad . \quad . \quad . \quad (223)$$

In this calculation the effect of the machine losses upon the currents carried by the two machines has been neglected. If the points A and B were joined, the action of the set would be even worse, as the result of a heavy load on one side would be to reduce the excitation of the machine functioning as a generator and to increase that of the other machine.

On the other hand, by cross-connecting the field windings, as indicated in Fig. 232, the action is greatly improved. A fall in voltage on the heavily-loaded side, and a corresponding increase on the other, now increases the excitation of the machine which functions as a generator and weakens that of the other. Since the two machines operate at the same speed, it follows that the induced e.m.f. will increase on the machine which is generating and decrease on the other machine, thus tending to maintain the point B at the correct potential.

Three-Wire Generator and Static Balancer. Consider two diametrically opposite conductors on the armature of a 2-pole generator, as indicated by A and B in Fig. 233. As the armature rotates, an alternating e.m.f. will be established between these two points, having its maximum value when A and B respectively pass under the positive and negative brushes. If a reactor be connected across A and B, the potential of the mid-point, X, of the reactor will always be midway between that of A and B. For any position of the conductors let the line AB be displaced by θ from the brush axis DE, and let $A'B'$ be constructed as shown displaced by θ on the other side of the brush axis. Then by inspecting Fig. 233 it is

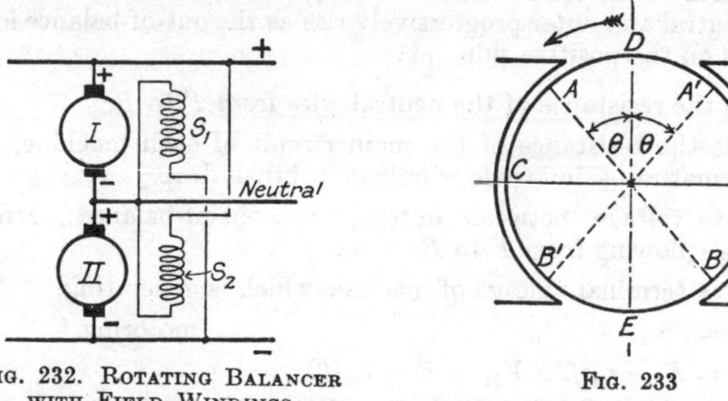

FIG. 232. ROTATING BALANCER WITH FIELD WINDINGS CROSS-CONNECTED

FIG. 233

apparent that the total e.m.f. induced in the portion of the armature between A and A', and between B and B', respectively, is zero. Hence the potential difference between A and B' is the same as that

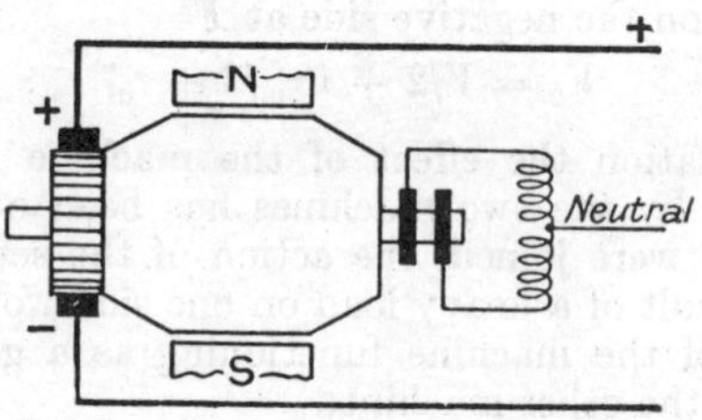

FIG. 234. SINGLE-PHASE STATIC BALANCER

between A and B. A conductor at the fixed point C will therefore have a potential midway between that of A and B, for all positions of the armature. Hence the potential midway between that of A and B is also the same as that midway between the brushes D and E. The neutral point of a 3-wire system can therefore be obtained by joining the neutral wire to the mid-point of a reactor connected across two diametrically opposite tappings on the armature of the main generator. Normally, the tapping points are connected to slip rings, and the connexions to the 3-wire system made as in Fig. 234.

With a lap-wound multipolar generator there will be as many tappings as there are poles, the tappings being one pole-pitch apart and connected alternately to the two rings.

With a 3-wire generator of this type, the currents at positive and negative brush sets have only the same value when the load is balanced. In addition, the armature conductors do not all carry the same current, but carry current at one time corresponding to the load on the positive side, and at the other to the load on the negative side.

To obtain more uniform conditions for the armature conductors, more than two slip rings are employed. For example, another pair

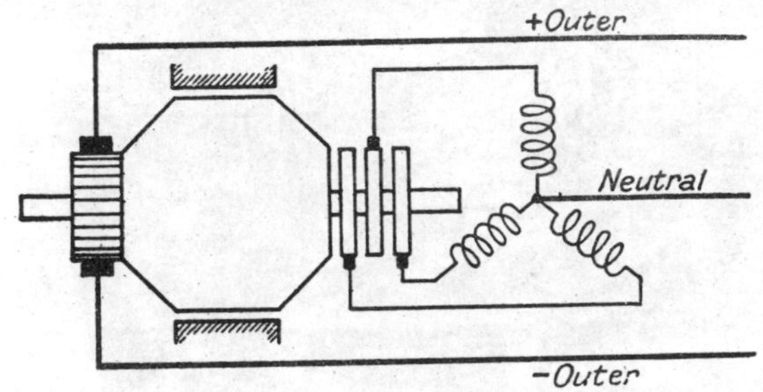

Fig. 235. 3-phase Static Balancer

of tapping points placed midway between A and B could be connected to a second reactor and the mid-point of this also connected to the neutral wire forming a 4-phase star connected system of coils; or three such pairs of tapping points could be used forming, with their reactors, a 6-phase star-connected system. Instead of using the complete 6-phase system, only three tapping points and three slip rings need be used, connected to a 3-phase star connected reactor, the star point forming the terminal for the neutral as indicated in Fig. 235. This is the arrangement commonly employed, except for very small machines. In Fig. 236 two such 3-wire generators are shown, coupled together for a Hopkinson test. The equalizing connexions can plainly be seen on one of the machines. The slip rings also serve as equalizers.

The machines have a cast steel frame and do not represent the most recent construction.

Special Characteristics. The normal power system is operated at a nominally constant voltage with the generators and loads connected in parallel. The constant-voltage system is most convenient when loads differ widely in power, and may sometimes be operated simultaneously and at others, separately. In some applications, however, there is only one load connected to the generator at all times, and when this is so it is often an advantage if the generator current is inherently limited to a safe value even on short-circuit. The desired generator characteristic is then as shown by curve b of

FIG. 236. 3-WIRE D.C. GENERATOR ARRANGED FOR HOPKINSON TEST
(Lancashire Dynamo and Crypto, Ltd.)

Fig. 237 in contrast with the ideal constant voltage characteristic shown by curve *a*. Single-operator welding generators require characteristics of this shape, so also do generators supplying the motors driving excavating machinery which may strike obstacles, and cause the driving motor to stall.

Again, for some applications it is advantageous to keep the current nominally constant. The generator and motor armatures

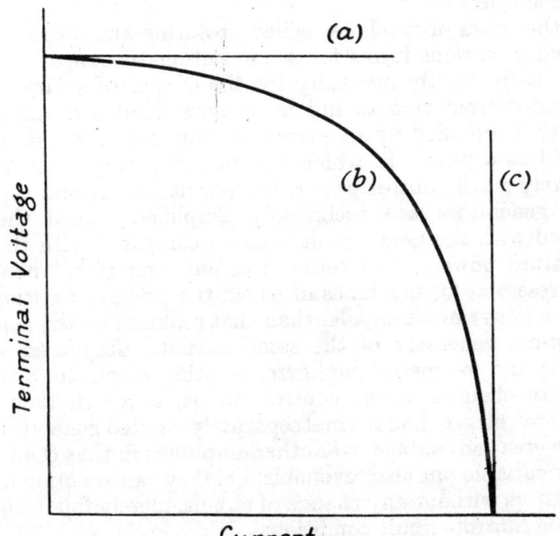

FIG. 237. GENERATOR CHARACTERISTICS, (*a*) IDEAL CONSTANT VOLTAGE CHARACTERISTIC, (*b*) PRACTICAL CHARACTERISTIC SHOWING INHERENT CURRENT AND VOLTAGE LIMITATION, (*c*) IDEAL CONSTANT CURRENT CHARACTERISTIC

are then all connected in series and carry the same current. Generators and motors are switched off by removing the excitation and then short-circuiting the armature. The voltage across them is then zero, and, since power is proportional to voltage when the current is constant, the power of the short-circuited unit is zero. The theoretical ideal characteristic of the generator would then be as shown in curve *c* of Fig. 237. A system employing series generators and motors connected in series was developed by Thury, between 1889 and 1906, for d.c. power transmission, since it is not practicable to generate a sufficiently high voltage in a single machine. In more recent applications, however, the purpose is to permit the use of constant-armature-current motors with their simpler wiring and controllers, and their desirable characteristics. The advantages of such an arrangement are: a motor can be stalled with safety and takes very little power although still maintaining its torque under this condition; the speed can be controlled over the whole range by field control alone, so that the controller has to handle only field-circuit power and not armature power, as with constant voltage motors under similar conditions; the wiring is simpler. This type of constant-current system was developed* for auxiliary drives on ships in about 1928, and has more recently been extended† to include also the main propulsion motors on special-purpose vessels such as dredgers and ferries.

A further class of machines called "rotating amplifiers" has been developed in various forms for use in automatic control systems in which it is frequently necessary for the output of a large machine to be varied from zero to full by a very small controlling power such as that provided by an electronic amplifier. Since an amplifier is defined as a device in which the output power is controlled by a relatively small control power, the normal shunt and separately-excited generators are technically amplifiers, since the power associated with the field circuit is small compared with the output or armature power. The term "rotating amplifier," however, is usually reserved for machines in which the primary exciting circuit requires a power much smaller than that required by the field circuit of a normal generator of the same output. Compared with the electronic and magnetic amplifiers, rotating amplifiers are slow to respond to changes in the control circuit, although they respond much more quickly than normal separately-excited generators. They have the great advantage over other amplifiers in that their power is not only variable but also reversible, i.e. they can function as motors or generators without any change of circuit, merely following change of load or control-circuit conditions.

Machines which differ from the normal nominal constant-voltage machines described earlier will be considered in the following pages.

Three-field (Kramer) Generator. The "three-field" or "three-winding" generator, first described by Kramer in 1909, is of conventional form but has a separately-excited winding, a cumulative shunt winding and a differential series winding. On short-circuit, there is little or no current in the shunt winding since the machine terminal voltage is zero. The m.m.f. of the separately-excited winding exceeds that of the series winding only by the amount required to generate an e.m.f. equal to the volt-drop in the generator itself. In other words, the short-circuit current must be such that the m.m.f. it produces in the series winding is less than the m.m.f. of the separately-excited winding. The short-circuit current is thus limited and is controlled by adjustment of the separately-excited field. On no load there is little or no current in the series winding and the e.m.f. is high under the cumulative effect of separate and shunt excitations.

* By Gilbert Austin, Ltd.

† By Metropolitan-Vickers Electrical Co., Ltd.

By varying the proportions of the three field windings, the shape of the external characteristic can be varied over a wide range as shown in Fig. 238. As the resistance of the shunt field circuit is reduced towards the critical value, the machine becomes almost

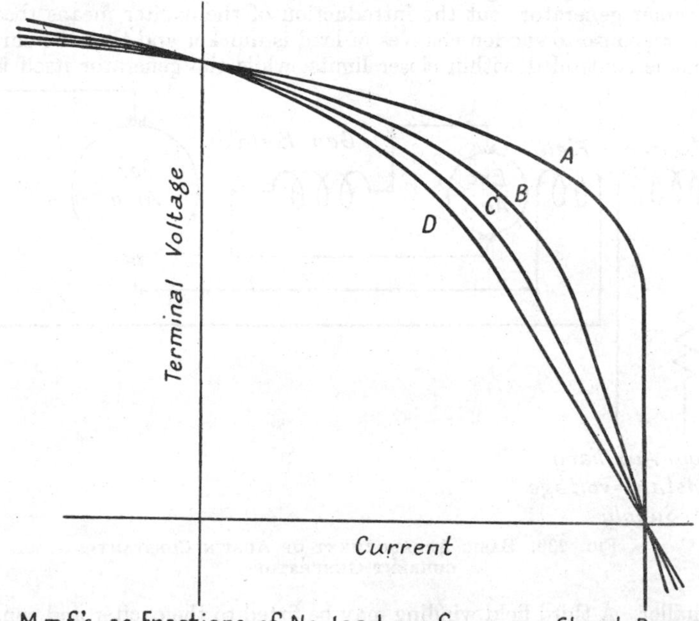

	M.m.f.'s as Fractions of No-load m.m.f.			Shunt Res. as Fraction of Critical Res.
	Separate	Shunt at No-load	Series at Short Circuit	
A	0·4	0·6	0·4	1·0
B	0·6	0·4	0·6	1·5
C	0·8	0·2	0·8	3·0
D	1·0	0	1·0	∞

(Neglecting armature resistance drop)

FIG. 238. THREE-FIELD-WINDING (*Kramer*) GENERATOR CHARACTERISTIC

self-exciting on the shunt winding, the series winding and separately-excited winding have almost equal and opposite m.m.f.'s., and the current is almost constant over the range of voltage for which the magnetization curve is approximately straight.

It is interesting to note that this is a very early example of a negative feed-back system of control.

Austin Constant-current Generator. The Austin constant-current generator resembles the Kramer generator in using a differential winding connected in series with the output, but it is a separately-excited generator, the exciter of which has basically two field windings, one separately-excited from a constant voltage supply, the other being the differential winding connected in series with the main generator output (Fig. 239). The action is thus similar to that of the Kramer generator, but the introduction of the exciter means that the response to sudden changes of load is quicker and that the current is controlled within closer limits, while the generator itself is

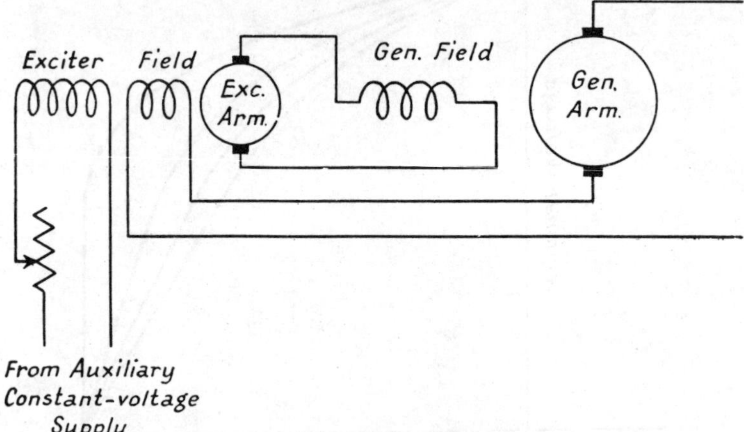

FIG. 239. BASIC ARRANGEMENT OF AUSTIN CONSTANT-CURRENT GENERATOR

smaller. A third field winding may be fitted to the exciter and connected cumulatively in series with the exciter output to compensate for the armature reaction of the exciter itself. If no constant voltage supply is available it is possible to use the exciter as a two-stage machine to provide its own separate excitation, but naturally this makes it more complicated.

Welding Generators. The principal requirement of a supply for electric arc welding is a limit to the current when a metallic circuit exists, either when the arc is struck, or when a droplet of metal bridges the gap. The voltage across the arc is almost independent of the current, except at very low currents, and thus has a straight horizontal characteristic for a given length of arc. The machine and arc characteristics (Fig. 240) should intersect at an angle as near a right angle as practicable, to give a steady current, i.e. the machine external characteristic should fall steeply at the arc voltage (about 20 V), and the open-circuit voltage must be reasonably high both to provide this slope and to facilitate striking the arc. The type of machine required depends on whether one or more operators use the machine simultaneously. When several operators are supplied by one generator, it must produce a constant voltage adequate to strike an arc, and the current limitation has to be provided by a resistor in the circuit to each individual welder. When each operator has his own generator, however, the current limitation is more efficiently provided by the inherent characteristic of the machine. The current limit must be adjustable to meet the requirements of different jobs, so also should the open-circuit voltage.

It is important that the current should be limited not only under steady conditions but also following sudden changes in circuit

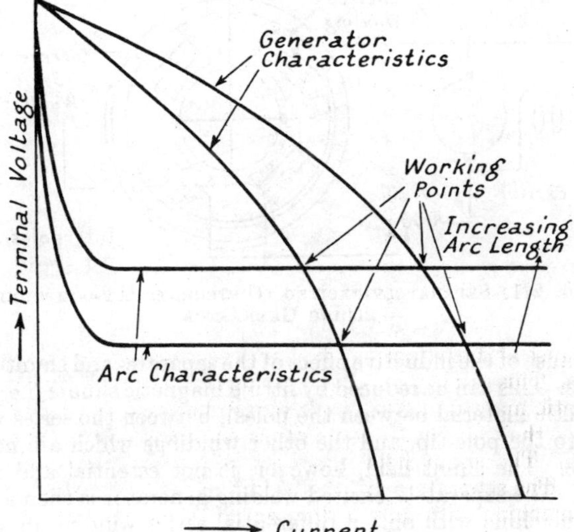

FIG. 240. CHARACTERISTICS OF METALLIC ARC AND OF WELDING GENERATOR

resistance. A full consideration of such "transient" performance is beyond the scope of this volume, but, in brief, the fall in voltage on short-circuit depends upon the collapse of the field flux. This collapse will be retarded by induced currents in any closed electric circuits linking the flux, and thus momentary excessive output currents may occur. Such circuits are formed by shunt- and separately-excited field windings, and by eddy current paths in the magnetic materials. The transient performance is improved, therefore, by laminating the field system and by reducing the time constants of any field coils linking the main flux. Additional circuits may also be provided to accelerate the flux changes, but they have no effect on the steady-state performance.

The required characteristics can be produced in various ways. Continental and American practice favour machines which are fundamentally cross-field generators, although the construction and arrangement of the windings vary greatly. British practice favours machines which are fundamentally normal machines compounded to give the required characteristics. Only a few basic types of generator can be considered from the wide variety which have been developed.

Separately-excited (differentially-compound) Welding Generators. The three-field generator* has a suitable steady-state characteristic for a welding generator. The transient performance, however, is

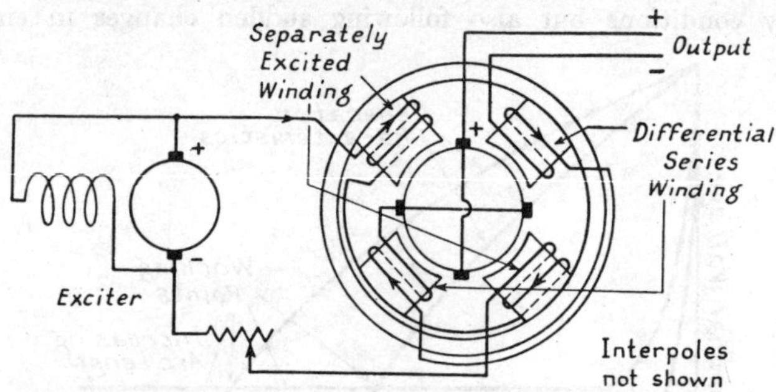

FIG. 241. SEPARATELY-EXCITED (DIFFERENTIALLY-COMPOUND) WELDING GENERATOR

poor because of the inductive effect of the separate- and shunt-excited windings. This can be reduced by fitting magnetic shunts (i.e. bridges of magnetic material between the poles), between the series winding nearest to the pole-tip, and the other windings which are nearer to the yoke. The shunt field, however, is not essential and may be omitted. The separately-excited welding generator is then a normal type of machine with only a differential series winding in addition to the separately-excited winding. It usually requires a d.c. supply or an exciter for the excitation, although permanent magnets have been used. Control of the open-circuit voltage is given by adjusting the field rheostat—or the position of the permanent magnets—and of the short-circuit current, by tappings on the series winding. A diverter can be used as an alternative to tappings but must be inductive to be effective on transients. To reduce the mutual interaction of the two field windings, they may each be wound on only one pole of each pair as shown in Fig. 241, but this causes considerable leakage with the possibility of bearing troubles due to currents within the bearings induced by stray fluxes.

Shunt Welding Generators. If a shunt generator has few field ampere-turns relative to its armature ampere-turns, the external characteristic has a shape as shown in curve *a* Fig. 242, with a maximum current which is satisfactory for welding purposes. The bottom part of the curve, however, is unstable, and operation has

* *See* page 310.

to be confined to the top part by using a series resistor, the resistance line for which is shown at *b*. The characteristic of the combination is then as shown in *c*. This method is obviously inefficient. The efficiency is improved in one design by increasing the inherent voltage fall of the machine. This is effected by dividing the field winding into two, connecting one part across the armature terminals,

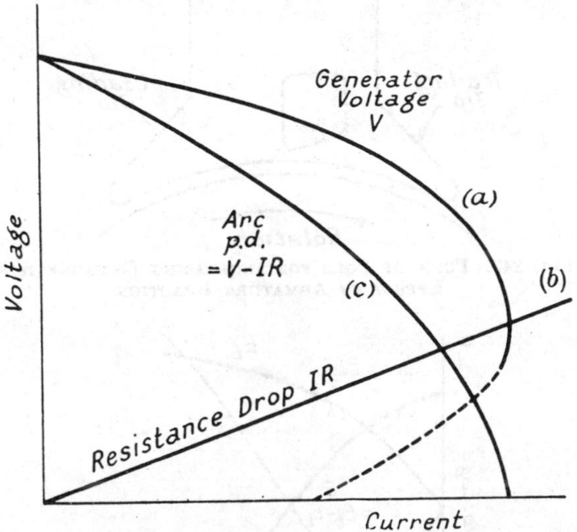

FIG. 242. SHUNT WELDING GENERATOR CHARACTERISTIC

and the other part across the load terminals so that it develops no m.m.f. on short-circuit. The resistance of the stabilizing series resistor can then be reduced, thus reducing the losses.

Cumulative-compound Welding Generators. The falling characteristic of the shunt machine is due partly to resistance and partly to armature reaction. If the effect of the latter were adequately increased, the series resistor and its losses could be avoided. This can be accomplished by restricting the flux path in the trailing half of the poles by a "tunnel" as shown in Fig. 243. Compared with the normal d.c. generator the field m.m.f. is small relative to the armature m.m.f. so that the flux in the leading half pole is not only weakened on load but actually reversed, and, since this part of the pole is not restricted, this flux is not limited by saturation. The reduced cross-section of the pole body for the trailing half causes saturation which prevents any great increase of flux in that half, in spite of the great increase of m.m.f. The e.m.f.'s E_l and E_t, corresponding to the fluxes of the leading and trailing halves of each pole are thus as shown in Fig. 244. The total e.m.f. E is their algebraic sum, hence the characteristic of the machine droops as shown in the figure. The shunt winding alone is now insufficient, since on short-circuit there would be no voltage to pass current through it. A *cumulative* series winding has, therefore, to be added to generate an e.m.f. equal to the resistance drops and brush drops

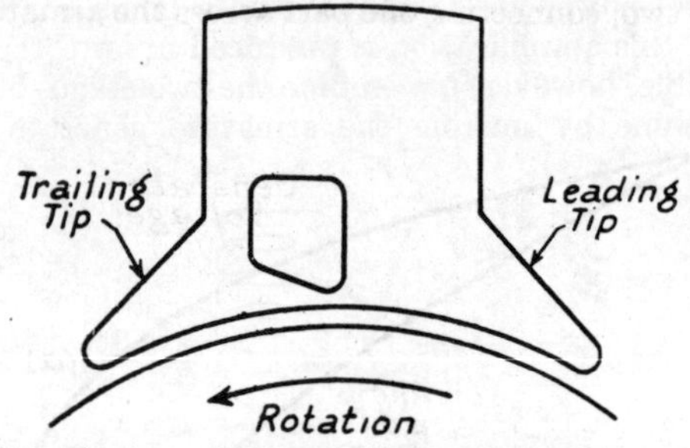

FIG. 243. FORM OF POLE FOR INCREASING DEMAGNETIZING EFFECT OF ARMATURE REACTION

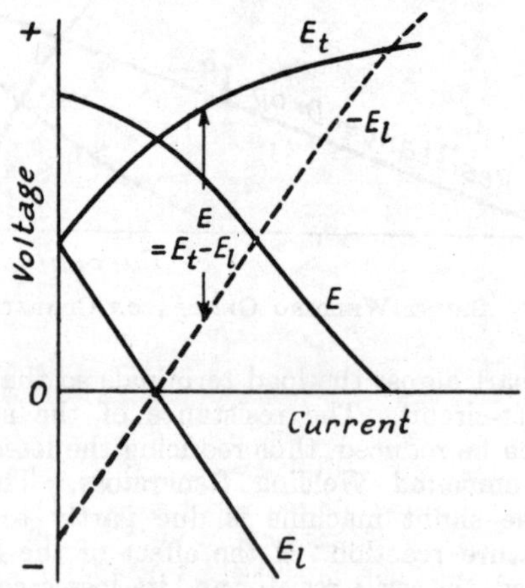

FIG. 244. CHARACTERISTIC OF CUMULATIVE-COMPOUND WELDING GENERATOR

of the machine. Tappings on this series winding will permit variation of the short-circuit e.m.f. and thus of the short-circuit current. The open-circuit voltage is adjusted by the shunt field rheostat.

Two-stage Machines: Rotating Amplifiers. When the controlling power available is insufficient to supply the field winding of a generator of the required output, an obvious sulution is to provide an exciter, that is, a smaller d.c. generator whose sole load is the field winding of the main generator. The controlling power is then required to supply only the field winding of the exciter as in Fig. 245.

Such a scheme is frequently used, even where ample control power is available to control the generator, merely to reduce the losses, size and cost of resistors in the control circuit or to increase the speed of response to changes. The combination of exciter and generator can be regarded as a two-stage amplifier, since changes in the control power are reproduced on a much bigger scale in the output, and this amplification is produced in two steps.

It is possible, however, to combine the two stages of amplification in one machine by making the armature generate two separate

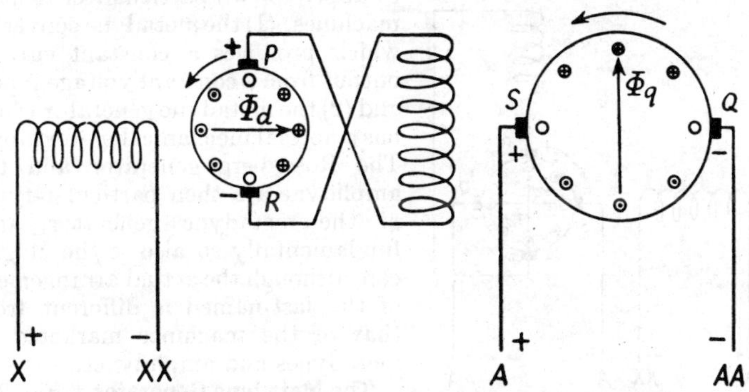

FIG. 245. GENERATOR WITH SEPARATELY-EXCITED EXCITER

currents, the first of which is used to provide the excitation for the second or output current. This double use of one armature winding may be effected in two ways—

1. By using two field fluxes electrically at right angles to each other and two corresponding sets of brushes, thus making a "cross-field" generator.
2. By using different numbers of effective poles for the two stages, e.g. two poles for the first or excitation stage, and four poles for the output stage. The armature and field system would be a normal four-pole arrangement for the output stage, but for the first stage only, two opposite poles would be used. The armature would then be chorded to half pitch for the first stage but full-pitch for the output stage.

Such two-stage generators are called "rotating amplifiers" but the term is not applied to a normal single-stage separately-excited generator, although of course its output power can be regarded as an amplified reproduction of the field power.

Cross-field Machines. The cross-field machine is not a recent invention, nor was it originally developed as an amplifier. The first to be widely used was the Rosenberg generator of 1904,* applied to

* See page 326.

train lighting, but a patent for a cross-field machine was granted nearly twenty years earlier. The next effective step was due to Pestarini, about 1930, and the biggest development, the cross-field amplifier, was produced about 10 years later.*

It is convenient to consider the metadyne generator first. Pestarini was not concerned with amplifiers, but with d.c. machines with additional sets of brushes generally, and to which he gave the generic title "metadynes." The cross-field machines are, therefore, particular kinds of the metadyne, but the term is normally taken to apply to two particular cross-field machines: (1) the metadyne converter which produces a constant current output from a constant voltage input, and (2) the metadyne generator which has a normal mechanical power input. The Rosenberg generator and the amplidyne are then particular types of the metadyne generator, and, fundamentally so also is the Magnicon, although the actual arrangement of the last-named is different from that of the machines marketed as metadynes and amplidynes.

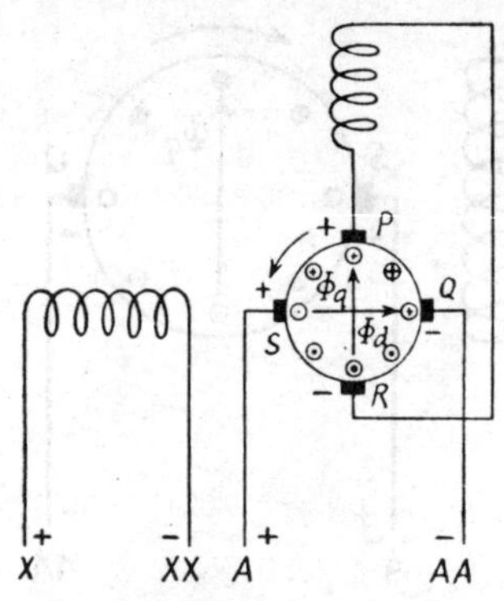

FIG. 246. GENERATOR ACTING AS OWN SEPARATELY-EXCITED EXCITER

The Metadyne Generator.† Fig. 245 shows a separately-excited generator and its exciter. The two field axes and the two sets of brushes are shown at right angles to demonstrate the similarity with the cross-field generator shown in Fig. 246, in which the functions of both exciter and generator are performed by one armature. The separately-excited winding *X–XX* produces the flux Φ_d, and the rotation of the armature in this flux generates an e.m.f. which circulates current through the second field winding between the brushes *P* and *R*. The flux Φ_q set up by the latter winding generates a second e.m.f. which produces the potential difference between brushes *Q* and *S*, which are connected to the load. In the cross-field machine, however, the flux Φ_q results not only from the current flowing in the coil between brushes *P* and *R*, but also from the passage of this current through the armature itself. In some circumstances the armature m.m.f. alone is used, and these brushes are then joined together so that the arrangement is as shown in Fig. 247.

Although the metadyne generator is basically the machine shown diagrammatically, this figure does not show the physical arrangement of the poles. If the normal type of d.c. machine poles are used to carry the control winding *X–XX*, the coils short-circuited by the

* By the General Electric Co. of America and the Macfarlane Engineering Co., Ltd. of Glasgow, independently.

† Metropolitan-Vickers Elec. Co., Ltd.

output brushes Q and S have e.m.f.'s induced in them by the direct-axis flux Φ_d set up by the control winding. To reduce the flux density at this position, the poles have to be cut away at the middle as shown in Fig. 248. Even then it is not possible to fit interpoles. which are desirable. In the more common form the pole is, therefore, divided into two separate parts, thus giving four polar projections in what is a two-pole machine, Fig. 249. To avoid confusion it has been proposed that the elemental machine with four pole pieces and a two-pole armature winding shall be described as a "one-cycle" metadyne (and so on) since there is one complete cycle of armature currents per revolution with such a system. The polar projections each have their own field coils and are excited in consequent pairs for each of the two

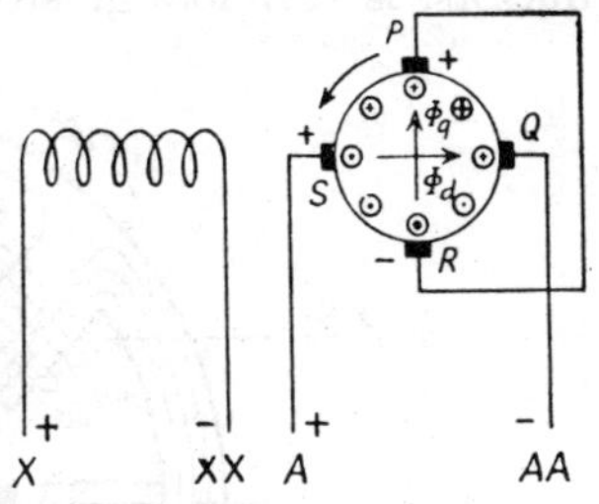

Fig. 247. Basic Crossfield or Metadyne Generator

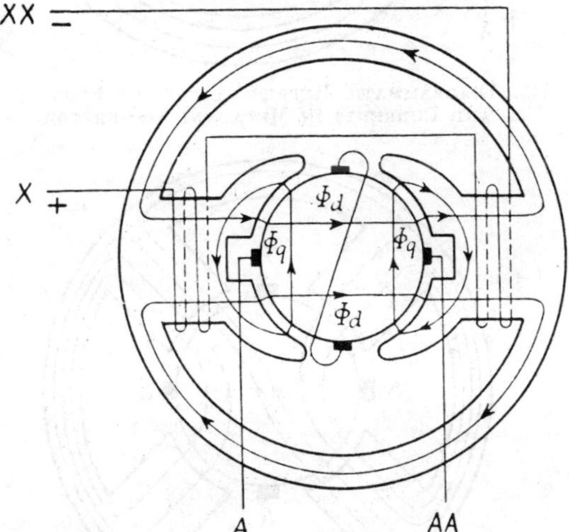

Fig. 248. Simple Form of Metadyne Generator

flux components. Interpoles are normally required only for the output brushes, although this arrangement provides space for interpoles on the quadrature axis if they were necessary.

It must be understood that the e.m.f.'s and currents exist simultaneously in the same armature conductors, and that it is only the position of the sets of brushes which "separate" them out. In particular, between the brushes P and Q, and also between R and S, the component currents are additive, while between Q and R, and also between S and P, they are in opposition. The armature heating, being dependent upon the r.m.s. value of the currents in the conductors, is therefore greater than it would be due to the output

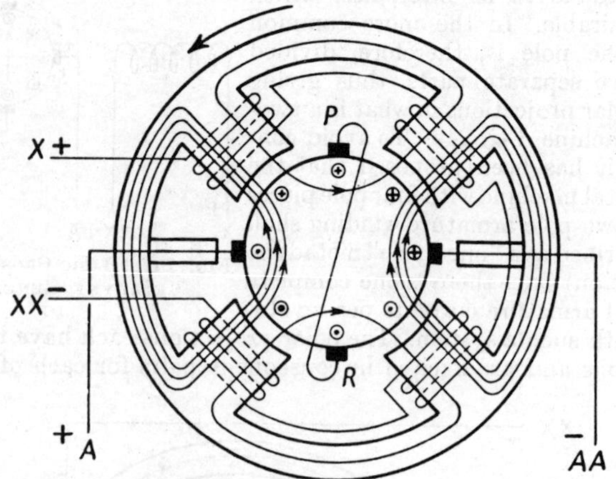

FIG. 249. DIAGRAMMATIC REPRESENTATION OF FLUXES, E.M.F.S AND CURRENTS IN METADYNE GENERATOR

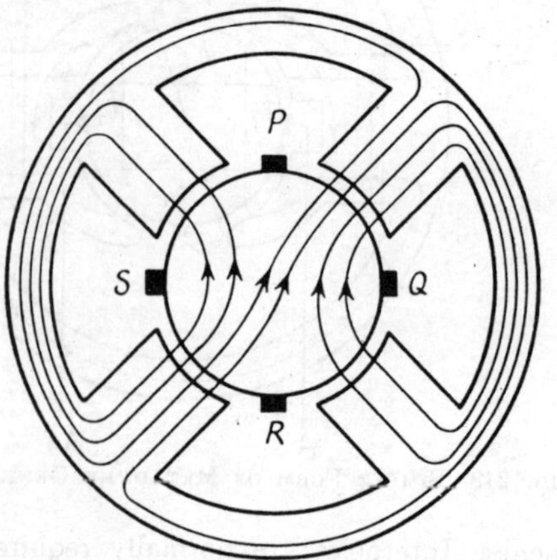

FIG. 250. FLUX PATHS IN ONE-CYCLE METADYNE

current alone. Similarly, the fluxes Φ_d and Φ_q cannot exist separately, but only as components of the total flux, as shown in Fig. 250.

In practice the metadyne generator may have a considerable number of windings on the poles—

CONTROL WINDINGS. These windings provide the initial excitation, and since only a small m.m.f. is required, will normally consist of many turns of very fine wire, or a few turns of relatively thick wire. Two windings may be fitted and connected in opposition, one to a control circuit and the other to the output of the complete system. The metadyne output is then proportional to the difference between the m.m.f.'s of these two windings which is a measure of the amount by which the system output differs from the desired value. The metadyne thus produces an output tending to reduce this difference or "error," until it has reduced it to a negligible value, since only a very small resultant m.m.f. on the direct- or control-axis is required under steady-state conditions.

Further control windings may be fitted to suppress unwanted oscillations which may develop in the output when the amplification is high, and yet others to provide an "over-riding" limit to the output to prevent damage from excessive current or voltage. The latter are inoperative until the limit is reached. These additional control windings are not shown in the diagrams.

QUADRATURE (AMPLIATOR) WINDINGS. The output of a metadyne is limited by armature heating just as is the output of a normal type of machine. The metadyne armature, however, has two component currents flowing in it, and the rated output will be less than that of the same armature in a normal d.c. generator. This output can be increased if the current flowing between the short-circuited brushes is reduced. This current can in fact be reduced if the number of turns through which it flows is increased. The extra turns cannot of course be accommodated on the armature, but they can be placed on the poles so as to magnetize in the same direction as the same current does in the armature. The short-circuit is therefore opened and a "quadrature" or "ampliator" winding connected across the brushes. This is a reversion from the diagram of Fig. 247 to that of Fig. 246, simplicity being sacrificed for economy.

COMPENSATING WINDING. The output current flowing in the armature produces an m.m.f. in direct opposition to that of the control winding. This will automatically impose a limit to the output current, even on short-circuit, in much the same way that the differential series winding does in the three-field generator. Although this has advantages at times, in that it protects against excessive currents and hence also against excessive torques in motors supplied by the metadyne, the limit thus imposed is too low. It can be raised by providing a "compensating" winding, magnetizing in the same direction as the control winding and carrying the output current. The compensating winding m.m.f. is thus in direct opposition to that of the output current flowing through the armature. It may have less turns than the armature or the same number, being under- or fully-compensated accordingly. Compensation has no effect on the no-load voltage of the metadyne since there is then no current in it. Increasing compensation increases the short-circuit current, and with a fully-compensated metadyne the short-circuit current is limited only by the resistance of the windings.

AUXILIARY SHUNT WINDING. If a winding on the direct (control) axis is connected in parallel with the load as a shunt winding, it will have no effect on short circuit since there will then be no current in it. If it is connected cumulatively, it will have little effect on no-load since the quadrature axis will be relatively saturated. At intermediate loads it will materially increase the output voltage, thus having the same effect as the shunt winding in the three-field

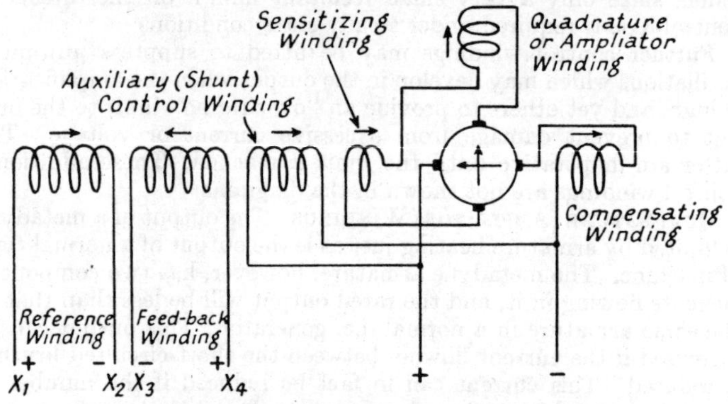

FIG. 251. SHOWING WIDE VARIETY OF FIELD WINDINGS POSSIBLE ON METADYNE GENERATOR

generator. By means of shunt and compensating windings, therefore, the external characteristic of the metadyne can be "tailor made" to suit almost any requirements.

SENSITIZING. A given output requires some definite current in the ampliator (primary) circuit, and since this circuit has resistance, there is a volt-drop in it necessitating a corresponding m.m.f. in the control winding to give the required output. If, however, the quadrature current is also made to flow through a winding of the correct number of turns on the direct axis, it will provide its own excitation and the output would be obtained with a much smaller input control power. Too many turns would make the metadyne self-exciting and the control would be lost. This modification is a "positive feed-back" around the first stage of amplification. Since ampliator coils already exist on all four polar projections the sensitizing does not require any extra coils but merely a slight increase in the number of turns of half the coils and a similar decrease for the others.

The essence of sensitizing is that the ampliator current shall itself produce flux on the direct axis. This can also be done without any differences in the coils if the air-gaps of the polar projections are made alternately bigger and smaller since some of the flux then passes diagonally and thus has a direct axis component.

Fig. 251 shows the various windings diagrammatically, the arrows indicating the normal steady-state directions of the m.m.f.'s. There may also be stabilizing windings on the control axis.

Metadyne Generator Ratings. Ratings of machines with a steeply-falling characteristic can be misleading. With a characteristic such as shown in curve *a* of Fig. 252, the output power has a graph such

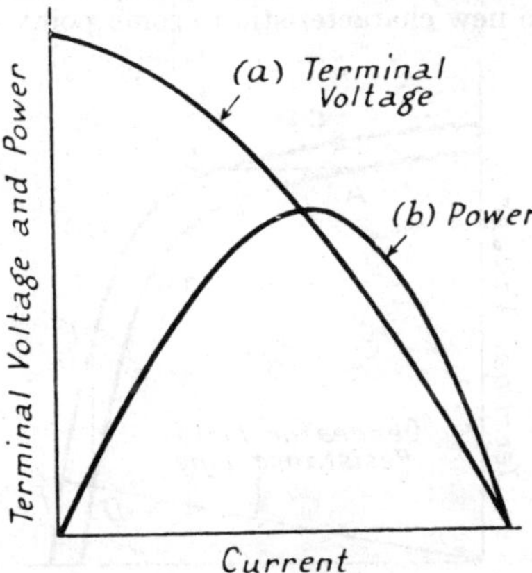

FIG. 252. CHARACTERISTIC AND OUTPUT OF METADYNE GENERATOR

as shown by *b*. The maximum power is not the product of maximum voltage and maximum current.

The rated output power of a metadyne is appreciably less than that obtainable from the same armature in a normal generator. First, the output current has to be reduced by about 10 per cent, because of the heating caused by the quadrature current flowing in the armature. Secondly, the provision of the commutating zones for two sets of brushes reduces the effective pole-face area, and hence the total flux and terminal voltage, by about a third. The rated output power of the armature is thus about $0{\cdot}9 \times 0{\cdot}667 = 0{\cdot}6$ of that of the same armature in a normal d.c. generator. For this reason it is not economical to use metadynes for large output powers, and, above 15 to 20 kW separately-excited generators are used with metadynes as exciters to obtain very similar results as from a metadyne generator.

Metadyne as Main Generator. A metadyne is frequently used as a main generator supplying the armature of a motor used for some control purpose. It is not an immediate large increase of e.m.f. that is required, following an increase of control current, but rather, a considerable increase in output current to provide the torque needed to accelerate the motor. Since the motor is separately excited its flux is substantially constant, there is no sudden increase of speed, and therefore no sudden increase of e.m.f. and no necessity for a sudden increase of applied voltage beyond the increased *IR* drop. The voltage will of course rise gradually as the speed increases. This result is most effectively obtained if the metadyne operates at some point such as *A* on the external characteristic *a* of Fig. 253. A sudden increase of the control current to correspond to a new characteristic such as *b* then causes a rapid change of the working point from *A* to *B* at almost constant voltage, followed by a slower change along the new characteristic to some point such as *C* where

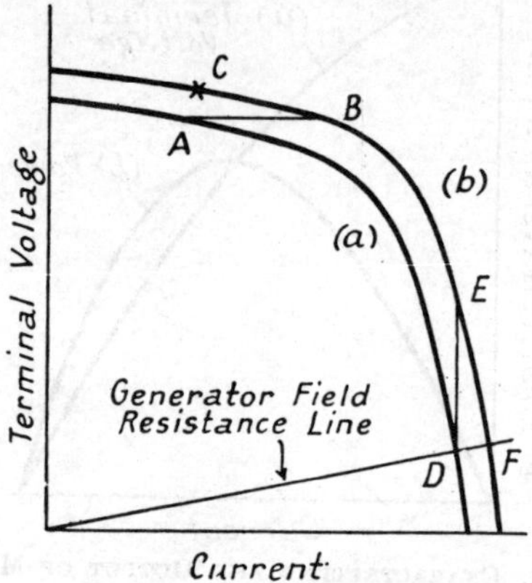

FIG. 253. RESPONSE OF METADYNE GENERATOR

the current is only slightly greater than that due originally to the higher motor torque usually required for a higher speed, corresponding to the increase in motor armature voltage.

Metadyne as an Exciter. When a metadyne functions as an exciter, the requirements are exactly opposite to those of a metadyne supplying a motor armature. When the control current is increased, tending to increase the field current of the generator a considerable voltage opposing the change is induced in the generator field winding owing to its enormous inductance. There can, therefore, be no immediate increase of current and the requirement from the exciter is a considerable and rapid increase in its terminal voltage to provide a so-called "field forcing" voltage to oppose the e.m.f. induced in the generator field.

If the metadyne operates at some point such as *D* on its characteristic *a*, Fig. 253, and there is an increase of control current, the working point moves to *E* on the new characteristic *b* with negligible change of current. It then moves more slowly along this characteristic as the current and resistance drop of the generator field increase to the point *F* which lies on the resistance line of the generator field circuit. The metadyne thus forms an ideal field-forcing exciter and thus compensates to some extent for the inductance of the generator field which tends to delay the response to control changes. The metadyne itself responds very quickly, since the change in direct axis flux required is relatively small, and the main flux is provided by the armature and ampliator windings which have a small time constant compared with the field of a normal machine.

Amplidyne. The amplidyne* is fundamentally identical with the metadyne generator already described, with the exception that normally it is fully compensated. To achieve the most effective compensation the compensating winding is distributed in a number of slots instead of concentrated in a single coil as previously described. The magnetic circuit may then consist of slotted punchings resembling those used for a.c. series motors. This form does not provide as much space for a variety of field windings as the form with polar projections, and the various control inputs are more usually combined in an electronic or magnetic amplifier providing the control excitation instead of in the machine itself.

Magnicon.† The distinction between metadyne and magnicon is the manner in which commutating zones are provided. In the magnicon the conductors of the coils undergoing commutation are removed from the field fluxes by short-chording the winding to half pitch. Polar projections are provided on both axes instead of being at 45° to them as with the metadyne. The short-winding pitch reduces the output voltage, of course, but has other compensating advantages, e.g. the length of mean turn of the armature winding, also the axial length of the armature is reduced; only half the number of field coils is required. A complete comparison of metadyne and magnicon is beyond the scope of the present treatment, but since each pole carries only one of the two component fluxes, the magnicon is very well suited to serve as a regulating exciter since the constant-control excitation required in a regulator can easily be provided by permanent magnets on the control axis, thus avoiding the necessity for a reference supply.

Metadyne Converter.‡ The metadyne converter is not a generator in that both input and output powers are electrical. If the input power is supplied at constant voltage, it is transformed to an output at constant current. The machine consists of a normal d.c. type of armature with two sets of brushes at 90° electrically to each other, mounted in a suitable field system, which, in the elementary form, need not carry any field windings. A small shunt or lightly-compounded motor is required to run the set up to speed and supply the fixed losses. The constant voltage supply is connected to one set of brushes P and R, Fig. 254, and the output taken from the others, Q and S. Since there are no field windings, the e.m.f. in each circuit is dependent entirely on the flux set up by the current flowing through the armature between the other set of brushes. Neglecting the resistance and brush-contact drops, the e.m.f. induced in the armature between the brushes P and R must be equal and opposite

* General Electric Co. of America and the British Thomson-Houston Co., Ltd.
† Macfarlane Eng. Co., Ltd.
‡ Metropolitan-Vickers Elec. Co., Ltd.

to the applied voltage, and hence constant. If the speed is constant then so also must be the flux set up by the current in brushes Q and S, and for the flux to be constant, the current must be constant. If the resistance or back e.m.f. of the load should increase, there must be an increase in the output voltage between brushes Q and S, in order to maintain the current constant. This in turn requires an increase of the corresponding flux, hence an increased current is taken from the supply. The input and output powers must be equal, neglecting losses, or the set will accelerate, and it is assumed that the speed is constant.

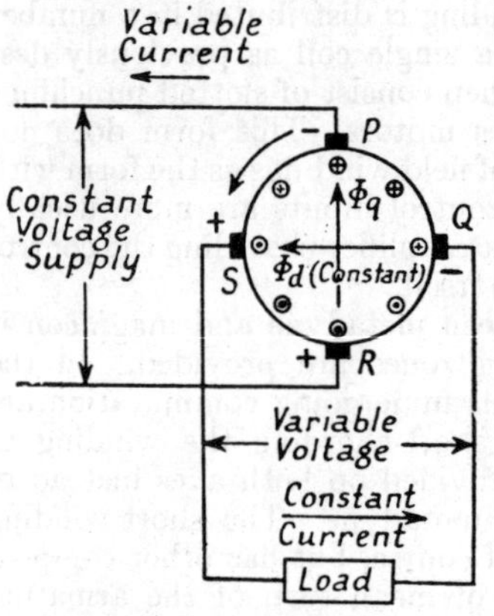

Fig. 254. Principle of Metadyne Converter

This machine has been applied to the acceleration of traction motors. If the current were maintained indefinitely at the accelerating value, it is apparent that the speed of the train would become excessive. To prevent this, it is necessary to provide field windings so that the output current can be reduced when the required speed is reached. Further complication arises from the interconnexion of motors and metadyne so that one generator can control four motors.*

Rosenberg Train-lighting Generator. The generator of a railway coach must be driven from the axle, since there is no prime mover on the coach. A battery must also be provided to supply the lighting when the coach is stationary, and this battery must be charged by the generator. Since the coach travels both forwards and backwards, the polarity of the generator must be independent of the direction of rotation. Again, since the speed of the coach varies widely, it is necessary for the generator to charge the battery at low speeds but not to overcharge it at high speeds, i.e. there must be an inherent current limit in the generator which will not be exceeded whatever the speed or the state of the battery. The cross-field generator is ideally suited to this purpose since reversal of rotation reverses the quadrature axis flux and the reverse rotation in this reversed flux maintains the same output polarity. The generator is an uncompensated metadyne generator of the form shown in Fig. 248 operating at variable speed. The current is, therefore, limited by the armature m.m.f. which is in direct opposition to the primary excitation. The latter is derived from the battery itself and is therefore

* Although over a hundred such equipments were installed on the London underground system about 1938 and have proved satisfactory, their otherwise desirable regenerative properties apparently produce an excessive rise of voltage on that system and has prevented their more widespread adoption.

automatically reduced when the battery e.m.f. is low. A rectifier in the armature circuit prevents the battery discharging when the generator e.m.f. is less than that of the battery at standstill, and at very low speeds, Fig. 255.

Two-stage Machines with Different Numbers of Poles in the Two Stages. The simplest case of two different numbers of poles for the

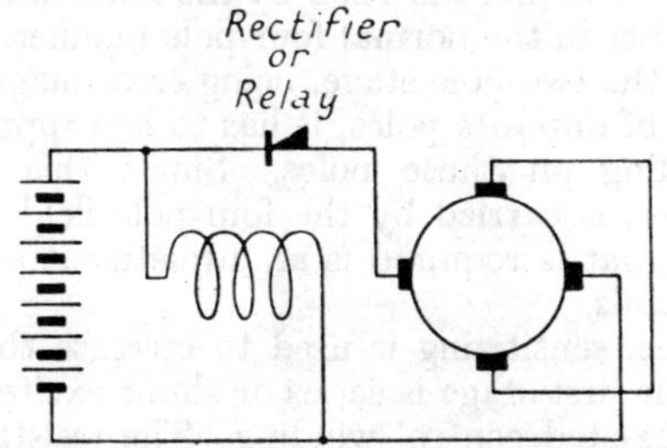

FIG. 255. ARRANGEMENT OF ROSENBERG TRAIN-LIGHTING GENERATOR

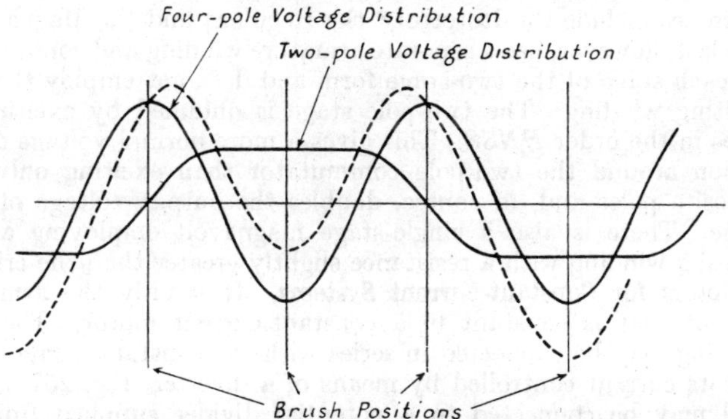

FIG. 256. VOLTAGE DISTRIBUTION AROUND COMMUTATOR

two stages is the combination two and four. Since a two-pole machine cannot operate as a four-pole one, it is apparent that the machine will be a normal four-pole machine and that its operation as a two-pole machine will be abnormal. For two-pole operation, two opposite poles are excited with opposite polarity, the other two being unexcited. If the winding is full-pitch as a four-pole machine, then it will be half-pitch as a two-pole machine. This renders the behaviour peculiar in that the potential distribution around the commutator is as shown in Fig. 256, which can be confirmed by drawing out a complete winding. Thus, with the brushes in the normal four-pole positions, there is an e.m.f. between two opposite brushes when operating as a two-pole machine, but no voltage between the other two. Four-pole operation does not produce any e.m.f. between opposite brushes. The two systems are thus independent and can be operated simultaneously.

Two-stage operation consists of using the two-pole output as the first stage to excite all poles of the four-pole output stage. One output terminal is then formed by the mid-point of the four-pole field winding, which has a potential mid-way between that of the two opposite brushes, and the other by the remaining pair of brushes connected together in the normal four-pole manner. Since the armature m.m.f. of the two-pole stage, being cross-magnetizing, acts on the second pair of opposite poles, it has to be suppressed by a compensating winding on those poles. Since, this same armature current, however, is carried by the four-pole field coils already on these poles, all that is required is an adjustment of the number of turns on these coils.

In most forms, sensitizing is used to increase the amplification. For example, the first stage is series or shunt excited in addition to the separately-excited control winding. The resistance of the self-excited winding is made slightly greater than the critical value so that the machine almost, but not quite, excites itself. Only a very small separate excitation is then required for full output. Such machines include the Rototrol,* the Regulex† and the Magnavolt.‡ The last, however, has a separate armature winding and commutator for each stage of the two-stage form and does not employ the self-exciting winding. The two-pole stage is obtained by exciting all poles in the order *NNSS*. This gives a more normal voltage distribution around the two-pole commutator than exciting only two opposite poles and, of course, doubles the output voltage of that stage. There is also a single-stage magnavolt employing a self-exciting winding with a resistance slightly greater than the critical.

Motors for Constant-current Systems. It is only the armature current that is constant in a constant-current motor. The field winding may be connected in series with the constant-current loop and its current controlled by means of a diverter, Fig. 257, or the field may be connected to a potential divider supplied from an auxiliary constant voltage supply. Such a simple method is, however, only satisfactory for such as fan motors which are permanently coupled to their loads. It is important to understand that with constant current in the armature, the torque and speed of the motor is zero with no flux, and increases with increase of field current. The danger is, therefore, not that a motor will run away with too weak a field, as with a normal constant-voltage motor, but that, on light load, it will run away with too strong a field current. For motors not coupled permanently to a big enough load, it is necessary to fit a small exciter on the motor shaft and connect it in

* Westinghouse Electric Corporation of America.
† Allis-Chalmers Corporation of America.
‡ English Electric Co., Ltd.

series opposition to the field supply. When the speed rises, so also does the exciter e.m.f., thus decreasing the field current, the flux, and the torque. With a non-overhauling load the speed is thus limited to that at which the exciter e.m.f. is equal to the voltage applied to the field from the potential divider. If the load overhauls,

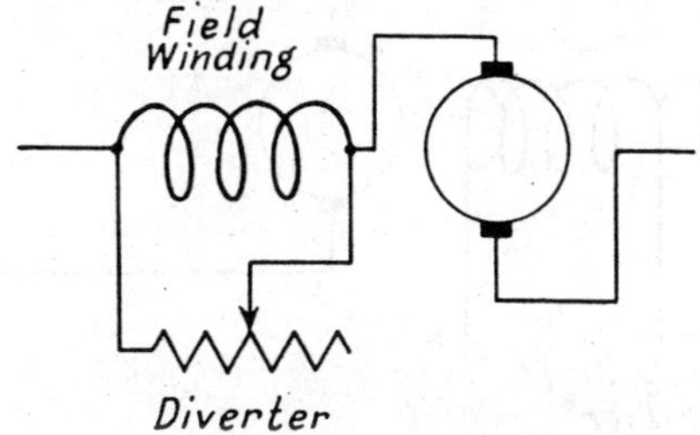

FIG. 257. CONTROL OF SMALL FAN MOTORS ON CONSTANT-CURRENT SYSTEM

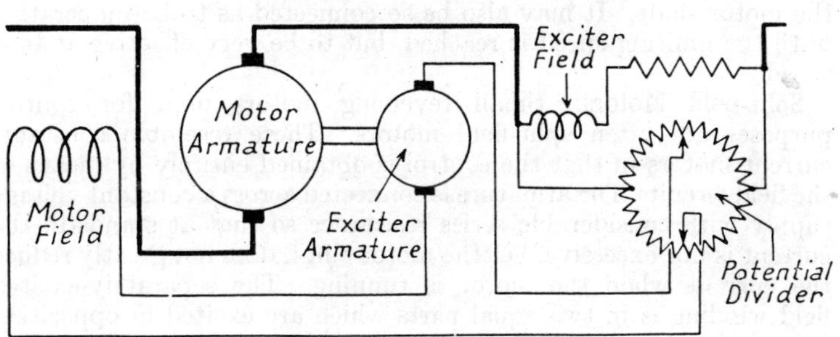

FIG. 258. CONTROL OF LARGE MOTORS ON CONSTANT-CURRENT SYSTEM

the exciter voltage exceeds that of the potential divider and the motor field current is reversed. Since the flux is reversed, the e.m.f. is reversed; the armature current, however, is unchanged. The motor, therefore, is acting as a regenerative brake. The terminal voltage of the main generators will be reduced, representing a reduction in their load. It is interesting to note that under motoring conditions, the exciter armature current flows against its e.m.f., and the exciter is motoring and adding its quota to the total torque. When the motor is regenerating, the exciter armature current is reversed, and the exciter is generating, thus adding to the braking torque.

The Austin system uses a double-potential divider shown in the off position in Fig. 258. Forward and reverse drives are obtained by opposite rotations of the contact arm, from the off position. The output of the divider is connected in series with the armature of the exciter on the motor shaft to supply the motor field winding.

The input to the divider is obtained by connecting it in parallel with the exciter field, and a resistor in series with the field.

When constant-current systems include the main propulsion motors, a rotating amplifier may be used to supply the motor field. The speed limit may then be derived from a tacho-generator on

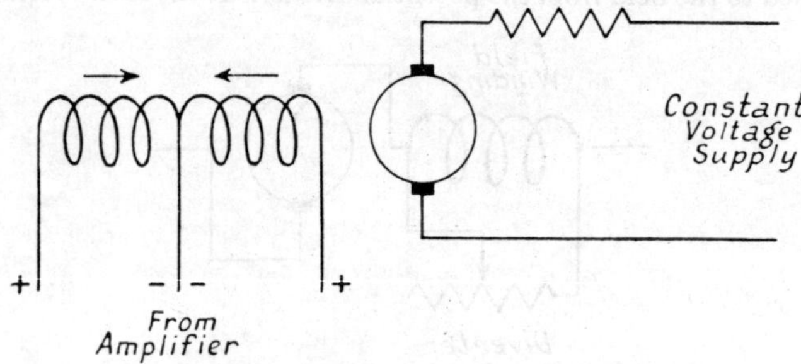

FIG. 259. SPLIT-FIELD MOTOR

the motor shaft. It may also be so connected as to be inoperative until the limiting speed is reached, but to be very effective at that speed.

Split-field Motors. Small reversing motors used for control purposes are often split-field motors. These resemble constant-current motors in that the control is obtained entirely by means of the field circuit. The armature is connected across a constant voltage supply with considerable series resistance so that at standstill the current is not excessive, yet the motor e.m.f. does not greatly reduce the current when the motor is running. The separately-excited field winding is in two equal parts which are excited in opposition from a "push-pull" electronic or magnetic amplifier output stage, Fig. 259. The motor is thus controlled in direction and torque as the currents in the two halves of the field winding become unbalanced under the influence of the input to the amplifier.

CHAPTER XIV

INSULATION: ARMATURE AND FIELD COILS

Classification of Insulating Materials. Most machine insulation includes, to a greater or lesser extent, organic material in the form of fibres or varnish. Such material is subject to change of properties with time, a process which is accelerated by raising the temperature, being doubled in rate for each 10 to 12°C increase of operating temperature. This change, called "ageing," is characterized particularly by increasing brittleness which leads to cracking under the mechanical stresses imposed by differential expansion, vibration, etc. It is therefore necessary to limit the maximum temperature which the insulation is permitted to reach, in order to obtain a reasonably long life. It is largely, but not entirely, the insulation which by imposing the temperature-rise limitation on the machine, thus limits the output of which a particular size of machine is capable. In view of the difficulty of measuring the temperature at the hottest place, and the variation of the temperature of the surrounding air, the limit for electrical machinery is expressed in the form of the maximum permissible temperature rise above that of the ambient air, and when the permissible temperature *rise* is decided a margin is provided to allow for the difference between the temperature of the hottest place and that actually measured.

What is considered a reasonable life will obviously vary according to the size, importance and duty of the machine. In addition, different materials have different ageing rates, hence the permissible temperature rises vary both with the insulating material and with the size, type and duty of the machine and the part of the machine being considered.

The British Standards divide insulation into seven classes with appropriate maximum permissible operating temperatures (*not temperature rise*) as follows*—

Class Y (90°C) insulation consists of materials or combinations of materials such as cotton, silk and paper without impregnation.

Class A (05°C) insulation consists of materials such as cotton, silk and paper when suitably impregnated or coated or when immersed in a dielectric liquid such as oil.

Class E (120°C) insulation consists of materials or combinations of materials which by experience or tests can be shown to be capable of operation at the Class E temperature.

* For detailed classified lists of materials, see B.S. 2757:1956 (*British Standards Institution, 2 Park Street, London, W.1*).

Class B (130°C) insulation consists of materials or combinations of materials such as mica, glass fibre, asbestos, etc., with suitable bonding, impregnating or coating substances.

Class F (155°C) insulation consists of materials or combinations of materials such as mica, glass fibre, asbestos, etc., with suitable bonding, impregnating or coating substances which by experience or tests can be shown to be capable of operation at the Class F temperature.

Class H (180°C) insulation consists of materials such as silicone elastomer and combinations of materials such as mica, glass fibre, asbestos, etc., with suitable bonding, impregnating or coating substances such as appropriate silicone resins.

Class C (above 180°C); insulation consists of materials or combinations of materials such as mica, porcelain, glass, quartz and asbestos with or without an organic binder.

Because of the hygroscopic nature of untreated fabrics and papers, Class Y is not used for machine insulation, since the presence of even a minute quantity of moisture lowers the breakdown strength of the insulation enormously.

The use of Class C materials for machine insulation is impracticable mainly due to their brittle nature. Classes A and B have long been established for machine insulation; Classes E, F and H are of more recent origin to cover developments in synthetic materials, especially in resins and varnishes.

Apart from varnishes and enamels, electrical machine insulation is composed principally of fibrous materials, either natural, such as cotton, silk, jute, wood, and asbestos, or artificial such as rayon, nylon, terylene and glass. These may be used as servings for wire covering, woven into tapes or wide fabrics, or felted into papers and boards. After the fibrous materials, the most important material is the natural laminar mineral mica. The properties of these various materials will be reviewed briefly before considering their application.

COTTON. When perfectly free from moisture, cotton affords quite good insulation, but, in common with all the usual forms of cellulose materials—paper cloth, etc.—it is very hygroscopic and absorbs large quantities of moisture if untreated. Treatment with varnish increases not only the resistance to moisture absorption, but also the heat resistance, as it does with most "dry" organic materials. Cotton is much used for covering wires, as woven tapes, and as the raw material for presspaper, pressboard, and, in the form of cloth, for bakelized fabric boards.

SILK. For many purposes silk is preferable to cotton since it is far less absorbent of both moisture and varnish and can be prepared much thinner, thus giving a better space factor. The supply is limited however, the price is high and hence the use is restricted.

JUTE, GRASS, WOOD, ETC. Many vegetable materials provide fibres which are coarse or short and therefore suitable only for papers and boards. These materials in general have ageing properties inferior to cotton.

ARTIFICIAL ORGANIC FIBRES. There are a considerable number of artificial organic fibres, such as rayon, nylon, terylene, which are suitable for wire coverings, tapes and other woven fabrics for insulation. The cost of these materials delayed their widespread use as insulators, but they can now compete successfully with natural materials in a number of applications. For example, nylon is widely used as an alternative to silk as a base for varnished fabric tape. Terylene is very suitable for use as a finishing tape for coils.

GLASS. Glass used for machine insulation is not of course in lumps, but is a fine fibre. Although glass wool has been available for many years, the application of glass to electrical machine insulation was delayed until the development of continuous filament glass fibres which are drawn out to a diameter of about 0·00023 in. These filaments are bunched together to form a yarn that can be applied to wire by roving or by brading, or woven into tapes, cloth and tubing.

Although the fibre itself is not absorbent, the breakdown value of untreated woven glass is only of the same order as that of untreated cotton. The material is therefore used in the same manner as other fabrics. The great advantage of glass is of course its immunity from ageing. Since, however, it is largely dependent for its insulating properties on the varnish between its fibres, there is still a limit to the temperature at which it can be operated. Woven glass can be produced just as thin as can silk, and hence gives good space-factors. It is less extensible than organic fibres, hence does not form as neat a surface when used as tape (except bias-cut varnished glass tape). The two main disadvantages of woven glass are that it is not as resistant to abrasion as natural fibres, and that, since the fibre is non-absorbent, varnish does not adhere as well.* Varnish treatment is satisfactory however, provided the fabric is not subsequently strained excessively on the bias. Varnish treatment considerably improves the resistance to abrasion.

ASBESTOS. Asbestos is a naturally fibrous mineral which can be reduced to a "wool" and spun. (A small percentage of cotton is sometimes added to assist in spinning.) It can be used as a wire covering, woven into tapes and cloth, or made into a paper. In addition to its well-known heat-resisting properties it is much more resistant to abrasion than woven glass. Its great disadvantage is that it cannot be applied so thinly as the organic materials or glass.

MICA. Mica is a natural mineral material which can be split into sheets and very thin flakes—less than 1 mil thick. It is an excellent insulating material; it can withstand high temperatures, does not appreciably absorb moisture, and has a high dielectric strength and

* The adhesion is dependent partly on the size etc. applied to the fibres to assist in weaving.

insulation resistance. Sheet mica can be used only in compression at right angles to the plane of cleavage and its price rises rapidly with the size of sheet due to the relative scarcity of large sheets. The cost and stiffness of sheet mica preclude its use for machine insulation except for the insulation between the segments of small commutators. Mica flakes, however, are flexible and relatively cheap, hence mica is widely used in machine insulation in the form of "micanite" sheets, mouldings, and tapes built up from flakes and a bonding resin or varnish.

It is convenient next to consider products made from the raw materials so far mentioned and also some others.

Varnishes and Enamels. A distinction is sometimes made between varnish and enamel in that the former consists of resins, either natural or synthetic, oils and solvents and thinners. Since these are all either liquid or soluble, varnish does not contain any solid matter in suspension. For some purposes, however, it is an advantage to add finely-divided insoluble solids to a varnish, thereby making it into an enamel. Enamels usually have a moisture resistance and thermal conductivity superior to varnishes, and, since the solids fill the interstices in fabrics, they give a better finish on surfaces which are inherently rough, as for example a taped coil.

Some varnishes consist of a solid resin dissolved in a solvent without any "drying oils," e.g. shellac dissolved in methylated spirit. Such varnishes dry by mere evaporation of the solvent, and produce a film which tends to crack subsequently. The varnishes formerly used most widely were compounded with drying oils such as linseed oil, and dry not only by solvent evaporation but also by chemical changes, including oxidation of the drying oil. There are now, however, numerous synthetic varnishes having superior properties such as toughness, slower ageing, etc., to those made entirely from natural resins and oils. Among the most important of these are the thermosetting resins of the phenol-formaldehyde (bakelite) type, the alkyd (glyptal) type, vinyl-acetal type and the silicone group.

The phenolic resin varnishes are excellent for providing a solid coil by bonding the fibrous insulation together and also in oil-modified forms for surface-finishing varnishes. The alkyd resins give varnishes suitable for bonding mica flakes for flexible sheets and also for finishing varnishes. The vinyl-acetal type is so far pre-eminent for the insulation of wires, although it has a more recent competitor in ethyoxyline resin which has superior heat resistance.

Silicone varnishes are of all types and are unsurpassed for heat resistance. Silicone is a generic title covering a wide range of types of chemical compound in the same way that "organic" does. Most varnishes incorporate oils or resins which are chemically classified as organic, i.e. they are composed primarily of carbon, hydrogen and oxygen, and depend for their chemical stability on the linkages between the carbon atoms. When these materials are exposed to high temperatures some of these linkages fail and the material is decomposed. If, however, the carbon atoms in these vulnerable linkages are replaced by silicon atoms, the result is a more stable compound which can be subjected to far higher temperatures without decomposition. The "silicones" thus constitute a vast group of compounds, some of which are suitable constituents for insulating varnishes of various types, having excellent temperature-resisting properties. A temperature (not temperature rise) of 180°C is considered appropriate for Class H insulation, but machines using only glass, mica and silicone have been operated at temperatures as high as 250° and 300°C. Such temperatures, however, produce other problems which limit the advantage which can be taken of the properties of silicones.

Insulated Wires. Round wires can be insulated with enamel (really varnish), or a fibrous roving. There have, therefore, long been available round wires insulated with a natural oil-based "enamel," cotton or silk. Cotton and silk are applied in the form of a roving wound helically on to the wire to form a complete covering. Two layers are usually applied, the second being wound in the opposite direction around the wire to the first. The material is thus d.c.c. or d.s.c., that is double cotton covered or double silk covered in contrast to the singly-covered wires known as s.c.c. and s.s.c. The use of enamelled wires for electrical machines was restricted to very small machines, since the insulating film was easily abraided, was not resistant to heat, and was liable to be softened by subsequent varnish treatment of the coils. Hence a fibrous covering was usual, either alone as in d.c.c., or, more rarely, in combination as in enamel with s.s.c. The earlier artificial organic fibres were not particularly satisfactory and were little used. The space "wasted" on insulation increases in the order, enamel, silk, cotton, hence it was unfortunate that enamel was so unsatisfactory and silk so expensive. These materials are all Class A, and for Class B, wires insulated with asbestos roving were developed. The thickness of asbestos covering however is even greater than that of cotton covering.

This position was completely changed by two developments: vinyl-acetal resins and continuous glass fibres. Vinyl-acetal-based varnishes produce a film so resistant to abrasion that the insulated wire can even be drawn down in size or rolled flat with negligible damage to the varnish film. It is, moreover, superior to oil-based varnishes in both heat and solvent resistance. Other synthetic materials have very similar properties. Continuous glass fibres can be applied to wires as a roving and provide a Class B, F or H covering far thinner than is possible with asbestos, although it is more susceptible to abrasion and other mechanical damage than is asbestos. This defect is to some extent remedied by varnishing the covering immediately after its application. Thus vinyl-acetal varnished wires have almost entirely replaced "oleo-resin enamelled" wires, and silk-covered wires, and, to a considerable extent even cotton-covered wires; whilst glass-covered wires have largely replaced asbestos-covered wires.

Square and ribbon wires could not formerly be enamelled and were, therefore, either double-cotton-covered or asbestos covered. It is practicable to apply vinyl-acetal varnish to these cross-sections, and, of course, glass-covering can be applied. To obtain an improved resistance to abrasion, glass is sometimes applied as a braid woven around the conductor. The abrasion resistance of glass is also improved not only by varnishing it after application, but also by bonding it to the copper with varnish.

The thickness of wire insulation varies both with the wire diameter and the material. With some materials several thicknesses may be available for each wire diameter. It is, therefore, not practicable to give complete tables. Moreover there are tolerances on both wire diameter and insulation thickness. The figures below, therefore, give the maximum amount in mils (0·001 in.) by which the nominal wire diameter can be exceeded by the diameter of the insulated wire and they include the tolerance on the wire diameter as well as the nominal thickness of, and tolerance on, the insulation.

ROUND WIRES

Nominal Wire Dia. in.	*Oleo-resin Enamel* (B.S. 156) Normal	Thick
0·002	0·5	0·7
0·028	2·3	3·4
0·160	6·1	—

	Vinyl-acetal Enamel (B.S. 1844) Fine	Medium	Thick	Extra-thick
0·002	0·4	0·6		—
0·028	2·3	3·2	4·1	5·7
0·160	5·0	6·2	7·5	10·0

	Ethoxylene Enamel Fine	Medium
0·002	0·4	0·6
0·028	2·3	3·2
0·080	3·6	4·7

	Silk and Rayon Covering (B.S. 2480) Single	Double
0·0016	1·3	2·3
0·028	2·3	3·3
0·064	3·6	4·6

	Cotton Covering (B.S. 1791) Single		Double	
	Fine	Ordinary	Fine	Ordinary
0·006	3·5	4·1	6·1	9·1
0·028	5·2	6·3	7·3	11·3
0·160	8·5	9·6	12·6	15·6

Impregnated Asbestos (B.S. 1497)

0·01	8 to 11
0·02	10 to 12
0·08	12 to 15

Varnish-bonded Glass Covering (B.S. 1933)

	Single	Double
0·01	5·1	8·1
0·08	5·8	8·8
0·16	—	13·6

Glass Braided

Nominal	14/17	18/21	28/32
0·08	17·8	21·8	32·8
0·160	18·6	22·6	33·6

Ribbon Wires

Vinyl-acetal Enamel

Fine	Medium	Thick
2·4	4·0	6·0

Cotton

Single	Double		Triple	
	Fine	Ordinary	Fine	Ordinary
10	16	20	25	30

Glass Lapped

Extra-fine	Fine	Ordinary
9	13	17

Tapes. Tapes can be applied to both straight and curved parts of coils, whereas sheet material can be applied only to straight parts, hence the wide use of tapes. Tapes can be used both for structural and for insulating purposes. For Class A insulation, cotton tape is usually employed for structural purposes and ranges from a smooth tape only 3 mil thick to heavy twill-weave webbings 23 mil thick. For insulation of Classes B, F and H glass tape or asbestor tape (both woven) may be used, the thickness ranging from 2 mil to 9 mil, and from 10 mil to 20 mil respectively. Glass tape has a very high tensile strength, whilst asbestos tape has a higher resistance to abrasion.

When tape or cloth is used for insulation purposes it has of course to be varnish treated as explained on page 340. If however this varnish treatment is carried out by a number of successive treatments on the material before it is applied to the coils, the result is a "varnished fabric," a smooth material in which all the interstices are filled and which has a breakdown-strength of the order of 1,000 V/mil, which is about four times as great as is obtained from fabrics whose only varnish treatment is subsequent to their application to the coil. In the manufacture of varnished fabrics the raw cloth is subjected to various preliminary processes and then passed through a bath of varnish and up a heated drying tower, the varnishing process being repeated a number of times. This varnished fabric has much less tendency to absorb moisture than fabrics which receive only a simple varnish treatment after application to the coil. Varnished fabric tapes are usually made by slitting varnished cloth. If the cloth is slit parallel to the warp, giving a "straight-cut" tape, it has too little elasticity to permit neat taping of the bends in a coil. Bias-cut tape, slit at an angle to the warp (usually 45°) overcomes this difficulty, but naturally comes in short lengths unless specially woven. The fabric base for Class A materials is usually cotton, but where thinner, or more elastic tape is required, silk or nylon is used. Varnished glass tapes are classified according to the heat resistance of the varnish used.

A Class B, F or H tape can be built up in the form of a sandwich of mica flakes between two layers of thin paper, or between a layer of paper and a layer of thin woven silk or glass. The mica flakes are bonded to each other and to the paper or fabric with varnish, the alkyd resin varnishes being very suitable for this purpose. Such "mica tapes" have very little elasticity and poor resistance to mechanical damage, and must be followed by a fabric tape to consolidate and protect them. In the manufacture of mica tape, a sheet of paper 1 mil thick, or thin fabric, is drawn slowly along over a table, being coated with varnish as it is drawn off the roll. Mica flakes are stuck on by hand and varnished. The top layer of paper is then applied from a second roll and the whole passed through an oven to dry the varnish before being wound on a roll. It is subsequently slit to the required width of tape, usually from ½ in. to 1 in.

Tape used as insulation is usually applied "half-lap", i.e. each turn overlaps the previous turn by almost half the width of the tape. The electrical breakdown value thus depends on the puncture of one thickness of the tape, or the creepage between the turns for a distance of half the width of the tape. The overall radial thickness per layer of half-lap tape is obviously twice that of the tape itself. Tapes applied purely for structural purposes (e.g. over the mica paper wrap of strap-wound armature coils) are not lapped but may be "butt" i.e. edges almost touching, or "staggered" i.e. in an open helix, according to circumstances.

Flexible Sheet Materials. Flexible sheet materials include papers, woven fabrics and built-up sheets incorporating mica flakes.

Papers vary widely in thickness, mechanical properties and ageing characteristics, according to the raw material used and the treatment during manufacture. Papers are essentially sheets of felted fibres, the base of which is usually vegetable, e.g. cotton, wood, jute, grass, although asbestos is also made into a paper for electrical insulating purposes. Papers are of course absorbent and, except for asbestos paper subject to ageing. A paper about 1·5 mil thick is sometimes used for insulating the laminations of armature cores, being pasted on the back of the complete sheets before punching. This paper is made from wood-pulp treated chemically to remove lignin and resins. The thin paper used in the manufacture of mica tapes was formerly Japanese tissue paper but an alternative was developed in America using old manilla rope as the raw material. These papers, although only 1 mil thick, are required to have a very high tensile strength along the length so that they can be drawn along the table when wet with varnish. The transverse strength is correspondingly very low.

Wide use is made of relatively thick papers called presspaper made from cotton, cotton and wood-pulp, or jute (the cotton having better ageing properties). The paper is highly compressed and calendered and ranges from 5 to 20 mil in thickness.

Another material called leather-paper or leatheroid resembles presspaper but is far tougher. It is made by treating the wet paper with chemicals which convert the cellulose into a gelatinous mass, so that on drying, the material is much more homogeneous and tougher than presspaper. It is also much more expensive and hence is used only where toughness is essential. It is available in thicknesses from 5 mil to $\frac{1}{32}$ in.

A most important group of flexible sheet materials is formed by the varnished fabrics which have already been mentioned under tapes. These materials, both tapes and sheet material, are sometimes referred to as "empire," this word however is the trade name of a particular manufacturer.*

Flexible sheets are made consisting of a layer of mica flakes bonded to a paper backing with varnish. Two kinds are used for armature coil insulation, differing primarily in the nature of the varnish employed. In one, a shellac varnish is used, so that heat is necessary when applying the material to the coil, but it results in a hard strong insulation. In the other, a varnish which produces a flexible plastic bond enables the material to be applied to the coil cold.

Boards. Boards include such materials as pressboard, bakelized boards and micanite Pressboard is fundamentally identical with presspaper; the distinction is merely that it is too thick to be made in rolls, and therefore has to be made in short lengths. The degree of compression in manufacture, hence the porosity, depends upon the conditions under which it is to be used. It is available from 20 mil to $\frac{3}{4}$ in. or more in thickness. Its name, and the alternative "fuller-board" are derived from its original use in finishing cloth.

Bakelized boards are usually built up from paper or fabric—usually cotton, but sometimes asbestos—treated with bakelite varnish and bonded together by heat under pressure. The paper board is superior in electrical properties, the cotton fabric board in mechanical properties, and the asbestos fabric board in ageing properties. Glass cloth can also be used, but is not completely satisfactory with bakelite varnish, and a better product results if a melamine-based bond is employed. The cost of fabric boards, especially glass, is relatively high.

An alternative to bakelized fabric board is provided by plywood,

* Micanite and Insulator Co. Ltd.

usually beech, bonded with bakelite resin and highly compressed in the curing process. This material has excellent mechanical properties and can to some extent be made anisotropic by varying the arrangement of the directions of the grain in successive lamina.

Micanite boards are made by bonding mica flakes together either with varnish or with a dry resin powder under heat and pressure. The properties of the product depend upon the size of mica flake and proportion and nature of the bond. High bond content in varnish form gives more flexibility, and low proportions of dry bond give rigid sheets, which can be finished to a specified thickness. Some forms, made from relatively large flakes, can be moulded under heat and pressure to form complicated shapes such as the micanite vee-rings of commutators, and also moulded on to poles and brush-holder arms.

Varnish Treatment. Treatment with varnish or some other material is necessary for all coil insulation. In the body of the coil it serves to bond the conductors and insulation into a mechanically sound structure and to fill the interstices, while on the surface it is essential to prevent the ingress of dirt and the absorption of moisture. Normally, varnish treatment takes the form of oven-drying, immersion in varnish until cool, draining and finally oven-drying the varnish. The treatment may be repeated once or twice to obtain better impregnation.

The initial drying, however, may be carried out under vacuum after preliminary heating. This heating cannot be economically performed under vacuum without unduly prolonging the process, since radiation alone is an ineffective means of heat transfer at the temperatures required. After vacuum drying the varnish is usually forced into the coil by positive pressure instead of relying on simple immersion. Again, after the removal of the varnish and draining of the coil, vacuum may be employed to remove the solvent prior to drying the varnish.

Instead of varnish a compound, usually of bitumen, may be used. It is liquified by heat and used in a vacuum-pressure process in the same manner as varnish. After draining, the coil is allowed to cool, when the compound in it solidifies.

Normal varnishes cannot fill the interstices of a coil and its insulation completely since they contain volatile solvents and thinners with the consequence that the solid residue after drying is smaller in volume than the varnish from which it is deposited. Complete impregnation is possible with so-called "solventless varnishes" or allyl co-polymers. These consist of two synthetic compounds or "plastics," one of which is initially a liquid in which the other is dissolved. On heating, the whole mass becomes solid without the evolution of any vapour. These materials are too expensive for general use and the alternative, bitumen compound impregnation, which is relatively cheap, is widely used for stationary windings, i.e. field coils. Since it is thermoplastic it, however, becomes soft when the machine is hot and is liable to be thrown off on to the poles if used for armature windings. It is possible to use compounds with higher softening temperatures, but impregnation then becomes more difficult.

With small machines, varnish treatment can most conveniently and economically be performed by dipping or impregnating the whole armature after the coils have been placed or wound into the slots, but with larger machines, handling complete armatures would be difficult and varnish penetration inadequate, so that the coils are treated before being placed in the slots.

Air-drying varnishes and enamels (i.e. not requiring heat for drying) are also used, but only to obtain a better surface finish on an impregnated coil, or on a coil which cannot be heated, e.g. on armature coils of which the slot portion is insulated with shellac-bonded mica, which would soften and loosen if heated.

Armature Coils. On relatively small machines the armature coils are wound from round wire, but as the size of the conductor increases, the poor space factor associated with round wire leads to the use of square and rectangular or ribbon wires. These have rounded corners so as not to cut the insulation, and rectangular wires are preferable to square since it is easier to avoid twists. With the largest rectangular conductors the width is several times the thickness and the term "strap" is used instead of "wire." With round and other wires of small cross-section it is possible to use insulated wires in making the coils, but with strap-wound coils it is necessary to use bare copper and to insulate the conductors after the coil has been shaped, since the relatively wide conductor stretches considerably on the outer edge when it is bent "on edge," and the insulation would be opened or unduly thinned if applied previously. Coils for fractional horse-power machines are made by winding insulated wires directly into the slots which have been lined with a "liner." With larger machines wire-wound coils are most easily made by "pulling." In this method loops such as that shown in Fig. 260 are first wound, and then placed in a pulling machine in which clamps grip the coil at the "noses" and at the four points which will become the ends of the straight or slot portion of the coil. The latter two pairs of clamps then move in opposite directions perpendicular to the plane of the loop until the required pitch is obtained, Fig. 261. This operation must not stretch the copper any more than can be helped and during the pulling process the clamps holding the noses are drawn towards one another. Using pneumatically-operated pulling machines reduces the time required to shape a coil to a mere fraction of the time required by "forming." The latter process consists in winding the coil by hand, conductor by conductor, around a "former" of the appropriate shape. This method still has to be employed for the manufacture of strap-wound coils (Fig. 262) which, however, are usually single-turn coils consisting of two or three conductors side by side in the slot. Since the copper strap is

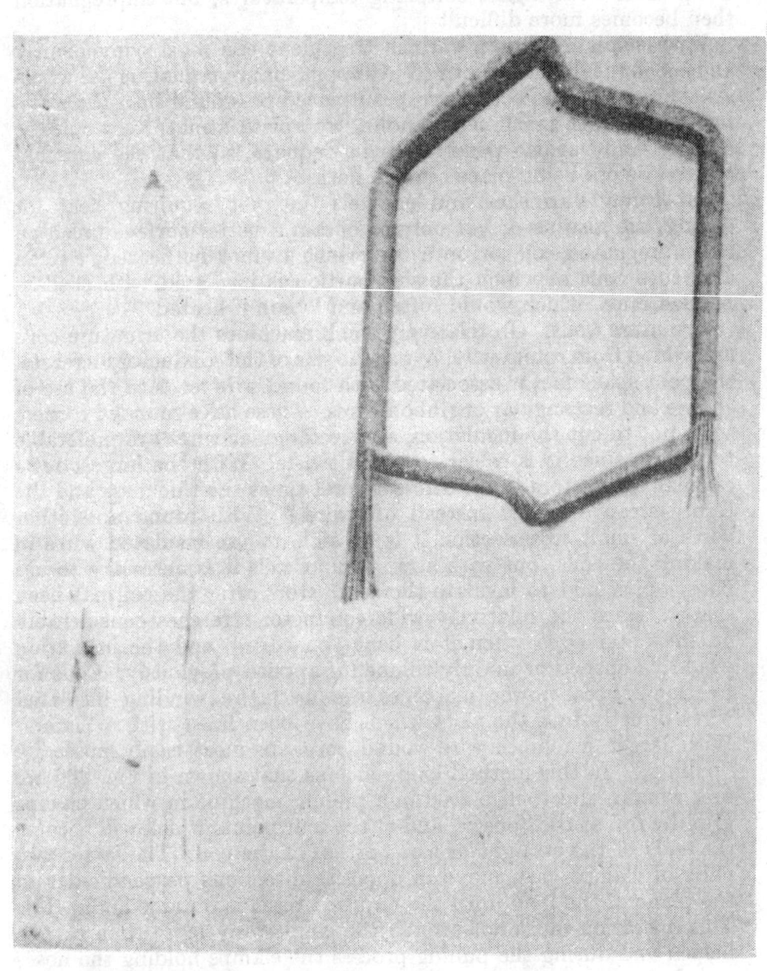

Loop ready for shaping by "pulling."

Finished "pulled" coil after insulating.

FIGS. 260 AND 261. COMPLETE CELL OF ARMATURE WINDING OF TRACTION MOTOR

5 coil-sides per layer and 3-turn coils

(*English Electric Co., Ltd.*)

relatively stiff, even though "soft" copper is used, considerable hammering into shape is required and the formers are frequently

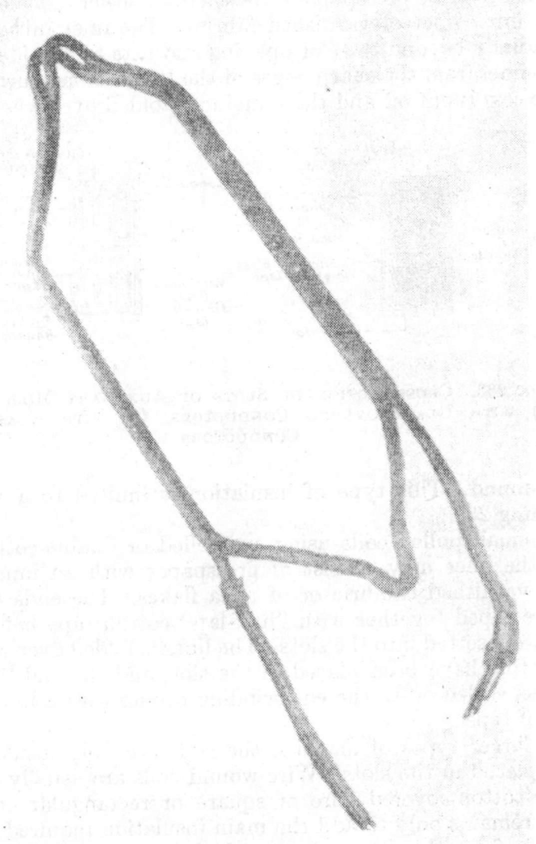

Fig. 262. Complete Cell of Armature Winding
3 coil-sides per layer; 2-turn coils
(*Lancashire Dynamo and Crypto, Ltd.*)

made of steel, although wood may be used if neither the number of coils required, nor the cross-section of the strap, is great.

Insulation of Armature Coils. It is convenient to consider first, only coils insulated with Class A insulation in order to reduce the number of alternatives to be discussed simultaneously.

For fractional horse-power machines with the armature coils

wound directly into the slots, vinyl-acetal enamelled wire is almost ideal. The enamel provides the insulation between turns whilst the insulation between the coil and core is in the form of a "slot liner." The liner may be presspaper or leatheroid, either alone or reinforced by an inner liner of varnished fabrics. The liner initially projects out radially beyond the slot opening and acts as a guide, protecting the enamel from the sharp edges of the laminations during winding. The excess is cut off and the remainder folded over when the coil is

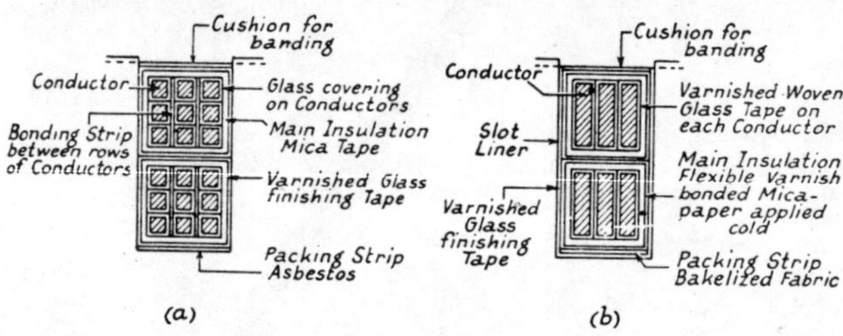

FIG. 263. CROSS-SECTION OF SLOTS OF AUXILIARY MILL MOTORS (a) WITH GLASS-COVERED CONDUCTORS, (b) WITH GLASS-TAPED CONDUCTORS

fully wound. This type of insulation is limited to a voltage not exceeding 250.

For small pulled coils using enamelled or double-cotton-covered wire, the liner may consist of presspaper with an inner lining of either varnished cambric or of mica flakes. The ends only of the coil are taped together with "half-lap" cotton tape before the coil sides are inserted into the slots. The liner is folded over when all the conductors have been placed in the slot, and the ends of the liner are then sealed on to the end winding taping with a few additional turns of tape.

For larger types of machine the coils are fully insulated before being placed in the slots. Wire-wound coils are usually made from double-cotton-covered wire of square or rectangular cross-section and it remains only to add the main insulation required to insulate the coils from the core, other earthed parts, and from each other. The periphery of the coil can be divided into two sections: the straight portions which lie in the slots, and the end-windings which include all the bends. The slot portions can be insulated by wrapping them with pieces of sheet material, whereas on the end-windings it is necessary to use tape to get around the bends. The sheet material most commonly used is a flexible mica wrap consisting of mica flakes stuck with flexible varnish on to a thin sheet of paper. For taping the end-windings a cotton tape, or varnished fabric tape followed by cotton tape is normally used, according to the voltage.

For strap-wound coils it is first necessary to insulate the individual conductors from each other. For the end-windings this is necessarily done by taping either all or alternate straps, usually with half-lapped cotton tape or varnished fabric tape. The slot portions can be similarly taped,* but this is not essential since they can be adequately insulated by interleaving the beginning of the mica paper wrap between the conductors before wrapping it around the coil to form the main insulation as shown in Fig. 265. This method has the advantage of taking up very little of the slot compared with taping, but is not suitable for voltages exceeding 1,500. The mica paper is, of course, at least, Class B, but, because of its outstanding reliability, it is normally used even when the remainder of the materials, including conductor insulation inside it, are Class A.

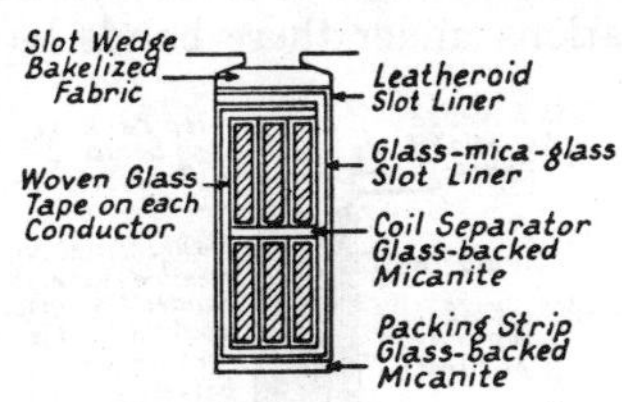

FIG. 264. CROSS-SECTION OF SLOT OF 225-V GENERATOR WITH SLOT LINER AS MAIN INSULATION

For Classes B, F and H insulation the processes are similar, merely the materials being different. Insulated wires are glass-covered, glass-braided or asbestos-covered instead of enamelled or cotton-covered; cotton tapes and varnished fabric tapes are replaced by mica tape when used solely for insulating purposes or by glass or asbestos tape when some mechanical strength is required. It is permissible to use small quantities of materials with lower permissible temperature rises in Class B, etc. insulation so long as it is only for structural purposes and the overall electrical and mechanical properties are not impaired at the higher temperatures. The paper backing of the mica-paper wrap and the paper and fabric used in the mica tapes are deemed to be permissible under this dispensation, although glass cloth can be used as a genuine but more expensive high temperature alternative.. Even then the varnish, being organic, may not be strictly Class B.

Even when the coils are fully insulated before being placed in the slots, it is common practice to line the slots to protect the coil insulation from the edges of the laminations. Normally the liner consists of presspaper 0·005 or 0·007 in. thick, but asbestos paper may be used for high temperatures. In addition, packing strips of pressboard or bakelized fabric etc. are used, one at the bottom of the slot to prevent the corners of the coil being damaged by the radii at the corners of the slot; one between the two coil sides to provide a space between the two layers at the ends of the straight portion; and one at the top of the slot to prevent damage to the coil when the slot wedge is driven in. For large machines the coils, wound in "open" slots, are invariably retained in the slot by wedges, usually of bakelized fabric or bakelized compressed wood, fitting into grooves in the side of the teeth. The wedges thus have to be driven in from the ends of the core. In addition to the top packing strip it is common practice to fold over the slot liner at the top of the slot

* Fig. 263*b* shows a slot section with Class B insulation of this type.

if space permits. With small machines using semi-closed slots a strip of pressboard is frequently used as a wedge; no wedge grooves are required in this case since the tips of the teeth are sufficient. Alternatively, with small machines, a series of circumferential bands of steel wire may be wound around the armature at intervals along the core to retain the coils. To accommodate them the laminations under these bands have a reduced outside diameter. With

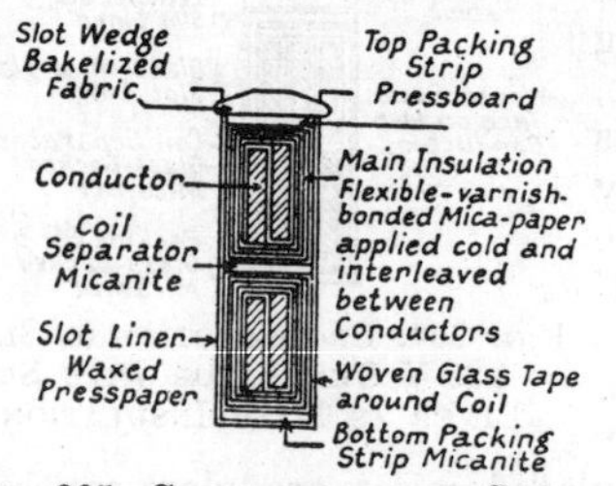

FIG. 265. CROSS-SECTION OF SLOT OF 1,100-V GENERATOR WITH CELL INTERLEAVED BETWEEN CONDUCTORS

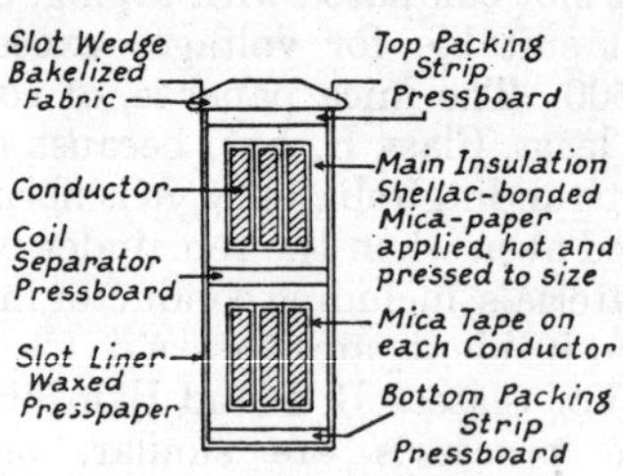

FIG. 266. CROSS-SECTION OF SLOT OF 7,000-H.P., 1,500-V MOTOR WITH SHELLAC-BONDED INSULATION

such banding, strips of insulating material must be placed over the coils to protect them where the bands bear on them.

Banding over the end-windings of all medium size and large machines is necessary to support them against centrifugal forces. Here again a tough material is required to protect the coil insulation, and presspaper is frequently used with varnished cambric or mica flakes to give an improved insulation. The presence of the band over the end-windings inevitably restricts ventilation and the consequent effect on the temperature rise must not be overlooked.

For large machines and others of particular importance or for arduous duty, more elaborate armature coil insulation may be employed. For example (Fig. 266), for a large rolling-mill motor each copper strap may be taped with mica tape, and the wrap, bonded with shellac varnish, applied hot and subsequently pressed to size on the slot portion, while the ends are taped with more mica tape and finished with glass tape. For the smaller auxiliary mill-motors, to obtain a high rating for weight and size, glass-covered wires, taped on both slot portion and ends with mica tape and finished with varnished glass tape, may be used, Fig. 263*a*. Traction motors, from which such mill motors were developed, may similarly be specially insulated.

Thickness of Insulation. Although, as might be expected, the thickness of insulation used increases with the working voltage of the machine, there is no simple relation between thickness and rated voltage. Since electrical failure of insulation in service is frequently preceded by its mechanical failure, it is apparent that machines working under arduous conditions (involving say repeated mechanical stressing due to reversing or other transient loadings, dirty conditions, high operating temperatures, etc.), require more substantial insulation than other machines of the same rated voltage working under more favourable conditions. Again, with a machine that is vital to the operation of the plant, it is economic to use thicker and better insulation to eliminate, as far as practicable, the possibility of failure. Where, however, a large number of similar motors are used for independent drives to individual machines, it may be more economical to use a cheaper construction, involving less insulation, and to have a number of spare motors. A motor which fails is then replaced immediately by a spare and is repaired at leisure.

It may be noted that the apparent factor of safety in insulation is very high since the manufacturer has to allow a reasonable margin to avoid failure when the insulation is tested at twice rated voltage + 1,000 V. It must be remembered, however, first, that there is a minimum thickness which can be applied without undue risk of damage during manufacture, and that this is more than adequate electrically for low voltage machines; secondly, that the test is applied to clean new insulation, whereas failure must be avoided under service conditions when the insulation has aged and become dirty.

The amount of insulation required for a given voltage and operating conditions is therefore governed largely by statistical information on failures in similar conditions. Entire absence of failures would suggest that the insulation used is unnecessarily and uneconomically thick; an appreciable proportion of failures on machines during their accepted working life obviously indicates that the insulation is deficient.

These facts, together with the wide variety of insulating materials available, makes it impracticable to specify here the insulation required for various voltages and conditions. A few typical cases are therefore given as an indication of actual practice.

Fig. 263*a* shows the slot cross-section of a motor used for auxiliary drives in steel mills for voltages of the order of 500 and horse-powers up to 200. The conductors are small enough to permit the use of glass-covered copper strips. The main insulation consists of mica tape on both slot portion and ends of the coil, thus avoiding a joint just beyond the end of the slot such as occurs in the other examples. The finished size of the slot is obtained by adding the following to the dimensions of the bare copper—

Width		cm
Glass covering on conductors	3 × 0·043 =	0·129
Bonding strips	2 × 0·025 =	0·050
Main insulation, mica tape		0·100
Glass finishing tape		0·025
Varnish		0·025
Tolerance and clearance		0·021
		0·350

Depth		cm
Glass covering on conductors	6 × 0·043 =	0·258
Main insulation	2 × 0·100 =	0·200
Glass finishing tape	2 × 0·025 =	0·050
Varnish	2 × 0·025 =	0·050
Packing strip		0·075
Banding		0·247
		0·880

Fig. 263*b* shows the slot arrangement of a motor which is similar to that of Fig. 263*a* except that the conductors are of such proportions that glass-covered strip cannot be used. Bare copper conductors are insulated from each other by taping each with a half-lap layer of varnished, woven-glass tape. The main insulation in this case is flexible mica-paper wrapped around the coil and secured with a varnished-glass, finishing tape. (Mica tape could have been used as in the previous example.) A slot liner is used to facilitate insertion of the coils into the slot. The dimensions are as follows—

Width		cm
Glass tape on conductors	3 × 0·030 =	0·090
Mica-paper wrap		0·100
Glass finishing tape		0·025
Varnish		0·025
Slot liner	2 × 0·018 =	0·036
Tolerance and clearance		0·024
		0·300

Depth		cm
Glass tape on conductors	2 × 0·030 =	0·060
Mica-paper wrap	2 × 0·100 =	0·200
Glass finishing tape	2 × 0·025 =	0·050
Varnish	2 × 0·025 =	0·050
Packing strip		0·075
Slot liner		0·018
Banding		0·247
		0·700

Fig. 264 shows the slot of a low-voltage (225 V) generator with a rating of the order of 200 kW. Here a semi-closed slot is used. The conductors are taped with woven-glass tape and inserted through the slot opening one at a time. The main insulation consists of a liner formed of flexible micanite with glass cloth on both sides. This, together with the leatheroid liner, is folded over when all the conductors are in position and the whole secured by the slot wedge.

The ends of the coil are then taped up to and over the projecting liners. The dimensions are as follows—

Width		cm
Glass tape on conductors	3 × 0·050 =	0·150
Glass-mica-glass liner	2 × 0·075 =	0·150
Leatheroid slot liner		0·025
Varnish		0·025
Tolerance and clearance		0·020
		0·370

Depth		cm
Glass tape on conductors	2 × 0·050 =	0·100
Glass-mica-glass liner	3 × 0·075 =	0·225
Packing and separator		0·125
Varnish	2 × 0·025 =	0·050
Leatheroid slot liner		0·038
Wedge		0·312
		0·850

Fig. 265 relates to a 1,100-V generator of the order of 2,000 kW. The bare copper conductors are insulated from each other by interleaving the beginning of the mica-paper wrap between them. The wrap is secured with a woven-glass tape. On the ends each has to be taped with mica tape. The dimensions are as follows—

Width		cm
Interleaved mica-paper wrap		0·230
Glass tape around coil		0·025
Varnish		0·025
Slot liner		0·036
Tolerance and clearance		0·034
		0·350

Depth		cm
Interleaved mica-paper wrap	2 × 0·230 =	0·460
Glass tape around coils	2 × 0·025 =	0·050
Varnish	2 × 0·025 =	0·050
Packing strips and separator		0·225
Slot liner		0·018
Wedge		0·477
		1·280

The insulation shown in Fig. 266 is expensive and is used on most important machines only. In this case the machine is rated at 7,000 h.p. at 1,500 V. The conductors are insulated from each other by mica tape. They are then bound together and insulated with shellac-bonded mica-paper applied hot and tightened in a special machine and then pressed to size. The coil ends are insulated with mica-tape throughout. The slot dimensions are—

Width		cm
Mica tape on conductors	3 × 0·035 =	0·105
Main insulation (including binding tape) . .		0·284
Slot liner		0·036
Tolerance and clearance		0·025
		0·450

Depth		cm
Mica tape on conductors	2 × 0·035 =	0·070
Main insulation	2 × 0·284 =	0·568
Packings and separator		0·310
Slot liner		0·018
Wedge		0·554
		1·520

Field Windings. For shunt coils, enamelled-, cotton-, asbestos- or glass-covered conductors are used. For small wires, round section is invariably employed, but with larger areas square or rectangular conductors are frequently used. Vinyl-acetal enamelled wires are used with small sections, being more reliable than single-cotton-covered wires. Double-cotton covering is suitable for larger Class A coils, and asbestos or glass covering for higher temperatures. Braiding may be used for the very large conductors required in the case of large low-voltage generators.

Small shunt coils are generally wound on formers, the conductors being held together by tapes and, later, by varnish. The pole must then be insulated before the coil is placed upon it. For this purpose leatheroid or presspaper is commonly used, the thickness ranging from 0·2 mm to 1 mm according to size and working voltage. Larger coils may be wound on bobbins or spools which may be made entirely of insulating material such as presspaper and pressboard or may be made of sheet steel and insulated from the coil with presspaper. For higher temperatures, poles and spools are insulated with micanite or asbestos paper. Thick washers of pressboard or micanite are used to protect the ends of the coil from the frame and pole tips, or from the spool flanges. Plywood is also used for this purpose.

Shunt coils are frequently protected by taping, the tape being applied toroidally to former-wound coils. Although providing protection against mechanical damage, taping reduces the cooling.

With fan-ventilated machines, and large high-speed machines, sectionalized—instead of solid—field coils can be used. With these coils, instead of having the conductors in one solid mass, they are wound in sections. These sections are separated by distance pieces so that ventilation spaces are formed between them. A further ventilation space is left between the coil and the pole. Such a ventilated coil is illustrated in Fig. 267.

In winding rectangular field coils, pieces of presspaper or strong cloth are inserted between layers at the corners, to prevent damage to the conductor insulation which is subjected to heavy pressure at

FIG. 267. SHUNT COIL
(*British Thomson-Houston Co., Ltd.*)

these points. In all cases, the field coils must be packed so as to fit tightly on the pole to avoid any movement.

Series and Interpole Windings. For small machines, insulated wires are used for the series and interpole windings as with shunt coils. When the sectional area permits, however, a good construction is obtained for interpole coils and for the series coils of series motors by using a single layer winding consisting of heavy copper strap wound on its narrow edge. When thin conductors are thus wound on edge, the turns are insulated from one another by inserting thin leatheroid or presspaper about 0·15 mm thick as spacers between them.

For large conductors, the coils may conveniently be wound so that there is a small air space between turns, as in Fig. 268, which illustrates an interpole coil constructed by the British Thomson-Houston Co. For very heavy currents it is a mechanical convenience to use a number of conductors in parallel instead of one very heavy conductor. Thus in Fig. 268 there are three copper straps in parallel. This arrangement also suits very well the construction of good joints between coils and connectors. For the series coils of compound generators very few turns are required in large machines. The conductor may again with advantage consist of a number of straps in parallel, and these straps may be wound on edge or on the flat, whichever is the more convenient.

Where the number of turns per pole is small, and the conductors large, considerable space is wasted owing to the conductor being wound in a spiral. Packing pieces must therefore be provided to take up this waste space, as shown at the top of the coil in Fig. 268.

Space Factor. For a coil, the space factor σ is given by

$$\sigma = \frac{\text{the total sectional area of copper in the coil}}{\text{total sectional area occupied by the coil}} \quad . \quad (224)$$

Neglecting slight effects such as the waste of one turn due to spiralling of the winding, if a coil of rectangular wire could be wound

Fig. 268. Interpole Coil
(British Thomson-Houston Co., Ltd.)

perfectly tight, the space factor would be simply the ratio of the area of the uninsulated wire to the overall area of the insulated wire.

For a coil wound with round wire, the space factor is less. If the conductors are arranged without any "bedding," the space factor for a wire of diameter d insulated to a diameter d_i is then, if there is no slack—

$$\sigma = \frac{\pi}{4} \times \frac{d^2}{d_i^2} = 0{\cdot}785 \left(\frac{d}{d_i}\right)^2 \quad . \quad . \quad . \quad (225)$$

For field coils, however, the layers may bed into each other, as in Fig. 269, arranging themselves at 60°. Under these conditions the ideal space factor, if there be no slack, is, cf. Fig. 269,

$$\sigma = \frac{\pi}{4} d^2 / \left(\frac{\sqrt{3}}{2} d_i^2\right) = 0{\cdot}906 \left(\frac{d}{d_i}\right)^2 \quad . \quad . \quad (226)$$

For small armature coils there is no definite arrangement of the turns, the conductors being dropped into the slot. Under these conditions the space factor is some 10 or 15 per cent less than in (226). If then the dimensions of a conductor and the thickness of its insulation be known, the space factor can be estimated with sufficient accuracy. For a coil wound with very small wire the space factor is extremely small owing to the relatively large thickness of the cotton covering. The factor increases as the diameter is increased. For example, a wire 0·2 mm dia, insulated s.c.c., has a dia of 0·29 mm; a wire 2 mm dia insulated s.c.c. has an overall dia of 2·15 mm.

For 60° bedding, the value of the space factor is—

For the wire 0·2 mm dia $\sigma = 0{\cdot}91(0{\cdot}2/0{\cdot}29)^2 = 0{\cdot}43$

For the 2 mm wire $\sigma = 0{\cdot}91(2/2{\cdot}15)^2 = 0{\cdot}79$

For a 2 mm wire insulated d.c.c., the overall dia is 2·26 mm and the space factor $\sigma = 0{\cdot}91(2/2{\cdot}26)^2 = 0{\cdot}715$

For a 6 mm wire insulated d.c.c., the overall dia would be 6·35 mm, and the space factor $\sigma = 0{\cdot}91(6/6{\cdot}35)^2 = 0{\cdot}815.$

These values are all "ideal" values and make no allowance for slackness, such allowance being in particular necessary in square and rectangular field coils.

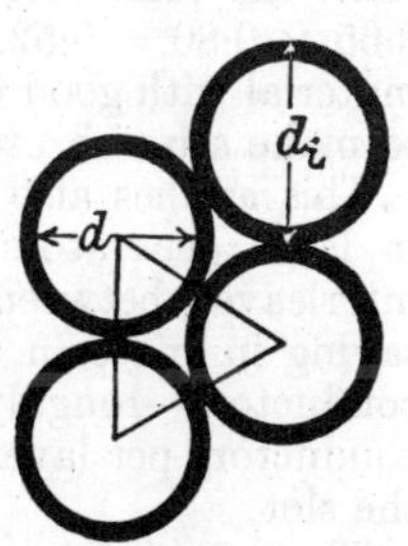

Fig. 269. Illustrating the Bedding of Round Wires

The gain obtained by using square conductors instead of round is 10 per cent where the conductors bed properly, the space factor for the square conductor being $(d/d_i)^2$ instead of $0{\cdot}91(d/d_i)^2$. It is where bedding is not possible that the square or rectangular conductor is at an appreciable advantage, the space factor then being increased, for a square conductor, from $0{\cdot}785(d/d_i)^2$ to $(d/d_i)^2$, i.e. by just over 27 per cent.

Space Factor of the Slots. For the slots, the space factor is given by—

$$\frac{\text{Total sectional area of copper in the slot}}{\text{Total area of slot}} \quad . \quad . \quad (227)$$

In this, the space factor not only includes the effect of the minor insulation, i.e. the insulation on the actual conductors, but also includes that of the major insulation, i.e. the insulation cell. Often, too, it is taken to include the effect of the retaining wedge.

In wire-wound armatures, there is in general not much that can be done to improve the space factor of the actual coils; for very fine wires, vinyl-acetal enamel covering can be used instead of cotton. For former-wound coils, and fairly large conductors, rectangular or square conductors may be used. So far as is practicable, the actual number of conductors per slot should be as low as possible for the number of slots chosen. That is, the winding should have only two paths in parallel, and where possible the conductor should consist of a single wire and not two wires in parallel. However, to employ conductors of stock dimensions, it often becomes necessary to use two small wires in parallel.

The reduction in the space factor due to slot insulation becomes less important as the slot width is increased. The use of very narrow slots is, then, to be avoided if possible. On the other hand, if the slot width be made too great, and the number of slots too few, other difficulties arise, and the slot width, even for large machines, is generally of the order of half an inch. For a slot width of, say 12 mm, and a depth below wedge of 45 mm, the total space taken up in width by the insulating cell, slot lining and slack, for a 500-V machine, will be of the order of 2·5 mm to 3 mm, and, in depth, about 5 to 6 mm. This reduces the space available for the actual windings from the total of 45×12 to about 9×39, i.e. to $351/540 = 65$ per cent of the total space available below the wedge. If the space factor of the actual conductors were as high as 80 per cent, the total factor for the slot below the wedge would be only $0{\cdot}65 \times 0{\cdot}80 = 0{\cdot}52$. It therefore is imperative to employ thin material with good insulating and mechanical properties rather than to make a lavish use of cheap, thick material.

This applies also to the insulation of the strap conductors used in large generators. Using bare copper straps with mica paper interleaved between adjacent conductors, results in a considerable saving in space in the width of the slot, as compared with taped conductors—roughly a saving of rather more than 1 mm for 4 conductors per layer, allowing space for 10 per cent more copper in the slot.

The actual value of the space factor for the slots may be as low as 0·1, in the case of very small motors, and rarely exceeds 0·55.

It must not be concluded that, if the space factor can be improved, a correspondingly increased output could be taken from the machine. If it be taken that for a given machine running at a given speed a certain definite armature loss can be dissipated, for a machine of definite proportions the armature copper loss would then be constant, and hence the copper loss in the slot would be constant, i.e. if A_s = the slot area,

$$\rho(\sigma \times A_s \times \alpha^2)L \text{ would be constant,*}$$

and therefore $\alpha^2\sigma$ would be constant.

This would indicate that the current density should vary inversely as the square root of the space factor. Against this, the increased insulation in the slot would give rise to a larger temperature gradient from the conductors to the walls of the slot. However this may be, it is indisputable that, to obtain maximum output from a given frame, it is necessary to employ such a winding and such insulating material that the space factor in the slot shall be as high as possible.

* α denotes the current density.

CHAPTER XV

MECHANICAL DETAILS

It is not proposed to deal at very great length with the subject of the mechanical design of d.c. machines. A careful study of the plates and other drawings given in the book, together with the following brief notes, should satisfy the needs of the student. A number of illustrations of relatively small machines has been given, as the average engineer will have to deal with relatively small motors rather than with large generators. The drawings may be taken to represent good modern practice.

Magnet System. Almost invariably, the frame of a d.c. machine constitutes also the yoke magnetically connecting the poles. In special cases with very small motors an entirely laminated field system is used, and in certain other machines the field circuit is entirely laminated.

Cast iron yokes were mainly used in early machines but cast steel yokes largely supplanted them. Because of their greater permeability, cast steel yokes were normally only about one-half the sectional area and weight of the corresponding cast iron yokes. The normal practice now is to use forged-steel yokes, and the majority of yokes are fabricated, although cast steel frames, and cast iron, are still employed occasionally. The drawing given in Fig. 270 is typical of a large machine having a cast steel yoke, and Fig. 283, p. 373, illustrates a large motor having a fabricated, rolled steel frame. These fabricated constructions have been made possible by the great developments that have taken place in the art of welding, the feet, etc., being welded to the main portion of the frame which is normally of simple cylindrical form as in Fig. 273. With fabricated yokes, there is no need for expensive patterns. All possibilities of blow holes are removed, resulting in improved magnetic and mechanical properties as compared with steel castings, so that it is possible to work with a rather smaller sectional area with the forged steel yoke.

With small machines the yoke may be in one piece, as in Fig. 273, but for larger machines it is usual to employ a split yoke, the two halves being bolted together, as in Fig. 283.

Main Poles. Solid poles, with separate pole shoes, were much used for earlier machines. Frequently, when cast steel frames were employed, the poles and frame formed part of the same casting. The use of such single castings for pole and yoke resulted sometimes in the poles not being homogeneous, and was abandoned by many manufacturers long before the adoption of the rolled steel yokes.

The type of pole construction most generally favoured is the

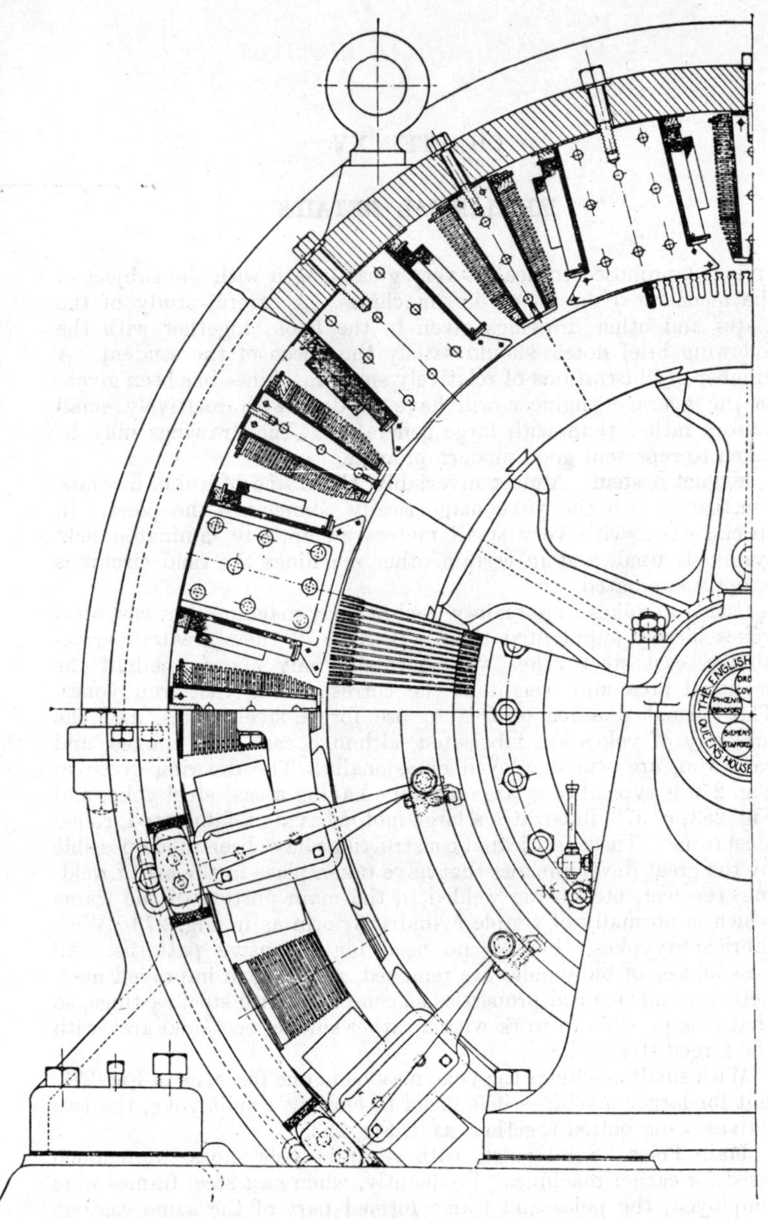

Fig. 270. Sectional Drawing of a Large 12-pole Generator

(English Electric Co., Ltd.)

laminated pole. The poles are built up of sheet steel laminations of the form shown in Fig. 270, the pole shoe and pole body forming part of the same laminations. The laminations are 1 to 1·5 mm thick, and are not insulated other than by the natural scale, etc. This type of pole construction lends itself admirably to commercial designs, where it becomes necessary to work a frame of given diameter with various values for the core length.

For small machines, stiffening plates of sheet steel are used at the two ends, and for large poles the laminations are held between two thick plates, the whole being riveted together as in Fig. 273. The poles are bolted to the yoke by means of set screws, seatings on the yoke being machined to receive them, and two or three sheet iron liners placed between the pole and the yoke, to permit of adjustments being made to the air gap. The poles may be drilled and tapped as if solid. Some firms prefer to insert a solid bar through a hole left in the punchings, this bar being tapped to act as a nut.

To increase the reluctance of the iron at the pole tips and so to avoid excessive field distortion, the pole laminations are sometimes shaped with one normal and one shortened tip. In assembling the laminations, the plates are arranged so that the normal tip falls alternately to the left and the right.

Sometimes the poles are made of solid steel forgings, bolted to the frame as shown. With such poles, separate pole shoes are necessary. These shoes are built up of sheet steel stampings lightly insulated by paper, etc., to minimize eddy currents due to the changes in the flux distribution set up by the motion of the armature teeth. The laminations are securely fastened together by steel rivets, and the whole shoe fixed to the pole body by means of countersunk screws.

Interpoles. These may be laminated or solid, and are built of wrought iron or mild steel. They are fastened to surfaces machined on the yoke by means of set screws. Sometimes the yoke, in addition, is slotted to receive the poles. The interpoles may be mounted centrally as in Fig. 273, or may be placed at one end of the core. The particular interpole shown in Fig. 273 is tapered towards the tip. This interpole has a uniform taper and is solid. Fig. 270 shows a tapered interpole built of laminations. Frequently the interpole is made rectangular, and of sufficient sectional area to avoid magnetic saturation at the root. Such interpoles are shown in Figs. 272, 275 and 276.

Armature. The armature core plates are from 0·4 to 0·5 mm thick and insulated with paper or by a thin coating of a suitable compound. For normal machines good sheet steel is used, but for totally-enclosed machines where it is essential to keep down the value of the iron losses it is common to employ silicon steel. This material may also be used with advantage when the frequency is high.

Except in small sizes the core is divided into a number of packets by ventilation spacers of the type illustrated in Fig. 274, this spacer

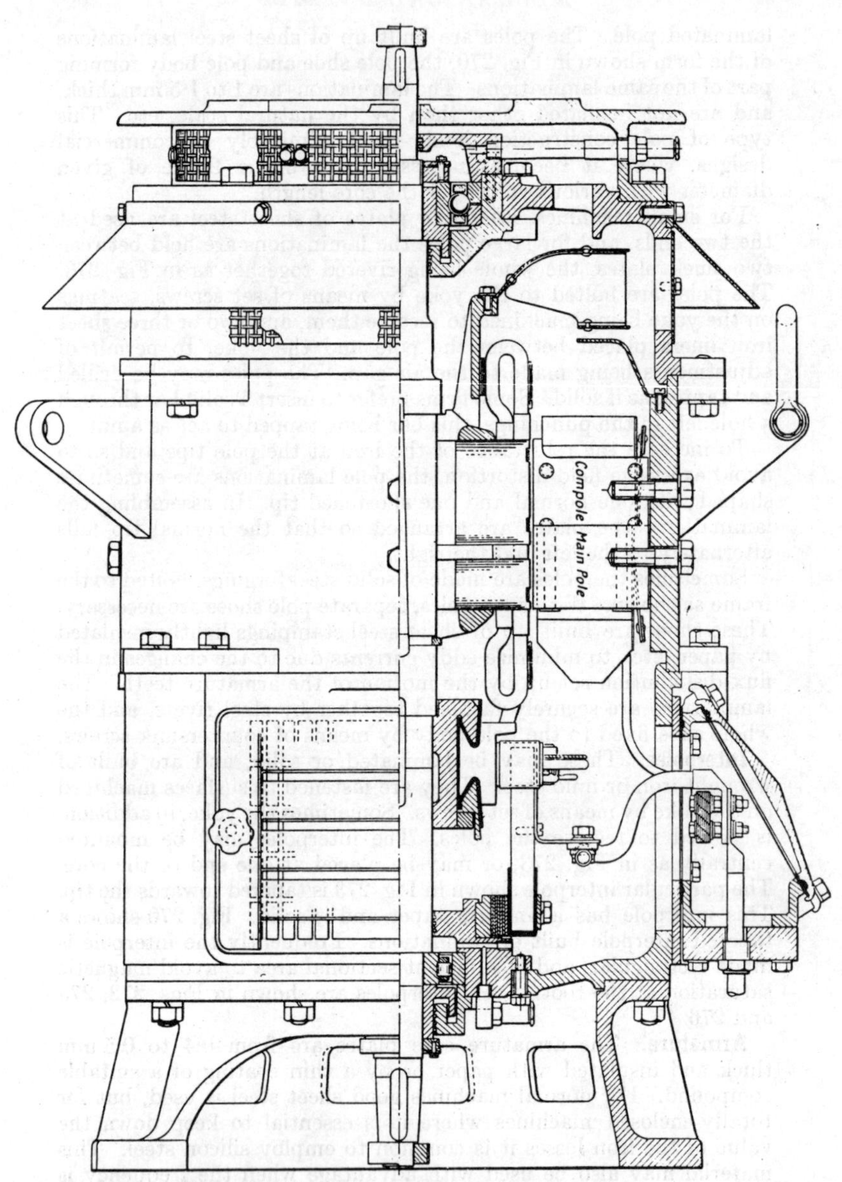

FIG. 271. 4-POLE D.C. MOTOR, VERTICAL TYPE

(Metropolitan-Vickers Elec. Co., Ltd.)

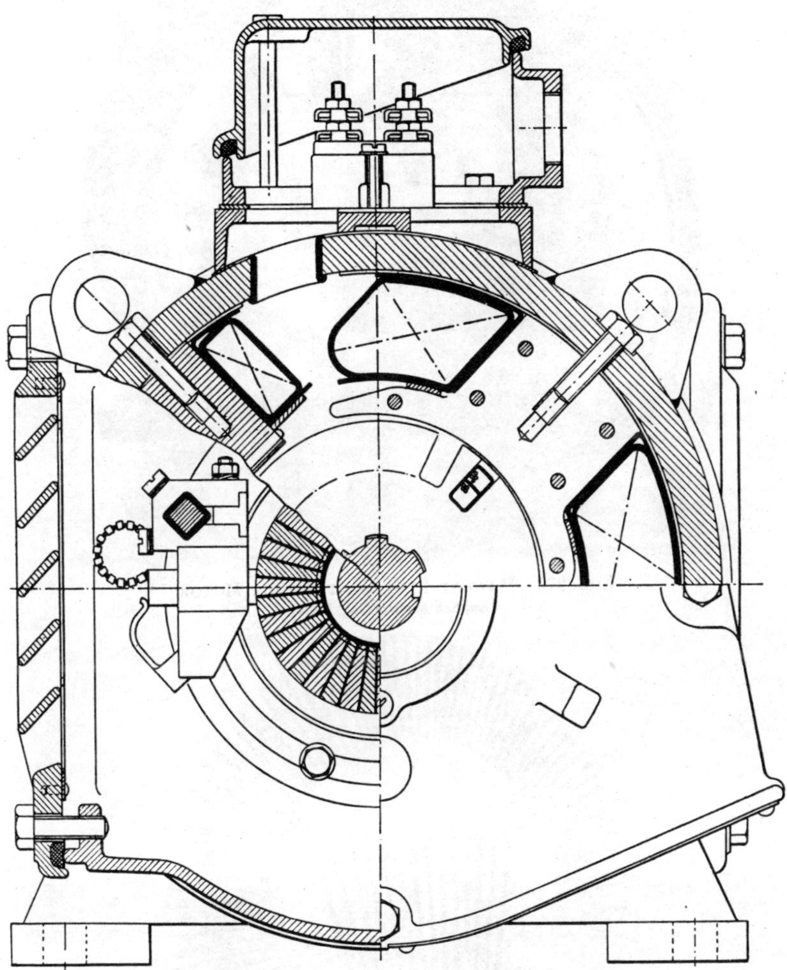

FIG. 272. 2-POLE D.C. MOTOR
(*Laurence Scott and Electromotors, Ltd.*)

being formed by spot-welding steel pieces of I-section to a thick core plate. The I-pieces pass centrally down each tooth, and there is ample room for cooling air to pass from the inner side of the core between these I-pieces to the outer surface of the armature. The vent spacers shown in Fig. 274 include one complete spacer for a machine of medium size, with the core plates in one single piece, and one segment of a vent spacer for a large machine in which the core is built up in segments.

Fig. 273. Magnet System of 4-pole Motor
(*English Electric Co., Ltd.*)

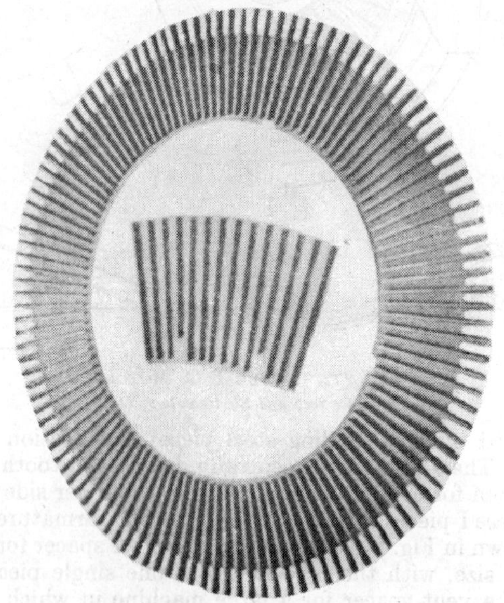

Fig. 274. Ventilation Spacers
(*British Thomson-Houston Co., Ltd.*)

Except in very small sizes, a vent spacer is used at each end of the core, separating the core from the end plate. This end plate generally forms part of the support for the end connexions of the armature winding—i.e. the "winding carrier." With small machines, the core plates are keyed direct to the shaft, or splined as in Fig. 276, which shows the construction of a small machine with the winding carrier butting against a shoulder on the shaft. During construction, the core plates and vent spacers are threaded on the shaft, the key or splines serving to "register" the plates in their correct position. At the commutator end another end plate is placed, and the core tightened up by hydraulic pressure. For the construction shown in Fig. 276, while under the action of the press, a split shrink ring is sprung into position in the circumferential slot made in the shaft. On removing the pressure, the core springs out slightly and the end plate is pressed firmly against the ring, which thus serves the same function as the fixed shoulder at the other end of the core. Alternatively, instead of using a split ring, the core is held tight by a lock-nut, and this lock-nut is prevented from slacking off by means of a grub-screw.

With small machines which have the core built up direct on the shaft, it is necessary to provide axial ventilation holes so that air can pass into the ventilation ducts and, in fan-ventilated machines, so that air can pass completely through the core. These ventilation holes are shown clearly in Fig. 276.

They, and the key way, are punched out at the same time as the complete disc, and the slots are punched out separately on a notching machine.

Machines having more than four poles, and in special cases 4-pole machines, have the core plates built up on a cast iron centre commonly called a "spider," as in Fig. 275. Such an arrangement permits the shaft being removed without interfering with the armature core plates and winding. With large machines it would naturally be uneconomical from magnetic considerations to employ more iron than is necessary. For comparatively small machines, up to about 100 cm dia, the core plates are in one complete disc, and driven from the spider by one or more keys.

Fig. 300, p. 389, shows an armature with a simple "sleeve" spider.

With large armatures it is not possible to blank out the core discs in one piece, and it becomes economical to construct the core of a number of segments—six or more, depending upon circumstances. These segments are then driven from the spider by means of dove-tailed keys, as shown in Fig. 270 for a 12-pole generator, or the segments themselves may have projecting dove-tailed tongues which fit into corresponding slots machined out in the spider.

In assembling such segmental cores, the plates are arranged so that the joints in one layer are intermediate between those of the adjacent layers. This is indicated in Fig. 270, where there are six segments, and the dotted radial line indicates the joint in the layer

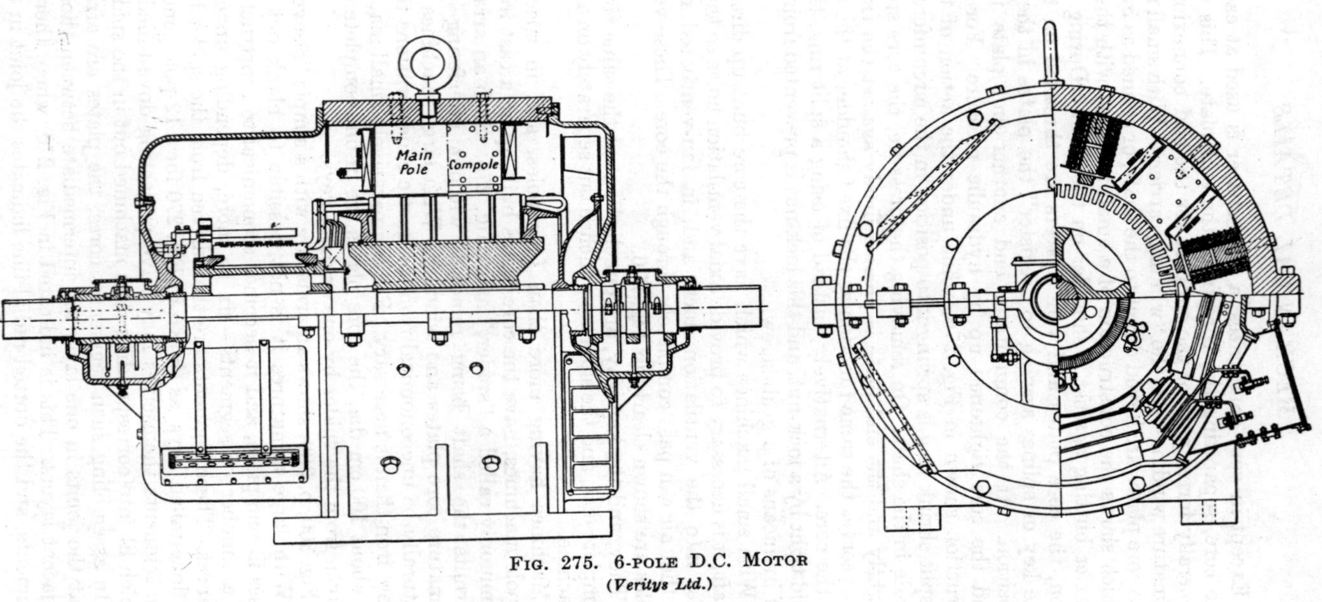

FIG. 275. 6-POLE D.C. MOTOR
(*Veritys Ltd.*)

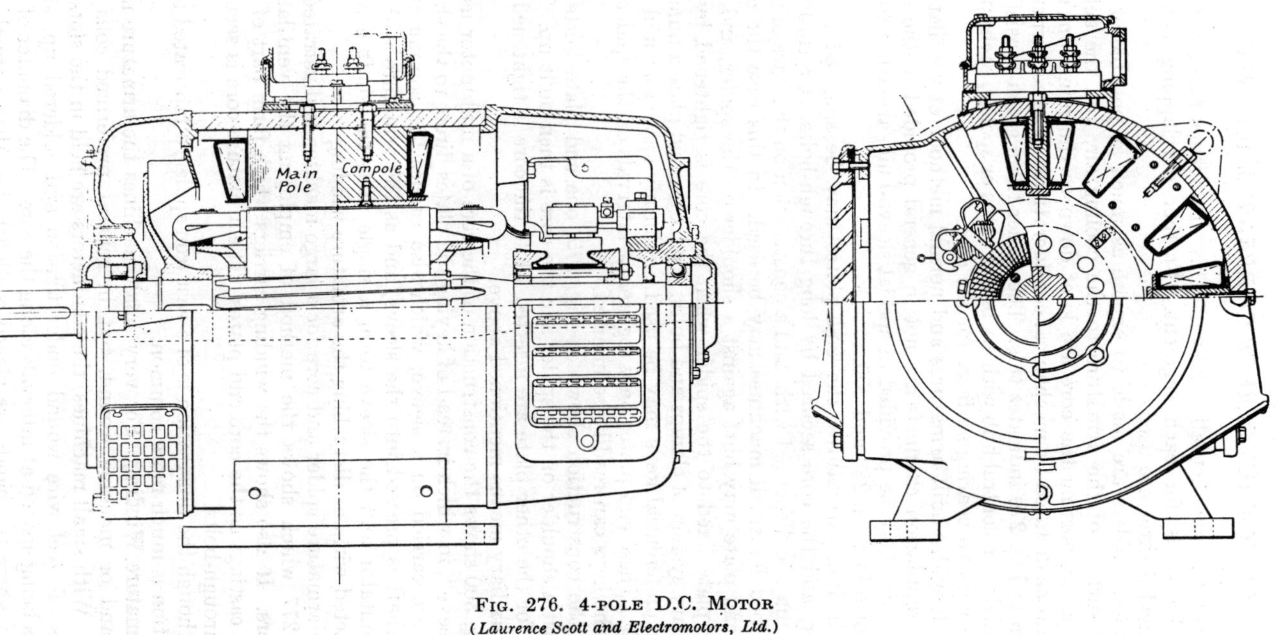

Fig. 276. 4-pole D.C. Motor
(*Laurence Scott and Electromotors, Ltd.*)

beneath the one actually shown. This arrangement necessitates two dove-tails for each segment. Further overlapping can be arranged if three are used.

Modern spiders are made to expose as much as possible of the inner surface of the armature to the cooling air, and the actual surfaces supporting the core are kept down to the minimum safe amount as dictated by mechanical considerations. The construction shown in Fig. 270 indicates this. The spider there given consists in essentials of a central hub with six hollow arms or spokes, each spoke providing two bearing surfaces for the core.

With spider-built armatures and normal methods of ventilation, axial ventilation channels are not in general provided in the core plates, but may be provided in special sleeve-built armature cores, as shown in Figs. 289 and 300.

With large armatures, the end plates may be spigoted to the spider, and the core secured by long through-bolts. Or through-bolts may be dispensed with, and a construction on the lines of that described for small machines may be used. In this case the one-core end plate may butt against a shoulder on the spider, and the other plate fixed to the spider when the core is tightened by an auxiliary system of clamps and bolts. To fix the end plate a number of circumferential keys may be used, uniformly spaced around the spider. These end plates only register at intervals on the spider, so that the keys can readily be inserted.

In the construction shown in Fig. 275, one end plate butts up against a shoulder on the spider and the core is then built up. The keys for the other plate are inserted while the core is tightened by the auxiliary means mentioned above.

Fig. 300 shows the construction of the core of a mill motor using a "sleeve", in which instead of keying the plates direct to the shaft, they are carried on a sleeve, which also carries the commutator. The shaft is pressed into the sleeve, and as the armature with its commutator and the sleeve form a single unit, the shaft can be extracted without disturbing the armature winding

An armature spider and core for a large machine is illustrated in Fig. 277 which shows the method of employing the ventilation spacers. It also shows the winding carriers which form part of the same castings as the core end plates. The armature core is secured by through-bolts.

Although the illustration is of a cast iron spider, a fabricated construction is much more common.

Armature Winding. For very small machines the armature may be hand or machine wound, but in general, preformed coils are used. With small machines, the conductors are held in the slots by bands of steel wire wound under tension and soldered up, such bands being spaced at intervals along the core. The diameter of the plates where the bands are placed is slightly less than normal. To restrain the end connexions from flying out due to centrifugal forces, the coils are bound tightly on to the winding carriers, similar bands of steel wire being used. Since centrifugal forces tend to cause the winding ultimately to move outwards from the carrier, these carriers are sometimes dispensed with in small machines, at the commutator end. The binding wire is put on over a layer of micanite,

FIG. 277. ARMATURE SPIDER AND CORE FOR LARGE MACHINE
(*British Thomson-Houston Co., Ltd.*)

or leatheroid, to prevent damage to the coils, as illustrated in Figs. 197 and 275.

With large machines it is usual to employ wedges of bakelized compressed wood or bakelized fabric, dove-tailed into the slots to hold the coils in place, and to use steel wire bands for the end coils.

Commutator. For high-speed machines, shrink-ring commutators are used. In these the segments are first clamped tightly together and, after turning, are insulated with hard micanite rings at the ends and at one or two places along the cylindrical surface. Steel rings are then shrunk on top of the micanite rings, forcing the segments inwards to form a complete circular arch. Such commutators were used in the early years of this century for d.c. generators for direct coupling to steam turbines. They are used for direct-coupled exciters for turbo-alternators, and for high-speed generators used as dynamometers for engine testing. Radial, or disc, commutators are shrink-ring commutators using only one shrink ring. As the name implies, the working surface is the radial and not the cylindrical surface. Again, commutators on these lines were sometimes used for d.c. turbo-generators many years ago. Their use is now mainly limited to direct-coupled exciters for turbo-alternators.

In practically all cases, therefore, commutators are built with Vee-shaped clamping rings, the segments having a form on the lines of that shown in Figs. 275, 276, and 300.

The segments are made of hard-drawn copper, and are separated by thin plates of mica or micanite, the usual thickness of which is about 0·8 mm. Mica is to be preferred, but cannot be used for large commutators because of the difficulty of obtaining large sheets, making the cost of large mica segments prohibitive. On account of cost, also, micanite is often used for small commutators. The copper segments are tapered, as in Fig. 272, and are given sufficient wearing depth to conform to standard requirements.

The copper is insulated from the Vee-rings and the hub by micanite carefully moulded to the exact shape required. Frequently the hub is not insulated, but sufficient clearance is left to avoid the need of this insulation.

For low speed machines, the Vee-rings may be of cast iron, but for high-speed machines steel is necessary.

Large commutators are mounted on hubs or spiders as in Fig. 270. In that shown in Fig. 270, both Vee-rings are loose and through-bolts are used. For medium sizes, one of the loose rings may butt against a shoulder on the spider, and the outer Vee-ring secured to the spider by set screws. The "risers" connecting the segments to the armature coils are made of copper strip for large machines. Before assembling the segments, these strips are riveted and soldered into saw cuts made in the segments. The outer end of the riser is shaped, as in Fig. 270, so as to form a clip into which the armature conductors are soldered.

For small commutators a lug is provided, to form part of the segment, and the armature conductors are soldered directly into the lug, separate risers thus being dispensed with.

For low-voltage generators of large output, the current to be collected becomes very great, necessitating an unduly long commutator. It then becomes necessary to construct the commutator in two parts, each held by its own pair of Vee-rings. The connectors joining corresponding segments are shaped so as to give the fan action necessary to cool the outer element of the commutator. In addition, projecting lugs may be left on the outer edge of the segments and shaped so as to draw air over the surface of this outer element.

Structurally commutators with Vee-rings may be "arch-bound" or "Vee-bound." The arch-bound commutator is similar in principle to the shrink ring commutator except that steel "Vee-rings" are used, insulated with micanite Vee-rings and fitted into a Vee-shaped groove turned in the ends of the segments. The Vee-rings are pulled towards one another by bolts *and effectively press only on the inner side of the Vee-shaped groove.* The reaction to this pressure is provided, as in the shrink-ring commutator, by the circular arch formed by the segments themselves.

The Vee-bound commutator is apparently the same as the arch-bound commutator, but the Vee-rings are a tight fit in the Vee-groove so that there is no tendency to force the segments radially inwards. This type of construction is generally considered inferior to the arch-bound commutator.

The ideal commutator is perfectly symmetrical about any diameter, is, and remains, truly cylindrical. Since the copper and mica segments both vary slightly in thickness, symmetry can be obtained in large commutators only by grading the segments according to thickness so that thick and thin ones are evenly distributed around the periphery. The maintenance of the commutator's shape requires the use of hard micanite, i.e. one in which a low proportion of bonding material between the mica flakes is consolidated under high pressure. The bolts holding the Vee-rings must be tightened gradually as the commutator is heated, to eliminate slackness due to any softening of the micanite under heat. When the commutator is cold the surface is turned. For large machines, the final turning operation may be carried out in the assembled machine using a diamond-pointed tool. For small commutators it is possible to use a soft grade of mica flake for the micanite segments so that they wear down at the same rate as the copper segments. Usually, however, hard mica is used and the mica segments are under-cut, i.e. the top of each mica segment is removed so as to form a shallow groove between the adjacent copper segments. This tedious operation may be carried out with a hack-saw blade or more speedily with a small hand milling-cutter. The edges of the copper segments should be rounded off after under-cutting to remove any burrs.

Fig. 278 represents a photograph of the armature of a 1,500-kW, 250-V, 6,000-A generator constructed by the Lancashire Dynamo and Crypto Co. Details of the design of this machine are given on page 469.

Brushes. The type of brush used is dependent upon the peripheral speed of the commutator, and upon the working voltage. For very high surface speeds, graphitic brushes are much used. For low-voltage machines such brushes are also necessary, otherwise medium hard electrographitic brushes may be used in most cases.

Fig. 278. Armature with Double Commutator for a 1,500-kW, 250 V, 6,000-A Dynamo
(*Lancashire Dynamo and Crypto, Ltd.*)

The brushes normally used are of relatively small width, a number of such small brushes being preferred to relatively few wide brushes. For very small machines, brushes $\frac{5}{8}$ in. in width are available for thicknesses of $\frac{1}{4}$ in., $\frac{5}{16}$ in., and $\frac{3}{8}$ in. The next standard width is $\frac{7}{8}$ in., in thicknesses of $\frac{3}{8}$ in. up to $\frac{3}{4}$ in. in steps of $\frac{1}{8}$ in. For larger machines, brushes $1\frac{1}{4}$ in. wide and of thicknesses from $\frac{3}{8}$ to $\frac{9}{8}$ in., in

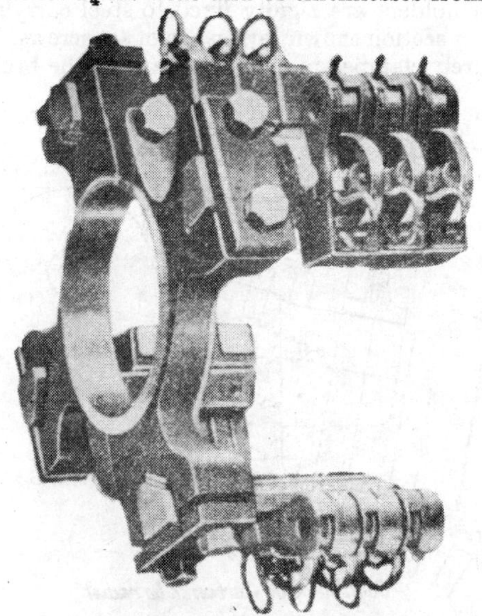

FIG. 279. BRUSH GEAR FOR 4-POLE MACHINE
(*Standard Carbon Brush Holder Co.*)

steps of $\frac{1}{8}$ in., are available. The standard lengths for these three brush widths are $1\frac{1}{4}$ in., $1\frac{1}{2}$ in., and $1\frac{3}{4}$ in. respectively.

To form a definite connexion between the brush and its holder a twisted copper flexible conductor is used, this conductor being securely fixed into the brush during its manufacture. Tags are provided to the flexibles and slotted so as to pass under the head of terminal screws on the brush holder.

Brush Holders. For all ordinary d.c. machines, the holders now employed are of the box type. The lever type of holder is not often used, and attention will therefore be concentrated upon the box type.

A very good type of brush holder is illustrated in Figs. 279 and 280. The brush is a good sliding fit in the box. Certain tolerances have to be permitted in the manufacture of the brush and the holder, and the clearance between brush and holder will total from one-hundredth to one-five-hundredth of an inch in width and about half these amounts in thickness.

The pressure of the brush upon the commutator is regulated by a clock-type phosphor bronze spring, the extension of which is arched to form the pressure finger. With this type of spring a uniform pressure is obtained throughout the life of the brush. The pressure is adjusted by moving the adjusting lever into a different notch.

The brush holders are rigidly fixed to steel carrying bars of the special form of section shown, by means of set screws. These bars in turn are securely clamped in machined jaws in the brush bracket by

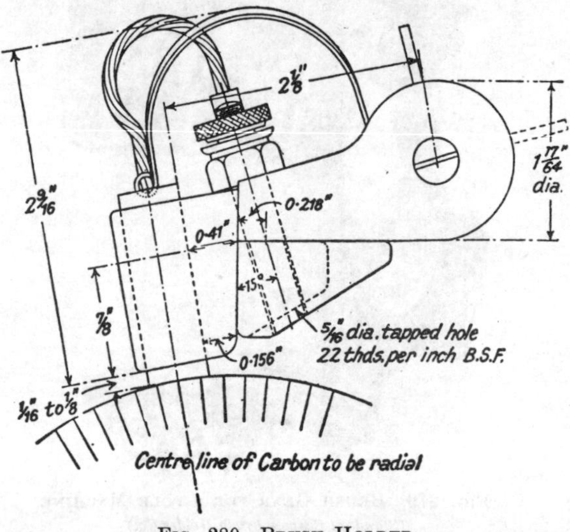

FIG. 280. BRUSH HOLDER
(*Standard Carbon Brush Holder Co.*)

a cover plate and two or four screws. In Fig. 279 the cover plate is clamped on by two such screws. Packing pieces are placed under the carrier bar and these are removed as the commutator wears. Such packing pieces can be seen in the bottom carrier bar in Fig. 279, and are also shown in Fig. 271.

The carrier bar is insulated from the brush rocker by an insulating sleeve moulded round one end of the bar, and projecting well beyond the cover plate.

The terminals to which the pigtails are joined are fitted to the top of the set screws which clamp the brush holders on to the carrier bar.

With small machines, the brush rocker is mounted on a seating turned on the end-shield and, where the commutator diameter is not too great, the rocker is often mounted on a seating turned on the pedestal bearing for larger machines. The rocker in such machines is constructed in two parts, as shown in Fig. 279, these two parts being held together by set screws, one of which is shown in Fig. 279, to the left of the upper brush set.

For larger machines the brushes are carried in brackets which are bolted to a brush yoke. This yoke is supported by the magnet frame in such a manner that the brush position can readily be adjusted when the machine is on the test bed, and then the brush

Fig. 281. Heavy Brush-holder for Auxiliary Mill Motor

(General Electric Co., Ltd.)

yoke can be locked in the desired position. Fig. 281 shows the heavy duty brush holder of a steelworks motor.

Where perfect collection of the current is difficult, and in particular for high-speed machines, better results are obtained by using two thin brushes instead of one thick brush. Fig. 282 illustrates such "duplex" brush gear. For each brush a separate spring is used, the pressure of the brushes on the commutator being adjusted by nuts and screws.

Fig. 283 is an illustration of a large motor for a rolling mill. It shows clearly the commutator and brush gear. It also illustrates the features of the fabricated construction for the frame.

For very long commutators, such as the double commutator shown in Fig. 278, the brush brackets are most conveniently mounted on an entirely separate yoke bolted direct to the bedplate, as shown in Fig. 284. This illustration shows two heavy-current generators operating in tandem. Fig. 285 shows the complete brush assembly for one of the machines.

Angle Between Brush and Commutator. With most motors the brushes are set so that their centre-line is radial to the commutator as in Fig. 272; this permits of operation under similar conditions for both directions of rotation.

Frequently, too, with large non-reversing machines, the brushes are radial. With such machines, however, it is sometimes the practice to have the brushes inclined, as indicated in Figs. 282 and 286.

The brushes are said to be trailing when the commutator is run in the same direction as the brushes are tilted, and reaction—or

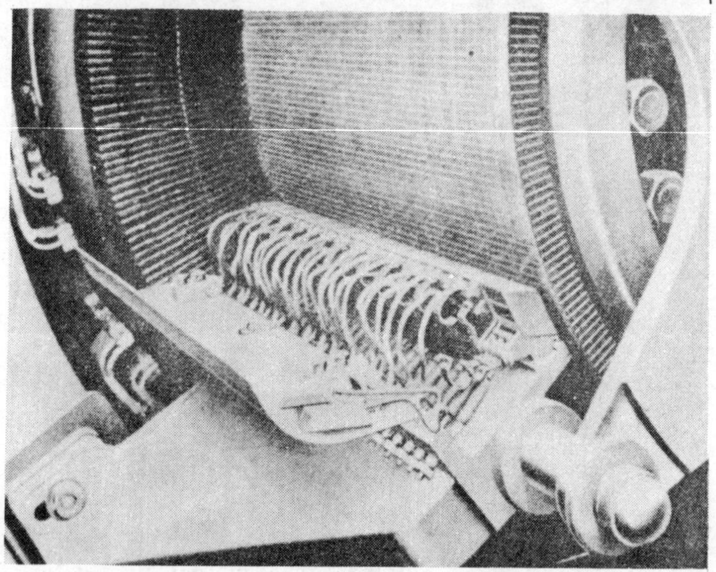

FIG. 282. DUPLEX BRUSHGEAR OF LARGE MOTOR
(*Metropolitan-Vickers Electrical Co., Ltd.*)

leading—when the commutator is rotated against the angle of tilt, as in Fig. 286.

In reaction brushes the angle of tilt is usually from 30° to 40°; for trailing brushes the angle is smaller and usually from 10° to 15°.

For machines which rotate in one direction only, reaction brushes are preferable. Ideally, the radial reaction force and the tangential friction force at the commutator surface have a resultant which is parallel to the length of the brush, so that there is no side-thrust in the brush box to produce friction and prevent the brush from moving up and down freely. For machines which may be used for either direction of rotation, reaction brushes cannot be used and radial brushes are preferred. Most motors of course are in this group.

Staggering the Brushes. To prevent ridges forming on the commutator, due to all the brushes following the same track, the

FIG. 283. ROLLING MILL MOTOR
(*British Thomson-Houston Co., Ltd.*)

FIG. 284. HEAVY-CURRENT GENERATORS, EACH 3,360 A, 205 V

(Lancashire Dynamo and Crypto, Ltd.)

brushes are staggered, as indicated in Fig. 287, each set of positive brushes being staggered with relation to the previous set of positive brushes, and the negative brushes similarly arranged. By this

FIG. 285. BRUSH GEAR FOR THE MACHINE IN FIG. 284
(*Lancashire Dynamo and Crypto, Ltd.*)

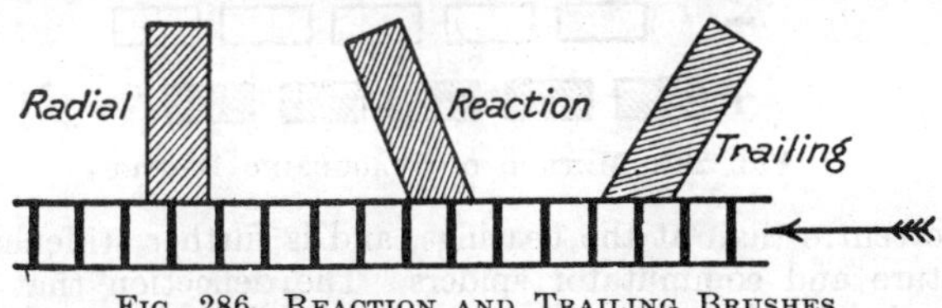

FIG. 286. REACTION AND TRAILING BRUSHES

means each track on the commutator is covered by equal numbers of positive and negative brushes.

The Shaft. In determining the diameter of the shaft, consideration must be given not only to the turning moment that has to be transmitted, but also to the bending of the shaft under the weight of the armature and, in addition, unbalanced magnetic pulls.

In designing the shaft, then, it is given sufficient rigidity to prevent the natural deflection exceeding a certain percentage of the air gap dimension. The deflection is frequently calculated approximately by assuming the shaft to have a constant diameter, equal to the minimum value, and that the whole of the weight of the armature and commutator is concentrated midway between the bearings.

Then if L = distance between bearings in cm

W = total weight of armature in kg

E = the modulus of elasticity for the shaft material in kg/cm²

δ = the deflection of the shaft at the centre, in cm

I = the moment of inertia for the shaft section

d = the diameter of the shaft

$$\delta = W \times \frac{L^3}{48EI} \quad . \quad . \quad . \quad . \quad . \quad (228)$$

For a normal shaft of circular section $I = \frac{\pi}{64} \times d^4 = \frac{\pi}{4} \times r^4$

and the deflection $\delta = \frac{4W}{3\pi} \times \frac{L^3}{d^4} \times \frac{1}{E}$

$$= \frac{0{\cdot}425\ WL^3}{d^4E} \text{ cm} \quad . \quad . \quad . \quad (229)$$

When calculated by this method the deflection should in general not exceed 3 or 4 per cent of the air gap dimension.

Actually the load is not concentrated at the centre, and the centre of gravity of the load is nearer the driving end. The shaft is thicker

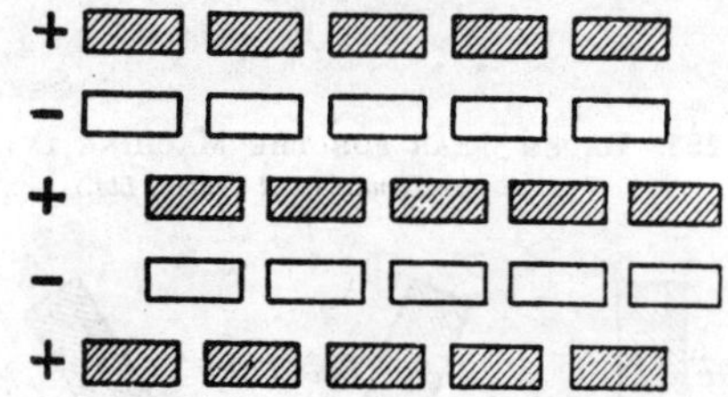

FIG. 287. METHOD OF STAGGERING BRUSHES

in the centre than at the bearings, and is further stiffened by the armature and commutator spiders. The deflection that actually occurs is therefore less than would be calculated by the above formula.

For ordinary shafting steel, the modulus E is $2{\cdot}1 \times 10^6$; in special cases steel having a higher value may be employed. In terms of the air gap l'_g, the diameter of the shaft for 3 per cent deflection must then not be less than,

$$d^4 = \frac{0{\cdot}425WL^3}{0{\cdot}03l_g \times 2{\cdot}1 \times 10^6}$$

$$d = 0{\cdot}05 \sqrt[4]{\frac{WL^3}{l'_g}} \text{ cm} \quad . \quad . \quad . \quad . \quad (230)$$

The shaft is not designed, then, solely from considerations of pure torsion, although sometimes formulae are used based upon such considerations.

The twisting couple on a shaft subjected to pure torsion, when transmitting load is

$$T = \frac{1}{2\pi} \times \frac{P^*}{n} \text{ newton-metre}$$

$$= 97{,}500 \frac{P^*}{N} \text{ kg cm} \quad . \quad . \quad . \ (231)$$

Where P^* is power in kW and N the speed in rev/min.

For a solid shaft of diameter d, the maximum intensity of stress q established is

$$q = \frac{2T}{\pi r^3} = \frac{16T}{\pi d^3} = \frac{16 \times 97{,}500}{\pi d^3} \times \frac{P^*}{N}$$

$$= \frac{0{\cdot}5 \times 10^6}{d^3} \times \frac{P^*}{N} \text{ kg/cm}^2$$

whence $$d = 79{\cdot}4 \sqrt[3]{\frac{P^*}{N} \times \frac{1}{q}} \text{ cm} \quad . \quad . \quad . \ (232)$$

If, then, the shaft could be designed solely from torsion considerations, the maximum working stress q would be constant and the diameter of the shaft could be obtained from a formula of the type

$$d = \text{const. } \sqrt[3]{P^*/N}$$

Occasionally such a formula is used, the constant being taken so as to correspond to a very low value for the working stress due to pure torsion, thus allowing a considerable factor of safety. A very rough value for the constant, giving the diameter of the shaft in cm, is from 15 to 18.

A better method is to work out the equivalent bending moment due to the combined effect of bending and twisting and to calculate the stress established in the shaft designed so as to be sufficiently stiff in accordance with (230).

For a load concentrated at the centre of the shaft the bending moment is

$$B = \frac{WL}{4} \text{ kg cm}$$

The actual bending moment will have a maximum value less than this owing to the weight being distributed.

The equivalent bending moment, M, due to the combined effect of bending and twisting, is

$$M = \tfrac{1}{2}\{B + \sqrt{B^2 + T^2}\} \quad . \quad . \quad . \ (233)$$

The moment of inertia of a circular section to bending is $I = \frac{\pi}{64} d^4$ and the maximum intensity of stress, at the distance $y = d/2$ from the axis,

$$f = \frac{My}{I}$$

$$= M \bigg/ \left(\frac{\pi}{32} d^3\right)$$

$$= 10{\cdot}2\ M/d^3 \quad . \quad . \quad . \quad . \quad (234)$$

For normal shafting steel the intensity of stress may, when calculated by the above method, be taken up to 700 kg/cm².

As shown in the various drawings given in the book, the shaft has its largest diameter at the armature spider, and the diameter is reduced in stages to a minimum at the bearings.

Bearings. Ring lubrication is employed for large machines and to a limited extent for comparatively small machines, but unless noise has to be reduced to a minimum, ball and roller bearings are now fitted as standard for all except comparatively large machines. The use of these bearings not only reduces tremendously the bearing losses, but eliminates a great source of trouble hitherto experienced with d.c. machines, namely oil throwing.

With small machines, ball bearings may be used at both ends. For larger machines, roller bearings are used at the driving end, and ball bearings may be used at the non-driving end, i.e. at the commutator end. The ball bearing also acts as a location bearing, but where excessive end-thrust is anticipated, thrust bearings also are fitted. Alternatively, where a limited amount of end-thrust is experienced, a double-lipped roller bearing may be fitted.

Examples of ball and roller bearings are given in Figs. 276 and 300.

To permit removal of the end-brackets of the machine without disturbing the bearings, the bearings are frequently carried in housings as shown in Figs. 288, 300 and 301. By the use of such "cartridge housings" the motor can be dismantled without interfering with the press-fit of the bearing on the shaft, and the bearing remains completely enclosed, so that there is no danger of dirt entering it.

Sleeve bearings, with ring lubrication, are used for motors when very silent running is required. The bearing housings form integral parts of the end bracket castings. For small bearings, the bearing bush is in one piece and is made of brass or phosphor bronze. Slots are made in the upper half, to receive the lubricating rings. These rings ride loosely on the shaft and dip into the oil reservoir. They are carried round at a comparatively low speed by the shaft, and carry oil to the top of the shaft. To prevent the rings from jumping on to the outside of the bush, guards are provided.

For large machines pedestal bearings are generally used, of the type illustrated in Fig. 270 and in Fig. 236. Such bearings normally have spherical seatings, and are self-aligning. The bearings are split horizontally and consist of cast iron shells lined with white metal (babbit metal), grooves being cut in the white metal lining so as to feed the oil over the whole bearing surface of the shaft.

When sleeve bearings are used, the axial length of the bearing surface is normally from $2\frac{1}{4}$ to $2\frac{3}{4}$ times the diameter of the shaft in

FIG. 288. ARMATURE OF A D.C. MACHINE HAVING BEARINGS ENCLOSED IN CARTRIDGE-TYPE HOUSINGS

the bearing. Larger ratios are used for high speed machines, and for very low speeds a smaller ratio suffices.

The actual dimensions of the bearing surface are determined from joint considerations of the total pressure load on the bearing and the peripheral speed of the shaft in the bearing.

If D_B = dia of shaft in the bearing
L_B = length ,, ,, ,,
v_B = peripheral speed ,, ,,
P_B = the total bearing load,

then the intensity of pressure is commonly expressed as

$$p_B = P_B/D_B L_B \qquad . \quad . \quad . \quad . \quad (235)$$

If the peripheral speed is expressed in m/sec and the pressure p_B in kg/cm², then for normal ring lubrication the product $p_B v_B$ will in general not exceed 20.

The lower the value of this product, under normal conditions, the less is the likelihood of the bearing overheating.

Normally the dimensions may be based upon $p_B v_B = 15$. For low speeds the use of this formula would result in the intensity of pressure, p_B, becoming too high. The dimensions should then be such that p_B does not exceed about 12 kg/cm^2.

Shoulders are provided on the shaft to limit the end play. For large machines this end-play is generally limited to from one-sixteenth to one-eighth of an inch.

Fig. 275 shows ring lubrication with sleeve bearings.

If the shaft of a machine becomes magnetized axially, each bearing becomes in effect a short-circuited homopolar generator. The resulting currents passing from shaft to journal rapidly destroy the bearing surfaces. The process is equally destructive of roller bearings. It is important, therefore, when connecting brushes, compole coils, series field coils and compensating windings, to ensure that the load current does not link the magnetic circuit formed by the shaft, bedplate, etc., thus making the armature shaft an electromagnet.

Traction Motors. For traction work the normal type of motor design is unsuitable, owing mainly to limitations of space.

Formerly, motors of the totally-enclosed type were used, but modern practice is to employ ventilated-type motors.

The general features of the mechanical design of a modern ventilated-type diesel-electric locomotive motor can be obtained from Figs. 289 and 290. These drawings represent a 200-h.p., 450 rev/min, 300-V traction motor constructed by the English Electric Co., Ltd.

The cast steel frame is of the box type. The six main poles and six interpoles are laminated and bolted to the frame. No radial ventilating ducts are used, but there are a number of circular ventilation ducts passing axially through the core. The core plates are built on to a sleeve spider, so that if necessary the shaft can be removed. Wedges are used to retain the armature winding in the slots and steel binding wire to support the end connexions.

The brush holders are of the box type and there is one brush block per spindle.

The construction of motors for electric railways is on the same general lines as that illustrated in Figs. 289 and 290. These motors are sometimes designed for a working voltage of 1,500; often, however, on 1,500 or 3,000-V lines, the motors are designed for 750 V and 2 or 4 connected permanently in series.

A diagrammatic longitudinal section of another railway motor is given in Fig. 291. This illustrates a 200-h.p., 775-V motor and is self-explanatory. For this motor there are two brush spindles—it is a 4-pole wave-wound machine, and there are four brush blocks per spindle, as in Fig. 292, the rated current being 218 A. This motor, driving wheels 42 in. dia through a 21/72 reduction gear, develops a tractive effort of about 2,500 lb at a speed of 29 m.p.h. when taking its rated current of 218 A at 775 V. The insulation on this machine is based on a line voltage of 1,600. The armature slots are increased in size at the core ends so that additional insulation can be used at these points. The ends of the armature are covered with impregnated cotton duct caps, held in position by the steel banding wire.

For surburban railway work the motors work on an intermittent load and are normally rated on a 1-hour basis. The permissible

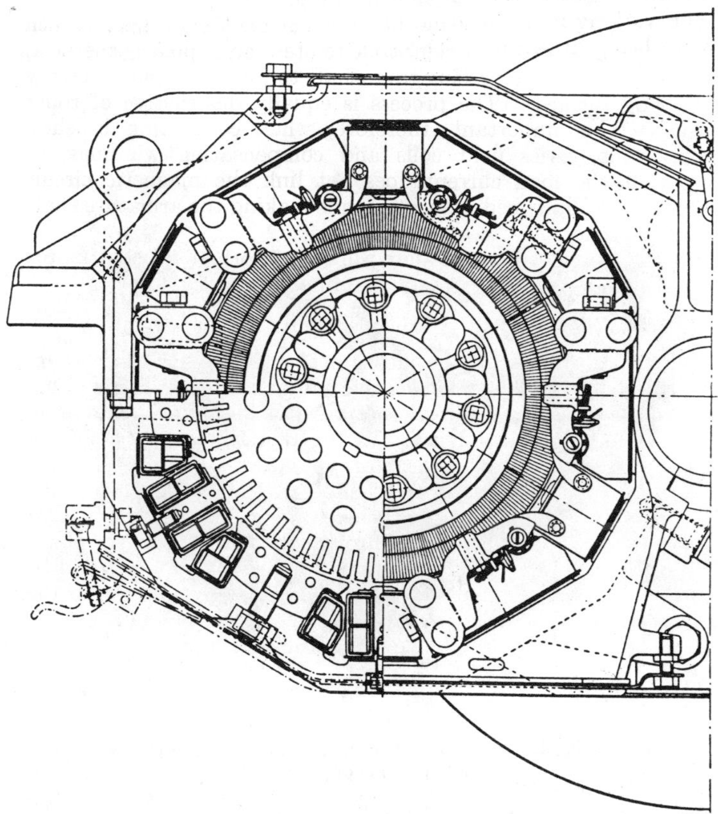

FIG. 289. SECTION OF 200-H.P., DIESEL-ELECTRIC, TRACTION MOTOR
(*English Electric Co., Ltd.*)

temperature rise for Class B armature windings is 90°C by thermometer and 120°C by winding resistance for both 1-hour and continuous ratings. The insulation used has to be of the heat-resisting type so that glass fibre, mica and asbestos are very largely employed.

Traction motors are invariably series wound.

Types of Industrial Motor. For very many purposes the *protected type* of motor construction is well suited. Fig. 293 illustrates such a machine. The internal parts of the motor are adequately protected from accidental contact without interfering with the ventilation of the machine.

For still further protection the *screen-protected type* is used, for which all the openings in frame and end-shields are fitted with perforated metal covers, as shown in Fig. 294.

For working in dusty atmospheres, *pipe ventilation* may be used, the air being drawn into the machine through a pipe fitted to one

FIG. 290. 200-H.P., TRACTION MOTOR SHOWN IN FIG. 289
(*English Electric Co., Ltd.*)

of its openings, and discharged either through another pipe or direct into the surrounding atmosphere. Fig. 295 shows a motor arranged for pipe ventilation.

It is often essential to make use of a *drip-proof* construction, for which purpose it is common practice to use louvred ventilating covers as illustrated in Fig. 296.

The photographs of the various types of construction show that the same general construction can be adapted to them all.

For dusty atmospheres, and in dangerous atmosphéres, totally-enclosed motors also are used, the construction of some being rendered flame-proof. As already discussed on page 260, the output that can be taken continuously from totally-enclosed motors is

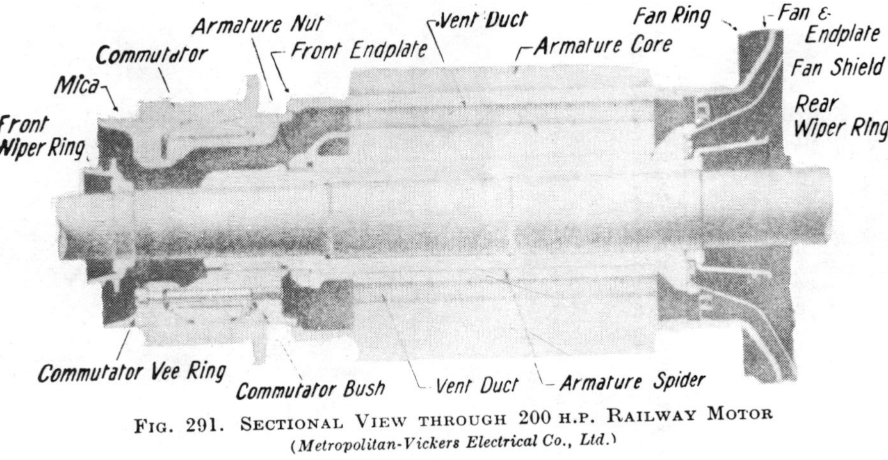

FIG. 291. SECTIONAL VIEW THROUGH 200 H.P. RAILWAY MOTOR
(*Metropolitan-Vickers Electrical Co., Ltd.*)

FIG. 292. BRUSH GEAR FOR 200 H.P. 775-V RAILWAY MOTOR
(*Metropolitan-Vickers Electrical Co., Ltd.*)

FIG. 293. PROTECTED-TYPE MOTOR
(*English Electric Co., Ltd.*)

FIG. 294. SCREEN-PROTECTED MOTOR
(The British Thomson-Houston Co., Ltd.)

FIG. 295. MOTOR ARRANGED FOR PIPE VENTILATION
(English Electric Co., Ltd.)

Fig. 296. Drip-proof Motor
(*English Electric Co., Ltd.*)

Fig. 297. Small Fan-cooled Enclosed Motor
(*Laurence Scott and Electromotors, Ltd.*)

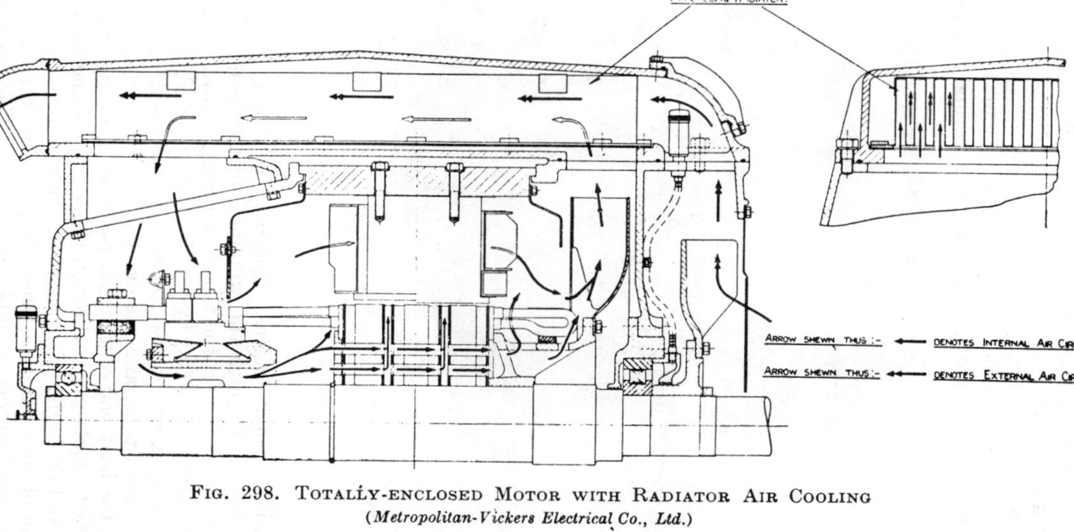

FIG. 298. TOTALLY-ENCLOSED MOTOR WITH RADIATOR AIR COOLING
(*Metropolitan-Vickers Electrical Co., Ltd.*)

relatively small, so that the machines are relatively heavy and expensive as compared with a protected type machine. To improve the radiating properties of these machines the yoke or end-covers are frequently ribbed as shown in Fig. 198, which illustrates a complete geared unit.

Frequently the system employed is much simpler, especially for small motors, an external fan being fitted to the shaft extension at the driving end and blowing air over the normal outside surface.

FIG. 299. EXTERIOR VIEW OF MACHINE ILLUSTRATED IN FIG. 298

(*Metropolitan-Vickers Electrical Co., Ltd.*)

Comparatively small internal fans may be fitted, in addition, to circulate the air through the windings on to the frame. Fig. 297 illustrates such a motor, the fan casing being shown, dismantled.

A more elaborate cooling system is indicated in Figs. 298 and 299. In this arrangement, hollow cooling fins are welded to the outside of the magnet frame, and the external fan—the casing of which is shown at the back end of the motor—blows air along these fins. An internal fan circulates air through the windings and over the commutator and through the inside of the hollow fins.

The use of such external fan cooling permits the construction of totally-enclosed motors of quite large ratings. The motor illustrated in Figs. 298 and 299 is a 6-pole machine with an armature 30 in. dia × 12·5 in. core length. At its rated output of 245 h.p., 810 rev/min its temperature rise is 43°C. Depending on the size of

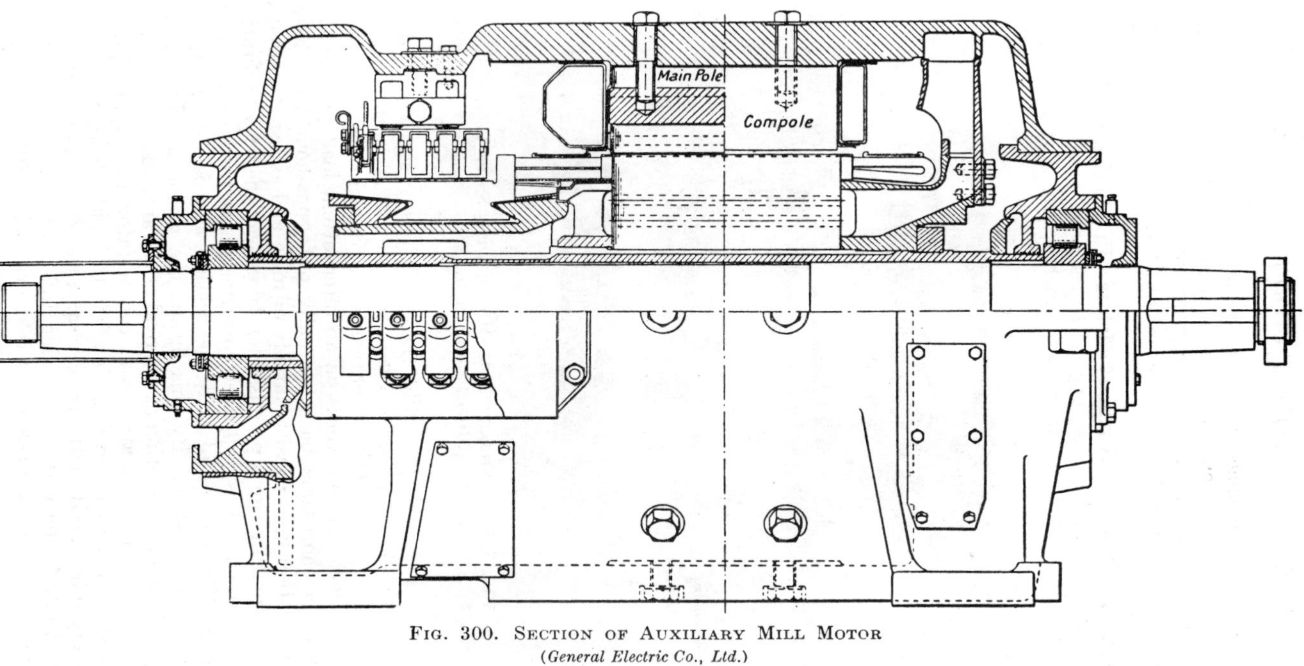

Fig. 300. Section of Auxiliary Mill Motor
(*General Electric Co., Ltd.*)

the motor, and its speed, the rated output of such machines may be as much as 70 per cent of the normal protected rating with small machines—or more under favourable circumstances. With large motors the relative value of the rating is as much as 50 or 60 per cent.

Iron- and Steel-mill Auxiliary Motors. In addition to the normal types of industrial motor, special types have been developed for various duties, such as for cranes, winches, etc. For the auxiliary

FIG. 301. AUXILIARY MILL MOTOR WITH SPLIT YOKE
(*General Electric Co., Ltd.*)

services in iron and steel mills the conditions of service are often very severe. Fig. 300 gives a sectional view of a motor designed especially to withstand the severe conditions of such duty. The motor is totally enclosed and works on an intermittent load, so that it is rated on a short-time basis, actually on a one-hour 75°C basis. The insulation used consists largely of mica and asbestos so that a high temperature rise is permissible. In view of the heavy vibration experienced, a robust all-steel construction is used as far as is possible. The yoke is of box-type section, and is split along the horizontal centre-line. By this means the motor can very easily be opened up for inspection or removal of the armature by removing the top half. The armature and commutator are mounted on a single steel spider as shown in Fig. 300, the spider being pressed on to a very substantial shaft. If necessary, this shaft can be extracted without interfering with the winding and commutator.

Another such steelworks motor is shown in Fig. 301.

CHAPTER XVI

DESIGN CONSIDERATIONS

Fundamental Output Relations. It has already been shown on page 9 that the total work done in one complete revolution of a d.c. machine is

$$W_r = 2p\Phi \times I_a Z_a \text{ watt-sec}$$

This work done thus depends solely upon two quantities—

1. *The Total Magnetic Loading*, $2p\Phi$, in webers . . (236)

and 2. *The Total Electric Loading*, $I_a Z_a$ in ampere-conductors (237)

The *total magnetic loading* gives simply the total number of magnetic lines cut by each conductor in one complete revolution. Sometimes the term "total flux" is used to denote the same quantity.

The *total electric loading* denotes the sum of all the currents in all the conductors on the armature. This quantity is sometimes called the *total current volume.*

For the active material of the armature to be fully utilized it is evident, for an armature of given dimensions, that the working values of the total magnetic and electric loadings will vary but little with minor changes in the values of the speed and working voltage. At very low speeds the electric loading may have to be reduced because of poor ventilation and cooling, and at relatively high voltages more space will be taken up by insulation due mainly to the increased number of active conductors. But, to a first approximation, the product (total electric loading × total magnetic loading) is constant for an armature of given dimensions, and thus an armature of given dimensions is capable of performing practically the same amount of work in one revolution, independent of minor changes in the actual working speed and voltage.

The dimensions of a d.c. armature, and hence also of the whole machine, except the commutator, are thus determined by the amount of work it has to perform in one revolution.

The output of a machine, as commonly defined, is the power output, which is given by the product of the work done per revolution and the number of revolutions per second, i.e. $P = W_r \times n$. The same machine will therefore be capable of developing an output of widely varying amount according to the speed of rotation. For example, a 25-h.p. 220-V 500 rev/min series motor can be utilized without modification to develop 50 h.p. 440 V at about 1,000 rev/min.

The work done in one revolution may be measured by the product of the power developed and the time for one complete revolution, i.e. the product $P \times 1/n$.

The dimensions of a machine are then not directly a function of the power output, but are a function of the work done per revolution, which may conveniently be expressed in the form P/n. (This latter may be regarded as giving the power output the machine would develop at a speed of 1 rev/sec.)

Since the work done in one complete revolution of the armature is numerically equal to $2\pi \times$ torque, it follows also that the dimensions of a machine may be regarded as being functions of the torque, rather than of the output. For example, a machine may have to develop 1,000 kW at the low speed of 60 rev/min. The dimensions will be great because the machine has to perform 1,000 kW sec, i.e. nearly 750,000 ft lb of work in one revolution. Another machine may have to develop 1,000 kW at 3,000 rev/min. This machine will be relatively small, as it has to perform only one-fiftieth of the amount of work in one revolution, i.e. 20 kW sec. (The corresponding values of the torque are 117,500 and 2,350 lb at 1 ft radius.)

Flux Density in Air Gap. In mechanical problems one is concerned as much with the *intensity* of loading as with the actual value of the load. Similarly in electrical engineering problems, the *intensity* of the loading affords at once a criterion as to its suitability. The figure for the current density in an armature conductor for example, at once is a criterion although not a complete one, as to the probable heating of the armature, whereas the mere knowledge of the total current is no criterion whatever.

With the magnetic loading of an armature similar conditions hold, although if the figure for the total magnetic loading is large, it is at once obvious that the machine must be large to accommodate the flux. Each square centimetre of armature surface is capable of receiving a certain magnetic flux, its value being but little influenced by the actual dimensions of the machine, except when the diameter is very small. The actual working value of the intensity of the magnetic loading is thus very important, as it affords at once a criterion as to the extent to which the active material is being employed.

This intensity may be expressed as the average value of the flux density at the armature surface, assuming the latter to be smooth. Thus the average density $B_{av} = \dfrac{\Phi}{\tau_p L} = \dfrac{2p\Phi}{\pi DL}$ may be used. Or the actual maximum value of the air-gap density, $B_g = B_{av}/K_f$, may be used as a measure of the intensity of the magnetic loading. Since B_g and not B_{av} determines the actual value of the iron losses in the teeth, the former affords the better criterion, but with machines all designed on standard lines it is immaterial which is used. The total magnetic loading is then given by—

$$2p\Phi = (\pi DL)B_{av} \quad . \quad . \quad . \quad . \quad (238)$$

or

$$2p\Phi = K_f(\pi DL)B_g \quad . \quad . \quad . \quad . \quad (239)$$

The value of the flux density in the air gap does not vary much in modern machines. For machines having an armature diameter of 100 cm or more, B_g will vary between 0·9 and about 1·05 Wb/m², i.e. 9,000 to about 10,500 gauss, although occasionally this latter figure may be exceeded, and even in quite small machines the value does not fall below about 0·65 Wb.

The corresponding values for B_{av} are about 0·65 to 0·75 Wb/m² for large machines and 0·45 as the lowest limit given above.

In (238) and (239) a relationship has thus been established between the actual dimensions of the armature and the total magnetic loading of the machine.

Armature Strength, or Specific Electric Loading. This is defined as follows—

$$\text{Armature strength} = \begin{array}{l}\text{Electric loading of the armature} \\ \text{per unit of circumference,}\end{array}$$

i.e.

$$q = \frac{I_a Z_a}{\pi D} = \frac{I_a Z_s}{\tau_s} \qquad . \quad . \quad . \ (240)$$

Again, it is obvious that the numerical value for the armature strength affords at once a criterion as to the suitability of the dimensions chosen for the machine. Although for large machines the value of the armature strength will not vary between wide limits, with small machines the space taken up by insulation is relatively greater, the taper of the teeth is more pronounced, and in general it is not possible to accommodate so many ampere-conductors per unit of the circumference as in larger machines. The manner in which it varies under different circumstances is discussed later. For the moment it may be taken that its value does not vary between very wide limits in large machines.

The total electric loading may then be expressed in terms of the armature strength and the circumference,

$$I_a Z_a = \pi D q \qquad . \quad . \quad . \quad . \ (241)$$

Esson Output Coefficient. The work done per revolution may be expressed in terms of the main dimensions of the armature, and the specific electric and magnetic loading constants. Thus,

$$\begin{aligned}\text{Work done per rev} &= 2p\Phi \times I_a Z_a \text{ watt sec} \\ &= (\pi D L B_{av})(\pi D q) \\ &= \pi^2 D^2 L q B_{av} \times 10^{-3} \text{ kW sec} \qquad . \quad . \ (242)\end{aligned}$$

Since this work per revolution is also equal to P^*/n where P^* is power in kW,

$$P^*/n = D^2 L(\pi^2 q B_{av} \times 10^{-3}) \qquad . \quad . \quad . \ (243)$$

Since it is more usual to express the speed of rotation as the number of revolutions per minute, it is more convenient to express the above equation in the form

$$\frac{P^*}{N} = D^2L\left(\frac{\pi^2}{60}\,qB_{av} \times 10^{-3}\right) \quad . \quad . \quad . \quad (244)$$

The right-hand sides of the above formulae are based on m.k.s. units,† i.e. D^2L in m^3, q in ampere-conductors per metre, and B in Wb/m^2.

In general, as first stated by Esson:

$$P^*/N = C_o \times D^2L \quad . \quad . \quad . \quad (245)$$

C_o is a coefficient first called by Esson the *Output-coefficient.*

The output coefficient thus varies in direct proportion to the product of the specific electric and magnetic loading constants. Its actual value is

$$C_o = \left\{\frac{\pi^2}{60} \times qB_{av}\right\} \times 10^{-3}$$

$$= 0{\cdot}164\,qB_{av} \times 10^{-3}\ \ddagger \quad . \quad . \quad . \quad (246)$$

In actual commercial design it is convenient to utilize the coefficient C_o as expressed above, as it gives at once a direct indication of the extent to which the dimensions of the machine are being employed. A large value for the coefficient indicates that a relatively large output is being obtained from the particular armature.

For the preliminary determination of the dimensions of a new design it is convenient to express (245) in the form

$$D^2L = \left(\frac{1}{C_o}\right) \times P^*/N \quad . \quad . \quad (247)$$

using the reciprocal of the output coefficient.

With C_o based on complete m.k.s. units of B_{av} etc. eqn. (247) gives D^2L in m^3.

For machines having an armature diameter of 100 cm and upwards, the variation in the value of the output coefficient is relatively small, since B_{av} is practically constant and q does not vary greatly. For small machines, the working value of q varies fairly largely with the value of the armature diameter, and hence the output coefficient also varies fairly largely. The working value of the coefficient is also dependent upon the value of the speed and upon the type of ventilation employed. With the introduction of fans of relatively large diameter it has been possible in most cases to increase considerably the output coefficient.

† If c.g.s. units are used D'^2L' is in cm^3, q' is in amp-cond/cm and B' in gauss and $P^*/N = D'^2L'(\pi^2q'B'_{av})/60 \times 10^{-11}$.

‡ In terms of the gauss, and amp-cond/cm the output coefficient giving D'^2L' in cm^3 is $C_o' = 0{\cdot}164\,q'B'_{av} \times 10^{-11}$.

In small machines, the actual working voltage affects the value of the coefficient, a larger value being possible for the lower voltage since the amount of space taken up by insulation is smaller than for relatively high voltages. With large machines, normal variations in the value of the working voltage have but little influence upon the working value of the output coefficient. If, however, the working voltage reaches the figure of 1,500 required in certain systems of electric traction, the coefficient has to be appreciably reduced.

For a fairly large armature the value of the coefficient, giving D^2L in m³, will be of the order of 4·5 to 5·5, for medium sizes, the value

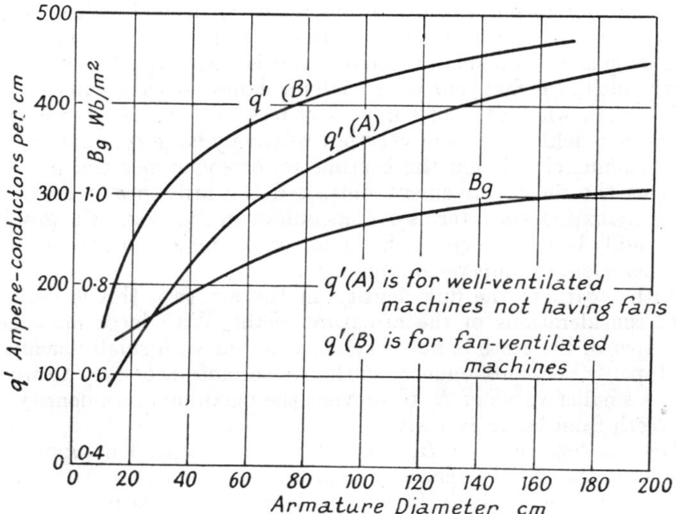

FIG. 302. LOADING CONSTANTS FOR D.C. MACHINES

will be from 3 to 4; and for small to medium sizes from 1 to 3. For very small machines, as used for table fans, etc., the value of the coefficient becomes extremely small. These figures and also the graphs given in Figs. 302 and 303 must be taken simply as affording a guide to help the student to obtain preliminary dimensions as a basis for the design of a machine. The data given by the graphs may be taken to represent keen modern practice for well-ventilated machines. For low speed machines lower constants must be taken both for the armature loading and for the output coefficient, and in any case the beginner would be well advised to base the preliminary dimensions upon more conservative constants. As the design is proceeded with it will soon become apparent whether the dimensions can be slightly reduced or need to be increased, as discussed later.

Variation in Output Coefficient with Dimensions. In Fig. 302 are given graphs showing the manner in which the values of the armature strength and the *maximum flux density*† in the gap B_g vary with the diameter of the armature. These figures must simply be taken as guides, and, in particular, so far as the armature strength is concerned, are subject to very appreciable variations depending upon the effectiveness of the ventilation. Normally the effect of progress will be to increase the value of the armature strength beyond the values given in the graph. Formerly the working value of the armature strength was often limited from considerations of armature reaction and commutation. The introduction of interpoles has removed this limitation, and the working value of the armature strength is now chiefly limited from heating considerations. The rapid decrease in the armature strength with small values of the diameter is due in part to the fact that it is not possible to employ such deep slots with small machines as it is with large, but it is also partly due to the fact that with small machines each slot will contain many small wires each of which is of necessity insulated, whereas the large machines contain very few relatively large conductors.

In commercial design the centimetre, or sometimes the inch, is used as the most convenient unit, and the armature strength is normally expressed in terms of that unit as in Fig. 302. The graphs can readily be used to give either gauss or Wb/m^2 and either ampere-conductor/cm or ampere-conductor/m.

With regard to the flux density in the air gap, this is limited from considerations of the armature teeth. With large machines the taper of the teeth is not very marked, but with small machines the taper is very pronounced, so that it becomes necessary to work with a smaller value of B_g to prevent the maximum flux density in the teeth from being excessive.

The working value of B_g is also sometimes dependent upon the speed. If the speed is such as to cause the frequency of the flux reversals to be relatively high—50 c/s or more—it is necessary to work with rather lower values of the air gap flux density. On the other hand, if the frequency of the flux reversals is extremely low, the flux density may be slightly increased. For a large proportion of d.c. machines the frequency of the flux reversals in a direct current machine is 25–50.

From Fig. 302 the manner in which the output coefficient varies with the diameter can be deduced. The form of the graph for C_o corresponds to the "product" of the graphs for q and B_g.

For the preliminary determination of the main dimensions D and L of a machine, Fig. 302 is not well suited as, naturally, the actual value of the diameter is not known. The author has therefore prepared Fig. 303, which gives the value of the coefficient C_o as a function of the work done per revolution by the armature expressed in the form P^*/N. The graphs correspond to average conditions of ventilation, and for machines of normal proportions.

Proportions of the d.c. Machine. In determining preliminary

† $B_{av} = K_f B_g$.

dimensions for a machine a sufficiently close value to take for the output coefficient can be obtained from Fig. 303, and the corresponding value of D^2L at once calculated.

It then remains to select suitable values of D and L to correspond with the value of D^2L. Broadly speaking, it is the designer's aim to select such dimensions as will result in the minimum cost of active material and manufacture, and yet lead to a design fully capable of fulfilling all guarantees.

To obtain suitable proportions for the machine, it is necessary to consider both the magnetic and the electric circuits. So far as

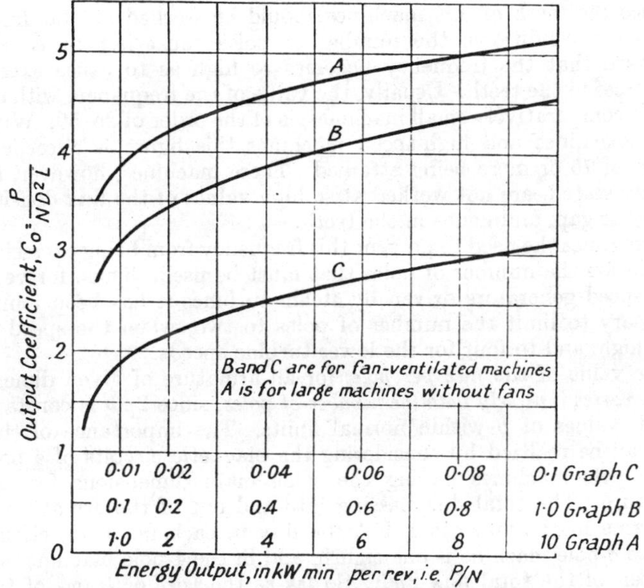

FIG. 303. OUTPUT COEFFICIENT OF D.C. MACHINES

the former are concerned, it is necessary to choose a suitable number of poles and also to choose suitable proportions for those poles. With regard to the electric circuits, it is desirable to choose such a design that a satisfactory winding and commutator are obtained.

Since the number of poles employed has an important bearing upon both the magnetic and electric circuits of the machine, considerations affecting the choice of the number of poles will first be discussed.

Number of Poles. With a d.c. machine the number of pairs of poles is not, as in a.c. generators, fixed from considerations of speed and frequency of currents to be supplied. Theoretically any number of pole pairs might be used for any particular machine. The need of producing a commercial, competitive design, usually narrows down the number of pole pairs that may be suitable to quite small limits.

Consider that the main dimensions of an armature are fixed and that the number of poles may be varied.

For an armature of fixed dimensions, and constant speed, the total magnetic loading, $2p\Phi$, and the total electric loading $I_a Z_a$, will not vary to any marked extent with comparatively minor changes in the number of poles. It must, however, be remembered that the frequency of the flux reversals in the armature teeth and core increases in direct proportion to the number of poles, being given by,

$$f = p \times n \qquad (248)$$

Since the teeth of d.c. machines should be worked not far from saturation conditions, the number of poles must not be so far increased that the frequency becomes so high as to cause excessive losses in the teeth. Usually, the value of the frequency, with all except comparatively small machines, is of the order of 25–50. With small machines and high-speed machines this figure is exceeded, values of 75 or more being attained. These machines, however, as already stated, are not worked at so high values of the flux density in the air gap, and hence in the teeth.

Sometimes the need to prevent this frequency from being excessive will decide the number of poles that must be used. Thus, for very high speed generators driven by steam turbines it has been found necessary to limit the number of poles to two, when the speed is very high, and to four for the lower turbine speeds.

The value of the *flux per pole*, for an armature of given dimensions, *varies inversely with the number of poles*, since $2p\Phi$ is constant for all values of p within normal limits. The importance of this fact can be realized by considering the magnetic circuits of 4-pole and 8-pole machines having the same main dimensions for the armature. The total flux passing into and out of the armature is the same in the two cases, but the flux in each magnetic circuit for the 8-pole machine is one-eighth, and in the 4-pole machine one quarter, of the total flux $2p\Phi$. So far as the *yoke* or *frame* of the machine is concerned, the flux is steady and the cross-sectional area can be chosen entirely from considerations of flux density. Hence the cross-sectional area of the yoke, and its weight, may be taken to vary inversely as the number of poles.

For the armature, the sectional area is determined partly by considerations of iron losses. Since the power wasted in eddy currents varies as the product $B^2 f^2$, it follows that, to obtain constant value for the losses, the cross-section of the armature core would be constant, and independent of the number of poles. Increasing the number of poles will then cause Φ, and hence B, to vary inversely proportional to p. So far as hysteresis losses are concerned, as the power wasted is proportional to $fB^{1 \cdot 6}$, for the same loss the core section can be reduced if the number of poles is increased.

For the armature core, then, it is not possible to reduce the cross-section of the core to the same extent as with the yoke, when the number of poles is increased, although it is possible to make considerable reductions. Where the working value of the frequency is low under any circumstances, it is possible with the armature, as with the yoke, to make the cross-sectional area vary inversely with the number of poles adopted, i.e. to work with a given value of the flux density in the core, quite independent of the number of poles.

The *total number of ampere-turns* developed by all the field coils will, for the case considered, be independent of the number of poles.

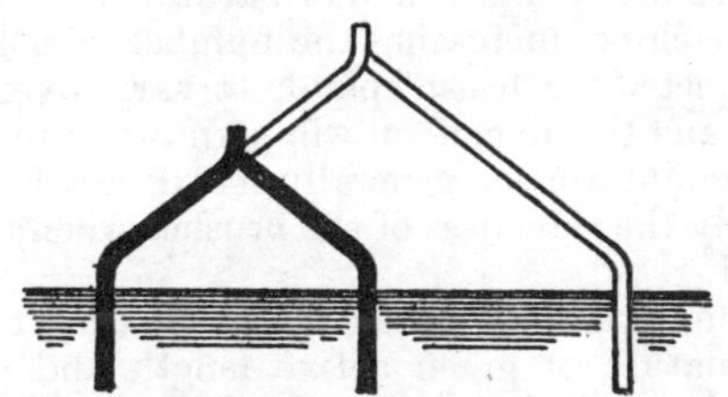

FIG. 304. EFFECT OF NUMBER OF POLES ON ARMATURE COILS

In logically-designed machines there is a fixed ratio between this total number of ampere-turns and the total number of ampere-turns developed by the armature, i.e. $I_a Z_a/2$. The ampere-turns developed by each field coil will thus vary inversely with the number of poles, and for the same depth of winding, the winding space required and hence the actual pole height will decrease as the number of poles is increased. *The weight of all the iron parts of the magnetic circuit will therefore decrease as the number of poles increases.*

With regard to the electric circuits, as already stated, the total electric loading will not vary appreciably with the number of poles, but the value chosen for this number of poles will have a marked influence upon the weight of "inactive" copper in each winding. This is illustrated by Fig. 304, showing the proportions of armature coils intended for two different numbers of poles, one twice the other. To a first approximation the weight of inactive copper, i.e. of the portions outside the slots, will vary inversely as the number of poles. Hence, increasing the number of poles will reduce the weight of copper on the armature, and at the same time reduce the outhang of the winding so that the overall length of the armature will be reduced, a point which sometimes is of considerable importance.

With the field windings, too, increasing the number of poles results in a smaller weight of copper being required.

As stated above, the total number of ampere-turns developed by all the coils is independent of the actual number of poles, as the design will be arranged so that the actual value of the ampere-turns developed by each field coil varies directly with the ampere-turns developed per pole by the armature, and hence varies inversely with the number of poles. The mean length of turn of the field winding will decrease as the number of poles is increased, and hence the total weight of copper can be reduced. To obtain the same losses, the cross-section of the conductor can be reduced, so that the saving in field copper due to increasing the number of poles is quite appreciable.

The weight of copper required for both field and armature windings will therefore decrease as the number of poles increases.

When the number of brush arms employed is equal to the number of poles, as it must with a lap-wound armature; and may be so with a wave-wound machine, increasing the number of poles causes the current to be collected per brush spindle to vary inversely with the number of poles, and this in general will permit of a reduction in the length of the commutator, as generally it will not be necessary to reduce appreciably the thickness of the brushes when the number of poles is increased.

In general, then, it may be taken that increasing the number of poles for an armature of given active length and diameter will decrease the weight of all the active material and reduce the overall length of the machine.

Further, armature reaction ampere-turns per pole $= 0{\cdot}5 I_a Z_a/$(no. of poles), and if too few poles are used this value may be excessive. Some designers endeavour to keep the value down to about 10,000 ampere-turns. Larger values can be used with compensated machines.

With a simple wave-wound armature, the total number of armature coils is not affected by the actual number of poles chosen, and there are two paths in parallel in all cases. But with a simple lap winding, the total number of armature coils will increase in direct proportion to the number of poles if the total magnetic loading is constant. The number of parallel circuits also increases in direct proportion to the number of poles.

Since single-turn coils are almost invariably used with multipolar lap windings, it follows also that the number of commutator segments will increase in direct proportion to the number of poles. For the same diameter of commutator the distance between brush sets varies in inverse proportion to the number of poles, so that the tendency towards flashing over from brush to brush will increase as the number of poles is increased, or alternatively, the diameter of the commutator will have to be increased as the number of poles increases.

The labour charges will increase as the number of poles increases —there are more armature coils to wind, insulate and solder to the commutator, there are more commutator segments to build up, more poles to assemble, more field coils to wind, and so on.

It may then be summarized that the larger the number of poles employed, the smaller will be the weight of both the active iron and copper in the machine, and hence the smaller will be the cost of material. On the other hand, the larger the number of poles the larger will be the labour charges, and if the number of poles is increased too much the saving in cost of active material will be completely outweighed by the increased labour charges.

For any particular case there will be one particular number of poles that will result in the cheapest machine, but this number will be determined largely by local manufacturing conditions and the price of copper and iron. Where labour is cheap and materials expensive it will be well to employ designs calling for more poles than when the opposite conditions prevail. It is to be expected, then, that considerable differences may obtain in the number of poles that would be employed by various firms for machines having the same main dimensions.

Further, when the best number of poles is fairly large—8 or more—it will generally be found that designs having either two more or two less than the best number will differ very little from the cheapest either from the point of view of cost or performance. The methods in use for costing, etc., have not been reduced to such an exact science that designers are yet in a position to state, for example, that for a particular case a 16-pole design is definitely better than a 14- or an 18-pole.

When designing a machine without limitations due to existence of patterns, etc., already made, the number of poles may be chosen so as to suit best the particular machine in question. Since the cost of the commutator is often a very large percentage of the total cost, the number of poles might then be chosen so as to give a satisfactory commutator. Under such conditions, the number of poles chosen for a 220-V machine and for a 550-V machine having the same output and speed, would then differ appreciably, since in both, the number of poles would be chosen to give as far as possible a reasonable value for the current per brush arm, and hence also for the length of the commutator.

For fairly large 500-V machines, a convenient value of the current per brush arm is 400 A, corresponding to 200 A per circuit in the armature winding. For very large machines this value might be increased a little. For smaller machines, the number of poles fixed from considerations of the iron dimensions would result in a smaller value for the current per brush arm. With low-voltage machines double the above figures, or more for very low voltages, may be necessary.

In commercial design it is necessary to base designs upon a relatively limited number of standards, and to choose for a given frame a number of poles which is best suited to average conditions. It then becomes necessary to arrange that the standard armature can be fitted with commutators of varying length. With d.c. machines it must be borne in mind that the dimensions of the commutator are settled largely by the current output, that is, largely by $(P^*/V) \times 10^3$, whereas the main dimensions of the armature are determined by the work done in each revolution, P^*/n. Hence low speed and normal values of voltage give rise to machines having relatively inexpensive commutators, whereas for high speeds and low voltages the commutator dimensions may become unwieldy, and difficulties arise.

Pole-pitch. In practice, then, each firm bases the number of poles used in standard machines upon its own experience as to costs, etc. For very small machines, or for high tension machines of modest output, bipolar machines are used as shown in Fig. 272. The exact diameter at which the change over from 2 to 4 poles takes place varies for different manufacturers, but for normal machines, an armature diameter of about 6 in. is about the largest that would be used for a bipolar design, as the pole-pitch then becomes 23·5 cm. Above this diameter, a 4-pole design is adopted until the pole-pitch becomes excessive and the poles unduly large, when a 6-pole design is used. In general, it may be taken that, for standard machines, when the number of poles is greater than four, it is chosen that the actual pole-pitch varies between comparatively narrow limits, although these limits will vary for different manufacturers. For 4-pole designs, the largest value of the pole-pitch that normally would be employed is from 35 to 40 cm. For 6-pole designs the pole-pitch may be taken to rather larger values, as the 6-pole machine being larger than the 4-pole, can well be worked with a rather longer core length. For a 6-pole machine, then, the pole-pitch may be taken up to a limiting value of from 40 to 45 cm, or even 50 cm in some cases. For larger machines, it may in general be taken that a pole-pitch of from 45 to 55 cm is suitable, and the number of poles to adopt is deduced accordingly. Armature reaction ampere-turns per pole, for a given value of q, are proportional to the pole-pitch, so that fact also, imposes a limit on the pole-pitch chosen. For low-voltage machines, a lower figure for the pole-pitch may be taken.

The above values of the pole-pitch suit well those for the axial lengths of armature core that experience has shown to be economical.

To sum up, a guide to the number of poles may be obtained from considerations of the pole-pitch dimension. For very small machines, two poles are often used until the pole-pitch reaches about 24 cm. Beyond this point 4 poles would be used until the pole-pitch becomes from 35 to 40 cm, after which 6 poles would be used until the pole-pitch reaches from 40 to 45 cm, or even 50 cm. For more than 6 poles, the pole-pitch will normally be from 45 to 55 cm, although in border-line cases a pitch of 40 may have to be used.

In machines specially designed for low-voltage heavy-current work the design, if possible, would be arranged to have more poles than the machine designed for more usual voltages. It must, however, clearly be understood that no hard and fast rules can be laid down with regard to the number of poles that should be adopted for any particular design. For example, two manufacturers may differ by as much as 50 per cent in the number of poles used for machines of the same output, e.g. the one may use an 8-pole design where the other uses 12 poles.

Core Length. One factor in determining the value to adopt for the core length is the ratio of the core length to the pole breadth as this determines the proportions of the pole.

The pole section which necessitates the smallest copper weight for the field winding is circular, since its periphery is the smallest possible for a given area, and hence the mean length of turn is smaller with circular than with any other form of pole section.

Most large manufacturers use laminated poles, and the square form of pole section will then give the most economical field winding. Since the pole breadth is normally about 40 to 55 per cent of the pole-pitch, the use of poles of square section would usually result in a comparatively short machine. Many manufacturers find that a rather longer machine is cheaper to construct, and build machines having the axial length of the poles up to as much as twice the pole breadth. It may then be taken that the core length of the machine will vary from about 75 per cent to 150 per cent of the value for the pole arc, i.e. from about 55 per cent to 110 per cent of the pole-pitch, the larger values as a rule holding for comparatively small machines. Some designers, as far as possible, aim at obtaining a core length equal in value to the pole arc. This gives a pole having an axial length rather greater than its breadth, and in general the student would find it well, for a normal case, to take for a preliminary design a core length differing but little from this value.

There is, however, a limit to the value of the core length, for if the core length be taken too great the e.m.f. induced per turn in the armature winding will be excessive, and the voltage between neighbouring commutator segments will become too high. Normally, it is not considered desirable that the *maximum* value of the voltage between adjacent commutator segments shall exceed 20 V on open circuit. In special cases it is necessary to exceed this figure and to employ compensating windings, but wherever possible steps are taken to keep within this limit.

For a normal single-turn coil, the maximum value of the induced e.m.f. is $2\ B_g Lv$ volts.

For a surface speed of 40 m/sec and a maximum flux density of 1 Wb/m², the total voltage between segments for each 1 cm core length is $2 \times 1 \times 0{\cdot}01 \times 40$ volts, i.e. 0·8 V, so that the maximum value of the core length, to keep within the above limit of voltage, is $20/0{\cdot}8 = 25$ cm.

Taking a lower value of the surface speed, e.g. 25 m/sec, it becomes possible to work then with a core length of 40 cm, and at 20 m/sec with a 50 cm core length without exceeding 20 V between commutator segments. It is important to check the value of the voltage between segments immediately the preliminary dimensions of the machine are settled.

Dimension of the Armature Diameter. The *peripheral speed* is sometimes a *limiting* factor to the value of the diameter. It is not, however, as a rule the *determining* factor, otherwise all machines operating at the same angular speed, i.e. the same rev/min, would be constructed with the same value for the diameter. Normally the peripheral speed should not exceed 6,000 ft/min, i.e. 30 m/sec. This speed does not call for any very special construction, but if exceeded, the bands of binding wire on the outhang of the armature winding will have to be especially strong. In absolute necessity, very much higher speeds are used, involving special methods of restraining the end connexions from flying out due to centrifugal action.*

In the preliminary determination of the dimensions of a d.c. machine, the value of the diameter may be obtained by assuming a reasonable value for the core length, and deducing the value of the diameter from the known value of D^2L. Better still, a number of alternative preliminary values of the main dimensions D and L should be worked out in tabular form, as shown later, and selection made from them.

Ratio of Pole Arc to Pole-pitch. For non-interpole designs this ratio may be taken as high as 0·75, for large machines. For interpole designs a slightly smaller value is necessary and generally the ratio taken is

$$\alpha = \frac{b_{ps}}{\tau_p} = 0{\cdot}67 \text{ to } 0{\cdot}7$$

* See later p. 452.

CHAPTER XVII

DESIGN OF D.C. GENERATOR: MAIN DIMENSIONS; ARMATURE

In this chapter a method of procedure in designing a d.c. generator will be given. The method will be illustrated by considering in detail the selection of the dimensions for a 1,000-kW, direct current, level-compounded generator.

Preliminary Determination of Main Dimensions. To illustrate the method of attack, preliminary dimensions will now be deduced for a 1,000-kW, 500-V, 300-rev/min, d.c. generator, to develop full load for 6 hours continuously with a temperature rise, by thermometer, not exceeding 40°C.

The work done per revolution = 1,000/300 = 3·33 kW min. A suitable preliminary value for the output coefficient is, therefore, from Fig. 303, about 4·5.

Hence
$$D^2L = \frac{3{\cdot}33}{4{\cdot}5} = 0{\cdot}74 \text{ m}^3$$

Full load current = 1,000 × 10³/500 = 2,000 A. For 400 A per brush arm this would necessitate 5 positive and 5 negative brush sets, i.e. a 10-pole machine. An average value for the length of the armature core for fairly large machines is 30 to 35 cm. It is well then to work out on a slide rule values of the diameter corresponding to these values and to others such as 25 and 40 cm for $D^2L = 0{\cdot}74$ m³.

The corresponding values are given below in Tabular form.

Core Length cm	Diameter cm	Circumference cm	Peripheral Speed m/sec
25	172	540	27
30	157	493	24·7
35	145·5	457	22·9
40	136	427	21·4

(The diameter for the first case, $L' = 25$ cm, is calculated,

$$D = \sqrt{\frac{0{\cdot}74}{0{\cdot}25}} = 1{\cdot}72 \text{ m} = 172 \text{ cm})$$

For the shortest machine the peripheral speed is 27 m/sec, a perfectly normal value so that all the designs are possible with normal construction.

The *maximum voltage between commutator segments* on open circuit for a simple lap winding is $2B_gLv$ volts. With $B_g = 1$ Wb/m², i.e. 10,000 gauss, and the longest core length of 0·4 m, the maximum voltage per segment is $2 \times 1{\cdot}0 \times 0{\cdot}4 \times 21{\cdot}4 = 17{\cdot}1$. For all the other alternatives its value will be lower, the lowest value being for the 25 cm length for which: max. volts/segt $= 2 \times 1{\cdot}0 \times 0{\cdot}25 \times 27 = 13{\cdot}5$. All the above designs are therefore permissible from this consideration.

Although a 10-pole design gives a very suitable value for the current per brush arm it would be well to consider also 8-pole and 12-pole designs, and to work out various important dimensions for each of these numbers of poles and the above values of D and L. This is most conveniently done in tabular form, as given below.

PRELIMINARY DESIGNS FOR 1,000-KW, D.C. GENERATOR, 500 V; 300 REV/MIN

Number of poles	Frequency f	Current per brush arm I_A/a A	Core length L' cm	Diameter D' cm	Pole-pitch τ'_p cm	Pole arc $b'_{ps} = \alpha\tau'_p = 0{\cdot}7\tau'_p$ cm	$\frac{L}{b_{ps}}$
			25	172	67·5	47·3	0·535
8	20	500	30	157	61·7	43·2	0·695
			35	145·5	57·2	40	0·875
			40	136	53·4	37·4	1·07
			25	172	54·0	37·8	0·66
10	25	400	30	157	49·3	34·5	0·87
			35	145·5	45·7	32·0	1·095
			40	136	42·7	29·9	1·34
			25	172	45·0	31·5	0·794
12	30	333	30	157	41·1	28·8	1·04
			35	145·5	38·1	26·7	1·31
			40	136	35·6	24·9	1·605

The table gives the more important dimensions of twelve alternative designs. So far as the proportions of the poles are concerned, it is at once apparent that for the smallest diameter, and hence the longest machine, 8 poles is very well suited, although some designers would regard 10 poles with equal favour. The pole-pitch for the 12-pole machine of this diameter is too small for normal conditions.

On the other hand, the largest diameter and shortest machine hardly seem suited for any of the three numbers of poles, except possibly 12 poles. A length of only 25 cm is rather smaller than is generally adopted for an armature greater than 100 cm diameter.

For the core lengths of 30 and 35 cm the pole-pitch is rather large for the 8-pole designs and smaller than usual for the 12-pole designs, although the proportions of the 12-pole machine with a

30-cm core length are quite reasonable. For the 10-pole design either of these lengths give rise to suitable proportions.

The frequency of the flux reversals ranges from 20 for the 8-pole machines to 30 for the 12-pole designs; neither of these limiting values is abnormal, so that the number of poles cannot in this case be decided solely from considerations of frequency; a frequency of 25 is quite a good value for a low speed machine.

Armature Reaction Ampere-turns. It is desirable to keep down armature reaction to a value not exceeding 11,000 AT/pole except for extremely large machines. Many designers prefer to keep it down to about 10,000.

With an armature strength of 400 ampere-conductors per cm this permits of a pole-pitch of $11{,}000/(0{\cdot}5 \times 400)$ i.e. 55 cm as a maximum and 50 as a quite safe limit. So all the designs are satisfactory in this respect.

Although, then, a design based upon any of the twelve preliminary designs outlined in the table could be built to operate satisfactorily, those having the most reasonable proportions are limited to—

The 8-pole machine of 40 cm length;
The 10-pole machine of 30 or 35, or possibly also 40 cm length.

The 8-pole machine has 500 A per brush set; although not impossible, this figure is somewhat on the high side so this machine may be ruled out in favour of the 10-pole design.

Having thus settled that the design shall be a 10-pole machine, it remains to decide on its diameter. Here, designers normally are guided by their experience as to the best ratio to adopt for L/b_{ps} and, as already stated, the beginner will not go far wrong with a normal machine if he takes this ratio about unity, or a little more. Nothing whatever is to be gained by working out fully two different designs for the same number of poles and slightly different core lengths, unless actual detailed calculations of costs are carried out, since the difference in the performance of the two machines would be insignificant as well as the difference in the weights of active material.

It must be remembered that the dimensions D and L are only preliminary, and that it may be found possible to shorten the machine, or possibly it may be necessary to increase the length.

For this machine, one might well take a diameter of 145 cm, and a length, provisionally, of 35 cm.

It will be noted that, for this particular machine, the number of poles which gives a convenient value of the current per brush arm also gives the best proportions to the magnetic circuit. This is not of necessity always so, and when time permits it is always well to consider a number of skeleton designs as outlined above, and to select the machine which seems the most satisfactory from all considerations.

Determination of Main Dimensions, given the Number of Poles. If the ratio $\beta = L/b_{ps}$ and also of $\alpha = b_{ps}/\tau_p$, be known from previous experience, the main dimensions can at once be determined if the number of poles is known.

Thus $$b_{ps} = \alpha\tau_p = \alpha \,.\, \pi D/2p$$

and $$L = \beta b_{ps} = \alpha\beta \,.\, \pi D/2p$$

Whence $$D^2L = \alpha\beta \,.\, \pi D^3/2p$$

and $$D = \sqrt[3]{D^2L} \times \sqrt[3]{\frac{2p}{\pi\alpha\beta}} \quad . \quad . \quad . (249)$$

As an example, the 1,000-kW machine just discussed may be taken, and the main dimensions obtained for a 10-pole design having $\beta = 1{\cdot}1$, and $\alpha = 0{\cdot}7$.

As already deduced, $D^2L = 0{\cdot}74$ m^3,

hence $$D = \sqrt[3]{0{\cdot}74} \times \sqrt[3]{\frac{10}{\pi \times 0{\cdot}7 \times 1{\cdot}1}}$$
$$= 1{\cdot}45 \text{ m}$$

and $$L = \alpha\beta \times \pi D/2p$$
$$= 0{\cdot}77 \times \pi \times 1{\cdot}45/10 = 0{\cdot}35 \text{ m}$$

or $$L' = 35 \text{ cm}$$

For the case of "square" pole shoes, and $\alpha = 0{\cdot}7$—a very common value—the diameter is given by—

$$D = \sqrt[3]{D^2L \times 2p} \div \sqrt[3]{0{\cdot}7\pi}$$
$$= 0{\cdot}766\sqrt[3]{D^2L \times \text{No. of poles}} \quad . \quad . (250)$$

and the length,

$$L = 0{\cdot}7\pi D/2p = 2{\cdot}2D/(\text{No. of poles})$$

As an example, preliminary dimensions may be determined for a 750-kW, 500-V generator operating at 375 rev/min.

The work done per revolution = 750/375 = 2 kW min.

From Fig. 303 a suitable trial value for the output coefficient is about 4·2.

Hence $D^2L = 2/4{\cdot}2 = 0{\cdot}475$ m^3.

The current output is $750 \times 10^3/500 = 1{,}500$ A. With 8 poles this would give 375 A per brush arm; with 6 poles, 500 A, and with 10 poles 300 A per brush arm. From this consideration the general choice could then be between 8 and 10 poles.

Taking 8 poles, for square shoes,

$$D = 0{\cdot}766\sqrt[3]{0{\cdot}475 \times 8} = 1{\cdot}2 \text{ m}$$

and $$L = 2{\cdot}2\, D/8 = 0{\cdot}33 \text{ m or } L' = 33 \text{ cm}$$

For the 10-pole alternative,

$$D = 0{\cdot}766\sqrt[3]{0{\cdot}475 \times 10} = 1{\cdot}295 \text{ m}$$

and $$L = 2{\cdot}2\, D/10 = 0{\cdot}285 \text{ m or } L' = 28{\cdot}5 \text{ cm}$$

For the 10-pole machine, the pole pitch is $\tau_p' = 3{\cdot}14 \times 129 \cdot 5/10 = 40{\cdot}7$ cm; for the 8-pole machine it is $\tau_p' = 3{\cdot}14 \times 120/8 = 47{\cdot}2$ cm. Either of the designs could then be adopted. The smaller number of poles would be preferred by many designers, especially for cases where labour charges are relatively high.

As preliminary dimensions for the 750-kW machine, a diameter of 120 cm, a length of 33 cm, and 8 poles may then be taken.

Reverting to the 1,000-kW machine:

The dimensions provisionally decided are then,

Diameter $D' = 145$ cm. Core length $L' = 35$ cm

Circumference $\pi D' = 3{\cdot}14 \times 145 = 455$ cm

Surface $\pi D'L' = 455 \times 35 = 15{,}900$ cm^2

Pole-pitch $\tau_p' = 455/10 = 45{\cdot}5$ cm

Peripheral speed $v_A = 4{\cdot}55 \times 5 = 22{\cdot}75$ m/sec

Design of Armature Stampings and Winding. From the preliminary dimension chosen for the armature diameter a suitable value for the maximum flux density in the gap can be selected by reference to Fig. 302. In any case B_g varies but very little with change in diameter, for large machines. Once the value of B_g is selected it automatically settles the electric loading intensity, q, for a machine in which the output and dimensions are fixed, since, as in (246) $C_o \propto qB_{av}$.

For machines of fairly large dimensions a reasonable figure for the maximum flux density in the air gap, B_g, is about 1 Wb/m^2. Some designers prefer a rather smaller figure of about 0·9.

Taking $B_g = 1{\cdot}0$ Wb/m^2, and $K_f = 0{\cdot}7$, the corresponding value of B_{av} is then $B_{av} = 0{\cdot}7 \times 1{\cdot}0 = 0{\cdot}7$ Wb/m^2.

The armature strength, in the 1,000-kW generator, will then have the value, cf., (246), given by

$$q = C_o^*/(0{\cdot}164 B_{av} \times 10^{-3}) = \frac{4{\cdot}5}{0{\cdot}164 \times 0{\cdot}7 \times 10^{-3}}$$

$$= 39{,}000 \text{ amp-cond/m}$$

$$q' = 390.$$

In all probability, these exact values of B_g and q cannot be used in the actual design, as it is necessary to select a number of conductors that give a perfectly satisfactory winding. This will only involve a slight modification of B_g, and a corresponding change in q.

Determination of Mean e.m.f. Induced per Conductor.

The mean e.m.f. $= B_{av} L v_A$

The peripheral speed $v_A = \pi D n = 3{\cdot}14 \times 1{\cdot}45 \times 5$

$= 22{\cdot}75$ m/sec

* $C_o = 4{\cdot}5$ for the 1,000-kW machine.

Hence the average e.m.f. $= 0{\cdot}7 \times 0{\cdot}35 \times 22{\cdot}75$
$= 5{\cdot}58$ V.

(Alternatively,

The total flux around the machine: $2p\Phi = B_{av} \times \pi DL$
$= 0{\cdot}7 \times 3{\cdot}14 \times 1{\cdot}45 \times 0{\cdot}35 \simeq 1{\cdot}10$ webers

and the mean e.m.f. $= 2p\Phi n = 1{\cdot}10 \times 5$
$= 5{\cdot}50$ V.)

Number of Conductors in Series on Armature. On full load, the e.m.f. induced in the armature winding exceeds the terminal voltage by an amount equal to the sum of the voltage drops in the armature winding, the interpole winding, the series winding, and the contact drop at the brushes. On a large machine these will aggregate about 2 per cent to 2·5 per cent of normal voltage in 500-V machines. For small machines the percentage will be from 5 per cent to 10 per cent or more for very small motors. For very low speed machines the percentage drop is relatively much greater.

Great precision in the estimation of the internal voltage drop is not necessary at this state of the design, and it will suffice to take the induced e.m.f. on full load as being 510 V, i.e. to allow a total drop of 10 V.

The number of conductors in series is then given by dividing the induced e.m.f. by the average e.m.f. per conductor—

$$Z_c = \frac{E}{2p\Phi n}$$

$$= 510/5{\cdot}58 = 91{\cdot}5.$$

The armature winding will then be arranged so that Z_c is the nearest convenient number to 91·5. If it is necessary to take a higher value for Z_c it simply means that, instead of working with $B_g = 1{\cdot}0$ Wb/m², a slightly smaller value is being taken. On the other hand, by taking a higher value, the armature strength is somewhat greater than that initially assumed.

For this machine, as discussed later, 90 is a most suitable number. Using 90 instead of 91·5 simply necessitates the working value of the flux being just a shade more than the assumed value. The revised value for the total flux is then

$$2p\Phi = 1{\cdot}10 \times 91{\cdot}5/90 = 1{\cdot}115,$$

and the number of conductors in series $Z_c = 90$.

The flux per pole $= 2p\Phi/2p$.

$$\therefore \qquad \Phi = 1{\cdot}115/10 = 0{\cdot}1115 \text{ weber.}$$

The value of Z_c largely determines the type of armature winding that is employed.

For a simple lap winding, Z_c represents the total number of armature conductors per pole, i.e. the number of conductors per slot × number of slots per pole.

For a simple wave winding—a 2-circuit winding—Z_c represents one half the total conductors on the armature, irrespective of the number of poles.

For large multipolar machines, the simple lap winding is used for all normal cases on account of the fact that the number of paths in parallel in the armature is equal to the number of poles. For the 1,000-kW design, a simple lap winding is obviously very well suited, as the current per circuit in the armature will then be 200 A, a very convenient current to handle. For small machines, unless the voltage is low, the simple wave winding is best suited as it is undesirable to have more than two paths in parallel when the total current is low. The actual type of winding to be adopted cannot be decided without reference to the commutator.

Normally it is not practicable to work with commutator segments less than 3 to 4 mm thick at the outer circumference. The thickness of the mica between segments is 0·8 mm or thereabout, so that the minimum value for the pitch of the segments under normal conditions is about 4 mm.

For a 500-V machine, at least 36 segments are necessary between adjacent brush sets, in order to keep down the maximum value of the voltage per segment on open circuit to 20 V, and under normal conditions it would not be considered desirable to work with more than twice that number of segments.

When possible, single-turn coils are used; for lap windings almost invariably, as a simple wave winding is preferred to a multi-turn lap winding. For small current, wave-wound machines, multi-turn coils must often be used, the number of turns per coil being chosen so as to give a suitable number of commutator segments.

For most cases, either a simple lap winding, or a simple wave winding will give suitable results.

For exceptional machines having a large number of poles and operating at low speed, it may be desirable to use a winding having 4, or may be 6, paths in parallel.

Examples of machines requiring a simple wave winding are given later.

For the 1,000-kW machine, with $Z_c = 90$ there will be in all $2pZ_c = 900$ conductors on the armature as there are 10 paths in parallel. This number has been chosen as it gives a convenient number of slots, i.e. 15 per pole, or 150 in all, with $Z_s = 6$ conductors per slot. The total number of single-turn coils will be one half of the number of conductors, i.e. $C = 900/2 = 450$.

There are thus 450 commutator segments, i.e. 45 between brushes.

With this number of segments and a commutator diameter of 100 cm—as shown later on page 432—a commutator of satisfactory proportions is obtained.

Number of Slots. With a lap-wound armature the number of slots must be such that equalizing connexions can be made to the winding. This involves the number of slots being a multiple of the number of pole pairs. Frequently the number of slots for lap wound machines is made a multiple of the number of poles, so that full pitch coils can be used. The various points connected to successive equalizers need not be equally spaced, but normally it is convenient that they should be.

From a purely theoretical point of view the larger the number of slots the better is the performance, especially so far as flux swinging is concerned. Further, since the armature conductors are cooled by heat conduction through the slot insulation to the teeth, it is desirable not to bunch too many conductors in one slot.

The value of the total current per slot, i.e. the product (current per conductor) × (number of conductors per slot), should normally not exceed 1,500 A. Higher values are sometimes taken, but the student would do well to base his designs, if anything, upon a rather lower limit.

From the point of view of cost, the smaller the number of slots the better, as the cost of punching the slots increases with their number. Further, each slot has a lining of mica paper or other insulating material about 1 mm thick, so that of the total slot width about 2·5 mm is taken up with this lining and the necessary allowance for "slack," this amount being practically independent of width of the slot. With a wide slot, relatively more space is therefore available for copper than with a narrow slot.

The smaller the number of conductors per slot, and hence the greater the number of slots, the better it is for commutation.

Normally, the value of the slot pitch will lie between 2 and 4 cm as extreme limits, and between 2·5 and 3·5 cm as usual limits for all except very small machines. The number of slots will not differ greatly from the figure giving the diameter of the armature in centimetres.

For small armatures the complete core disc can be blanked out in one piece. This holds for diameters up to about 100 cm. Beyond this, the width of the sheet steel as normally rolled necessitates the building up of the core in segments as already discussed in Chapter XV. It then becomes necessary to choose a number of slots which will permit of this, it obviously being desirable that the total number of slots shall be divisible by the number of segments employed. For the 145 cm diameter machine six segments would suffice.

So far as the 1,000-kW machine is concerned, 150 slots has tentatively been selected.

This number is divisible by 10, the number of poles, and, therefore, is suitable for equalizing connexions. It is divisible by 6, the number of segments into which the core is divided. The slot pitch is $\pi \times 145/150 = 3{\cdot}04$ cm, which is quite a good value.

The total current volume per slot, $I_a Z_s = 200 \times 6 = 1{,}200$ A conductors, which is quite normal for a large machine.

Dimensions of Armature Conductor. The current density at which the armature conductors are worked is determined from considerations of temperature rise, and hence depends very largely upon the effectiveness of the method of ventilation employed. In turn, this will depend to a very large extent upon the peripheral speed of the armature and, of course, upon the nature of the fans, if any.

Considering the loss occurring in all the conductors of one slot, the power wasted for the total active length is, for Z_s conductors each carrying I_a amperes—

$$Z_s I_a^2 \times \rho \frac{L}{A_a}$$

The slot pitch being τ_s, the loss per unit area of the armature surface becomes,

$$p_a = \frac{Z_s I_a^2 \rho}{A_a \times \tau_s}$$

$$= \frac{Z_s I_a}{\tau_s} \times \frac{I_a}{A_a} \times \rho = q\alpha\rho \quad . \quad . \quad . \ (251)$$

i.e. the loss per unit area of surface is proportional to the product of the current density and the armature strength.

If m.k.s. units are used eqn. (251) will give the loss per m², and if c.g.s. units—the loss per cm².*

The current density is rarely expressed in A/m², but quite often in A/mm². Taking as an example $q' = 400$ ampere-cond/cm, $\alpha' = 450$ A/cm² and $\rho' = 2{\cdot}05 \times 10^{-6}$ (for hot copper) the ohmic loss per cm² of armature surface is $450 \times 400 \times 2{\cdot}05 \times 10^{-6} = 0{\cdot}37$ watt.

Other things being equal, then, machines working with a small value of the armature strength can be worked at larger current densities than those in which the armature strength is relatively great. This is borne out in practice; in particular the current density with small wire-wound armatures may safely exceed the values that are admissible with large strap wound machines.

For our present purpose it will suffice to take the following as being suitable values—

$\alpha = 4{\cdot}5$ A/mm² for large strap-wound armatures and very good, normal ventilation.

$\alpha = 5{\cdot}0$ A/mm² for small wire-wound armatures and very good normal ventilation.

$\alpha = 6{\cdot}0$ to $7{\cdot}0$ A/mm² for fairly high speed, fan-ventilated machines.

* A convenient expression based on mixed units is: Watts lost/m² of surface = 2·05 × ampere-conds/cm × (A/mm²).

If the peripheral speed is relatively low, the values must be reduced, and for fan-ventilated machines it should be realized that to work at the figure given, and in addition with a high value of the ampere-conductors per centimetre, the fans must be both large and efficient.

For the 1,000-kW generator, for example, the peripheral speed being normal, a current density of about 450 A/cm² may be taken as being suitable. The current per circuit is 201 A* approx. so that the sectional area required is

$$A_a = \frac{I_a}{\alpha}, \quad A_a' = \frac{201}{450}$$

$$= 0{\cdot}446 \text{ cm}^2.$$

For six conductors per slot, and a sectional area of this magnitude, the most convenient method is to employ copper strap for the conductors—a practice to be used wherever possible, as it results in a very sound armature winding, easy to insulate and of good mechanical properties.

The maximum depth of conductor that can be employed in direct current armatures, where the frequency of the flux reversals is about 25, is about ¾ in. or 19 mm. If this dimension is much exceeded, as stated later, appreciable eddy e.m.f.'s may be induced and the copper losses increased. The corners are normally rounded off, so that the conductor width should be rather greater than that given by 0·446/1·9 cm, i.e. than 0·235 cm. The conductor may then be taken to be 2·4 mm wide by 19 mm deep, its area being 0·45 cm² when allowance has been made for rounding off the corners.

Where the frequency of the flux reversals is relatively high, the conductor depth should be taken rather less than 19 mm—about 15 mm.

Design of the Slot. Various methods are in use for determining *ab initio* the slot dimensions. The method favoured by the author is to select first a suitable dimension for the slot depth. The depth employed is naturally of considerable importance. By increasing the depth more copper can be accommodated and the total current loading of the machine increased. To obtain the maximum possible output from a machine of given dimensions it is therefore imperative that the depth of the slots should be taken as great as experience has found to be desirable. Slots that are too deep result in the reluctance of the teeth becoming excessive, and would necessitate the use of smaller values of the flux density in the gap than can be used with slots of more normal dimensions. In addition the tooth losses would become excessive in many cases. For strap-wound armatures the copper losses are increased owing to eddy e.m.f.'s induced in the deep conductors, in particular when the slots approach

* Including a small allowance of 0·5 per cent for the current in the shunt winding.

and leave the poles. The "eddy current" losses so caused depend very largely upon the depth of the conductor, so that it does not become expedient to increase the depth of the conductor beyond the dimensions given in the previous paragraph, unless the frequency is low. Since the normal winding has but two layers, for normal large machines the slot depth is then simply large enough to accommodate a 2-layer winding in which the conductor depth is about $\frac{3}{4}$ in.

For medium-sized and small machines the taper of the slots is more important, and the slot depth must further be limited. To be precise, for small machines the best slot depth should be a function of the armature diameter. The following table will serve as a guide for small machines, but need not, of course, be adhered to rigidly—

Diameter of Armature cm	Slot depth cm
15	2·25
20	2·75
25	3·25
30	3·75
40	4·25
50	4·5

These figures would not apply to exceptional cases, such as, for example, very low speed machines.

In non-interpole machines the maximum permissible slot depth was often limited from considerations of the leakage reactance of the armature coils, i.e. from commutation considerations.

Having settled upon the slot depth, the width of the slot can then be determined from considerations of the conductor dimensions, slot insulation, etc. A good method of procedure for a large machine may be illustrated by considering that of the 1,000-kW machine under discussion. For this machine the conductor dimensions have been fixed at 19 × 2·4 mm. A good method of insulating the conductors, which is at the same time very economical of space, is to employ mica paper wrapped in between and around the bare conductors as already discussed in Chapter XIV. The group of three conductors is then insulated with $2\frac{3}{4}$ turns of mica paper wrap. This is bound with one layer of cotton tape (not overlapped). To protect the coils against damage during insertion, the slot is lined with a thin trough of leatheroid—or press paper. Between the two layers a strip of mica paper is placed, and in the event of there being any undue slackness at the winding in the depth of the slot, a strip of pressboard is placed at the bottom. The winding is retained against centrifugal forces by means of a bakelized fabric wedge 4 mm thick. It is important to remember that allowance must be made for roughness of the slot walls and a certain degree of slackness in the width chosen for the slot. In the slot depth a fairly liberal allowance should be made lest the winding, as built up, is too tight a fit in the width of the slot, in which case the coils can be compressed in width at the expense of an increase in their other dimension. A suitable allowance for slack with a strap wound armature is about 0·2 mm per conductor; in depth an allowance of from 2 to 3 mm total is ample.

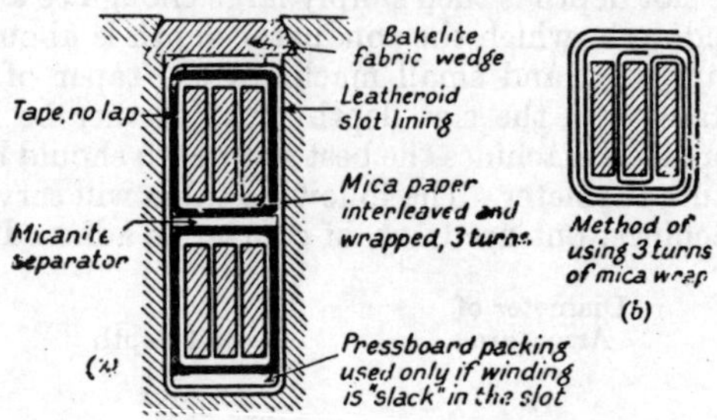

FIG. 305. ARRANGEMENT OF CONDUCTORS IN SLOT (NOT TO SCALE)

A drawing of the slot for the 1,000-kW machine is given in Fig. 305, the dimensions being compiled as follows—

Slot width—	mm
Bare copper, 3 × 2·4 mm	7·2
Mica wrap, 0·3 mm thick; total of 8 thicknesses, 8 × 0·3	2·4
Tape, no lap, each side 0·2 mm	0·4
Leatheroid lining, 0·2 mm thick	0·4
Allowance for slack, etc.	0·6
Total slot width	11·0

Slot depth—	mm
Bare copper, 2 × 19	38·0
Mica wrap 6 thicknesses total per cell, or 12 in all, 12 × 0·3	3·6
Tape, no lap, for 2 cells	0·8
Leatheroid lining, 3 × 0·2	0·6
Bakelized Fabric wedge	4·0
Micanite separator between layers	0·5
Slack	2·5
Total slot depth	50·0

The slot dimensions are then 11 × 50 mm. These dimensions would be considered unsuitable for a non-interpole machine, as the ratio slot depth to slot width is rather large.

It is important that the winding shall be a fairly tight fit in the slot, so that heat can travel from the conductors through the insulation to the walls of the slot. Any air pockets would seriously impair the heat conduction.

Having finally settled upon the number and dimensions of the slots and the total number of conductors, it remains to decide finally upon the main dimensions L_i and L of the armature. The dimensions chosen at the start were based upon an assumed value of the output coefficient and served to determine a suitable winding and value for the working flux. It is now necessary to determine the total tooth section and hence to obtain a final figure for the core length of the armature.

Final Dimension for Core Length. The core length finally adopted is chosen to give the desired value for the flux density in the teeth. As already discussed in Chapter II, the calculations for the teeth are based upon conditions holding one third of the distance from the bottom of the teeth. For a large machine the apparent flux density may be from 2·15 to 2·2 Wb/m², this figure representing the maximum value and occurring for the teeth under the centre of the pole shoe.

The average flux density in the teeth—over the whole pole-pitch —is given by

$$B_{t_{av}} = \text{Total flux per pole} \div \text{total tooth section per pole}$$

$$= \frac{2p\Phi}{Sb_{t_{ef}}L_i} = \frac{\Phi}{\dfrac{S}{2p} \times b_{t_{ef}} \times L_i}$$

This may also be written

$$B_{t_{av}} = \frac{2p\Phi}{S(\tau_{s_{ef}} - b_s)L_i} = \frac{2p\Phi}{(\pi D_m - Sb_s)L_i}$$

$\dfrac{S}{2p}$ being the total number of teeth per pole. The maximum density is

$$B_t = \frac{1}{K_f} \times B_{t_{av}}$$

$$= \frac{1}{K_f} \times 2p\Phi \times \frac{1}{(\pi D_m - Sb_s)L_i}$$

For the machine under consideration, the external diameter of the armature is 145 cm and the slot depth 5 cm. The diameter at the root of the teeth is then $145 - 2 \times 5 = 135$ cm, and D_m the diameter at one third of the distance from the root of the teeth is $135 + 10/3 = 138{\cdot}3$ cm.

The tooth pitch at this point is $\tau'_{s_{ef}} = \pi \times 138{\cdot}3/150 = 2{\cdot}9$ cm

The effective tooth width is then $b'_{t_{ef}} = \tau'_{s_{ef}} - b_s'$

$$= 2{\cdot}9 - 1{\cdot}1 = 1{\cdot}8 \text{ cm}$$

The number of teeth per pole is $150/10 = 15$

As there is a certain degree of magnetic saturation in the teeth, the flux coefficient may be taken to be 0·72, instead of 0·7 (the value of α). The effective area of the teeth, per pole, is then

$$A_t = K_f \times \frac{S}{2p} \times b_{t_{ef}} \times L_i$$

and $$A_t' = 0{\cdot}72 \times 15 \times 1{\cdot}8 \times L_i',$$

where L_i is the actual active length of iron.

Taking then $B_{appt} = 2{\cdot}15$ Wb/m² = 21,500 gauss for the apparent flux density in the teeth,

$$A_t = \Phi/B_{appt} = 0{\cdot}1115/2{\cdot}15 \text{ m}^2 = 0{\cdot}0518 \text{ m}^2$$

and $$A_t' = 518 \text{ cm}^2$$

Hence the necessary value for the active iron length is

$$L_i' = 518/(0{\cdot}72 \times 1{\cdot}8 \times 15) = 26{\cdot}7 \text{ cm}$$

For an iron factor of 0·89, the total axial length of the packets is

$$L_n' = 26{\cdot}7/0{\cdot}89 = 30 \text{ cm}$$

This length could conveniently be divided into six packets, each about 2 in. wide, so that five ventilating ducts would suffice, each 0·9 cm wide.

The total core length necessary thus becomes,

$$L' = L_n' + n_v b_v' = 30 + 5 \times 0{\cdot}9 = 34{\cdot}5 \text{ cm}$$

The value which has provisionally been adopted is 35 cm, indicating that the output coefficient initially assumed fits in very well with the flux and current densities, and the method of insulation assumed.

The actual core length first adopted will now be revised to $L' = 34{\cdot}5$ cm.*

It may be well at this point to recapitulate the method employed. At the start, a value was *assumed* for the output coefficient. Normally it is not to be expected that this assumed value will fit in *exactly* for the particular machine being designed, but its value will be sufficiently exact to permit of a suitable diameter being selected, and an *approximate* value for the core length determined. From these provisional dimensions a reasonable value for the working flux was obtained, and a suitable number of armature conductors and slots chosen. The dimensions of these conductors were chosen from considerations of current density, and the conductors were accommodated in slots of as large a depth as experience has shown to be desirable, and the necessary slot width thus obtained. The core length was then finally adjusted so that it just sufficed to carry the working flux.

* Normally it may be expected that relatively greater modification will be necessary. Actually a change of 0·5 in 34·5 is insignificant; it is only made here to illustrate the general case.

With this procedure, if the various constants are properly chosen to fit in with the method of ventilation, etc., all doubts as to the suitability of the output coefficient initially chosen are dispersed, and there is no question of the machine finally decided upon being too "tight" or too "slack."

Depth of Armature Core. The core below the slots is usually worked at a density of about 1·25 Wb/m² = 12,500 gauss, except in abnormal cases. For very low values of the frequency, the density may exceed this, and, for higher values than normal, a lower flux density may be advisable.

Each path in the core has to carry one half of the useful flux per pole, i.e. 0·1115/2 Wb for the 1,000-kW generator.

The area required is then $A_c' = \frac{1}{2} \times 0{\cdot}1115 \times 10^4/1{\cdot}25 \text{ cm}^2$

$$= 446 \text{ cm}^2$$

The necessary depth below the slots is then,

$$d_c' = A_c'/L_i' = 446/26{\cdot}7$$
$$= 17 \text{ cm approx.}$$

The internal diameter of the armature stampings is then,

$$D_i' = D' - 2(h_s' + d_c') = 145 - 2(5 + 17) = 101 \text{ cm}$$

This corresponds to $A_c' = 17 \times 26{\cdot}7 = 454 \text{ cm}^2$

The dimensions of the core plates are then—

External diameter: $D' = 145$ cm

Internal diameter: $D_i' = 101$ cm

No. of segments: 6

No. of slots: $S = 150$

Slot dimensions: $b_s \times h_s = 11 \times 50$ mm

Gross length of core: $L' = 34{\cdot}5$ cm

No. of ducts, each 0·9 mm: $n_v = 5$

Net length of core: $L_n' = 30$ cm

Active iron length: $L_i' = 26{\cdot}7$ cm

Armature Winding. The type of winding to be adopted has already been discussed. For this large current machine it is a simple lap winding.

Hence: the commutator pitch $y_C = 1$.

Full-pitched coils will be used, the coil will then span 150/10 = 15 teeth.

Equalizing Connexions. Taking a tapping from every other slot would necessitate 30/2 rings as there are 30 slots per pair of poles.

With a tapping from every third slot, only 10 rings would be required. Better results would be expected with the larger number of rings, although 10 would probably suffice.

Taking 15 rings, the connexions will be,

Ring 1 joined to coils	1	91	181	271	361
" 2 " "	7	97	187	277	367
" 14 " "	79	169	259	349	439
" 15 " "	85	175	265	355	445

The sectional area necessary for the rings decreases as the number of rings is increased. The area of the conductors is 0·45 cm², and for this large number of rings about 40 per cent to 50 per cent of this value would suffice for the rings. The rings will then be taken to have a sectional area of 0·2 cm².

The total number of tapping points is 5 per ring, or 75 in all.

Length of Armature Conductors. To obtain accurately the length of the conductors, reference should be made to an actual drawing of the coil. But sufficiently accurate results can be obtained for present purposes by the following method: Assume that the machine has a large number of poles, so that considerations of curvature can be neglected, and that the coil ends are bent at an angle θ, as indicated in Fig. 306. Then, to a close enough approximation, the length of the free portion at each end of the coil may be taken to be the sum of the lengths of the straight lines, AC, CE, and EK. The insulating cell should extend for the upper coil-side by about 2 cm beyond the core, i.e. 2 cm from H. (HA is the space between the end of the laminations and the core end plates, and actually contains a vent-spacer, stiffening up the ends of the teeth.) Allowing for the fact that the coil is not bent sharply, the distance AC may then be taken to be 4 cm approximately; for EK we may approximate 3 cm.

The dimension CD being one-half of the pole-pitch, and the slant side CE being $CD/\cos\theta$, the total length of the free portion at each end may be taken to be,

$$\tfrac{1}{2}l_{fr}' = 7 + \tau_p'/(2\cos\theta) \text{ cm}$$

and $$l_{fr}' = 14 + \tau_p'/\cos\theta \text{ cm approximately, for large machines.}$$

Actually, the dimension CD should be based upon the mean diameter holding for the coil, but it suffices to take the pitch corresponding to the external diameter of the armature, as by this means some allowance is made for the actual buckle in the conductor at EK.

The angle at which the coils are bent is determined by the ratio of the slot width to the slot pitch at the root of the teeth, and by the distance that the coils are spaced apart.

Thus, if the width of the insulated coil group be represented by PF, as in Fig. 306, the spacing between the groups by FN, and the slot pitch by OP; then $\sin\theta = PN/OP$.

Often the coils are packed tightly against each other with no ventilating space provided, and then the dimensions ON may be taken to be the slot width, and $\sin\theta$ is simply the ratio of the slot width to the slot pitch, if no allowance be made for the ends of the winding drooping below the bottom of the slot.

Overhang of the Winding at each end is, from Fig. 306, given by $L_o' = 7 + \tan\theta \times \tau_p'/2$ cm for large machines.

For large machines, with no allowance for spacing of the coils, and for a slot width equal to 40 per cent of the slot pitch at the root of the teeth, $\sin\theta = 0{\cdot}4$ and $\theta = 23{\cdot}5°$. By increasing θ to 30° about 3 mm space between coils is normally obtained.

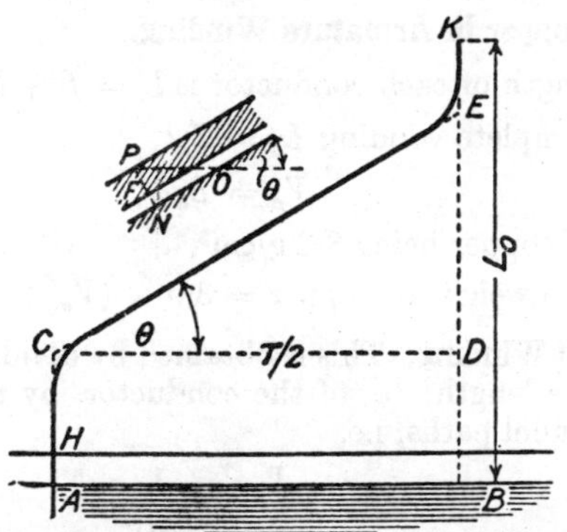

FIG. 306. END PORTION OF ARMATURE CONDUCTOR

For $\theta = 30°$, $\cos\theta = 0{\cdot}87$, and

$$l_{fr}' = 14 + \tau_p'/0{\cdot}87 = 14 + 1{\cdot}15\tau_p' \text{ cm approximately} \quad . \ (252)$$

The outhang, for this case, with $\tan\theta = 0{\cdot}58$

$$L_o' = 7 + 0{\cdot}3\tau_p' \text{ cm approximately} \quad . \quad . \ (253)$$

With the coils packed tight, the corresponding values are

$$l_{fr}' = 14 + 1{\cdot}09\tau_p' \text{ cm approximately,}$$

and $$L_o' = 7 + 0{\cdot}23\tau_p' \text{ cm approximately.}$$

These dimensions include, in large machines, the portion of the conductor that is soldered into the commutator riser.

For small machines, the ratio of slot width to slot pitch at the root of the teeth is much larger, and the outhang of the cell has to be smaller.

For such machines, a liberal allowance is obtained by taking

$$l_{fr}' = 5 + 1{\cdot}35\tau_p' \text{ cm} \quad . \quad . \quad . \ (254)$$

For the 1,000-kW generator, the slot pitch at the base of the slots is approximately 27 mm, the width of the insulated bundle of conductors will be about 10·5 mm, so that with 3 mm spacing between bundles, $\sin\theta = 13{\cdot}5/27$, and $\theta = 30°$.

The pole-pitch is 45 cm approximately.

Hence as in (252) $l_{fr}' = 14 + 1{\cdot}15 \times 45$,

$= 60$ cm approximately,

and $L_o' = 7 + 0{\cdot}3 \times 45 = 20{\cdot}5$ cm approximately

Weight of Copper in Armature Winding.

The total length of each conductor is $l_a = L + l_{fr}$,

and, for the complete winding $L_a = Z_a l_a$

The volume $V_a = L_a A_a$

The density of copper being 8·9 g/cm³,

$$\text{the weight of copper} = 8{\cdot}9 \times (V_a' \times 10^{-3}) \text{ kg} \quad . \ (255)$$

Resistance of Winding. This is obtained by dividing the resistance of the complete length, L_a, of the conductor, by the square of the number of parallel paths, i.e.

$$r_a = \rho \frac{L_a}{A_a} \times \frac{1}{(2a)^2} \quad . \quad . \quad . \quad . \ (256)$$

where ρ is the specific resistance of copper. For normal "cold" conditions, ρ may be taken to be $1{\cdot}7 \times 10^{-6}$ ohms, in cm measure, and allowing for 50°C rise in temperature, by increase in resistance, the resistance when hot may be taken to be 20 per cent greater than when cold.

Hence, for the 1,000-kW machine—

The free length per conductor . . .	$l_{fr}' = 1{\cdot}15 \times 45 + 14$	$= 60{\cdot}15$ cm
The total length . .	$l_a' = 34{\cdot}5 + 66$	$= 100{\cdot}5$ cm
The total length of all the 900 conductors .	$L_a' = 900 \times 100{\cdot}5$	$= 90{,}500$ cm
The total volume. .	$V_a' = 90{,}500 \times 0{\cdot}45$	$= 40{\cdot}7 \times 10^3$ cm³
The total weight . .	$W_a = 8{\cdot}9 \times 40{\cdot}7$ kg	$= 362$ kg
	$= 2{\cdot}2 \times 362$	$= 796$ lb
The resistance of the complete 905 metres of conductor . .	$= 1{\cdot}7 \times 90{,}500 \times 10^{-6}/0{\cdot}45$	
	$= 0{\cdot}342$ ohm, cold	
The number of parallel paths . . .	$2a = 10$	
The total armature resistance . . .	$r_a = 0{\cdot}342/10^2$	$= 0{\cdot}00342$ ohm, cold
	$= 1{\cdot}2 \times 0{\cdot}00342$	$= 0{\cdot}0041$ ohm, hot

The voltage drop in the winding . . . $= 2{,}010 \times 0{\cdot}0041$

$= 8{\cdot}24$ volts, approx. when hot,

and the power wasted, neglecting eddy currents

$= 2{,}010 \times 8{\cdot}24$ watts

$= 16{\cdot}6$ kW, when hot

As the conductors are rather deep, eddy current losses* will be appreciable and may be assumed to increase the total loss by 25 to 33 per cent. Taking the latter figure we have—

Power wasted, including eddy currents $= 1{\cdot}33 \times 16{\cdot}6 = 22$ kW

* Eddy current losses are considered in detail in Ch. XX.

CHAPTER XVIII

DESIGN: THE FIELD SYSTEM

Field Magnet. To prevent the effects of armature reaction from being excessive, the magnet system has to be designed so that the m.m.f. developed by the exciting coils is sufficiently powerful in comparison with that developed on full load by the armature. In d.c. machines, designed on a logical basis, there will thus be a certain minimum value for the ratio field ampere-turns/armature ampere-turns, under full load conditions, that is regarded as a safe limit. This minimum value will depend upon the duty the machine has to perform.

For non-interpole designs, the magnetic circuit should be designed so that the total ampere-turns required for the armature teeth and air gap are greater than the full-load ampere-turns developed by the armature. With such machines, the total ampere-turns developed by the field winding on full load may be taken to be from 30 per cent to 60 per cent in excess of the armature ampere-turns. For shunt motors intended to work over a wide range of speed, by field control, still more powerful field windings would be necessary for non-interpole designs.

The introduction of interpoles permitted the use of much less powerful field windings, and certain generators now work with the ratio field ampere-turns/armature ampere-turns of about unity, and motors not intended for speed control work with a much smaller ratio. In particular, with highly ventilated motors the ratio may be brought within the region of 0·6. The dimensions of the air gap have now, in fact, been brought down to a value which is considered a reasonable minimum from mechanical considerations, although still much greater than would be adopted for a polyphase induction motor.

It is well, however, to bear in mind that the larger is the ratio adopted for F_f/F_a, the less is the distorting effect produced by armature reaction, and, in a generator, the less is the liability to flash over from brush to brush during sudden violent overloads. In motors, the larger the value F_f/F_a the greater is the amount that the field strength can be reduced, and the speed thereby increased, without risk of flashing over. On the other hand, the larger the ratio F_f/F_a, the greater is the amount of copper required for the field winding, the greater is the winding space required, and the greater the cost of the machine.

Depending, then, upon the duty the machine has to perform, the designer can decide upon the value that he considers desirable for the ampere-turns to be developed by the field windings. In general, the student would do well to base the design of large d.c. machines upon $F_f/F_a = 1{\cdot}1$ to $1{\cdot}25$ on full load.

For motors where speed control is not required, many machines are being built with much smaller values. Where a very wide range of speed control is required, the ratio should be made as large as possible, up to 2 or more.

For the 1,000-kW generator the armature ampere-turns per pole are,

$$F_a = I_a Z_a / 4p \qquad (257)$$
$$= 900 \times 200/20 = 9{,}000$$

The magnet system will then be designed so that the field ampere-turns on full load are between 10,000 and 11,000.

The pole has then to be designed of sufficient section to carry the useful flux + the leakage flux, and of sufficient height to accommodate between 10,000 and 11,000 ampere-turns for the field coils.

Sectional Area and Type of Pole. Except in special cases the most usual form of pole is the laminated type, as in Fig. 270, bolted to the main poles. The pole plates may be relatively thick—1 mm to 1·5 mm—and the iron factor may be taken as 0·95.

The leakage factor, as already shown, is not constant, but may safely be taken as not exceeding 20 per cent on full load. For laminated poles the flux density on full load may be taken to be about 1·6 to 1·7 Wb/m², i.e. 16,000 to 17,000 gauss for machines of medium and large size.

For the example considered, taking $\nu = 1{\cdot}2$,

the pole flux on full load $= 1{\cdot}2 \times 0{\cdot}1115 = 0{\cdot}134$ Wb

For a working flux density of 1·6 Wb/m², i.e. 16,000 gauss, the area required

$$= 0{\cdot}134/1{\cdot}6 = 0{\cdot}0838 \text{ m}^2$$
$$= 838 \text{ cm}^2$$

Taking an iron factor of 0·95, the cross-sectional area of the pole becomes 838/0·95 = 880 cm².

To permit end play and to avoid magnetic centring of the armature, the axial pole length may be taken rather less than that of the core, 1 to 1·5 cm, less being sufficient normally.

For an axial length of 34·5−1·5 = 33 cm the pole limb breadth becomes $b_p' = 880/33 = 26{\cdot}5$ cm approx.

Height of Pole. As stated, the height must be chosen to permit sufficient room for the exciting coils. In normal practice, standard proportions for the pole punchings are developed so that sufficient field copper can be carried, not only for comparatively high-speed, well-ventilated machines, but for machines in which the ventilation is not quite so good.

For large machines, the coils are cylindrical in form, whereas for small machines they are more of the proportions of a disc. In the former the cooling surface offered by the ends is relatively small, whereas with the latter type the end surfaces are relatively important.

In a coil of cylindrical form, neglecting the cooling surface at the ends, the total cooling surface of the inner and outer cylindrical surfaces is $= 2 \times l_e$ per unit height of coil, where l_e is the mean length of turn.

Hence, if p_e is the permissible loss per unit area of cooling surface for normal temperature rise, the permissible loss per unit height of coil is

$$2l_e p_e \quad . \quad . \quad . \quad . \quad (258)$$

If now the winding depth is d, and the copper space factor of the coil σ, the total volume of copper is

$$V_e = \sigma l_e d \text{ per unit height of coil,}$$

and the power wasted, for a current density α,

is $\alpha^2 \rho V_e$ per unit height of coil . . (259)

Equating (258) and (259),

$$2l_e p_e = \alpha^2 \rho \times \sigma l_e d$$

hence $$\alpha = \sqrt{\frac{2p_e}{\rho \times \sigma d}} \quad . \quad . \quad . \quad (260)$$

and the ampere-turns per unit height $= \alpha\sigma d$

hence the ampere-turns per cm height

$$= 1{,}000\sqrt{p_e'\sigma d'} \quad . \quad . \quad . \quad (261)$$

for $\rho' = 2 \times 10^{-6}$ and d' in cm.

For a given temperature rise, then, the ampere-turns that can be developed per unit height of the exciting coil are directly proportional to the product of the square root of the permissible loss per cm of surface, the square root of the winding depth, and the square root of the space factor of the copper. For small wires the space factor σ may be as low as 0·4, increasing to about 0·65 for large round wires and to about 0·75 for large conductors of square section.

The winding depth should not be made too great or the temperature difference between the middle and outside layers will be excessive. As a normal value 4 or 5 cm may be taken, although larger values have been used.

As a rough approximation the total cooling surface of the coil should be about 2¼ to 2¾ in.² per watt lost. Great precision on this point is unnecessary in determining a suitable height for the pole. If only the two cylindrical surfaces are taken into account, it will suffice for our present purpose to take an equivalent figure of 14 or cm² per watt. Assuming then that $p_e' = 0{\cdot}07$ and the space factor = 0·6, for a winding depth of 5 cm the ampere-turns per cm height is:

$$1{,}000\sqrt{0{\cdot}07 \times 0{\cdot}6 \times 5} = 460 \text{ approx.}$$

or 92 per cm of winding depth.

For a winding depth of 4 cm the corresponding figure is

$460\sqrt{4/5} = 410$ ampere-turns per cm height of winding space or 102·5 per cm of winding depth.

Based on an approximate winding depth of 4 cm the total net height of exciting coils to develop 10,500 ampere-turns is then 10,500/410 = 25·5 cm approx. Allowing for insulation between the shunt and series coils, and between the coils and the frame, and also for the space wasted due to the curvature of the yoke, the height of the pole above the shoe may then provisionally be fixed as about 30 cm.

The total distance from the pole face to the yoke may thus be taken to be 33 cm, for a "shoe" thickness of 3 cm at the centre.

The Yoke. For a cast steel yoke, the normal value of the flux density on full load is about 1·2 Wb/m² = 12,000 gauss although occasionally it may be necessary to exceed this figure.

On full load, the leakage coefficient for the yoke is not likely to exceed 1·25, so that for the 1,000-kW generator the flux in the yoke becomes

$$\Phi_y = 1{\cdot}25 \times 0{\cdot}1115 = 0{\cdot}14 \text{ Wb approx.}$$

The flux in each yoke path is thus 0·14/2 = 0·07 Wb, and the sectional area necessary for a flux density of 1·2 Wb/m² is

$$A_y = 0{\cdot}07/1{\cdot}2 = 0{\cdot}0583 \text{ m}^2$$

and $$A_y' = 583 \text{ cm}^2.$$

Total Ampere-turns for Iron Parts of Magnetic Circuit. The dimensions of the complete magnetic circuit have now been determined, with the sole exception of the air gap.

To determine the gap dimension it is desirable first to calculate the value of the ampere-turns necessary to establish full working flux in the whole of the iron parts of the magnetic circuit.

For the *teeth* the effective area $A_t' = 518$ cm²; the height l_t' = 5 cm and the apparent density $B_{appt} = 0{\cdot}1115/0{\cdot}0518 = 2{\cdot}15$ Wb/m².

The real density is determined by the coefficient k_s. To obtain this we have: $L' = 34{\cdot}5$, the iron length $L_i' = 26{\cdot}7$, the tooth pitch $\tau'_{s,f} = 2{\cdot}9$ cm, and the width $b'_{t,f} = 1{\cdot}8$ cm.

Hence, $k_s = 2{\cdot}9 \times 34{\cdot}5/(1{\cdot}8 \times 26{\cdot}7) = 2{\cdot}1$.

From Fig. 22 the ampere-turns per cm $H_t'' = 390$, and the real density $B_{real} = 2{\cdot}09$ Wb/m² = 20,900 gauss.

The total ampere-turns for the teeth is then

$$F_t = H_t'' \times l_t' = 390 \times 5 = 1{,}950$$

For the *poles* the area $A_p' = 838$ cm². The effective density with a leakage factor of 1·2 on full load is

$$B_p = \frac{1 \cdot 2\Phi}{A_p} = 1 \cdot 2 \times 0 \cdot 1115/0 \cdot 0838 = 1 \cdot 6 \text{ Wb/m}^2$$

From Fig. 8 the ampere-turns per cm $H_p'' = 30$.

The length of the path is $l_p' = 30$ cm, and the total ampere-turns for the poles

$$F_p = H_p'' \times l_p' = 30 \times 30$$
$$= 900$$

For the *armature core* the area $A_c' = 454$ cm²

The flux density $B_c = 0 \cdot 1115/(2 \times 0 \cdot 0454) = 1 \cdot 23$ Wb/m²

From Fig. 8 the ampere-turns per cm $H_c'' = 10$.

The half length of mean line $l_c' \quad = \pi(135 - 17)/20 \ldots$

$= 18 \cdot 5$ (cf. p. 38), say $= 19$ cm

and the ampere-turns $F_c = H_c'' \times l_c' = 10 \times 19 = 190$

For the *yoke* the area $\quad A_y' = 583$ cm²

The flux density with a leakage coefficient of 1·25 on full load is

$$B_y = 0 \cdot 1115 \times 1 \cdot 25/(2 \times 0 \cdot 0583)$$
$$= 1 \cdot 2 \text{ Wb/m}^2.$$

From Fig. 6 the ampere-turns per cm for the cast steel yoke is then $\quad H_y'' = 7$.

The half length of mean line is obtained:

For a yoke thickness of 12 cm the mean diam is approx.

$$145 + 1 + 2 \times 33 + 12 = 224 \text{ cm}$$

Hence, $l_y' = \frac{1}{2}\{\pi \times 224/10 + 12 - 13\} = 35$ cm . . . (cf. p. 45)

The total ampere turns for the yoke thus become

$$F_y = H_y'' \times l_y' = 7 \times 35 = 250 \text{ approx.}$$

The total ampere-turns required per pole for the iron parts of the magnetic circuit, neglecting the effect of armature reaction, then aggregate to,

Teeth, 1,950; Pole, 900; Core, 190; Yoke, 250;

i.e. 3,300 ampere-turns in all, approx.

Of this amount the teeth require the greater part, i.e. just over 60 per cent, the poles roughly 27 per cent, and the effect of the armature core and the yoke is extremely small.

Air Gap. As already stated, the air gap must be chosen so that the main field is sufficiently stiff, that is, so that armature reaction shall not distort the field to an undesirable extent. With a compound-wound machine it is possible to employ relatively weaker exciting coils, as with momentary overloads the main field is stiffened by the heavy current in the series coils. For the case under consideration, l_g will be chosen so that the total ampere-turns developed by the field winding under full-load conditions shall be between 10,000 and 11,000.

Of the total ampere-turns so developed, however, a certain portion is used up in compensating for the effects of armature reaction. With an interpole machine, the brushes are set in the neutral position, and it may be taken that, to compensate for armature reaction, extra excitation equal to from 10 per cent to 20 per cent of the armature reaction ampere-turns is necessary. The smaller figure corresponds to when there is but little magnetic saturation in the teeth, and the latter to when the teeth are highly saturated. For the 1,000-kW generator, armature reaction $F_a = 9{,}000$ ampere-turns per pole, and it may be taken that to compensate for the effects of this reaction some 1,800 ampere-turns as a maximum will be required.

Hence, the length of the air gap, l_g, may conveniently be adjusted so that the total ampere-turns necessary to induce the e.m.f. of 510 V on open circuit is about 10,500 − 2,000 = 8,500 ampere-turns, approximately.

Of this 8,500, about 3,200 is expended on the iron parts, so that the gap dimension l_g must be chosen to make $F_g = 5{,}300$ approximately.

Until the dimension l_g is known the precise figure for the gap coefficient K_g cannot be determined. For large d.c. machines however, it is not likely to differ greatly from 1·15.

The effective area $A_g' = K_f \tau_p' L'$

$$= 0{\cdot}72 \times 45{\cdot}6 \times 34{\cdot}5 = 1{,}130 \text{ cm}^2$$

and $A_g = 0{\cdot}113 \text{ m}^2$

The flux density $B_g = 0{\cdot}1115/0{\cdot}113 = 0{\cdot}985$

For an assumed value of $K_g = 1{\cdot}15$,

the ampere-turns per cm $= 1{\cdot}15 \times 0{\cdot}796 \times 0{\cdot}985 = 9{,}000$

Hence, 5,300 ampere-turns will necessitate an air gap of 5,300/9,000, say = 0·6 cm.

This value will therefore be taken, i.e. $l_g' = 0{\cdot}6$.

Determining now the actual value of the coefficient K_g.

The ratio slot width/gap width = 11/6 = 1·8.

The slot pitch $\tau_s' = 3{\cdot}04$ cm.

From Fig. 17 Carter's coefficient = 0·73.

Hence, the correction factor for the slots is

$$\begin{aligned} K_{gs} &= \tau_s'/\{\tau_s' - b_s'(1 - \delta_s)\} \\ &= 30{\cdot}4/(30{\cdot}4 - 11 \times 0{\cdot}27) \\ &= 1{\cdot}105. \end{aligned}$$

There are 5 vent ducts each of width 9 mm.
The ratio vent width/gap width = 9/6 = 1·5.
From Fig. 17 Carter's coefficient = 0·77.

The correction factor for the vent ducts is thus

$$\begin{aligned} K_{gv} &= L'/\{L' - n_v b_v'(1 - \delta_v)\} \\ &= 34{\cdot}5/(34{\cdot}5 - 5 \times 0{\cdot}9 \times 0{\cdot}23) \\ &= 1{\cdot}03 \end{aligned}$$

The total gap coefficient is then $K_g = K_{gs} \times K_{gv}$
$= 1{\cdot}105 \times 1{\cdot}03 = 1{\cdot}14.$

The total ampere-turns for the air gap, on full load, is then

$F_g = 1{\cdot}14 \times 0{\cdot}796 \times 9{,}850 \times 0{\cdot}6 = 5{,}380$

For the iron parts $F_i = 3{,}300$

The total is then 8,680 AT

If, then, the allowance for armature reaction be taken as 20 per cent of the armature ampere-turns, i.e. 0·2 × 9,000 = 1,800, the total ampere-turns to be developed by the field coils on full load are 8,680 + 1,800 = 10,500 approx.

Open Circuit, or Magnetization Curve. The magnetization curve can now be calculated. The calculations are given in tabular form on page 47, and the actual curve is given in Fig. 28, page 48.

Calculation of Iron Losses, etc. For the teeth it suffices to base calculations upon the density obtaining at a distance of $l_t/3$ from the root of the teeth and upon a volume of a tooth of uniform width $b_{t_{\frac{1}{3}}}$.

Then the effective breadth = 1·8 cm
The number of teeth = 150
The area of all the teeth = 1·8 × 26·7 × 150 cm²
= 7,200 cm²
The volume of all the teeth = area × height
= 7,200 × 5 = 36,000 cm³

For a density of 7·8 gm per cm³ the tooth weight is

Weight of teeth = 7·8 × 36 kg = 281 kg
The real flux density (page 427) = 2·09 Wb/m² = 20,900 gauss
The frequency (page 406) = 25.

From Fig. 95 the loss due to hysteresis at $B = 2{\cdot}09$ Wb/m², $f = 25$ is 6·7 watt/kg.

For 0·5 mm plates, silicon content 0·5 per cent—

The eddy loss = $0{\cdot}0064 \times 2{\cdot}09^2 \times 0{\cdot}5^2 \times 25^2$ watt/kg
(*see* equation (112)) = 4·45 watts/kg.

Hence, the total tooth loss $= 6{\cdot}7 + 4{\cdot}45 = 11{\cdot}1$ watts/kg

and the total tooth loss for 281 kg $\Big\} = 11{\cdot}1 \times 281 = 3{,}120$ watts

For the core, the area, as on page 419, is 454 cm².

The mean dia of the core below slots is $(135 + 101)/2 = 118$ cm

Hence, the volume of the core $= \pi \times 118 \times 454 = 168{,}500$ cm³

and the weight $= 7{\cdot}8 \times 168{\cdot}5$ kg $= 1{,}315$ kg

The flux density on full load is 1·23 Wb/m².

From Fig. 95 the hysteresis loss at 1·23 Wb/m² and $f = 25$ is 2·15 watts/kg.

The eddy current loss is $0{\cdot}004 \times 1{\cdot}23^2 \times 0{\cdot}5^2 \times 25^2$ (*see* equation (112)) $= 0{\cdot}95$ watts/kg.

Hence, the core loss $= 2{\cdot}15 + 0{\cdot}95 = 3{\cdot}10$ watts/kg

and the total core loss (for 1,315 kg) $= 3{\cdot}1 \times 1{,}315 = 4{,}100$ watts

Hence, neglecting the effect of distortion, the total iron loss for the core and teeth is $= 3{,}120 + 4{,}100 = 7{,}220$ watts.

The "extra iron loss" on full load may be taken as one half of the tooth loss.

The total loss on full load then becomes

Iron loss + stray iron loss $= 7{,}220 + 1{,}560 = 8{,}780$ watts

Total Losses in Armature. On full load the total copper loss, when hot, in the armature windings has been calculated to be

$I_A{}^2 r_a$ + eddy current loss $= 22$ kW

The total iron losses + stray iron losses $= 8{\cdot}8$ kW

Hence, the total armature losses $= 22 + 8{\cdot}8 = 30{\cdot}8$ kW

Estimated Temperature Rise of Armature. Including the stray iron and copper losses, the total armature loss on full-load is approximately 31 kW.

The outhang of the winding has been calculated to be 20·5 cm.

Hence, the overall length $= 34{\cdot}5 + 2 \times 20{\cdot}5 = 75{\cdot}5$ cm

The external barrel surface $= 3{\cdot}14 \times$ dia $\times$ overall length

$= 3{\cdot}14 \times 145 \times 75{\cdot}5$

$= 34{,}500$ cm² approx.

The loss is thus $31{,}000/34{,}500 = 0{\cdot}9$ watts per cm².

The peripheral speed of the armature is 22·8 m/sec.

According to formula (188), an approximation to the temperature rise is $\theta = (250 \text{ to } 300 \times p')/(1 + 0{\cdot}09 v_A{}^{1{\cdot}3})$.

This machine is well ventilated, and the lower figure could be taken. Stray losses have been liberally allowed for. Taking, then, the figure of 275, the temperature rise should be well less than

$$\theta = 275 \times 0{\cdot}9/(1 + 0{\cdot}09 \times 22{\cdot}8^{1{\cdot}3})$$
$$\simeq 40°\text{C}$$

(With the lower figure, i.e. 250, the temperature rise works out to 36°C, which is more likely to be realized.)

Design of Commutator. In designing the commutator of large machines the pitch of the segments is rarely taken to be less than 4 mm, corresponding to a thickness at the surface of 3·2 mm and 0·8 mm mica separators. For small motors this thickness has sometimes to be reduced. With these machines the taper of the segments is quite appreciable, and considerations of the thickness at the inner edge of the segment must not be overlooked.

With large machines, the distance between neighbouring brush spindles for a 500-V machine should not be less than 11 or 12 in., and in every case the commutator surface must suffice to dissipate the heat without the temperature rise exceeding the normal limits.

Taking 30 cm as a normal distance between brush spindles for a 500-V machine, the commutator circumference for this 10-pole machine will become $10 \times 30 = 300$ cm. This would give a diameter of $300/\pi = 95$ cm. Taking the round figure of 100 cm for the commutator diameter,

Commutator diam	= 100 cm
Peripheral speed	$= 3{\cdot}14 \times 1 \times 5 = 15{\cdot}7$ m/sec
Number of segments	= 450
Pitch of segments	= 314/450 = 0·698 cm

This value is quite satisfactory. The commutator diameter is about 70 per cent of that of the armature—a fairly usual ratio.

Brushes. At normal full load, carbon brushes of medium hardness can be worked at a density of about 40 A/in.^{2}.

The current per brush arm = 402 A approx., including the shunt current.

The total sectional area of the brush blocks per spindle is then for 40 A/in.^{2}, 402/40 = 10 in.^{2} approx.

Standard brushes are available in various sizes. It is undesirable to use brushes that are too wide; it is better, that is, to use a large number of relatively narrow brushes per spindle rather than a few wide brushes cf. Fig. 282. The brush must also not be too thick, or the number of coils simultaneously undergoing commutation will be excessive. Practice varies in regard to the number of segments covered by the brush; normally a brush thickness of $\frac{3}{4}$ in., provided the number of segments covered is not more than 3, may be considered sufficient for large machines. Thicker brushes are sometimes used. A brush width of $1\frac{1}{4}$ in. is suitable for most cases.

Taking brushes $\frac{3}{4}$ in. × $1\frac{1}{4}$ in., this being a standard size—

Brush thickness $= 1{\cdot}89$ cm

No. of segts. covered $= 1{\cdot}89/0{\cdot}698 = 2{\cdot}72$

Area of brush $= 15/16 \text{ in.}^2 = 6{\cdot}05 \text{ cm}^2$

Using 10 brush blocks per spindle, the current density would be about 6 per cent greater than 40 A/in.²—a relatively unimportant amount.

Taking this number, the total brush area per spindle

$$= 60{\cdot}5 \text{ cm}^2$$

The full load current density $= 402/60{\cdot}5 = 6{\cdot}65 \text{ A/cm}^2$

The brush width $= 1\frac{1}{4}$ in. $= 31{\cdot}8$ mm

Allowing 6·2 mm for the thickness, etc., of the brush boxes, the total space taken up by the brush boxes is

$$10 \times (31{\cdot}8 + 6{\cdot}2) = 38 \text{ cm approx.}$$

Allowing 5 cm for staggering of the brushes and clearance at the ends, the total axial length of the commutator, excluding the risers, may then be taken to be 43 cm, say 47 cm overall, including the riser. The general proportions of the commutator differ but little from those indicated in Fig. 270.

Brush Friction Loss.

The peripheral speed = 15·7 m/sec

The coefficient of friction may be taken to be 0·22 for a medium hard carbon brush.

The total brush area = 605 cm² per 10 spindles.

The brush pressure,

at 120 g/cm² $= 120 \times 605 = 7{\cdot}25 \times 10^4$ g

and the power wasted $= 0{\cdot}22 \times 7{\cdot}25 \times 10^4 \times 15{\cdot}7 \times 981 \times 10^{-5}$

(cf. page 244) $= 2{,}450$ watts

Brush Contact Loss. For a medium hard carbon brush the total volt drop for both sets of brushes may be taken to be 2·0.

The total brush contact loss on full load is then

$$= 2 \times 2{,}010 = 4{,}050 \text{ watts approx.}$$

Total Losses at Commutator. The commutator losses on full load thus aggregate

$$2{,}450 + 4{,}050 = 6{,}500 \text{ watts}$$

The barrel surface of the commutator

$$= \pi \times 100 \times 43$$
$$= 13{,}500 \text{ cm}^2$$

The loss per cm² is then $6{,}500/13{,}500 = 0{\cdot}48$ watts

This omits all considerations of the surface of the risers, as in the case of (192), page 258.

The temperature rise estimated is then

$$\text{(cf. (192), page 258)} = 120 \times p'/(1 + 0{\cdot}1\, v_C)$$
$$= 120 \times 0{\cdot}48/2{\cdot}57$$
$$= 22{\cdot}5°\text{C}$$

The actual temperature rise at the end of the commutator remote from the risers is likely to exceed this value, as the commutator is rather long.

Design of Field Windings. The shunt winding has to be designed so that it can develop continuously the ampere-turns necessary to give rise to an induced e.m.f. of 500 V on open circuit.

The series winding has to develop the additional ampere-turns necessary to maintain approximately constant voltage under all conditions of load—the machine being level compounded.

The total ampere-turns for full load have been calculated to be 10,500, and on no-load at 500 V the value is approximately 8,000. On full load, then, the series winding has to develop 2,500 ampere-turns.

Shunt Winding. The total net winding space on the poles has been taken to be 25·5 cm. Taking about 80 per cent of this for the shunt winding, the space for this winding is 20 cm.

To allow for a certain amount of voltage regulation, the winding should be designed so that it will develop its normal ampere-turns when at its normal working temperature and with about 80 to 85° per cent of full terminal voltage—the remainder being absorbed by the field regulating resistance.

The winding will then be designed to develop 8,000 ampere-turns when hot, with a total applied voltage of $0{\cdot}85 \times 500 = 425$ V.

MEAN LENGTH OF TURN. The pole laminations are secured by steel rivets between two end plates each 3 cm thick as indicated in Fig. 270.

Allowing 1·0* cm for the spool and insulation, and rounding off the corners with a radius of 4·0 cm for the inner layers—the end plates are smaller in breadth than the pole—the inner dimensions of the coil are then approximately 28·5 × 41 cm gross. For a winding depth, assumed, of 4 cm the mean length of one turn will then be

$$= 2 \times (20{\cdot}5 + 33 + \pi \times 6)$$
$$= 145 \text{ cm approx.}$$

COOLING SURFACE

This is given by

$$\text{Surface} = 2l_e'(\text{Height of coil} + \text{winding depth})$$
$$= 2 \times 145(20 + 4) \text{ cm}^2 \text{ per coil}$$
$$= 6{,}950 \text{ cm}^2 \text{ per coil}$$
$$= 69{,}500 \text{ cm}^2 \text{ for the ten coils}$$

* A very liberal allowance.

PERMISSIBLE LOSS. The peripheral speed of the armature, v_A is 22·8 m/sec. For this speed the necessary cooling surface to obtain a temperature rise of 40°C is (cf. (191), page 258),

$$= 35/(1 + 0{\cdot}07 \times 22{\cdot}8) = 13{\cdot}5 \text{ cm}^2 \text{ per watt.}$$

It is not desirable to calculate the winding so that we have nothing in hand, as precise calculation of the hottest measurable temperature is not possible. It would then be well to base the calculations upon a rather larger surface, say upon 16 cm² per watt, which would, according to the formula, correspond to a temperature rise of 34°.

The permissible loss in all the ten coils is then

$$\text{Permissible loss} = \text{Surface/(surface per watt)}$$
$$= 69{,}500/16$$
$$P_e = 4{,}350 \text{ watts}$$

EXCITING CURRENT. This current is given at once by dividing the power wasted by the total voltage applied to the winding, and becomes

$$I_e = 4{,}350/425 = 10{\cdot}2 \text{ A approx.}$$

NUMBER OF TURNS. This is given by dividing the ampere-turns by the current, hence

$$N_e = 8{,}000/10{\cdot}2 = 784 \text{ approx.}$$

RESISTANCE OF WINDING. The resistance drop divided by the current gives the value of the resistance. Thus

$$r_e = 425/10{\cdot}2 = 41{\cdot}7 \text{ ohms, hot}$$

The resistance, hot, may be taken to be 20 per cent greater than the value when cold.

Hence, the resistance, cold, is $= 41{\cdot}7/1{\cdot}2 = 35$ ohms approx.

TOTAL LENGTH OF CONDUCTOR.

The total length = No. of poles × turns per pole × (length of mean turn)

$$= 10 \times 784 \times 145 \text{ cm}$$
$$L_e = 11{,}400 \text{ m approx.}$$

SECTIONAL AREA OF CONDUCTOR.

$$A_e = \rho \frac{L_e}{r_e}$$

Taking ρ' to be 1·7 microhms in cm measure,

$$A_e' = 1{\cdot}7 \times 10^{-6} \times 11{,}400 \times 10^2/35$$
$$= 0{\cdot}0555 \text{ cm}^2$$

The nearest s.w.g. No. to this is s.w.g. 12, having a diameter of 2·65 mm and a sectional area of 0·0546 cm². This wire, insulated d.c.c., has an external diameter over insulation of 3·00 mm.

For this diameter it is possible to wind nearly 20/0·3, i.e. 65 turns approximately in a winding length of 20 cm.

NUMBER OF LAYERS. With 65 turns per layer, the number of layers is

$$784/65 = 12$$

Taking 12 layers, the number of turns is then 780.

Winding depth, neglecting allowance for slackness and bedding

$$= 3{\cdot}6 \text{ cm}$$

This latter figure is near enough to the assumed value of 4 cm to do away with the need for re-calculating the mean length of turn, etc.

AMENDED DATA FOR SHUNT WINDING. The data finally settled are—

Number of turns per pole	780
Number of layers	12
Number of turns per layer	65
Conductor area	0·0546 cm²
Conductor diam bare	2·65 mm
Conductor diam insulated	3·00 mm

$$\text{Total resistance of winding} = 1{\cdot}7 \times 10^{-6} \times (1{\cdot}14 \times 10^{6})/0{\cdot}0546$$
$$= 35{\cdot}5 \text{ ohms, cold}$$
$$\text{or} = 1{\cdot}2 \times 35{\cdot}5 = 42{\cdot}5 \text{ ohms, hot}$$

From the magnetization curve on open circuit, the actual ampere-turns required from the shunt winding are 7,840, so that the shunt current may be taken to be 10 A.

$$\text{The volt-drop in the field winding} = 10 \times 35{\cdot}5 = 355 \text{ V, cold}$$
$$\text{and } I_e r_e = 10 \times 42{\cdot}5 = 425 \text{ V, hot}$$

The power wasted, with the winding hot, is

$$P_e = 10 \times 425 = 4{,}250 \text{ watts}$$

corresponding to 4,250/69,500 = 0·061 watts per cm² of total surface.

According to (190), the temperature rise may then be expected to be 1,400 × 0·061/(1 + 0·07 × 22·8)

$$= 33{\cdot}7^\circ\text{C}$$

WEIGHT OF COPPER.

$$\text{The total volume of copper} = A_e{}'L_e{}'$$
$$= 1{\cdot}14 \times 10^{6} \times 0{\cdot}0546 \text{ cm}^3$$
$$= 0{\cdot}0624 \times 10^{6} \text{ cm}^3$$

At 8·9 gm/cm³ the weight becomes

$$= 8{\cdot}9 \times 0{\cdot}0624 \times 10^3 \text{ kg}$$
$$= 555 \text{ kg}$$
$$= 2{\cdot}2 \times 555 = 1{,}220 \text{ lb}$$

This figure may be compared with that necessary for the armature winding, i.e. 754 lb. Further, there is the weight of the series coil.

CURRENT DENSITY IN SHUNT WINDING. It is well to check this figure which, in general for a fairly well ventilated machine, will work out to be in the neighbourhood of 2 A/mm². The actual figure is

$$= 10/5{\cdot}46 = 1{\cdot}83 \text{ A/mm}^2$$

As a check on the other calculations the loss with the winding when hot is $\alpha^2\rho' = 1{\cdot}2 \times 1{\cdot}7 \times 10^{-6} \times 183^2$ watts per cm³ where α is the current density in A/cm². The total loss, in terms of the weight in kg, is

$$I^2r = \rho' \times \alpha^2 \times \frac{\text{Wt. in kg} \times 10^3}{8{\cdot}9} \text{ watts}$$
$$= \frac{2{\cdot}04}{0{\cdot}89} \times \left(\frac{\alpha}{100}\right)^2 \times (\text{Wt. in kg}) \text{ watts, when hot}$$
$$= 2{\cdot}3 \times (\text{c.d. in A/mm})^2$$
$$\times (\text{Wt. in kg}) \ldots \text{ watts, hot}$$

Applying this to the 1,000-kW machine,

$$I_e^2r_e = 2{\cdot}3 \times 1{\cdot}83^2 \times 555 = 4{,}300 \text{ watts}$$

This figure accords with the previous calculation, so that it may be taken that errors have not crept in.

ALTERNATIVE METHOD OF DETERMINING CROSS-SECTIONAL AREA OF SHUNT CONDUCTOR. As stated in (149), page 178, the area, in terms of the m.m.f., is given

$$A_e = \frac{\rho_e F_e l_e}{v_e} \text{ where } v_e \text{ is the voltage per coil}$$

Taking ρ' to be 2·04 microhms in cm measure, and l_e' in cm

$$A_e' = 2{\cdot}04 \times 10^{-6} \times 8{,}000 \times 145/42{\cdot}5 = 0{\cdot}0556 \text{ cm}^2,$$

which checks with the figure already obtained.

Series Winding. This winding has to develop about 2,500 ampere-turns on full load, and the full-load current output is 2,000 A. Great precision in regard to the number of series turns for a compound generator is not necessary, as minor adjustments of the voltage can be made by giving the brushes a very slight lag or lead, and, in general, absolutely constant voltage is not insisted on. It is then desirable to arrange that the series winding shall develop not less than 2,500 ampere-turns, and if it becomes necessary to exceed this value it is of no great moment.

In the present case, by putting two circuits in parallel, the current per circuit becomes 1,000 A, and to produce 2,500 ampere-turns, $2\frac{1}{2}$ turns per pole will be necessary. The extra half turn can readily be provided by making connexions from coil to coil alternately at the two sides of the machine, as indicated in Fig. 307.

Owing to the large current, the series conductor will be of large section, and cooling conditions are therefore considerably better

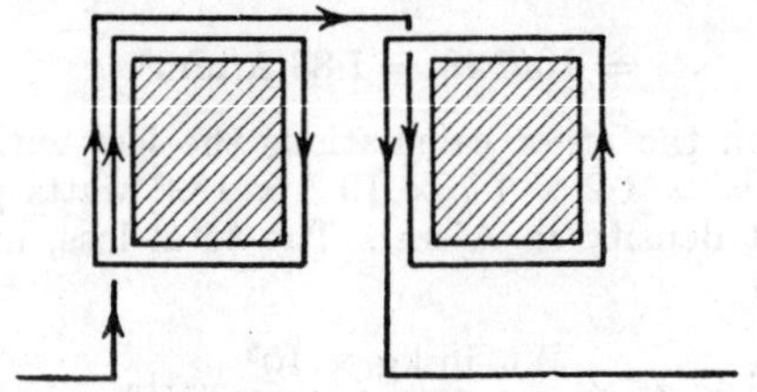

FIG. 307. WINDING WITH $1\frac{1}{2}$ TURNS PER POLE

than for the shunt coils. It is sufficient then to determine the sectional area of the conductor by taking a current density somewhat greater than that used for the shunt winding—a very suitable figure is 2 to 2·2 A/mm².

The series conductor may then consist of solid copper 12 × 40 mm wound on its narrow edge, or it may be built up of a number of smaller strips wound in parallel, e.g. 8 strips 12 × 5 mm.

The height of pole taken up by the series winding is then $3\frac{1}{2}$ × 12 mm net = 42 mm or 55 mm total, allowing for spacing between the turns. (This space, owing to the spiralling of the conductor, corresponds to one more than the actual number of turns.)

DETAILS OF SERIES WINDING. The data are then—

Number of turns per pole	$2\frac{1}{2}$
Area of conductor	480 mm
Current per conductor	1,000 A
Current density	2·08 A/mm²
Length of mean turn	145 cm
Total length	3,600 cm, neglecting connectors

The total resistance for 2 paths in parallel is then

$$r_{se} = \tfrac{1}{4}(1{\cdot}7 \times 10^{-6} \times 3{,}600/4{\cdot}8) \text{ ohms cold}$$
$$= 0{\cdot}00033 \text{ ohm, cold}$$

and
$$= 1{\cdot}2 \times 0{\cdot}00033 = 0{\cdot}0004 \text{ ohm hot}$$

The resistance drop $Ir_{se} = 2{,}000 \times 0{\cdot}0004 = 0{\cdot}8$ volt, hot

The power wasted I^2r_{se} = 2,000 × 0·8 = 1,600 watts, hot
The weight of copper = 8·9 × 3·6 × 4·8 kg = 154 kg
= 340 lb

Design of Interpole and its Winding. So that every coil shall be under the interpole shoe during the whole of commutation, the width of shoe should correspond to the distance the armature moves through during the commutation of the complete slot group, i.e. to β_{xA}, page 158.

For the 1,000-kW machine—

The brush thickness	$t_b' = 1{\cdot}89$ cm
The segment pitch	$\tau_C' = 0{\cdot}698$ cm
The number of coil-sides per layer	$x = 3$
The armature diam	$D' = 145$ cm
The commutator diam	$D_C' = 100$ cm

Hence $\beta_{xA}' = \{1{\cdot}89 + (3 - 1) \times 0{\cdot}7\} \times 145/100 = 3{\cdot}29 \times 1{\cdot}45$
$= 4{\cdot}8$ cm approx.

For straight-line commutation, then, the effective width of the shoe should be not less than 4·8 cm.

Flux Density in Air Gap under Interpole. The data necessary for calculating the permeance of the slot leakage paths, etc., are—

Slot width $b_s' = 1{\cdot}1$ cm. Height above copper $h_t' = 0{\cdot}55$ cm
Height of copper overall $h_c' = 4$ cm
Active length approx 34·5 cm. Free length $l_{fr}' = 66$ cm

The periphery of the total 3 conductors per layer is 2[1·9 + 0·9], i.e. $b = 5{\cdot}6$ cm.

The air gap under the interpole shoe may conveniently be first assumed to be about 50 per cent greater than that for the main pole and, in this case, 0·8 cm approx.

Then

$$\lambda = \left\{\frac{4\pi}{10}\left(\frac{h_c}{3b_s} + \frac{h_t}{b_s} + \frac{b_{cp}}{6l_{g_{cp}}}\right) + \frac{l_{fr}}{L}\left(0{\cdot}23 \log \frac{l_{fr}}{b} + 0{\cdot}07\right)\right\} \times 10^{-6}$$

$$= \left\{1{\cdot}257\left(\frac{40}{33} + \frac{5{\cdot}5}{11} + \frac{48}{48}\right) + \frac{66}{34{\cdot}5}\left(0{\cdot}23 \log \frac{66}{5{\cdot}6} + 0{\cdot}07\right)\right\} \times 10^{-6}$$

$$= \{3{\cdot}4 + 0{\cdot}6\} \times 10^{-6} = 4{\cdot}0 \times 10^{-6}$$

The total ampere-conductors per slot $Z_sI_a = 6 \times 200 = 1{,}200$, Hence, as in (141), page 157, the flux density in the gap is

$$B_{cp} = \frac{L}{L_{cp}} \times \frac{2Z_sI_a\lambda}{\beta_{xA}}$$

$$= \frac{L}{L_{cp}} \times \frac{2 \times 1{,}200 \times 4}{0{\cdot}048} \times 10^{-6}$$

$$= 0{\cdot}2\,\frac{L}{L_{cp}}$$

The axial length of the interpole may be taken to be about 0·5 to 0·67 of the length of the armature core, the actual ratio is largely a matter of the individual designer's fancy.

Taking provisionally $L'_{cp} = 21$ cm

$$B_{cp} = 0{\cdot}2 \times \frac{34{\cdot}5}{21} = 0{\cdot}33 \text{ Wb/m}^2$$

Interpole Ampere-turns. The iron parts of the magnetic circuit for the interpole flux are quite unsaturated, so that it suffices to calculate the ampere-turns for the air gap only. The air gap being relatively long, the coefficient K_{g_c} will be a little less than for the main poles. Estimating its value as 1·1, the net resulting ampere-turns to establish the flux density of 0·33 Wb/m² are $0{\cdot}796 \times 0{\cdot}33 \times 1{\cdot}1 \times 0{\cdot}008, \times 10^6$ i.e. 2,300 ampere-turns.

The total ampere-turns per pole for the interpole is then

$$F_{cp} = F_a + 0{\cdot}796\, K_{g_{cp}} B_{cp} l_{g_{cp}}$$
$$= 9{,}000 + 2{,}300 = 11{,}300$$

The nearest convenient value for F_{cp} is 12,000, corresponding to 6 turns per pole; the necessary adjustment can be made by increasing the air gap, or by reducing the axial length of the pole shoe.

Reducing this latter dimension to 17 instead of 21 cm increases the value required for the flux density B_{cp} to $0{\cdot}33 \times 21/17$, i.e. to 0·408 Wb/m² i.e. 4,080 gauss. For $K_{g_{cp}} = 1{\cdot}1$, this flux density corresponds to a magneto-motive force of 3,580 ampere-turns/cm.

The dimensions finally decided upon for the interpole shoe will then be: axial length 17 cm; air gap 0·85 cm.

With these values, $F_{cp} = 9{,}000 + 0{\cdot}796 \times 1{\cdot}1 \times 0{\cdot}0085 \times 0{\cdot}408 \times 10^6 = 12{,}000$ approx.

The main current is 2,000 A, hence 6 turns per pole, carrying the full current will be required. For this heavy current, bare copper strap, having good cooling properties, can be used. The current density may be taken from 2 to 2·5 A/mm². At 2·2 A/mm² the sectional area required is 2,000/2·2, i.e. 910 mm².

Using bare copper strap wound on its narrow edge, as illustrated in Fig. 268, a copper strap 3 × 1 cm² could be used, three in parallel. Allowing for one turn wasted due to spiralling, and 2 mm air space between conductors, the total winding height is $(6 + 1) \times (3 \times 1{\cdot}2) = 25{\cdot}2$ cm.

The depth of 3 cm will give sufficient clearance between the interpole windings and those on the main pole.

The interpole itself will be tapered to avoid the possibility of the iron parts at the root becoming saturated at normal overloads. If the thickness at the root be made 6 cm and the leakage coefficient be as high as 3·5 at 25 per cent overload, the flux density at the root would then be $\{0{\cdot}408 \times 1{\cdot}25 \times 3{\cdot}5\} \times 4{\cdot}8/6$,
i.e. just over 1·4Wb/m², a very conservative figure.

The interpole will then taper from 6 cm to 3 cm, and a shoe be fitted having an effective width of 4·8 cm.

For an approximation to sinusoidal commutation, a narrower shoe would be used—in this case, the tip could have a width of 3 cm, and the actual shoe be dispensed with. To compensate for the use of this smaller shoe, B_{cp} would have to be increased by about 40 per cent, as stated on page 159.

The mean width of the interpole is 4·5 cm and the axial length 17 cm. Taking the liberal allowance of 5 mm for insulation and using coils with semi-circular ends, the mean length of turn for a winding depth of 3 cm becomes

$$2(17 + 3{\cdot}14 \times 4{\cdot}3) = 61 \text{ cm}$$

Resistance of the Interpole Winding.

Total length per pole	$= 6 \times 61 = 366$ cm
Total per 10 poles	$= 3{,}660$ cm
Total sectional area	$= 6 \times 150 = 900$ mm²
Total resistance, cold	$= 1{\cdot}7 \times 10^{-6} \times 3{,}660/9$
	$= 0{\cdot}00069$ ohm
,, ,, hot	$= 1{\cdot}2 \times 0{\cdot}00069 \quad = 0{\cdot}00083$ ohm
Resistance voltage drop, hot	$= 2{,}010 \times 0{\cdot}00083 = 1{\cdot}66$ V
Power wasted, hot	$= 1{\cdot}66 \times 2{,}010 \quad = 3{,}330$ watts
Current density	$= 2{,}010/900 = 2{\cdot}22$ A/mm²
Weight of copper	$= 8{\cdot}9 \times 3{\cdot}66 \times 9 = 294$ kg
	$= 646$ lb

The cooling surface of the coil, neglecting the effect of the space between layers, is $2 \times 61(25 + 3) = 3{,}400$ cm²

The total surface for 10 coils is then 34,000 cm²

The cooling surface as thus determined is therefore

$$34{,}000/3{,}330 = 10 \text{ cm}^2/\text{watt}$$

This value is sufficient for a winding of this type, the interpole winding being well ventilated.

Bearing Friction and Windage Loss. These losses may be estimated by means of the formulae given in Chapter XI.

The bearings for the 1,000-kW, 300-rev/min machine may be taken to have the following approximate dimensions—

Commutator end: dia 7·5 in., length 18 in.

Engine end: dia 9 in., length 22 in.

The bearing loss, according to the formula (173), page 244, can then be calculated for a speed of 300 rev/min.

For the two bearings, respectively

$$D_B' = 19 \text{ and } 22{\cdot}8 \text{ cm}$$
$$L_B' = 45{\cdot}6 \text{ and } 56 \text{ cm}$$
$$v_B = 2{\cdot}99 \text{ and } 3{\cdot}59 \text{ m/sec}$$

The loss for the bearing at the commutator end is then:

$$\text{Bearing loss} = 0{\cdot}3 \times 19 \times 45{\cdot}6 \times 2{\cdot}99^{1{\cdot}5} = 1{,}350 \text{ watts},$$

and for the other:

$$\text{Bearing loss} = 0{\cdot}3 \times 22{\cdot}8 \times 56 \times 3{\cdot}59^{1{\cdot}5} = 2{,}600 \text{ watts}$$

The total bearing loss may then be taken to be 4,000 watts

For the *windage loss*, applying formula (174) on page 244,

$$\begin{aligned}\text{Windage loss} &= 0{\cdot}8 \text{ armature diam} \times \text{overall length in metres} \times v_A^2 \times 10^{-1}\\ &= 0{\cdot}8 \times 1{\cdot}45 \times 0{\cdot}85 \times 22{\cdot}6^2 \times 10^{-1} \text{ watts}\\ &= 5{,}000 \text{ watts.}\end{aligned}$$

Total Losses and Efficiency. At full load, with the machine hot, the total losses have been computed as follows—

	kW	kW
1. Losses in the main series circuit:		
Armature winding	22·0	
Interpole winding	3·3	
Series winding	1·6	
		26·9
2. Shunt excitation loss:		
Shunt winding	4·3	
Field regulator	0·7	
		5·0
3. Iron losses:		
Iron loss	7·2	
Stray loss	1·6	
		8·8
4. Commutator losses:		
Brush contact	4·0	
Brush friction	2·5	
		6·5
5. Bearing and windage:		
Two bearings	4 0	
Windage	5·0	
		9·0
Total full load losses, including stray losses		**56·2**

The full load output is 1,000 kW

Hence the full-load input $= 1{,}000 + 56{\cdot}2 = 1{,}056{\cdot}2$,

and the full load efficiency $= 1 - 56{\cdot}2/1{,}056{\cdot}2 = 1 - 0\ 0532$

$$= 94{\cdot}7 \text{ per cent}$$

To determine the efficiency at other loads, for the level compounded generator the losses in the shunt circuit and the various frictional losses may be taken to be constant. The iron loss of 7·2 kW may also be taken to be nearly constant, as the induced e.m.f. only changes by 2 per cent from no-load to full load. The stray losses may be taken to vary in proportion to the square of the load current; in addition, the losses in the main series circuit vary as the square of the load current.

With sufficient accuracy, the brush contact losses may be taken to be proportional to the load current. This will result in the losses being over-estimated at loads less than full load, and slightly under-estimated for over-loads.

It then appears that the full load losses may be divided into:

1. Constant losses $= 5{\cdot}0 + 7{\cdot}2 + 2{\cdot}5 + 9{\cdot}0 = 23{\cdot}7$ kW
2. Loss proportional to the output $= 4$ kW
3. Losses proportional to the square of the output $= 26{\cdot}9 + 1{\cdot}6 = 28{\cdot}5$ kW

The losses at other loads can then rapidly be deduced, as indicated in tabular form below—

CALCULATION OF EFFICIENCY CURVE FOR 1,000 kW. GENERATOR

Load	Losses				Input	Losses/Input	Efficiency
kW	(1) kW	(2) kW	(3) kW	Total kW	kW	%	%
0·25 = 250	23·7	1	1·8	26·5	276·5	9·6	90·4
0·5 = 500	23·7	2	7·1	32·8	533	6·15	93·8
0·75 = 750	23·7	3	16·0	42·7	793	5·4	94·6
F.L. = 1,000	23·7	4	28·5	56·2	1,056	5·3	94·7
1·25 = 1,250	23·7	5	44·5	73·2	1,323	5·5	94·5
1·5 = 1,500	23·7	6	64·2	93·9	1,594	5·9	94·1

Maximum efficiency occurs at just below full load. At full load the constant losses 23·7 kW differ but little from those proportional to the square of the load, 26·9 kW.

The total resistance of the main circuit is then:

For the armature $r_a = 0{\cdot}0041$ ohm, hot

„ main series windings $r_{se} = 0{\cdot}0004$ ohm, hot

„ interpole windings $r_{cp} = 0{\cdot}0008$ ohm, hot

Hence $r_m = r_a + r_{se} + r_{cp} = 0{\cdot}0053$ ohm hot

The total voltage drop in the windings is then, with full load armature current of 2,000 + 10 = 2,010 A—

$$I_A r_m = 2{,}010 \times 0{\cdot}0053 = 10{\cdot}85 \text{ V}.$$

Including the 2-V drop at the brushes, the total volt-drop on full load will then be about 13 V, and the induced e.m.f. 513. This is near enough to the value assumed, i.e. 510.

Having completed the calculations for the machine, the designer, before finally putting the machine in hand, would naturally review the whole design, not merely from the point of view of the particular machine under design, but from the point of view of possible future orders for machines which could be fitted into the same frame. In the present case, to make the frame suited for machines operating at much lower speeds, and therefore with much inferior cooling conditions, the designer would very likely feel inclined to increase the height of the pole by 4 or 5 cm.

CHAPTER XIX

FURTHER EXAMPLES OF DESIGN OF MOTORS AND GENERATORS

THE calculations for the 1,000-kW, 10-pole, 500-V, 300 rev/min generator, given at length in the two preceding chapters, can readily be adapted with but little modification for other designs. The machine has an output of 100 kW *per pole*, and the speed corresponds to a frequency of 25 flux reversals per sec. With but very minor modifications the calculations for the 10-pole machine can be used for other designs corresponding to the same output per pole, and to the same frequency, e.g. a 1,200-kW, 12-pole, 250-rev/min generator, or an 8-pole, 800-kW, 375-rev/min generator. For all such ratings the preliminary designs for 500-V generators can be based on the following data—identical with the data for the 1,000-kW machine—

Pole pitch = 45·5 *cm, core length* = 34·5 *cm, peripheral speed* = 22·75 *m/sec,* 15 *slots per pole, slot depth* = 50 *mm, width approx.* = 11 *mm,* 6 *conductors per slot,* 19 *mm* × 2·4 *mm approx., simple lap winding,* 45 *commutator segments per pole, average voltage between segments* = 11·2 *V, current per brush arm* = 400 *A, armature reaction* = 9,000 *AT per pole.*

The magnetic circuits of the pole-pitch of all the machines differ but very little. The most important difference arises out of the decrease in the taper of the teeth as the diameter is increased with increasing number of poles. To counter-balance this, the space between field coils is greater for the small diameter machines with few poles than for those having large diameters and many poles.

The external diameter of the armature will have to be increased by $2\tau_p/\pi$ for each pair of poles additional to 10—the original number —i.e. by 29 cm approximately. This will necessitate the same increase in the internal diameter of the core plates and of the yoke. The diameter of the commutator would be increased in proportion to the number of poles. Based on the same loading constants as the 1,000-kW generator the main dimensions of a range of 500-V machines from 800 kW to 1,600 kW are then as given on page 446. The pole and air-gap dimensions are identical with those for the 1,000-kW machine.

The other data for the design, with the exception of details of the field windings, have been given above. The table could be extended to include 1,800-kW and 2,000-kW machines at 167 and 150 rev/min respectively, having armatures of 261 cm and 290 cm diameter respectively. But if the range of diameters is too great,

MAIN DIMENSIONS OF A RANGE OF 500-V GENERATORS

kW	800	1,000	1,200	1,400	1,600
Rev/min	375	300	250	214	187·5
No. of poles	8	10	12	14	16
Core plates:					
D' cm	116	145	174	203	232
D_i' cm	72	101	130	159	188
Number of slots	120	150	180	210	240
Commutator diameter cm	80	100	120	140	160
Number of segments	360	450	540	630	720
Yoke:					
Internal diameter cm	183	212	241	270	299

the loading constants would have to be modified somewhat, as will be discussed later.

To a very close approximation the ampere-turns required for the shunt coils will be the same for all the machines. The sectional area of the conductor for these coils will vary in direct proportion to the number of poles, the field current will be proportional to the number of poles, and the number of turns inversely proportional to the number of poles.

For the series coils, the ampere-turns required will be the same for all numbers of poles. The nearest convenient number of turns will be chosen; if necessary the coils will be grouped in two parallel circuits.

So far as performance is concerned, all the copper losses will increase in direct proportion to the number of poles. This same proportionate increase will also hold to a very close approximation for the iron loss, the brush friction and the other mechanical losses. The full-load efficiency will therefore have practically the same value of 94·7 per cent for all the designs. The peripheral speed is the same for all the designs, and the cooling surface is proportional to the number of poles, so the temperature rise should be much the same for them all.

As stated already, the taper of the teeth is somewhat different for the various machines. As a result, in comparison with the 10-pole machine, the armature of the 8-pole machine will work out slightly "tight" and of the 12-, 14-, and 16-pole machines a little "slack." It may be well to examine this point in detail. The slot depth is 50 mm so that the slot pitch at "one-third up from the root" differs in cm from that at the circumference by $6{\cdot}67\pi \div$ number of teeth, i.e. by 21 cm $\div$ number of teeth.

This difference is 1·75, 1·4, 1·17, 1·0, and 0·875 mm respectively for the 8-, 10-, 12-, 14-, and 16-pole machines. The effective tooth width for the 10-pole machine, cf. page 417, is 18 mm, and the net iron length of the core is 30 cm. For the 16-pole machine the effective tooth width, for the same slot width of 11 mm, is 1·4–0·875, i.e. 0·525 mm greater. To retain the same flux density in the teeth the net core length of the 16-pole machine could be reduced to 30 × 18/18·525 cm, i.e. to 29 cm approx. Alternatively, the iron dimensions may be unaltered and the slot width increased by 0·525 mm. This latter alternative permits of the total copper width in the slot being increased from 7·2 to 7·725, an increase of about 8 per cent. This increased width of the conductors would permit of more current being carried, with an increase in electric loading and output. This relatively increased output per pole for the 16-pole large-diameter machine is small ; it is still smaller for the 14-pole machine and insignificant for the 12-pole design. For the 8-pole machine the diameter is only 116 cm and yet it is necessary to increase the core length only by about 0·5 cm to retain the same density in the teeth as for the original 10-pole design.

Dealing now with the influence of the number of poles upon the space available for the field windings. At the armature circumference the pole-pitch is the same for all numbers of poles. But whereas with the 8-pole machine there is a very considerable increase in the pitch at the yoke, the increase is very much less for the 16-pole generator. From the data given on page 446, the pole-pitch at the inner periphery of the yoke is 72, 66·5, 63, 60·5, and approximately 58·5 cm, respectively, for the 8-, 10-, 12,- 14-, and 16-pole machines. Thus the machine with the small number of poles is at a great advantage as compared with the 16-pole generator, as the total distance between main poles at the yoke is no less than 13·5 cm greater, leaving 6·75 cm greater clearance at the yoke between the main pole and the interpole. Not only does this give more room for the field coils and improved ventilation, but it results in a smaller magnetic leakage.

All the machines have the same core length and the same pole-pitch, so that the ratio L/τ_p is constant for them all. But obviously the ratio L/D is much less for the larger than for the smaller output. So the tendency is to work with rather a larger pole-pitch and relatively fewer poles for the larger outputs, i.e. to work with a larger value of the output per pole at a given frequency. This results in a larger armature reaction, a larger current per brush arm, and usually a rather longer core length. By this means, too, ample space is left for the field coils.

Further Designs Based on the 1,000-kW, 300 rev/min Generator. The actual design for the 1,000-kW, 500-V machine, and those given in skeleton form in the preceding pages, all work with an average value of 11·2 V between adjacent commutator segments. This figure is below the figure of 14 volts which is generally regarded as a comfortable limit, giving a reasonable safety factor.

Taking this figure of 14 V between segments, each of the machines previously considered could be driven at a speed increased in the ratio of 14/11·2, i.e. 1·25, with a corresponding increase in output and voltage. The peripheral speed would also be increased by 25 per cent to 1·25 × 22·75, i.e. 28·4 m/sec. a speed which presents no particular constructional problems. Alternatively, if the same speed of rotation were retained, the core lengths could be increased by 25 per cent to about 43 cm.

With the same main dimensions as those already given, and the same armature winding data, we therefore have at once available the designs for two further ranges of 625-V machines—

(1) With the 43 cm core length:

kW	1,000	1,250	1,500	1,750	2,000
Rev/min	375	300	250	214	187·5
Poles	8	10	12	14	16

(2) With the original core length of 34·5 cm

kW	1,000	1,250	1,500	1,750	2,000
Rev/min	468	375	287	267	234
Poles	8	10	12	14	16

For both these ranges the main dimensions and design data are as given on page 445 and page 446. The field windings must be altered to suit the new voltage, and for the longer machines the mean length of turn is altered.

For 500-V operation, the value of Z_c would be reduced from 90 to 90 × 500/625 in each case, i.e. to 72. This would involve using 18 slots per pole with 4 conductors in each; there would be 36 commutator segments per pole. The current per brush arm would rise to 500 A. The original brush dimensions could be retained and a brush capable of working at 50 A/in.^{2} employed instead of the 40 A/in.^{2} used in the original 1,000 kW design. The commutator dimensions corresponding to the original machine could thus be retained.

The efficiency of the machines can very readily be deduced from the calculations given for the "type" machine, i.e. the 1,000 kW at 300 rev/min.

Taking, for example, the 1,250-kW, 625-V, 375-rev/min generator, having, for the most part, identical design data as for the "type" machine. The electrical losses will be identical with those for the "type" machine. The eddy current loss will be increased in proportion to the square of the increase in speed, the hysteresis loss will be increased in proportion to the speed, the brush friction will be increased in proportion to the speed, the bearing friction in proportion to the 1·5th power of the speed, and the windage in proportion to the square of the speed. The speed increase is 25 per cent. The losses thus become—

	kW
Losses in the main series circuit	26·9
Shunt excitation loss	5·0
Iron loss	9·8
Stray loss	2·2
Brush contact loss	4·0
Brush friction	3·1
Bearing loss	5·5
Windage	7·8
Total	64·3

The efficiency is thus 1 — 64·3/1,314·3,

i.e. Efficiency = 0·951

This figure will, to a very close approximation, hold for all the 625-V machines having the core length of 34·5 cm.

In addition to the ranges of machines deduced above, for an output per pole of 100 kW at $f = 25$, 125 kW per pole at $f = 33{\cdot}3$, and 125 kW per pole at $f = 25$, designs for other intermediate values of the kW per pole follow at once from the data in the table on page 446. For example, for an output of 1,100 kW at 300 rev/min, i.e. 110 kW per pole at $f = 25$, 1,320 kW at 250 rev/min, and so on, the core lengths given in the table will have to be increased by 10 per cent to 38 cm approx. The winding data will need no modification for 550 V.

Other ranges for other values of the pole-pitch can also be worked out very quickly once a single machine of comparable pole-pitch has been worked out in detail.

Preliminary Design Based upon Output per Pole. Since the proportions of the d.c. machine are determined by the pole-pitch rather than by the diameter, there is much to be said for basing the preliminary design directly upon the pole-pitch, rather than upon the diameter. Thus—

The ampere-conductors per pole $= q\tau_p$

The flux per pole, $\Phi = B_{av}\tau_p L$

Hence the work done per pole when the armature moves through one pole-pitch $= \tau_p{}^2 L \times qB_{av}$ watt-sec.

If f is the frequency of flux reversals, the time for the armature to move through one pole-pitch is $1/2f$ sec.

Hence the power developed per pole $= 2f\tau_p{}^2 L \times qB_{av}$ watts

and
$$\tau_p{}^2 L = \frac{P^*}{2p} \times \frac{10^3}{2fqB_{av}} \tag{262}$$

For example, if the number of poles and speed correspond to

50 c/s
$$\tau_p{}^2 L = \frac{P^*}{2p} \times \frac{10}{qB_{av}}$$

For 25 c/s
$$\tau_p{}^2 L = 2\,\frac{P^*}{2p} \times \frac{10}{qB_{av}}$$

Similarly for other frequencies†

Limiting Value of Output per Pole with Normal Design. The power developed for each pole-pitch is obviously the product of the average e.m.f., and the ampere-conductors for the pole-pitch,

$$\text{i.e. power per pole} = e \times q\tau_p \text{ watts} \tag{263}$$

† The above relations are for m.k.s. units of q etc.

With c.g.s. units $\tau_p'^2 L' = \dfrac{P^*}{2p} \times \dfrac{10^{11}}{2fq'B_{av}'}$

This holds for both c.g.s. and m.k.s. units of q and τ_p.

Taking the normal conservative limits of $e = 7$ and $q' = 400$ for large machines,

$$\text{Power per pole} = 7 \times 400\tau_p' \times 10^3$$
$$= 2{\cdot}8\tau_p' \text{ kW}$$

To obtain larger values it is necessary to employ less conservative loading constants, by the use of compensating windings and taking $e = 10$ or special armature windings as already stated.

The armature m.m.f. in ampere-turns per pole is

$$F_a = 0{\cdot}5q\tau_p$$

Hence, for $\tau_p' = 45$ and $q' = 400$, $F_a = 9{,}000$

$\tau_p' = 50$ $q' = 400$, $F_a = 10{,}000$

$\tau_p' = 55$ $q' = 400$, $F_a = 11{,}000$

A large armature reaction is undesirable for normal designs, and most designers like to keep it down to 11,000—some prefer a lower limit.

The limiting outputs for simple lap windings are then—*

For pole-pitch = 45 cm maximum power per pole = 126 and 180 kW

,, = 50 cm ,, ,, = 140 and 200 kW

,, = 55 cm ,, ,, = 154 and 220 kW

To illustrate this method of approach to the design of a large machine, preliminary dimensions for a 1,500-kW, 600-V, 250-rev/min generator will be obtained.

Current output = 1,500/0·6 = 2,500 A.

With 12 poles this would require a commutator handling just over 400 A at each brush arm. The frequency with 12 poles would be 25 c/s, so 12 poles will be tried first. Then,

$$\text{Power per pole} = 1{,}500/12 = 125 \text{ kW}$$

From the preceding paragraph this value is quite practicable without exceeding normal limits.

$$\text{Taking } q' = 400, \text{ and } B_{av} = 0{\cdot}7$$

$$\tau_p{}^2L = 125 \times \frac{10^3}{2 \times 25 \times 40{,}000 \times 0{\cdot}7}$$
$$= 89{,}300 \times 10^{-6} \text{ m}^3$$
$$\tau_p'^2L' = 89{,}300 \text{ cm}^3$$

From equation (263) page 449 the minimum value for the pole-pitch,

* The former values are for non-compensated machines and the latter for machines with compensating windings.

if the average voltage per conductor is not to exceed 7, is approximately 45. At 25 frequency the peripheral speed is $50\tau_p$, and becomes 22·5 m/sec for a pole-pitch of 45 cm, 25 m/sec for a pole-pitch of 50 cm, and so on. Evidently the peripheral speed is not a limiting factor. A few slide rule operations give the four skeleton designs below—

τ_p' cm	L' cm	L/τ_p	$\tau_p L$	Φ	e	Z_c
45	44	0·98	0·198	0·139	6·95	88
47·5	40	0·84	0·190	0·133	6·65	92
50	36	0·72	0·180	0·126	6·3	97
52·5	32·5	0·62	0·170	0·119	5·95	103

The figures are approximate only.

Of the four designs the last one is definitely rather short, the first maybe is a shade long. Probably the two most likely are the second and third. For the second, 15 slots would be used with 6 conductors per slot for each pole; for the third, 16 slots per pole would be used.

Taking the second machine, the diameter is 12 × 47·5/3·14, i.e. 182 cm.

Physical Limits of d.c. Machines. The very large outputs possible with a.c. generators, e.g. 60,000 kW at 3,000 rev/min, are not possible with d.c. machines. The limitation is chiefly due to the commutator.

The total current loading is

$$\pi D q = I_A Z_c$$

Hence, if e is the average e.m.f. induced in each armature conductor

$$\pi D q e = I_A Z_c e = I_A E$$

= the total power developed by the armature.

Hence $$P^* = \pi D q e \times 10^{-3} \tag{264}$$

(P^* denotes $EI_A \times 10^{-3}$, and not the actual output.)

and $$D = \frac{P^* \times 10^3}{\pi q e} \tag{265}$$

It will be noticed that the speed of rotation does not enter into these equations. By inserting the limiting values of q and e in the equations, we have:

(1) the maximum possible output obtainable for any given armature diameter, and

(2) the minimum possible diameter for any given output.

Taking, for example, $q = 450$ and $e = 7$ as limits giving an ample factor of safety, we have for large machines, without compensating windings:

$$\text{Maximum kW} = \pi \times 450 \times 7 \times D' \times 10^{-3} = 10D',$$

where D' is the dia in cm.

Hence the minimum diameter which can be used, to conform to the above limits is

$$D_{min}' = 0{\cdot}1 \times P^*$$

For example, a 2,000-kW generator working with a specific electric loading of 450 ampere-conductors per cm and an average e.m.f. of 7 V in each conductor *must* have a diameter of 0·1 × 2,000 cm, i.e. 200 cm. This is quite independent of the actual speed in rev/min and of the air-gap flux density. If the angular speed is high, the core length perforce must be short; if the speed be low, the core length will be long.

At 31·4 m/sec the angular speed is 300 rev/min approx. For an average flux density $B_{av} = 0{\cdot}7$ Wb/m² the core length is 32 cm very nearly. At 200 rev/min the rating of 2,000 kW can still be obtained by increasing the core length to 40 cm; at 160 rev/min and with the same loading constants the core length will have to be 50 cm.

(This should be contrasted with an air-cooled turbo-alternator having a stator bore of 100 cm for which a rated output of 60,000 kVA is possible.)

For a machine fitted with *compensating windings* the average voltage between commutator segments may be taken up to 20 or a little more, i.e. the e.m.f. per conductor raised to 10 or a little more. With very large "compensated" machines forced ventilation may be used and the armature strength taken up to about 500 ampere-conductors per cm. Hence for such a machine, with a simple lap winding:

$$\max P^* = \pi \times 500 \times 10D' \times 10^{-3} = 15\ D' \text{ approx.},$$

and the minimum diameter that can be used becomes

$$D'_{min} = P^*/15$$

If the armature strength is kept down to 450, the relation is: $P^*_{max} = 13{\cdot}5\ D'$, and $D'_{min} = P^*/13{\cdot}5$.

With smaller machines the working value of the armature strength may be much less, depending on the diameter and max $P^* = 12\ D'$, or less.

It then becomes apparent that to obtain relatively large output a high peripheral speed may be necessary to obtain the necessary minimum diameter.

Peripheral Speed. If a high peripheral speed is necessary a suitable mechanical construction of the armature and commutator is designed. A peripheral speed of 30 m/sec presents no difficulty, 40 m/sec is quite usual for medium-sized armatures, and much higher speeds can be used with suitably designed mechanical construction—up to 60 or 70 m/sec being employed with very large diameter armatures.

Moreover, high-speed d.c. generators are in use as dynamometers for engine testing and their maximum surface speed may reach as high as 100 m/sec. Formerly high-speed d.c. generators were direct coupled to steam turbines, but geared drives have long since superseded direct coupling for this purpose.

Maximum Core Length. Since the average e.m.f. that can be generated in a single armature conductor is limited, 7 V for a non-compensated simple lap-wound machine and to 10 V for a compensated machine, the maximum core length in metres that can be used is limited by the peripheral speed to $7/(B_{av}v_A)$ and $10/(B_{av}v_A)$ respectively, i.e. to about $12/v_A$ and $17/v_A$ respectively, assuming that $B_{av} = 0{\cdot}6$, as the frequency will be fairly high.

By using duplex lap windings, for the same voltage between adjacent commutator segments these limiting core lengths could be doubled, but in practice somewhat shorter lengths would be used. (As already stated in Chapter III the use of these windings is very limited.)

Speed-Output Relations. An alternative approach tc the problem of maximum output is

$$P^* = \pi Dqe \times 10^{-3} = qe \times 10^{-3} \times 60v_A/N$$

i.e.

$$P^* \times N = 6qev_A \times 10^{-2} \qquad (266)$$

This expression holds without modification for both m.k.s. and c.g.s. values of q and v_A.

Taking for large machines limiting values† of $q' = 500$ amp cond/cm, $e = 10$ and $v_A = 60$ m/sec we have; for the maximum possible output under these conditions:

$$P^* \times N = 6 \times 500 \times 10 \times 6{,}000 \times 10^{-2}$$
$$= 1{\cdot}8 \times 10^6$$

This relation will only hold for large diameters permitting of an armature strength as high as 500.

Illustrating this by a given case, it follows from the above that (a) a 3,000-kW generator working to the above limits would have to be driven at 600 rev/min, and that 3,000 kW is the maximum possible output that can be obtained at that speed. The diameter would be about 190 cm; and (b) a 6,000 kW generator would have to be driven at 300 rev/min and a simple lap wound generator cannot be designed for a larger output than 6,000-kW at that speed.

More often than not, overload conditions with these large machines are very onerous, so that the normal rated output is less than would correspond to the limiting conditions given above.

† These values must not be assumed to be the absolute limits. Inevitably the opinion of designers on these matters differs a little.

For *duplex lap windings* the working value e of the e.m.f. in the single armature conductor can be increased considerably, and under most favourable conditions, i.e. no violent very heavy overloads, the limiting output could be raised maybe to $P^* \times N = 3{\cdot}6 \times 10^6$.

When the duty includes such momentary violent overloads the volts/segt with duplex lap windings would be kept low, and e limited to 6 or 7 volts. Under these conditions the output possible is given by $P^* \times N = 2{\cdot}4 \times 10^6$.

For smaller machines, the expression $P^* \times N = 6qev_A \times 10^{-2}$ indicates that for a given peripheral speed and mean e.m.f. induced in a single conductor, the product $P^* \times N$ is simply proportional to the armature strength. Thus, for simple lap-wound machines, it follows that—

for machines with compensating windings and $e = 10$

with $\quad v_A = 50$ m/sec $\quad P^* \times N = 3{,}000\ q'$

and $\quad v_A = 40 \quad P^* \times N = 2{,}400\ q'$

(For non-compensated machines $P^* \times N = 1{,}600\ q'$ and $1{,}200\ q'$ with $v_A = 40$ and 30 m/sec, respectively.)

Thus, for a machine with compensating windings—

At 1,000 rev/min and 40 m/sec, $D' = 76{\cdot}5$ cm

For $q' = 350$: $P^* \times 1{,}000 = 2{,}400 \times 350$, i.e. $P^* = 840$ kW

At 1,000 rev/min and 50 m/sec, $D' = 61$ cm

For $q' = 325$: $P^* \times 1{,}000 = 3{,}000 \times 325$, i.e. $P^* = 975$ kW

Design of Motors. In the two preceding chapters, the design of a 1,000-kW generator has been worked through in detail. Exactly the same method of calculation can be applied to machines of all sizes, whether generators or motors.

One important point should be remembered in connexion with the output coefficient. If reference be made to page 393 it will be seen that eqn. (242) connects the main dimensions and the loading constants with the *total* work done by the armature in one complete revolution. The formula is based on the product of the armature current and the induced e.m.f., and *not* on the product of the line current and terminal voltage.

To be logical, therefore, the coefficient should be expressed in the form—

$$C_o = \frac{EI_A \times 10^{-3}}{N} \times \frac{1}{D^2L}$$

where I_A is the armature current, and E the induced e.m.f. In the case of large machines, the efficiency is very high, and it is of no great moment in practice whether we deal with the terminal voltage instead of the induced e.m.f., or with the line current instead of the armature current. But with small and other inefficient machines it is of considerable moment.

Consider, for example, a small series machine which at a given speed and its rated current of 10 A generates an e.m.f. of 100 V; the resistance of the windings aggregates 2 ohms, and the iron and friction losses are 100 watts—fairly reasonable figures for a machine of this size.

When working as a generator, the terminal voltage is 80 V, since the volt-drop in the windings is 20. The output is therefore $80 \times 10 = 800$ watts.

When working as a motor, on the other hand, the terminal voltage, for the same conditions of speed and current, will have to be 120 V. The input is therefore $120 \times 10 = 1{,}200$ watts. The power output is less than the product EI_A by an amount corresponding to the iron and friction losses, and is therefore $1{,}000 - 100 = 900$ watts.

Under the given conditions, the machine can therefore be rated as a generator to develop 800 watts, or as a motor to develop 900 watts, i.e. 1·21 b.h.p. Its loading constants, losses and temperature rise will be identical; as a generator the terminal voltage will be 80, as compared with the 120 necessary for motor operation at the same speed. The torque on the armature—measured by the product EI_A—is the same in the two cases.

This simple example serves to show that, when designing motors, other than highly efficient ones, it is not wise to base the design upon that of generators having an output equal to the input of the motor.

The safest plan, for all machines, whether generators or motors, is to base the design constants upon the power developed by the armature, i.e. the product of the armature current and the induced e.m.f.

The example given above is for a series machine; similar considerations hold for shunt and compound machines. With these machines, the difference between the terminal voltage and the induced e.m.f. is relatively smaller than for a series machine, but this is counterbalanced by the fact that the shunt current in the case of a motor does not pass through the armature, and in the case of a generator is not available for useful output.

If then P_e = the loss in the shunt winding,

P_o = the iron and friction loss.

When working as a generator, the output is—

$$EI_A - I_A^2 r_m - P_e$$

When working as a motor, the output is—

$$EI_A - P_o$$

The input in this latter case is—

$$EI_A + I_A^2 r_m + P_e$$

The difference between the output when working as a motor and that when working as a generator under precisely the same conditions with regard to loading constants, is then $I_A{}^2r_m + P_e - P_o$ in favour of the motor. In the motor, the ohmic losses are entirely provided by the supply system, but the iron and friction losses absorb some of the power developed by the armature.

It is the contrary with a generator; the iron and friction losses are provided by the prime mover, but the ohmic losses absorb some of the power EI_A developed by the armature.

In the numerical example considered above it was shown that the rating of the series machine there considered was 12·5 per cent greater as a motor than when working as a generator. This is by no means abnormal; with very small machines, the difference in favour of motor operation is very much greater, since the ohmic losses with such machines are normally much greater—in fractional horse-power machines often many times greater—than the iron and friction losses. Where, as very rarely happens, the iron and friction losses exceed the ohmic losses, the permissible output for generator operation would exceed that for motor working.

For large and efficient machines, these points are of no moment in practice, and in such, the rated output for generator working may be taken to be the same as that for motor working.

The design of d.c. motors can proceed on the same lines as those given in detail for generators. (The generated e.m.f. is $E = V - I_Ar_m$ instead of $E = V + I_Ar_m$ as in generators.) In choosing preliminary dimensions the loading constants and output coefficient can be based upon the output—expressed by kW—and by following the procedure given, the core length can be revised as the design progresses.

In the commerical design of d.c. motors, however, excepting those of large dimensions, the problem that arises is not to design *ab initio* the particular machine required, but to select the most suitable size from the "line" of machines that has been standardized. In deciding upon the standard values of the armature diameter and core length, an endeavour is made to arrange that the values of D^2L increase in sufficiently small steps. For a line of machines ranging from 5 to 25 in. diameter, D^2L will range from about 75 in.^{3} to 7,500, i.e. a ratio of about 100 to 1. If each value of D^2L is 25 per cent greater than that of the size smaller armature, the number of sizes of machine required is given by log 100/log 1·25, and is approximately 20. Actually, rather larger steps can be tolerated at the smaller end of the range, but smaller steps are desirable for the large machines. To avoid the need for adopting too many different sizes of core plates, it is a common practice for all except the smallest diameters to employ two, or sometimes three, different core lengths as standard for a given diameter. This, of course, is a counsel of expediency, and not of perfection, as the short-core machine is of necessity more expensive and less efficient than the one of more normal proportions.

Having selected the best frame-size available, it then remains to design the windings. Again, the number and size of slots is also standardized. With the smallest diameters it suffices to have but one slotting as a standard, but for others two or three alternative slottings are generally available. It follows that if the number of slots is standardized, and the normal value of the working flux is adhered to, it is not always possible to arrange the design for operation at a given speed at any particular voltage. It is, therefore, the practice of manufacturers to give lists showing the various output and speeds that can be obtained from their standard machines. If a suitable standard machine is not available, it becomes necessary to employ non-standard slotting and a non-standard commutator.

The number of poles also is standardized on the lines already given on page 402.

Design data for a number of small machines are given at the end of this chapter. The selection of suitable armature windings will be considered in detail for three typical motors.

Examples of the Choice of Suitable Armature Windings for Motors. 1. DIAM OF ARMATURE, 20 CM; CORE LENGTH, 14 CM, 4 POLES. From Fig. 302 this small armature can be worked with a maximum flux density in the air gap of 0·7, i.e. B_{av} 0·49 Wb/m².

$$\text{The pole pitch } \tau_p' = 15{\cdot}7;\ \tau_p'L' = 15{\cdot}7 \times 14$$

$$= 220 \text{ cm}^2$$

$$\text{The normal flux } \Phi = \frac{0{\cdot}49 \times 220}{10^4} = 0{\cdot}0108 \text{ Wb.}$$

Hence the average e.m.f. generated in each conductor at 1,000 rev/min is 4 × 0·0108 × 1,000/60

$$= 0{\cdot}72 \text{ V}$$

A normal diameter for the commutator of a 20 cm armature is about 14 cm. Taking 4 mm as the minimum pitch for the segments, the maximum number that can be used is about 110. This would suit very well a 37-slot arrangement with 3 coil-sides per layer. For 440-V machines the average volts between segments would be 15·8. With 37 slots the slot pitch is only 20 × 3·14/37 = 1·67 cm. Smaller values than this are sometimes used on small diameters like the one under review. With 33 slots and 99 commutator bars, the voltage between segments is nearly 18. With too many slots it would not be possible to work up to 230 ampere-conductors per cm. With too few the number of commutator segments becomes too few unless 27 slots are used with 4 coil-sides per layer. The taste of the designer would largely decide the issue. By choosing 35 slots a good machine will be possible.

With 3 coil-sides per layer, which would be standardized for the 440-V machines of this size, there will be 105 commutator segments.

With 3-turn coils the no-load speed is $440 \times 1{,}000/(3 \times 105 \times 0{\cdot}72)$
$= 1{,}940$ rev/min.

,, 4-turn	,,	,,	,,	$1{,}940 \times \frac{3}{4} = 1{,}455$
,, 5-turn	,,	,,	,,	$1{,}940 \times \frac{3}{5} = 1{,}165$
,, 6-turn	,,	,,	,,	$1{,}940 \times \frac{3}{6} = 970$

and so on.

2. Armature Diam, 30 cm; Core Length, 20 cm, 4 Poles. $\tau_p{}' = 23{\cdot}55$ cm. $D^2L = 0{\cdot}018$ m³. From Fig. 302, the maximum flux density in the air gap for this diameter may be taken up to $0{\cdot}75$ Wb/m², corresponding to $B_{av} = 0{\cdot}7 \times 0{\cdot}75 = 0{\cdot}525$.

The normal, or "rated," value of the flux is

$$\Phi = B_{av}\tau_p L$$
$$= 0{\cdot}525 \times 0{\cdot}2355 \times 0{\cdot}2$$
$$= 0{\cdot}0247 \text{ Wb}$$

It is a convenience to adopt 1,000 rev/min as a unit of angular speed, and to work out the average e.m.f. in each conductor at this speed. Hence, at 1,000 rev/min

$$e = 4 \times 0{\cdot}0247 \times \frac{1{,}000}{60}$$
$$= 1{\cdot}65 \text{ V}$$

From this figure the number of conductors that must be connected in series at any speed to generate a given e.m.f. can at once be deduced.

The normal operating speed of a motor of the dimensions under review would be about 1,000 rev/min or less. Without special construction it could operate up to about 1,800 rev/min and with strengthened mechanical construction at appreciably higher speeds. Taking 1,000 rev/min as typical, the number of conductors necessary for each circuit for 440 V is $440/1{\cdot}65 = 267$. (The ohmic drop has been neglected.)

For a simple lap winding, as there are 4 parallel paths, the total number of conductors on the armature will be about 1,000; for a simple wave winding there are only 2 parallel paths and the number of conductors will be about 500. A usual value for the commutator diameter would be about 20 cm although a larger diameter could be employed. For 20 cm diameter and 0·4 cm for the pitch of the segments, the maximum number of segments that can be built into the commutator is about 160.

For the simple wave winding, therefore, it would be necessary to use 2-turn coils, e.g. 125 commutator segments with 2-turn coils (this gives $Z_a = 500$). For the simple lap winding the alternatives would be to use 160 segments, approx., and 3-turn coils, or about 120 segments and 4-turn coils. For the lap winding, too, equalizing connexions would be necessary, and it would be necessary to employ 4 brush spindles, whereas only 2 are essential for the wave-wound motor.

It may be taken that where a wave winding can be used without exceeding the permissible voltage between commutator segments, this type of winding is generally preferred to the lap winding.

The Number of Commutator Segments. With comparatively small motors the number of commutator segments cannot always be made large enough to permit of working with an average of 14 V between commutator segments, and a higher value of 16 or so must occasionally be used. Where a higher value is used for this voltage between commutator bars, only a limited range of speed control by field variation can be permitted. In some a 99-segment commutator is used for 440 V, 111 segments are often used, but it is desirable to employ 120 or more if possible.

The Number of Slots. Of necessity, the slot pitch with small machines must be taken lower than for large generators, and for very small diameters the slot pitch may have to be taken as low as 1·5 cm. The greater the number of slots for each pole, the more silent is the operation of the machine, but the large values of the electric loading coefficient given in Fig. 302 cannot be attained unless the slots are reasonably wide. For 4-pole machines about 40 slots is a reasonable number. But if the pole-pitch is small, it may not be possible to use more than about 30. On the other hand, when the pole-pitch is large the number of slots may be taken up to about 50.

Reverting to the 30 cm armature, the pole-pitch is 23·55 cm, so that rather more than 40 slots in all would give a slot pitch of about 2 cm.

To obtain various operating speed and outputs from the same frame it is necessary to be able to arrange for various values of Z_c.

This can be done for any given slotting by altering the number of turns in each coil and retaining the number of commutator segments. In addition, the number of coil-sides per layer can be changed, involving a corresponding change in the number of commutator bars. Thus 3 coil-sides per layer, or 4, can be arranged for. By providing alternative slottings with a ratio of about 1·15 for the two numbers of slots, the steps in the standard speeds and outputs for a given frame are made sufficiently small.

For 440 V, 41 slots with 3 coil-sides per layer give an average of 14 V between segments.

The alternative number of slots to this might well be 1·15 × 41 = 47.

These slottings suit wave windings but are not suitable for lap windings.

Speeds and Winding Data for 41-slot Armature. Based on the normal flux of 0·0247 Wb, the induced e.m.f. was 1·65 V per conductor at 1,000 rev/min. Hence, neglecting the very slight ohmic drop, the no-load speed of 440-V motors will be

$$\text{Speed} = 440 \times 1{,}000/(1{\cdot}65\, Z_c) \text{ rev/min}$$

The following table of no-load speeds can then be drawn up for the 41-slot armature, 30 cm diam × 20 cm core length with simple wave windings.

Coil-sides per Layer	Turns per Coil	Conductors in Series (Z_c)	No-load rev/min
3	1	123	2,170
3	2	246	1,085
3	3	369	720
3	4	492	540
4	1	163	1,635
4	2	326	820
4	3	489	545

For the first four there would be 123 commutator segments and for the others 163. There would be no dummy coils for the first four, but one would be necessary for the 4 coil-sides per layer group. For the speed of 540 rev/min the 123 commutator segment design would normally be preferred to the one requiring a dummy coil.

Five coil-sides per layer would require more commutator segments than could be accommodated by the commutator.

Speeds and Winding Data for 47-slot Armature. For simple wave windings this slotting is suitable with 1 dummy coil for use with a 93-segment commutator and 2 coil-sides per layer. This would suit quite well 220-V operation, but is not suited for 440 V. With 3 coil-sides per layer and no dummy coils it is very satisfactory for either 440-V or 220-V machines, the number of segments being 141. Four coil-sides per layer would require too large a commutator for a normal machine.

At normal flux the following no-load speeds are obtained with the windings indicated.

Coil-sides per Layer	Turns per Coil	Conductors in Series (Z_c)	Rev/min	Volts
2	1	93	1,435	220
3	1	141	1,890	440
3	1	141	945	220
3	2	282	945	440
3	2	282	470	220
3	3	423	630	440
3	4	564	470	440

The two slottings thus together give a very closely graded choice of speeds between 500 and 1,000 rev/min. A motor of this size would normally not operate above about 1,800 rev/min without special construction.

For 220-V operation the speeds available with both slottings are: 1,435, 1,085, 945, 820 rev/min, etc.

As alternatives to the choice of 41 and 47 as the numbers of slots, 37 and 43 can be chosen; the drawback to this pair is that with 3 coil-sides per layer there are only 111 commutator segments, corresponding to 16 V per segment. The advantage is that more standard speeds are available between 1,000 and 1,800 rev/min.

The calculations just given for the armature 30 cm diameter and 20 cm long can immediately be modified to suit other core lengths. The working flux will be proportional to the net core length so that the operating speeds on no-load for a given winding scheme will be inversely proportional to the core length. If, therefore, the other standard core length for this diameter were 15 cm, the speeds given in the tables on page 460 would all be increased by one-third.

It should be pointed out that with the motor on full load the induced e.m.f. will be less for the motors designed for low speeds than for the high speed machines. (Actually for constant current density the voltage drop in the armature winding will increase in direct proportion to the number of conductors in series if the output be taken as proportional to the speed.)

3. ARMATURE DIA, 55 CM, CORE LENGTH, 25 CM. This machine would have 6 poles as the pole-pitch for 4 poles would be unduly large.

As an upper limit the commutator could be given a diameter, about 40 cm or more if necessary, to accommodate rather more than 300 segments.

For 440 V it is desirable with 6 poles to have not less than 6 × 440/14, i.e. about 190 segments, but a slightly smaller number could be used.

From Fig. 302 the value of B_g can be taken up to about 0·825 and B_{av} to 0·575 Wb/m².

The pole-pitch $\quad \tau_p' = \pi \times 55/6 = 28{\cdot}8$

$$\therefore \quad \tau_p L = 0{\cdot}288 \times 0{\cdot}25 = 0{\cdot}072$$

$$\Phi = 0{\cdot}575 \times 0{\cdot}072 = 0{\cdot}0413 \text{ Wb}$$

The volts per conductor at 1,000 rev/min are

$$e = 6 \times 0{\cdot}0413 \times \frac{1000}{60} = 4{\cdot}13$$

For a simple wave winding with single-turn coils, the mean volts between commutator bars would be 6 × 4·13 = 24·8. Obviously the simple wave winding could not be used at this speed, which speed is actually rather higher than would normally be used for this diameter. But at 500 rev/min—a very common speed for a machine of these dimensions—the single-turn wave winding would give 12·4 V between adjacent commutator segments. At 600 rev/min this winding would give about 15 V between segments and could be used for normal working.

For speeds above 600 rev/min it is essential to use a simple lap winding. For speeds lower than 600 rev/min either the lap or the wave winding can be used. This machine is of fairly large dimensions, and in most cases the slotting could be chosen to suit best the particular duty required. Examples of various duties will now be considered.

(*a*) *Winding for* 440-*V Motor with Light-load Speed of* 750 *rev/min.* The required value of Z_c is $\frac{400 \times 1000}{4{\cdot}13 \times 750} = 142$. For a simple wave winding with single-turn coils, Z_c is equal to the number of coils and commutator bars. With such a winding the voltage between segments would average $6 \times 440/142$ or 18·6 volts, which is too high.

A simple lap winding with single-turn coils would require 3×142 segments approximately. This number exceeds the capacity of the commutator. But by using 2-turn coils the number is reduced to one-half—to about 210—which is quite practical.

A convenient value to take for Z_c would therefore be 144. There would be $6 \times 144 = 864$ armature conductors and 216 coils and commutator segments. The pole-pitch is 28·8 so that 12 slots could be used for each pole, 72 in all, with 3 coil-sides per layer, 2-turn coils, 12 conductors per slot. The number of slots suits very well 12 equalizers, 1 for every alternate slot.

(*b*) *Winding for* 440-*V Motor at* 500 *rev/min.* For these conditions

$$Z_c = \frac{440 \times 1000}{4{\cdot}13 \times 500}$$
$$= 213$$

With single-turn coils, and a wave winding, the voltage between segments will average just over 12, which is a very conservative figure, so that the wave winding can be used.

Probably 70 slots would be a more standard number to use than 71. The wave winding would then be based on 70 slots, 3 coil-sides per layer, single-turn coils, 209 active coils and commutator segments, 1 dummy coil.

To avoid the use of the dummy coil with 6 poles, either 2 or 4 coil-sides must be used for each layer. With 4, the number of slots would only be 53, a number which is small for this pole-pitch, whilst with 2 coil-sides per layer no less than 106 slots would be necessary and the pole-pitch is not large enough to accommodate about 18 slots.

The alternative to the simple wave winding with 70 slots is a simple lap winding. For this the total number of conductors necessary is $6 \times 213 = 1{,}280$ approx. Since the number of commutator segments should lie between 200 and 300, single-turn coils are not possible. The most suitable arrangement would appear to result from the use of 216 commutator segments and 3-turn coils.

This involves the use of 72 slots, 3 coil-sides per layer, 18 conductors per slot. Again 12 equalizers would be satisfactory.

Compared with the wave winding, the chief difference is that there are three times as many conductors in each slot. The wave winding would be bar-wound and on the lines of Fig. 305—an ideal arrangement.

(c) *Winding for* 440-*V Motor at* 400 *rev/min.* The number of conductors in series would have to be $213 \times 5/4 = 266$.

This would suit very well a wave winding arranged in 67 slots with 4 coil-sides per layer and 268 commutator bars. No dummy coils would be necessary for this winding.

If the alternative core length were 30 cm, the above wave winding without modification could be used for 440 V at $400 \times 25/30$, i.e. 333 rev/min.

Details of Design of 350-kW and 150-h.p. Machines. When designing a motor it is necessary to assume a reasonable value for the efficiency so that the full load current can be determined. Full calculations for a 150-h.p. motor and for a 350-kW generator are given below. The generator is shunt wound and the motor has a very weak series winding—a stability winding—which roughly neutralizes the tendency of armature reaction to weaken the field, and gives a slightly drooping speed-load characteristic.

The generator, although well ventilated, is not provided with a fan; the motor is intended to be provided with a fan of fairly large diameter. Neither design is pushed to its limits.

OUTPUT			
Generator	kW	350	
Motor	h.p.		150
Voltage	V	440	500
Speed	N	600	450
Terminal current	I	795	245
Shunt current	I_s	5	2
Armature current	I_A	800	243
Internal drop (estimated)	$I_A r_m$	11	28
Induced e.m.f. (estimated)	E	451	472
Armature power	$P^* = EI_A \times 10^{-3}$	361	115
Work per rev	P^*/N	0·601	0·256
MAIN DIMENSIONS			
Arm. diam	D'	75	55
Length, gross	L'	31	25
Volume, D^2L, in m³		0·175	0·076
Circumference	$\pi D'$	235·5	173
Speed	n	10	7·5
Peripheral speed	v_A	23·55	13
Number of poles	$2p$	6	6
Frequency	f	30	22·5
Pole-pitch	τ_p'	39·3	28·8
,, arc	b_{ps}'	27·5	20·2
Pole arc/pole-pitch	b_{ps}/τ_p	0·7	0·7
Length/pole arc	L/b_{ps}	1·125	1·24
Number of brush arms		6	6
Current per arm		267	81

OUTPUT CONSTANTS			
Output coef. $C_o = \dfrac{P^*}{N \times D^2L}$		3·44	3·36
Flux density	B_g	0·904	0·854
Armature strength	q'	326	342
MAGNETIC FLUX			
Field form coef.	K_f	0·71	0·7
Area Surface per pole	$\tau_p L$	0·122	0·072
Gap area	$K_f \tau_p L$	0·0865	0·0504
,, density	B_g	0·904	0·854
Flux per pole	Φ	0·0784	0·043
Total flux	$2p\Phi$	0·47	0·258
Av. e.m.f. per bar	e	4·7	1·935
Condrs. in series	Z_c	96	244
Max. e.m.f. per turn	$= 2e/K_f$	13·2	5·52
Max. volts bet. segts.		12·9	17·6
ARMATURE WINDING			
Type		lap	wave
Number of circuits	$2a$	6	2
Total number of condrs.	Z_a	576	488
Number of slots	S	72	61
Condrs. per slot	Z_s	8	8
Turns per coil		1	1
Current per condr.	I_a	133·3	121·5
Condr. dimensions, mm		18 × 1·7	16 × 1·5
,, area in cm²	A'_a	0·3	0·237
Current density, A/cm²	α'	444	512
Total current volume	$I_a Z_a$	76,800	59,300
Armature AT's per pole	F_a	6,400	4,950
Armature strength	q'	326	342
$\alpha' q'$ cm units		145,000	175,000
Type of insulation		bare condr. mica wrap	bare condr. mica wrap
Slot dimensions, below wedge, mm		11 × 44	10 × 44
Space factor, below wedge		0·495	0·43
Embedded length	L'	31	25
Free length	l_{fr}'	58	45
Length per condr.	l_a'	89	70
Total length, m	$L_a = Z_a l_a$	513	341
Volume, cm³		15,400	8,080
Weight, kg		137	72
Resis. of length L_a, cold		0·29	0·245
,, bet. brushes, cold		0·0081	0·061
,, ,, ,, hot	r_a	0·0097	0·073
Volt-drop, hot	$I_A r_a$	7·8	17·8
Power wasted	$I_A^2 r_a$	6,200	4,310
MAGNETIC CIRCUIT			
Arm. Core Plates			
Diam. ext.	D	75	55
,, int.	D_i'	38	28
Length, gross	L'	31	25
Number of ducts	n_v	5	3
Width of ducts	b_v'	0·9	0·9
Length, net	L_n'	26·5	22·3
Iron length	L_i'	23·6	19·8
Depth below slot	d_c'	13·7	9·1

MAGNETIC CIRCUIT—*contd.*			
Area of core	A_c'	323	180
Length of path	l_c'	13	10
Flux density, Wb/m²	B_c	1·21	1·2
Magnetizing Force, AT/m	H_c	1,000	1,000
M.m.f. for core, AT	F_c	130	100
Armature Teeth			
Number of slots	S	72	61
Slot pitch	τ_s'	3·27	2·84
,, dimensions	$b_s' \times h_s'$	1·1 × 4·8	1·0 × 4·4
Diam at root	D_t'	65·4	46·2
,, at $l_t'/3$	D_m'	68·6	49·1
Effective pitch	τ_{sef}	2·99	2·53
,, width	b_{tef}'	1·89	1·53
Number of teeth per pole	$S/2p$	12	10·17
Effective ,, ,,	$K_f S/2p$	8·5	7·1
,, tooth area	A_t'	380	215
Appt. density	B_{appt}	2·06	2·00
$K_s = \tau_s'/b_t' \times L/L_i$		2·07	2·09
Real density	B_{real}	2·015	1·965
Magnetizing Force, AT/m	H_t	26,000	20,000
Length of path	l_t'	4·8	4·4
M.m.f. for teeth, AT	F_t	1,250	880
Poles			
Axial length, cm	L_p'	30	24·5
Breadth	b_p'	21	14
Area	A_p'	600	326
Ht. above shoe, cm	h_p'	22·5	18
Leakage coef. F.L.	ν_p	1·2	1·2
Effective flux	Φ_p	0·094	0·0515
Flux density	B_p	1·57	1·58
Magnetizing Force, AT/m	H_p	2,600	2,600
Length of path	l_p'	22·5	18
M.m.f. for pole, AT	F_p	590	470
Yoke			
Area, cm²	A_y'	400	220
Leakage coef. F.L.	ν_y	1·25	1·25
Flux	Φ_y	0·098	0·054
Flux density	B_y	1·225	1·225
Magnetizing Force, AT/m	H_y	800	800
Length of path	l_y'	32	26
M.m.f. for yoke, AT	F_y	260	210
Summary of M.M.F. for Iron			
Core	F_c	130	100
Teeth	F_t	1,250	880
Pole	F_p	590	470
Yoke	F_y	260	210
Total	F_i	2,230	1,660
AIR GAP			
Length	l_g'	0·45	0·4
Slot pitch	τ_s'	3·27	2·84
Slot width	b_s'	1·1	1·0
Slot/gap	b_s/l_g	2·45	2·5
Coef. K_{gs}		1·13	1·13
Vent width/gap	b_v/l_g	2	2·25
Coef. K_{gv}		1·045	1·04

Item	Symbol		
AIR GAP—*contd.*			
Gap coef. K_g	$K_{gs} \times K_{gv}$	1·18	1·17
Flux density	B_g	0·904	0·854
Magnetizing Force, AT/m	$0{\cdot}796 K_g B_g \times 10^6$	850,000	795,000
Length of path	l_g'	0·45	0·4
M.m.f. for gap	F_g	3,830	3,180
TOTAL EXCITATION			
M.m.f. for iron	F_i	2,230	1,660
,, gap	F_g	3,830	3,180
Total		6,060	4,840
Armature Reaction	F_a	* 18%=1,140	* 15%=760
Total m.m.f. on F.L.	F_e	7,200	5,600
Field m.m.f./Armature m.m.f.	F_e/F_a	1·125	1·13
Shunt m.m.f.		7,200	4,870
Series m.m.f.		Nil	730
SHUNT WINDING			
M.m.f., F.L.		7,200	4,870
Voltage available	V_e	374	500
Mean length of turn	l_e'	120	95
Area of condr. reqd., mm²		2·83	1·13
Actual area used, mm²	A_e	2·83	1·13
Condr. diameter	mm	1·9	1·2
Actual m.m.f. developed		7,200	4,870
Distance B.C. approx., cm		18	12
Winding depth ,, cm		3·5	3·8
Surface per coil ,, cm²		5,100	3,000
Total surface ,, cm²		30,600	18,000
Watts lost		1,985	1,085
Surface per watt, cm²		15·4	16·5
Current	I_e	5·3	2·17
Current density, A/cm²		187·5	192
Number of turns	N_e	1,360	2,240
Total length in m	L_e	9,800	12,800
Resis. hot	r_e	70·5	230·5
,, cold		58·8	192
Volume of copper, cm³		27,800	14,450
Weight of copper	kg	248	128·5
SERIES WINDING			
Number of turns	N_{se}	—	3
Sectional area, cm²	A_{se}'	—	1·1
Current density, A/cm²		—	222
Mean length, cm	l_{se}'	—	95
Total length, m	L_{se}	—	17·1
Resis., hot	r_{se}	—	0·0032
Resis. cold		—	0·0026
Volt-drop		—	0·63
Watts lost		—	160
Volume, cm³		—	1,880
Weight of copper	kg	—	17
COMMUTATOR			
Diameter	D_c'	55	40
Circumference	$\pi D_c'$	172·5	125·5

* The allowance normally ranges from 15 per cent to 20 per cent according to the degree of saturation in the teeth; great precision in estimating the allowance is not necessary.

COMMUTATOR—*contd.*			
Periph. speed	v_C	17·25	9·4
Number of segments	C	288	244
Pitch of segments	τ_C	0·599	0·514
Pitch of brushes cm		28·75	20·9
Volts between segments, max.		12·9	17·5
Total current collected	I_A	800	243
Number of brush arms		6	6
Current per arm, amperes		267	81
BRUSHES			
Thickness, cm	t_b'	1·59	1·27
Width, cm		3·17	3·17
Area, cm^2		5·05	4·02
Number, per arm		8	4
Area, per arm, cm^2		40·4	16·1
Current density, A/cm^2		6·6	5·02
Volt-drop, total		2	2
Contact loss . . . watts		1,600	490
Total contact area, cm^2		121	96·6
Brush pressure, g/cm^2		120	140
Brush friction loss . . . watts		540	280
Total com. losses . . . watts		2,140	770
INTERPOLES			
Brush thickness, cm		1·59	1·27
Segt. pitch, cm		0·6	0·51
Coil-sides per layer		4	4
Arm. diam/com. diam		75/55	57/40
Distance β_{zA}' (*see* page 158)		4·6	4·0
Shoe width, cm		4·5	4
„ length, axial, cm	L_{cp}	16	14
Pole dimensions, ends rounded, cm		4·5 × 16	4 × 14
Air Gap, cm	$l'_{g_{cp}}$	0·8	0·6
λ_i		3·3	3·6
λ_{fr}		0·6	0·5
λ		3·9	4·1
Current per slot	$I_a Z_s$	1,070	970
Flux density	B_{cp}	0·35	0·352
Gap coef. approx.	$K_{g_{cp}}$	1·10	1·10
Net amp-turns	$F_{cq} - F_a$	2,440	1,850
Arm. amp-turns	F_a	6,400	4,950
Interpole amp-turns, actual	F_{cp}	8,800	6,800
Turns per pole	N_{cp}	11	28
Conductor area	cm^2	3·5	1·1
Current density, A/cm^2		229	221
Mean length of turn, cm		49·5	42
Total length, m		32·7	70·4
Resistance, cold	r_{cp}	0·00159	0·0109
„ hot		0·00192	0·0131
Volt-drop, hot		1·54	3·18
Power wasted	watts	1,230	770
Volume, cm^3		11,500	7,750
Weight in kg		102	69
TOTAL VOLT-DROP IN SERIES CIRCUIT			
In armature wdg.		7·8	17·8
„ interpole wdg.		1·5	3·2
„ series wdg.		—	0·6
„ brush contacts		2	2
Total		11·3	23·6

IRON LOSSES			
Core			
Diam int, cm		38	28
Depth, cm		13·7	9·1
Mean diam, cm		51·7	37·1
Area, cm²		323	190
Volume, cm³		52,500	22,100
Weight, Kg		410	172
Flux density		1·21	1·13
Frequency		30	22·5
Hyst. loss per kg		2·6	1·75
Eddy ,, ,,		1·64	0·81
Total ,, ,,		4·24	2·56
Total core loss		1,740	440
Teeth			
Effective tooth width, cm		1·89	1·53
Volume of teeth, cm³		154	81
Weight ,, kg		120	63
Flux density		2·015	1·965
Hyst. loss per kg		5·8	4·2
Eddy ,, ,,		7·3	3·9
Total ,, ,,		13·1	8·1
Total teeth loss		1,570	655
Total Iron Loss		3,310	1,095
Stray losses on F.L.		1,660	550
TOTAL ARMATURE LOSS			
Copper loss		6,150	4,310
Stray loss		1,660	550
Iron loss		3,310	1,100
Total loss on F.L.		11,120	6,960
EFFICIENCY			
Losses on Full Load (in kW)			
Arm. copper		6·15	4·31
,, iron		3·31	1·10
Stray loss		1·66	0·55
Shunt wdg.		1·99	1·09
,, regulator		0·35	—
Interpole wdg.		1·23	0·77
Stability wdg.		—	0·16
Brush contact		1·60	0·49
,, friction		0·54	0·27
Bearing Friction and Windage		2·80	1·60
Total losses		19·63	10·34
Output, F.L.	kW	350·0	111·90
Input, F.L.		369·63	122·24
Losses/Input	%	5·3	8·5
Efficiency	%	94·7	91·5
F.L. Current (terminal)		795	244·5

Further examples of designs are given below, full data being given from which the detailed performance can be worked out.

DESIGN DATA FOR 1,500-kW HEAVY CURRENT GENERATOR, BY THE LANCASHIRE DYNAMO AND CRYPTO, LTD.

This machine is used for electrolytic work; a photograph of the armature is given in Fig. 278, page 368.

Output:

1,500 kW; 250/275 V, 6,000/5,450 A; 300 rev/min.

Armature Core Plates:

Ext. diam 165 cm; int. diam 117 cm; in 10 segments.
Core length 45 cm, including 5 ducts each 1·25 cm.
Core plates 0·455 mm thick.
No. of slots 220, each 1·17 × 6·0 cm (wedge 9·5 mm thick).

Armature Winding:

Simple lap, 10-circuit; 22 equalizers each 9·5 × 1·6 mm.
Number of conductors per slot, 2.
Each conductor consists of 2 straps 4·45 × 22·6 mm in parallel.
Insulation, 1 mm tube moulded round each conductor.

Commutator:

Two parts in tandem, each 110 cm diam 47 cm brush face.
Brush thickness 2·54 cm, width 3·18 cm.
Number of brushes per spindle, 24.
Number of segments, 220.

Main Poles:

Number of poles, 10; material, cast steel (corners rounded).

Pole breadth, 28 cm; axial length, 43·5 cm; area, 1,150 cm² approx.; height above shoe, 30 cm.

Pole shoes laminated, held together with 4 W.I. rivets 1·6 cm diam and secured to pole by 6 screws, 2·2 cm diam; thickness of shoe at centre 3·8 cm; air gap, 4·8 mm.

Shunt Winding.

Number of turns per pole, 350; conductor, 3·64 mm diam d.c.c.
Resist. cold approx. 10 ohms.

Series Winding:

One turn per pole consisting of 6 straps in parallel, each 1·1 × 3·8 cm; a diverter was used on site to adjust the degree of over-compounding to suit the local conditions.

Interpoles:

Body of pole, 3·25 × 33 cm, with rounded ends, bevelled off to 5 × 25·5 cm at the armature face.

Air gap, 4·8 mm; a sheet brass liner, 0·8 mm thick was used between the interpole and yoke.

Number of turns per pole, 3, consisting of 6 straps in parallel, each 1·1 × 3·8 cm.

Design Data for 600-kW Generator, by the Lancashire Dynamo and Crypto, Ltd.

Output:

600 kW; 220 V; 2,730 A; 1,000 rev/min.

Armature Core Plates:

Ext. diam, 76·5 cm; int. diam, 44·5 cm.
Core length, 38 cm, including 5 ducts 9·5 mm wide.
Number of slots, 99, each 1·13 by 4·75 cm.

Armature Winding:

Lap, 6 circuits.
Number of conductors per slot, 2.
Two conductors in parallel, each 12·7 mm by 4·2 mm at slot mouth and 2 conductors in parallel 19·3 mm by 4·2 mm at slot bottom, i.e. remote from air gap.
Eleven equalizers each 9·5 mm by 1·6 mm.

Note. The type of winding may be regarded in some quarters as a form of Duplex lap winding, as there are 198 commutator bars. From a design point of view, however, the manufacturers consider it a simple lap winding, as the 2 conductors in each layer are solidly connected together in parallel at the end remote from the commutator, although at the commutator end each conductor enters a separate bar.

Insulation:

0·18 mm tape on 1 conductor of each pair in parallel, with 0·75 mm micanite tube moulded round each layer.

Commutator:

Two parts in tandem, each 55 cm diam by 28·5 cm brush base.

Brushes:

Thickness, 2·2 cm; width, 3·18 cm.
Total number of brushes per spindle, 14.
Number of commutator segments, 198.

Main Poles:

Number of poles, 6; material, forged steel.
Pole breadth, 19·6 cm; axial length, 35·6 cm.
Area. 695 cm²; pole shoes, laminated.
Thickness of shoe at centre, 2·54 cm.
Pole arc, 26 cm; air gap, 4·8 mm.
Yoke, 51 cm wide; area 392 cm².

Shunt Winding:

Number of turns per pole, 650.
Conductor, 2·65 mm. d.c.c.
Resistance (cold), 15·6 ohms.

Series Winding:

Two circuits in parallel of 2 turns per limb.
Conductor, 2 straps in parallel 33·3 mm by 9·5 mm.

Interpoles:

Interpole body, 23 cm axial length tapered in width from 8·9 cm nearest yoke to 4·45 cm at armature face.

Air gap, 6·35 mm with one brass liner 3·2 mm thick between yoke and interpole body.

Interpole Winding:

Eight turns per limb, consisting of 4 conductors in parallel, 35 mm by 47·5 mm.

The interpoles are connected to form 2 circuits in parallel.

Design Data for 6-pole Generator by Veritys, Ltd.

A drawing of this machine is given in Fig. 275.

Output:

175 kW; 225 V; 778 A; 350 rev/min.

Armature Core Plates:

Ext. diam 66 cm; int. diam 40·6 cm, sleeve; core length 40·6 cm including 4 ducts, each 0·95 cm; 84 slots, 1·05 × 3·81 cm.

Armature Winding:

Simple lap—six circuit.

Number of conductors per slot 6; turns per coil 1.

Area of conductor 31·2 mm^2; resistance of armature winding 0·0081 ohm.

Commutator:

Diam 40·6 cm; 23·5 cm face; 6 brush spindles; 8 brushes per spindle each 15·9 × 22·2 mm; number of segments 252.

Main Poles:

Electrical sheet steel laminations; width 16·5 cm; axial length 20·1 cm; air gap 3·2 mm.

Shunt Winding:

Number of turns per coil 630; area of wire 3·243 mm^2; resistance of winding, cold, 24·2 ohms.

Series Winding:

Number of turns per coil 2; area of conductor 321·5 mm^2; resistance of winding, cold, 0·001 ohm.

Interpoles:

Dimensions of pole, 5·08 × 30·5 cm; of shoe 3·81 × 30·5 cm.

Number of turns per coil, 9½; area of conductor, 321·5 mm^2; resistance of winding, cold, 0·0093 ohm.

Design Data for 4-pole Motor by Laurence Scott and Electromotors, Ltd.

This motor is illustrated by Fig. 276.

Output:

30 h.p.; 1,500 rev/min.

Input:

440 V; 57 A.

Armature Core Plates:

Ext. diam, 22·8 cm; core length, 20·3 cm; 31 slots.

Armature Winding:

Simple wave—2 circuit; coil span, 7, i.e. slot 1 to slot 8.
Number of conductors per slot, 18; turns per coil, 3.
Conductor, 3·55 × 2·03 mm; resistance of armature winding, hot, 0·191 ohm.

Commutator:

Diam 15·2 cm; width of track, 6·4 cm; number of brush arms, 4; 2 brushes per arm each, 1·27 × 2·22 cm.
Number of segments, 93.

Shunt Winding:

Number of turns per coil, 2,930; diam of wire, 0·61 mm; resistance of winding, hot, 550 ohm.

Series Winding:

Number of turns per coil, 9; conductor 3·05 × 7·72 mm; resistance of winding, hot, 0·0094 ohm.

Interpole Winding:

Number of turns per coil, 42; conductor, 3·05 × 7·72 mm; resistance of winding, hot, 0·065 ohm.

Design Data for 2-pole Motor by Laurence Scott and Electromotors, Ltd

Output:

5 h.p.; 1,500 rev/min.

Input:

440 V; 10 A.

Armature Core Plates:

Ext. diam, 14 cm; core length, 14 cm 24 slots, each 0·8 × 2·3 cm.

Armature Winding:

Simple lap—2 circuit; coil span, 11, i.e. slot 1 to slot 12.
Number of conductors per slot, 60; turns per coil, 10; conductor diam, 1·02 mm; resistance of armature winding, hot, 3·5 ohm.

Commutator:

Diam, 9·8 cm; width of track, 4·13 cm; number of brushes per arm, 2; each 1·27 × 1·59 mm.
Number of Segments, 72.

Shunt Winding:

Number of turns per coil, 6,400; diam of wire, 0·376 mm; resistance of winding, hot, 1,360 ohm.

Series Winding:

Number of turns per coil, 52; conductor, 1·14 × 2·54 mm; resistance of winding, hot, 0·415 ohm.

Interpole Winding:

Number of turns per coil, 225; conductor, 1·14 × 2·54 mm; resistance of winding, hot, 0·93 ohm.

CHAPTER XX

EDDY CURRENT LOSSES IN ARMATURE CONDUCTORS

THIS chapter, which is in the nature of an appendix, discusses in more detail the eddy current losses in armature conductors to which reference was made in Chapter XVII.

The current in the armature conductors of a d.c. machine alternates at a frequency $f = np$, and, for straight-line commutation, has the trapezium wave-shape shown in Fig. 308 in which 2β is the

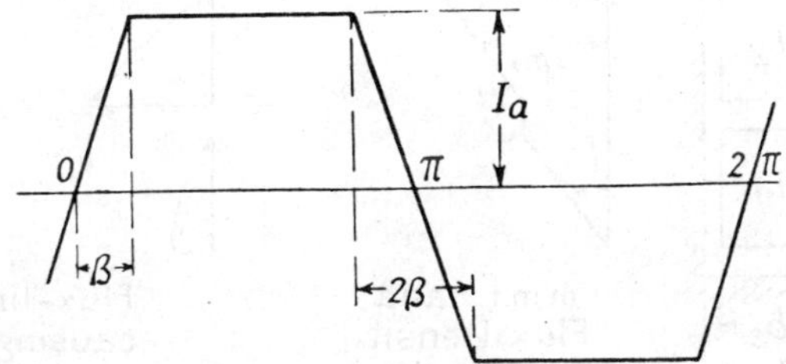

FIG. 308. WAVE-SHAPE OF CURRENT IN ARMATURE CONDUCTOR

angle corresponding to the time of commutation. This wave-shape is far from sinusoidal, the lower harmonics—third, fifth and seventh —being very prominent. The analysis of this waveform gives

$$i_a = \frac{4}{\pi} I_a \left\{ \frac{\sin \beta}{\beta} \sin \omega t + \frac{1}{3} \frac{\sin 3\beta}{3\beta} \sin 3\omega t + \ldots \right\} . \quad (267)$$

where $\omega = 2\pi f$.

Eddy Current Losses with Sinusoidal Currents. So far as the eddy currents in the conductors are concerned the effect of the total current can be determined by considering each of the harmonics separately. It is therefore desirable to consider first sinusoidal currents in the "embedded" portion of the conductors, i.e. the part in the slots. The diagram Fig. 309a, shows a winding with four similar conductors arranged, for convenience, in one vertical layer and carrying equal currents. The m.m.f. acting across the slot varies progressively from zero at the bottom to $(i_1 + i_2 + i_3 + i_4)$ at the top. Assuming initially that the current density is uniform, the m.m.f. will increase linearly, except for the small "steps" at the spaces between the conductors as shown. The flux density will be proportional to the m.m.f. since the reluctance of the slot is far greater than that of the iron. The flux density thus varies in a similar manner to the m.m.f. and the portion of this slot leakage flux linked with each conductor thus varies progressively, but not uniformly, from the top to the bottom of the conductor. As a result the e.m.f. induced in the conductor by this sinusoidally

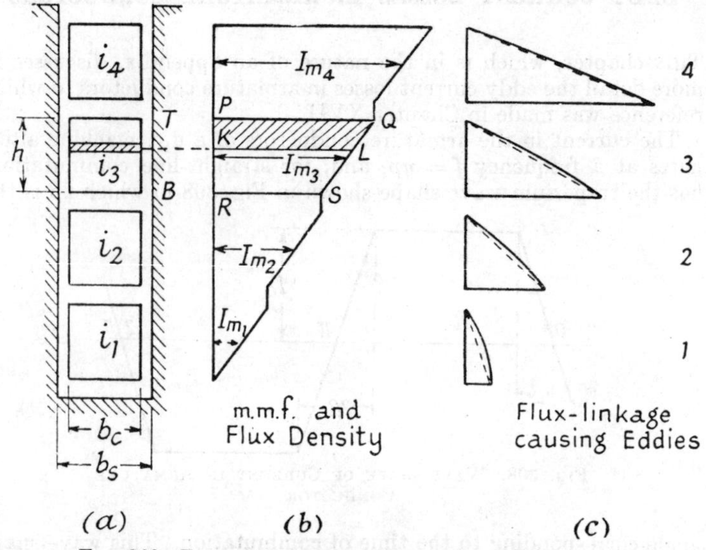

FIG. 309. FLUX-DENSITY AND FLUX-LINKAGE DUE TO CURRENTS IN ARMATURE CONDUCTOR

alternating flux also varies over the depth of the conductor so that eddy currents are produced.

The flux density at the top of the slot is

$$\frac{4\pi}{10^7}\,\frac{\text{Ampere-conductors per slot}}{\text{Slot width in metres}}\ \text{Wb/m}^2$$

and for a fairly large d.c. machine this will be of the order of 0·1 Wb/m² or 1,000 gauss. Considering one of the conductors—say No. 3—the flux linked with its bottom layer exceeds that linked with its top layer by the flux actually traversing the conductor No. 3 itself, i.e. by the flux represented by the trapezium *PQSR* in Fig. 309*b*. For the elemental strip shown the corresponding flux-linkages are represented by the area *PQLK*. The m.m.f. acting at the bottom B is $F_B = (i_1 + i_2)$, i.e. the total m.m.f. due to the conductors below the one under consideration, and the m.m.f. at the top T is $F_T = (i_1 + i_2 + i_3)$. The actual distribution of the current in the lower conductors 1 and 2 has no influence whatever on F_B or F_T. Further, any current in the conductors above 3 establishes no flux whatever in this conductor. As however the actual current density in 3 itself is not uniform, the m.m.f. and flux density will not increase uniformly from B to T, for the moment however, this variation will be neglected and the results of a rigid solution deferred until later.

Fig. 309c illustrates the manner in which the flux linked with the elemental strip of the conductor—as represented by the area of the trapezium $PQKL$—varies over the depth of the four conductors. The actual graphs are parabolic, but their deviation from straight lines is trivial except in the case of conductor No. 1 at the bottom of the slot. For this conductor the leakage flux itself is small and the eddy currents relatively unimportant. For the moment therefore the graphs will be assumed to be linear and then, as a rough approximation, we can take the eddy current loss as proportional to the square of the flux density at the mid-section of the conductor, it being understood that this will result in some error for conductor No. 1. This approximation leads to the conclusion that for similar conductors carrying equal currents the ratios of the eddy current losses in conductors 1, 2, 3 . . are $1^2, 3^2, 5^2$. . . . This relationship is in fact substantially correct for the depths of conductor normally used in a.c. machines, except for No. 1 conductor in which the losses are relatively about 7 per cent greater. With large d.c. machines the most common case is of a two-layer winding with deep conductors. For such machines, with sinusoidal current, the above approximation indicates that the ratio of the eddy current losses in the upper and lower layers of the winding is $3^2 : 1^2$ or, more nearly $3^2 : 1{\cdot}07$, i.e. 8·4, again, this is not far from the actual value.

The actual eddy current loss per unit volume could be calculated, to a rough approximation, in terms of the average flux density permeating the conductor. This would lead to precisely the same expression as given in equation (110), page 124, although by rather different reasoning. Thus the loss per unit volume is $1{\cdot}64\, B^2_{max} f^2 h^2/\rho$ watts if h is small and m.k.s. units are used. Again this result is sufficiently accurate to use with normal a.c. windings. There is no serious error in the loss so computed unless its value is too large to permit the use of such a conductor. Expressed in this form the eddy current loss is seen to vary directly with the square of the height h of the conductor and, in general, to conform to expectations. This tends to be obscured in the more usual method of expressing these losses in terms of a loss factor K_o such that the total losses, including those due to eddy currents, are given by $K_o \times$ (nominal loss) $= K_o I_a^2 r$, where r is the actual resistance.

Based on the above value for the loss per unit volume it can readily be deduced that, if K_e is the ratio of the eddy current loss to the nominal loss

$$K_e = \frac{1}{12}\left(\omega \frac{\mu_0}{\rho}\frac{b_c}{b_s}\right)^2 h^4 \left(\frac{I_m}{I_a}\right)^2 = \frac{1}{12}(ah)^4\left(\frac{I_m}{I_a}\right)^2 \quad . \quad (268)$$

where b_c/b_s is the ratio of conductor width to slot width, I_m is the nominal m.m.f. at the mid-section, I_a the actual current in the conductor, and $a^2 = \dfrac{\omega\mu_o b_c}{\rho b_s}$, so that a is a constant.

Thus for conductor 2, 3, 4 . . ., $I_m/I_a = 1{\cdot}5$, $2{\cdot}5$, $3{\cdot}5$. . . . Hence for conductor No. 1, $K_e = (ah)^4/48$ and for conductor No. 2, $K_e = 9(ah)^4/48$.

The above are in m.k.s. units hence $\mu_o = 4\pi \times 10^{-7}$ and at normal working temperature $\rho = 2{\cdot}04 \times 10^{-8}$.

Hence $a = 19{\cdot}65\sqrt{fb_c/b_s}$ and $K_e = 1{\cdot}25 \times 10^{-4}\left(\dfrac{b_c}{b_s}\right)^2 f^2h^4(I_m/I)_a{}^2$ where h is in metres. If the height is expressed in centimetres $K_e = 1{\cdot}25 \times 10^4\left(\dfrac{b_c}{b_s}\right)^2 f^2h'^4(I_m/I_a)^2$.

For conductors carrying equal currents the factor for the nth conductor or layer is thus

$$K_e = 0{\cdot}31 \times 10^4(2n-1)^2\left(\frac{b_c}{b_s}\right)^2 f^2h'^4 \quad . \quad . \quad (269)$$

where n is counted from the bottom of the slot.*

The drawback to any approximate method, such as that given above for determining K_e, is that it gives no indication of the limits of its application. These limits are determined in this case by the secondary effects produced by the eddy currents which cause the losses to be less than those computed above. For all ordinary purposes the error in K_e as given by equation (268) is of no practical importance if ah does not exceed 1·6, corresponding to conductors about 1·9 cm deep at a frequency of 25 and about 1·3 cm at a frequency of 50. The important higher harmonics in the current wave in Fig. 308 may however correspond to much higher values of ah, so that to deal with them a more precise method is necessary.

For computing accurately these eddy current losses in the armature conductors of a.c. machines, designers employ loss curves calculated and published by A. B. Field in 1905. Since Field's classic paper others have been published, notably by W. V. Lyon.† Using Lyon's results the loss curves published by Field can be extended to cover the higher values of ah more expeditiously than with Field's original formulae. It is not proposed to outline Lyon's development of the subject here, but merely to state that he calculates the *total loss factor* for the nth conductor in a single-layer winding in the form:

$$K_o = M_r + N_r n(n-1)$$

where M_r and N_r are functions of ah tabulated on page 478‡ and given in Fig. 310.

* As already stated this value is less accurate for $n = 1$ than for values of n greater than 1.

† *Trans. A.I.E.E.* Vol. 40, p. 1361.

‡ As given by E. A. Hanney: *J.I.E.E.* Vol. 71, p 282.

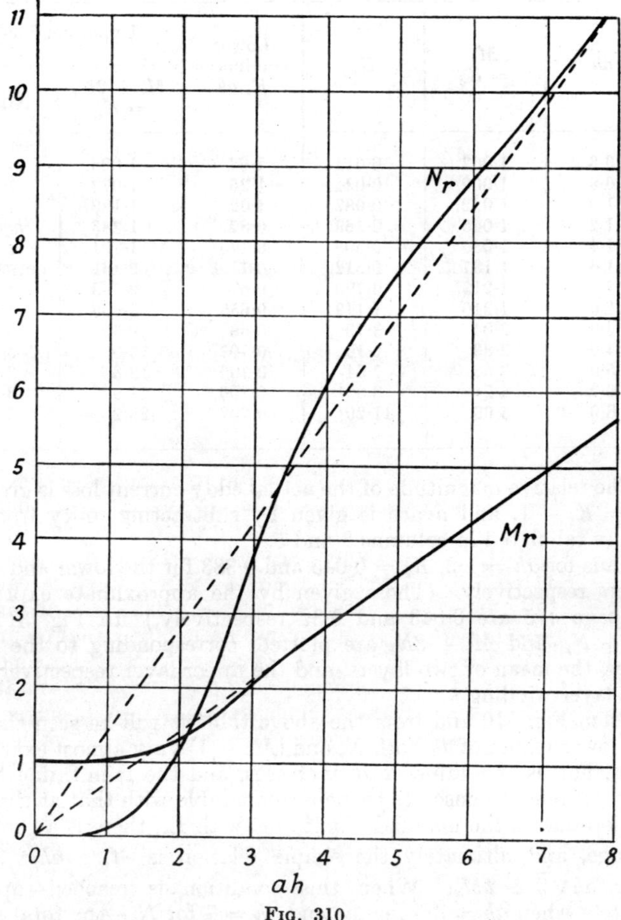

FIG. 310

For conductor No. 2 $K_o = M_r + 2N_r$
" " No. 3 $K_o = M_r + 6N_r$
" " No. 4 $K_o = M_r + 12N_r$

and the average value with Z_s conductors per slot in a single-layer winding is

$$K_o = M_r + \frac{(Z_s^2 - 1)}{3} N_r \quad . \quad . \quad . \quad (270)$$

The value $K_u = M_r + 2N_r$ corresponds to the upper layer of a two-layer winding and is also given in the Table. For the lower layer the loss factor is $K_e = M_r$.

ah	$M_r = K_e$	N_r	Lower conductor K_e/ah	Upper conductor	
				$M_r + 2N_r = K_u$	K_u/ah
0·6	1·003	0·011	1·67	1·025	1·71
0·8	1·009	0·034	1·26	1·077	1·35
1·0	1·022	0·082	1·02	1·186	1·19
1·2	1·045	0·169	0·87	1·383	1·15
1·4	1·085	0·308	0·775	1·701	1·21
1·6	1·137	0·512	0·71	2·161	1·35
1·8	1·213	0·790	0·67	2·793	1·55
2·0	1·310	1·149	0·655	3·608	1·80
3·0	2·04	3·73	0·68	9·50	3·17
4·0	2·83	6·10	0·707	15·0	3·75
5·0	3·53	7·64	0·706	18·8	3·76
6·0	4·24	8·82	0·706	21·9	3·65
8·0	5·66	11·29	0·707	28·2	3·53

The relative magnitude of the actual eddy current loss is given by $K_e = K_o - 1$, and hence is given by subtracting unity from the values tabulated in columns 2 and 5.

Thus for $ah = 1{\cdot}2$, $K_e = 0{\cdot}045$ and $0{\cdot}383$ for the lower and upper layers respectively. (Those given by the approximate expression on page 475 are 0·043 and 0·39 respectively.) In Fig. 311 M_r, $M_r + N_r$, and $M_r + 2N_r$ are plotted, corresponding to the lower layer, the mean of two layers, and the upper layer respectively of a two-layer winding.*

From Fig. 310 and from the above table it will be seen that for the lower values of ah both N_r and $(M_r - 1)$ vary almost exactly as $(ah)^4$, but as the value of ah increases, and the m.m.f. due to the eddy currents themselves become comparable with that at the mid-section due to the nominal m.m.f., i.e. with I_m, there is a profound change, and ultimately the simple relation is $M_r \simeq ah/\sqrt{2}$ and $N_r \simeq ah\sqrt{2} \simeq 2M_r$. When this condition is reached—approximately when $ah = 3{\cdot}5$ for M_r and $ah = 7$ for N_r—the total loss is not affected by any further increase in the conductor depth; increasing the depth results in an increase in the eddy current loss just compensating for the reduction in the nominal loss.

Since the nominal loss $\propto 1/h$, K_o/ah is a measure of the actual total loss and its value has been tabulated for both the upper and lower conductors in columns 4 and 6. From the values there given it is apparent that for the lower conductor the total loss changes very little for all values of ah from 1·8 upwards, the actual minimum

* They are in effect Field's curves extended to cover larger values of ah, it being noted that ah is used instead of mh as used by Field, a being $m\sqrt{2}$. From Fig. 310 or the table, the reader can readily construct graphs for $M_r + 6N_r$, etc., should he so desire.

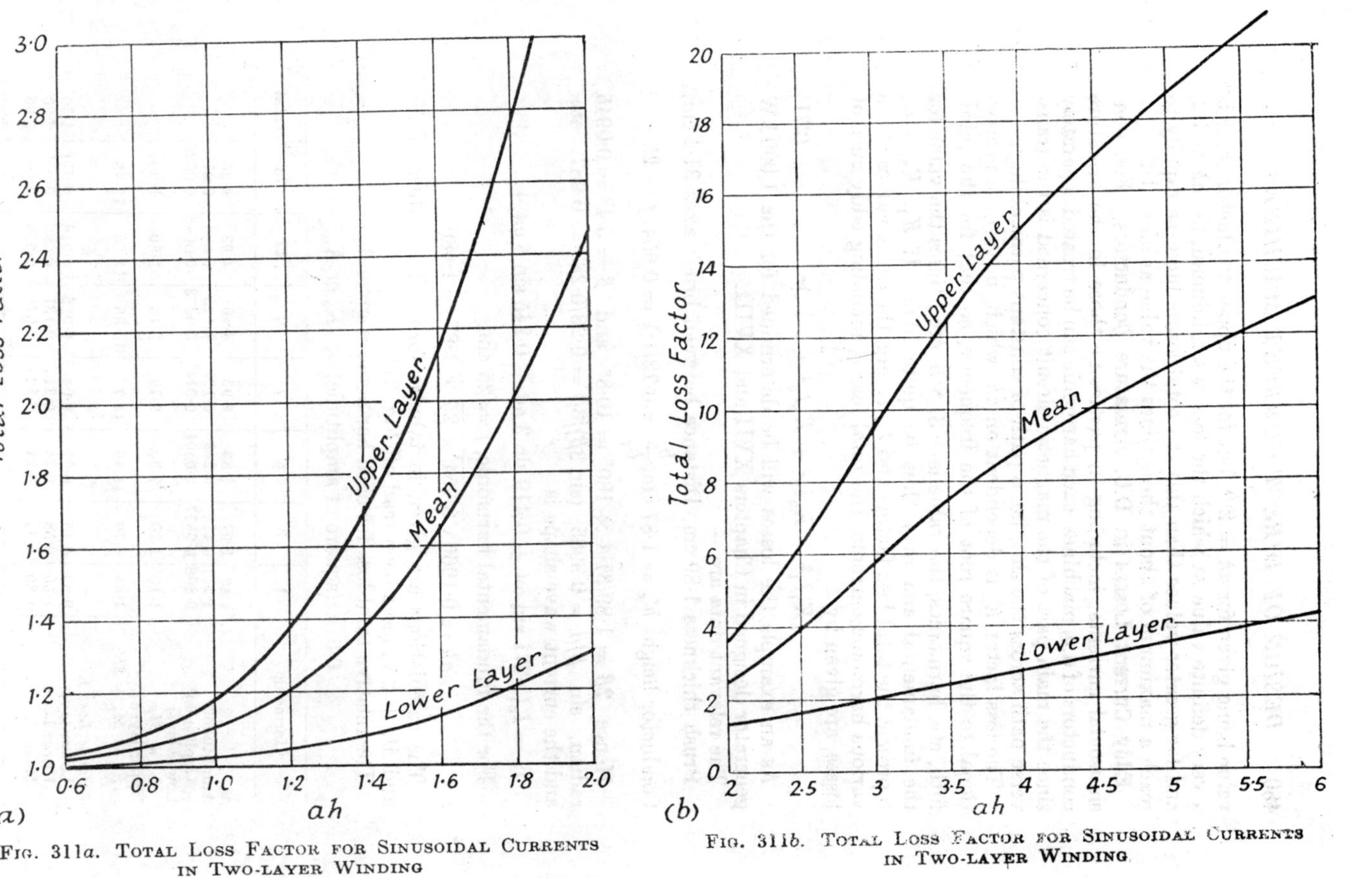

FIG. 311*a*. TOTAL LOSS FACTOR FOR SINUSOIDAL CURRENTS IN TWO-LAYER WINDING

FIG. 311*b*. TOTAL LOSS FACTOR FOR SINUSOIDAL CURRENTS IN TWO-LAYER WINDING

value being given for $ah = 2{\cdot}0$, but for the upper conductor there is a very definite value at which the loss is a minimum, i.e. $ah = 1{\cdot}2$, and for greater values than this the total losses increase rapidly to reach a maximum of about three times the value at $ah = 1{\cdot}2$.

Eddy Current Losses in D.C. Armature Conductors. For non-sinusoidal currents, including in particular those in the armature conductors of d.c. machines, each harmonic can be treated separately since the main part of the magnetic circuit concerned is the transverse path across the slot, i.e. a path of constant permeability.

The loss factor K_o is dependent on ah, which, in turn, is proportional to the square root of the frequency, so that for the third, fifth, etc. harmonics, the value is $\sqrt{3}$, $\sqrt{5}$, etc., times the value for the fundamental and may thus be quite high. If K_1, K_3, etc., represent the total loss factors and I_1, I_3, etc. the r.m.s. values of the various harmonic currents, the total losses, including eddy current losses, are given by

$$(K_1 I_1^2 + K_3 I_3^2 + K_5 I_5^2 + \ldots)r \qquad (271)$$

As an example the losses will be determined for the 1,000 kW generator designed in Chapters XVII and XVIII.

The relevant data are—

Brush thickness 1·89 cm, Distance between brush axes 31·4 cm, Conductor height $h'_c = 1{\cdot}87$ cm, $\frac{b_c}{b_s} = 0{\cdot}72/1{\cdot}1 = 0{\cdot}654$, $f = 25$.

Hence $2\beta = 1{\cdot}89/31{\cdot}4 \times 180° = 10{\cdot}8°$ and $\beta = 5{\cdot}4° = 0{\cdot}0945$ radian, $\sin\beta/\beta = 0{\cdot}9985$ $(\sin 3\beta)/3\beta = 0{\cdot}28/0{\cdot}283 = 0{\cdot}987$, etc. and the current wave-shape is

$$i_a = I_a\{1{\cdot}271 \sin \omega t + 0{\cdot}419 \sin 3\,\omega t + 0{\cdot}245 \sin 5\,\omega t + \ldots\}$$

For the fundamental harmonic, $f = 25$ and

$$ah = 0{\cdot}1965\sqrt{0{\cdot}654 \times 25} \times 1{\cdot}87 = 1{\cdot}486$$

The calculations are given in tabular form. The values of M_r and $M_r + 2N_r$ have been read off Fig. 311.

The relative total loss for each harmonic is given by

$$0{\cdot}5 \times (\text{square of amplitude}) \times (K_e \textit{ or } K_u)$$

Harmonic	1	3	5	7	9	11	13	Total
ah	1·49	2·57	3·32	3·93	4·46	4·93	5·35	
Amplitude	1·271	0·419	0·245	0·169	0·125	0·096	0·075	
½(Amplitude)²	0·808	0·0877	0·0301	0·0143	0·0078	0·0046	0·0028	
Lower layer $K_e = M_r$	1·105	1·70	2·30	2·79	3·15	3·50	3·80	
Upper layer $K_u = M_r + 2N_r$	1·88	6·70	11·45	14·7	16·88	18·60	19·95	
Relative Loss K								
Lower layer	0·89	0·15	0·07	0·04	0·025	0·016	0·011	1·20
Upper layer	1·50	0·58	0·34	0·21	0·131	0·085	0·056	2·90
Mean	1·20	0·36	0·21	0·12	0·078	0·050	0·033	2·05

In the lower conductors the total loss is thus $1{\cdot}2I^2r$, and in the upper conductors $2{\cdot}9I^2r$, the mean value being $2{\cdot}05I^2r$. The actual eddy current losses are $0{\cdot}2I^2r$ and $1{\cdot}9I^2r$ respectively, so that the eddy current loss for the upper conductors is about 9·5 times as large as for the lower conductors. This holds, of course, only for the embedded part of the coils. If the conductor is of uniform sectional area throughout, the total loss factor for the complete winding becomes

$$K_t = 1 + (K_o - 1) \times L/(L + 2l_{fr}) \qquad (272)$$

For the machine being investigated $L' = 34{\cdot}5$ and $L' + 2l_{fr}' = 100{\cdot}5$ (both in cm) whence

$$K_t = 1 + (2{\cdot}05 - 1)\ 34{\cdot}5/100{\cdot}5 = 1{\cdot}36$$

Due to the eddy current losses in the embedded conductors the copper loss on full load is thus increased from its nominal value of 16·6 kW to 1·36 × 16·6 kW = 22·6 kW.*

The actual mean square value of the current, for the trapezium wave is $(1 - 2\beta/3\pi)I^2$ which for this case is equal to $0{\cdot}96I^2$.

The very great effect of the harmonics of the lower orders—up to the ninth—is apparent, but it is also evident that, for all the harmonics higher than the third, the value of ah corresponds to the second part of the curves of Fig. 311 and not to the parts for which the approximate method given previously holds. The example has been worked out in detail, without any short cuts, to illustrate the principles involved. It has been based on a single conductor in each layer, whereas normally there are two or more coil-sides in each layer. In such cases the currents in adjacent coil-sides do not pass through zero at the same instant, the difference in phase corresponding to the number of commutator segments per pair of poles. For example, with 45 segments *per pole* the phase difference between currents in adjacent coil-sides is 180°/45 = 4° for the fundamental and $4n°$ for the nth harmonic. Due to this phase displacement the total m.m.f. acting transversely across the slot is slightly reduced, being due to the vector sum of the currents in the layer. On the other hand a radial field component—parallel to the slot axis—is established, due to the vector difference in currents in adjacent coil-sides. Neither effect is of major importance and the reduction in eddy currents due to the transverse field is offset by eddy currents established by the radial field. No error of practical importance can then arise by the use of the above method. Further, minor changes in β have but little effect on the total losses and it is therefore sufficient to employ loss curves calculated for an average value of β. Such loss curves† based on $2\beta = 12°$ are given in Fig. 312 for values of ah ranging from 0·8 to 2·0, where ah is based on the

* For very many years the full importance of these losses was not realized, and they were simply included as "stray load losses" and often underestimated.

† Such a loss curve was first given by R. E. Grime, *J.I.E.E.* Vol. 54, p 392 *See also* E. A. Hanney, *J.I.E.E.* Vol. 71, p. 274.

fundamental frequency. Supplementary scales are indicated from which the loss factor can at once be read off for any given depth of conductor, the scale being based on a normal working temperature (70°C) and a normal value of the slot width space factor b_c/b_s of 0·72.

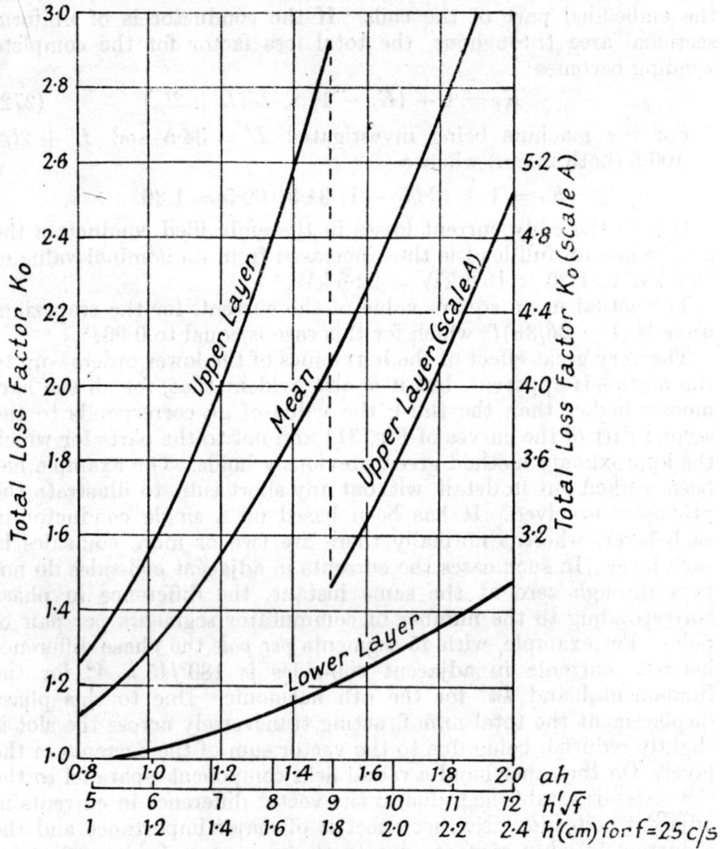

Fig. 312. Total Loss Factor for Embedded Portion of Two-layer Winding of D.C. Machine

The Variation in Total Losses with Conductor Depth. It is evident from the above that it is the upper layer of conductors in which the eddy current losses predominate, so that layer will be discussed first. In Fig. 312 the total loss factor for the upper conductor is 2·94 when the depth is 1·8 cm and $f = 25$, but is only 1·3 with a depth of 1·0 cm. For this latter conductor the nominal loss is 1·8 times its value for the deeper conductor, so the total "embedded" loss actually falls when the smaller conductor is used to 1·8

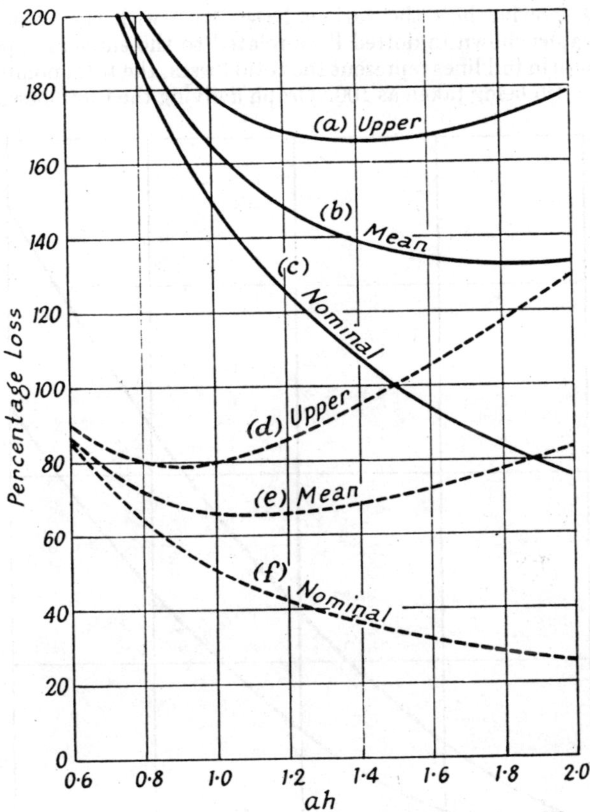

FIG. 313. PERCENTAGE LOSSES FOR COMPLETE TWO-LAYER WINDIN

$\times$ 1·3/2·94 = 0·8 of its value with the deeper conductor. Minimum loss in the embedded portion will occur when K_o/h is a minimum. For the upper conductor this condition is reached when $ah = 0{\cdot}9$ approximately, corresponding to a depth of just over 1 cm when $f = 25$, the depth varying inversely with frequency. The effect of the third and fifth harmonics is largely responsible for this value of ah being much less than for sinusoidal currents.

This minimum depth is too small for use in practice, especially if the same sectional area is used for both the embedded and "free" portions of the conductor, as the losses in the free portion would be far too great. With the smaller conductor the temperature gradient across the slot insulation would be greater. If the sectional area of the conductor is constant throughout, minimum total loss occurs when K_t/h is a minimum, and this depends upon both K_t and L/l_a. To illustrate this matter Fig. 313 has been prepared giving graphs showing how the various losses vary with the value of ah. The graphs shown in dotted lines related to the embedded portion, and those in full lines represent the total losses, the total nominal loss at $ah = 1{\cdot}5$ being taken as 100. Graph d giving the total losses in the

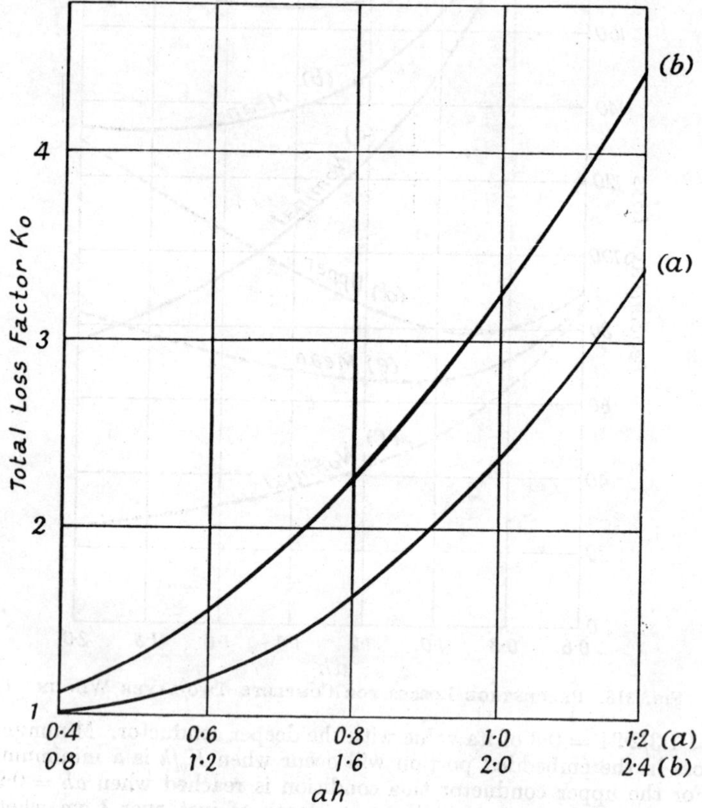

FIG. 314. TOTAL LOSS FACTORS FOR (*a*) LAMINATED WINDING WITH TWO LAMINATIONS PER CONDUCTOR, (*b*) UN-LAMINATED CONDUCTOR

buried portion of the upper conductor indicates that minimum loss takes place when $ah = 0{\cdot}9$. Graphs a, b, and c representing the total losses are based on $L/l_a = 0{\cdot}34$, corresponding to the example worked out above. Minor changes in this ratio have no practical effect on the general form of the graphs concerned. From a, minimum total loss in the upper conductor corresponds to $ah = 1{\cdot}4$, but there is little change in the losses from $ah = 1{\cdot}2$ to $1{\cdot}7$.

Graphs b and e represent the mean of upper and lower layers and therefore represent the total losses in the winding in terms of those given by *c* and *f*. From *b*, minimum total loss obtains at $ah = 1 \cdot 85$, but there is little change in the losses from $ah = 1 \cdot 6$ to $2 \cdot 0$.

The total embedded loss in the complete slot is a minimum at $ah = 1 \cdot 1$ and varies but little from $ah = 0 \cdot 9$ to $1 \cdot 4$. The losses in the lower conductor have not been plotted for clarity of the diagram. Their value can readily be deduced from the *upper* and the *mean* graphs.

Sometimes with very large machines two conductors are used in parallel instead of a single conductor of twice the depth. The conductors are then insulated throughout their length and parallelled only at the commutator. With a coil formed in the usual way with a "nose" at the back end this results in the one coil consisting of the top conductor in one slot and the bottom in the other, whilst the second coil consists of the two middle conductors, so that the e.m.f.'s induced by slot leakages fluxes in the two coils are approximately equal. There are thus four conductors arranged in one radial layer, each carrying one half of the total current. The reduction in eddy current loss thereby obtained depends on the nominal value of *ah*, but, owing to the higher harmonics in the current wave, the benefit is appreciably less than in a sinusoidal current of the main frequency. Fig. 314 gives in graph (*a*) the mean loss factor for such a four conductor arrangement and in (*b*) the corresponding graph for the ordinary two-layer winding. The scales are such that an immediate comparison is possible of the loss factors for equivalent single conductor and two conductors in parallel arrangements.

EXAMPLES

THE questions marked * are based on or taken from London University Papers and those marked † are taken from I.E.E. Papers and are reproduced by kind permission of these Authorities. The Authorities concerned are, of course, in no way committed to approval of the answers given to numerical problems.

CHAPTER I

1. Calculate the maximum value of the e.m.f. induced in the conductors of an armature 120 cm diam and 30 cm axial length, rotating at 400 rev/min. The maximum flux density in the air gap is 0·95 Wb/m^2.

2. A 12-pole machine is driven at 250 rev/min; determine the average value of the e.m.f. induced in each armature conductor when the useful flux per pole is 0·096 Wb.

3. Calculate the number of conductors required on the armature of a 4-pole motor, to generate an e.m.f. of 480 volts at a speed of 750 rev/min, when the flux per pole is 0·018 Wb. There are two paths in parallel in the armature winding.

4. Determine the value of the induced e.m.f. for a 10-pole armature winding having 1,200 conductors and 10 paths in parallel. The flux per pole is 0·09 Wb, and the speed 350 rev/min.

5. Determine the value of the force on the armature conductors in Ex. 1 when the current carried is 240 A.

6. If the total current carried by the armature of Ex. 4 is 2,000 A, calculate the work done, in ft lb, during one complete revolution.

7. Determine the value of the torque established by the armature of a 4-pole motor having in all 774 conductors, two paths in parallel, 0·024 Wb per pole, when the total current is 50 A.

*8. The 2-circuit armature of a 4-pole generator has 51 slots, each containing 20 conductors. What will be the voltage generated in the machine when driven at 1,500 rev/min, assuming the useful flux per pole to be 0·007 Wb?

*9. Find from first principles the voltage of a 4-pole lap-wound d.c. generator having a flux of 0·02 Wb and 560 conductors, when driven at 980 rev/min. How is the terminal voltage affected by load?

CHAPTERS II AND III

10. Compare the magnetic and electric circuits, pointing out the differences as well as the analogies.

*11. Discuss a method of estimating the reluctance of the air gap in the case of slotted armatures. Work out the ratio of the reluctance to that of a smooth-core machine in the case where the slot pitch is 55 mm, the slot width 20 mm, and the air gap 4 mm.

12. How is the apparent flux density in a highly saturated tooth on an armature core corrected in order to obtain the true density? Determine the apparent density in the case when the true density is 2·15 Wb/m^2 the effective slot pitch 2·8 cm, the slot width 1 cm, the gross length 35 cm, the number of ducts 5, each 1 cm wide, and the iron space factor 0·89.
(Use the magnetization curve in Fig. 8.)

*13. Explain the method of calculating the open-circuit magnetization curve of a direct current generator.

14. Explain what is meant by magnetic leakage, and indicate a method by which it can be estimated.

*15. Choose a suitable number of conductors, and draw a developed view of a 4-pole, 2-circuit armature winding, showing the positions and polarity of the poles, the position of the brushes, the direction of rotation, and the direction of the induced e.m.f.

16. Obtain a suitable commutator pitch for the following 2-circuit simple wave windings—

(a) 4 poles: 129 commutator segments.
(b) 6 poles: 128 commutator segments.
(c) 8 poles: 129 commutator segments.
(d) 10 poles: 351 commutator segments.

*17. Show that a d.c. armature with a 2-circuit wave winding, suitable for running in a field frame with an odd number of pole-pairs, can be wound so as to fill completely an odd number of slots, with the conductors grouped in fours, or multiples of four.

*18. A 40-h.p. 500-volt traction motor has four poles. There are only two brush arms. Assuming that there are 41 slots and 123 commutator segments. make a diagram sufficient to show how the armature winding may be arranged and connected to the commutator. It is sufficient to show the connexions of three consecutive coils.

19. The total number of conductors in series for a 2-circuit armature winding is required to be approximately 380, a tolerance of 5 per cent more, or less, being permitted. The number of slots is to be between 39 and 45, and of commutator segments between 117 and 135. Determine and give particulars of a suitable winding. There are 4 poles.

20. What is the object of equalizing connexions on a d.c. generator? To what class of winding are they applied? Give a diagram to indicate the manner in which these equalizers are connected to the armature winding.

21. The number of conductors in each circuit of an 8-pole simple-lap winding is to be within 5 per cent of 98; the number of slots per pole must lie between 12 and 18. Give details of a suitable winding, and give the full scheme of connexions to not less than 10 equalizing rings.

*22. Give a winding scheme for a 6-pole, 4-circuit, d.c. armature winding having a number of conductors as nearly as possible 600, arranged in a number of slots lying between 70 and 80. State whether the winding you give is made up to two separate circuits or not.

23. Determine a suitable winding scheme for a 4-circuit, 8-pole winding having between 340 and 360 commutator segments and single-turn coils; the number of slots is to be approximately 120.

Indicate the connexions that must be made to 12 equalizing connectors.

*24. Design a symmetrical 4-circuit wave winding for an 8-pole, 500-volt generator, assuming that the mean voltage between commutator segments is approximately 12 volts, and that each slot contains 6 conductors.

25. Determine a suitable winding scheme for a 4-circuit, 12-pole armature having between 340 and 360 commutator segments and single-turn coils, the number of slots to be approximately 180. Indicate the connexions that must be made to 12 equalizers.

26. Determine a suitable winding scheme for a 6-circuit, 12-pole winding having between 520 and 540 commutator segments, the number of slots to be approximately 180. Indicate the connexions that must be made to 11 equalizers.

27. Describe the use of Carter's coefficient in calculating the gap ampere-turns for a d.c. generator with a slotted armature.

Make a flux plot (as accurately as freehand sketching will permit) to scale, four times full size, showing the distribution of the no-load gap flux over half a slot pitch for a machine having a gap length 0·25 in., slot pitch 2 in., slot width 1 in., slot depth 2·5 in. The iron boundaries may be assumed to be equipotentials.

From the plot estimate Carter's air-gap coefficient corresponding to this slotting.

*28. Explain how the ampere-turns required for the air-gap, armature core, and teeth can be estimated in the case of a d.c. machine, the flux per pole

being given. Assume the slots to be parallel-sided and the armature core to be assembled with radial ventilating ducts.

***29. Explain with the aid of diagrams how the individual coils on a 4-pole d.c. armature may be connected to form *either* (i) a simplex lap winding, *or* (ii) a two-circuit wave winding.**

Deduce for the wave winding an expression for the commutator pitch.

Determine suitable numbers of conductors, slots and commutator segments for the armature of a 4-pole, 460-V, d.c. motor, the no-load speed of which is 720 rev/min. The flux per pole is approximately 0·024 Wb. Use a two-circuit wave winding.

†30. Explain the need for equalizing connexions used with d.c. armature windings. Draw a developed diagram of part of the winding of a 4-pole machine showing suitable equalizing connexions.

Show with the aid of sketches, the direction and the effect of currents that may be set up in these connexions.

†31. Explain and justify a method of determining the necessary ampere-turns per pole required for that portion of the magnetic circuit of a machine corresponding to the armature slots and teeth, given the necessary dimensions, the B/H curve for the iron, and the average gap density (flux per pole)/(area of pole shoe).

Show how a satisfactory allowance is made for the flux passing through the slot.

CHAPTERS IV AND V

32. Illustrate by means of graphs, and explain the manner in which the flux distribution in the air gap of a generator is dependent upon the value of the armature currents.

33. The voltage induced in a single turn of an armature winding is measured by connecting the coil to a pair of slip rings. Give graphs indicating the manner in which the voltage changes: (*a*) on no-load, and (*b*) on full load. Deal separately with the cases of generator and motor action.

34. The ampere-turns developed by the field coils of a generator are 3,000 per pole; armature reaction is equivalent to a maximum of 2,500 ampere-turns per pole. The air gap under the pole has the constant value of 0·5 cm, and the pole arc is 70 per cent of the pole pitch. Neglecting fringing and the reluctance of the iron parts, determine the flux density in the air gap at the two tips and in the centre.

35. Why are compensating windings used on certain d.c. machines? Explain their action, and the method of connecting up these windings, and give instances where their use is desirable.

36. Determine the value of the armature ampere-turns per pole for: (*a*) a wave-wound armature having 43 slots, 12 conductors per slot, 4 poles, the total current being 50 A.; (*b*) an 8-pole simple lap winding having 128 slots, 8 conductors per slot, the total current being 1,200 A.

*37. Explain the origin of "flashing over" in commutating machines, and show how the distortion of the field increases the possibility of its occurrence.

An armature develops 13,000 ampere-turns per pole; the pole arc is 70 per cent of the pole pitch; give the value of the m.m.f. that must be developed by the compensating winding to prevent distortion over the whole pole arc.

38. A single-turn coil carrying a current of 200 A develops a flux of 5×10^{-4} Wb. Calculate its inductance.

The current is completely reversed in 0·002 sec. Calculate the average value of the reversing e.m.f. (The resistance may be neglected.)

39. A coil has 1,500 turns. When 4 A are passed through it the total magnetic flux threading the coil is 0·06 Wb. The resistance of the coil is 20 ohms. Find an expression for the current in the coil immediately after it is switched on to a 100-volt d.c. supply. Assume that the inductance of the coil is constant.

40. Determine an expression for the energy stored in the magnetic field

established by a coil of N turns, wound uniformly upon an iron ring of mean diameter D and sectional area A, for a current I.

41. Calculate the value of the energy stored in the magnetic field established in the air gap of a dynamo, the pole face being 30×35 cm², the air gap 0·5 cm, and the ampere-turns absorbed by the air gap, 4,000.

*42. What is "magnetic hysteresis"? Show that when any material is subjected to cyclic changes of magnetism, a loss of energy is involved proportional to the area of the hysteresis loop. Calculate the loss of energy in one hour in 1 kg of iron when subjected to cyclic magnetic changes. The frequency is 25, the area of the loop represents 2,400 ergs per cm³, the density of the iron is 7·8.

43. Explain the difference between "alternating" and "rotating" magnetization, and give and explain graphs showing how the loss due to magnetic hysteresis varies with the maximum flux density for the two forms of magnetization.

44. If the coil in Ex. 39, when carrying a current of 5 A, has a steady reversing voltage of 50 applied, calculate the value of the current 1·2 sec after applying the reversing voltage.

45. As for Ex. 44. but calculate the time taken for the current to be reduced exactly to zero.

46. The total iron losses occurring in direct current machines are found to exceed considerably the values that are obtained on test samples when subjected to alternating magnetization under similar conditions of maximum flux density and frequency. Give full reasons for this.

47. Deduce an expression for the loss of power that takes place due to eddy currents in a laminated core, when subjected to uniform alternating magnetization. State briefly the advantages of silicon steel alloys over ordinary iron for use in cores of totally-enclosed machines.

CHAPTER VI

48. Explain what is meant by "straight-line commutation," "accelerated commutation," and "retarded commutation." How is the current density under the "leaving edge" of the brush affected by the "type" of commutation?

49. The current in a coil undergoing commutation changes according to a parabolic graph represented by—

$$\text{From } t = 0 \text{ to } t = 0{\cdot}5T_c, \quad i = I\left\{1 - 4\left(\frac{t}{T_c}\right)^2\right\}$$

$$\text{From } t = 0{\cdot}5T_c \text{ to } t = T_c, \quad i = I\left\{(T_c - t)^2 \times \frac{4}{T_c^{\,2}} - 1\right\}$$

Calculate and plot graphs showing the manner in which the current density varies over the leaving and entering portions of the brush, respectively.

50. Discuss in detail the various factors that contribute towards the reactance of the coils undergoing commutation on a d.c. machine.

51. Discuss the reason for giving the brushes of non-interpole machines a "lead." What type of commutation curve will result if the brushes of such a machine are fixed at the geometric neutral axis?

*52. What are the functions of interpoles? How are they wound in (*a*) the case of motors, and (*b*) the case of generators?

*53. Sketch approximately graphs showing the resulting field distribution over two complete pole pitches for the case of a loaded interpole generator. Explain in detail the reasons for the shape of the flux distribution curve.

*54. How is calculation made of the number of ampere-turns necessary for the interpoles of a direct current generator?

55. Discuss magnetic leakage in so far as it affects interpoles. Why is the Hopkinson leakage coefficient very much greater for the interpoles than for the main poles?

Draw a diagram of a 4-pole interpole generator showing the paths of the magnetic flux in the yoke, main poles and interpoles, and the armature core. Show also some of the leakage lines.

*56. Explain the separate functions of (a) commutating poles, and (b) compensating windings when employed in d.c. machines. Draw diagrams showing the arrangement of the windings and their connexions and polarities. Under what circumstances would both be employed in a d.c. machine?

Outline the method of calculating the ampere-turns required for the commutating poles in an uncompensated machine.

*57. Distinguish between the functions of (a) commutating poles and (b) compensating windings in a d.c. machine.

Outline a method of calculating the required flux density in the air-gap of the commutating poles.

Determine the number of turns on each commutating pole of a 6-pole machine, if the flux density in the air gap of the commutating pole is 0·5 Wb/m² (5,000 gauss) at full load and the effective length of the air gap is 4 mm. The full load current is 500 A and the armature is lap wound with 540 conductors. Assume the number of ampere-turns required for the remainder of the magnetic circuit to be one tenth of that of the air gap.

*58. Explain with diagrams the process of commutation in a d.c. machine having more than two coil-sides per slot. In a machine with six coil-sides per slot the leaving edges of every third commutator segment are frequently burnt while those in between are not. What is the cause?

A 400-kW, 460-V, 6-pole d.c. generator has a lap-connected armature with 90 slots and 6 conductors per slot.

Determine the number of turns required on each interpole if the effective interpole air-gap is 1·0 cm and the flux density in the air gap under the interpole is 0·22 Wb/m² (2,200 gauss).

CHAPTER VII

*59. The relation between the excitation current and the e.m.f. generated by a d.c. shunt-wound generator, running on open circuit at 850 rev/min, is—

Exciting current, amperes	2	3	4	5	6
Induced e.m.f., volts	68	87	100·5	109	112·5

The resistance of the shunt field winding is 22·2 ohms. Find the voltage at the terminals of the machine when it runs at 850 rev/min self-excited.

60. The machine in the above example is driven at half speed, i.e. 425 rev/min, and its field coils grouped in two parallel circuits. A regulating resistance is included in the field circuit, and the machine builds up to a voltage of 50. What is the valu of the resistance in the regulator?

*61. Describe, with the aid of diagrams, the variations of terminal voltage which are observed when a shunt generator is started from rest, gradually brought up to a high speed, and then gradually brought to rest again Explain what determines the open-circuit terminal voltage of a shunt generator driven at its normal speed.

*62. Given the magnetization curve of a self-excited shunt generator, show how to determine the open-circuit voltage, with a given total resistance in the shunt circuit. Show how to determine the rise in voltage when the speed is increased by 10 per cent.

63. A shunt generator, when driven at its normal speed, fails to self-excite. Discuss briefly the most likely reasons for this, and state how you would proceed to remedy the fault.

64. Having given the necessary data, show how to calculate the sectiona area of the conductor to be used for the shunt winding of a generator. What considerations determine the numbe of turns to be employed on a shunt coil?

65. Give a reasoned estimate of th resistance of the shunt winding of—

(*a*) A 5-kW 500-volt shunt generator.
(*b*) A 5-kW 100-volt shunt generator.
(*c*) A 1,000-kW 500-volt shunt generator.
(*d*) A 1,000-kW 220-volt shunt generator.

Only an approximate figure is required; the machines in every case are to be assumed to be operating at good normal speeds.

66. As for the above example (*a*) and (*b*), but for a series machine.

CHAPTER VIII

67. A direct current generator is separately excited. Explain carefully the manner in which the terminal voltage will change when the load on the generator is increased from zero to a maximum of 150 per cent of normal full load current. The excitation is maintained at a constant value. Discuss the effect of the brush position upon the external characteristic of the generator.

68. A shunt generator is loaded so as to maintain a constant load current. Draw and explain a graph showing the manner in which the terminal voltage varies with the value of the exciting current.

69. Explain in detail the manner in which the terminal voltage of a shunt generator will change when the load resistance is reduced steadily in stages from infinity to zero. The speed is maintained constant, and the resistance in the field circuit is unaltered after the initial adjustment.

70. As for Ex. 69, excepting that the machine is a series generator, and no field diversion is used after the initial adjustment, if any.

71. Explain fully the action of a level-compounded generator. If such a generator were worked with its initial voltage 10 per cent above normal, by strengthening the shunt excitation, what type of load characteristic would result?

72. Two shunt generators are connected in parallel. Discuss in detail the manner in which they will share the load.

73. A d.c. turbo generator provided with compensating windings is intended to operate in parallel with an existing non-interpole shunt generator, which latter works normally with the brushes, given a slight lead. How may it be arranged that the machines shall share uniformly a load which is liable to rapid fluctuations?

74. Explain in detail the function of the equalizing bus bar used when compound generators operate in parallel.

*75. A d.c. generator has the following field windings—

(*a*) a separately-excited winding of 1,000 turns per pole and resistance 80 ohms supplied from a constant-voltage source of 110 V.

(*b*) a shunt winding of 500 turns per pole and resistance 200 ohms.

The open-circuit magnetization curve at 500 rev/min is given by—

Armature voltage (V)	154	302	396	458	505	538
Field AT per pole	500	1,000	1,500	2,000	2,500	3,000

Determine the no-load terminal voltage at a speed of 600 rev/min, both windings being connected to their respective sources.

Determine also the changes in the no-load terminal voltage due to the insertion of (*a*) a resistance of 120 ohms in series with the separately-excited winding and (*b*) a resistance of 150 ohms in series with the shunt field winding.

*76. The magnetization curve of a level-compound wound 100-kW, 500-V generator is given below—

Ampere-turns per pole	6,000	7,000	8,000	9,000
E.M.F. (volts)	455	510	540	555

The total armature circuit resistance is 0·075 ohm and the brush drop 1 V per brush.

On full load the armature reaction has a demagnetizing effect equal to 1,000 ampere-turns per pole.

Determine the number of series turns per pole required for level compounding.

If new series coils are fitted having three extra turns per pole, calculate the terminal voltage at the same value of full load armature current neglecting any change of resistance.

CHAPTER IX

77. Deduce from first principles an expression for the total torque developed by the armature of a d.c. motor.

*78. Draw and explain curves showing the manner in which the torque exerted by a motor depends upon the magnitude of the current in the case of (*a*) a series motor, and (*b*) a constantly excited shunt motor. What influence has the set of the brushes upon the curves?

*79. The magnetization curve of a 4-pole d.c. series motor, with a 2-circuit winding, was obtained by separately exciting the field with currents of the values given below, and loading the armature to take the same current, the speed being maintained constant at 750 rev/min.

Current, amperes	10	20	30	40	50	60	70	80
Voltage between brushes, volts	160	295	375	425	460	485	505	525

Determine the speed-torque curve for this motor over the whole range of loading for a line voltage of 550, having given that the total number of conductors is 660, the resistance of the armature circuit is 0·28 ohm, and that of the field coils 0·29 ohm.

*80. A 4-pole series motor has a magnetization characteristic given by the following figures—

Magnetizing current, amperes	10	20	30	40	50	60	70	80	90
Flux per pole, Wb $\times 10^{-2}$	0·29	0·55	0·74	0·85	0·92	0·97	1·01	1·06	1·08

If the armature is wave-wound and has 820 conductors, and the resistance of the field windings and armature is 0·15 ohm, estimate the speed at which the motor will run when supplied at a constant voltage of 230 when taking a current of (*a*) 45 A, (*b*) 85 A. Calculate also the torque in lb ft that the motor will develop when carrying these currents.

81. Explain, with the aid of graphs, the load characteristics of cumulative compound and differential compound motors. State, with reasons, the class of work for which each type is best suited.

82. Two motors are connected in series and mechanically coupled to the same load. Having given the necessary data with regard to flux, number of conductors, etc., show how the value of the working speed of the combination can be determined, and also of the load carried by each motor.

83. Two shunt motors are connected in parallel and mechanically coupled to the same load. It is found that on heavy overloads one of the motors carries by far the greater portion of the load. Explain fully the reason for this, and state what steps must be taken in order to obtain satisfactory working.

*84. The open-circuit magnetization curve of a 500-V d.c. compound-wound motor at its rated speed is as follows—

Shunt field current A	2	4	6	8	10	12
Generated e.m.f. V	180	340	460	540	600	640

On full load the series field contributes one third of the total excitation and the resistance drop is 20 volts. Determine, in terms of the corresponding values at full load, the gross torque and speed when the armature current is (*a*) 0·5, (*b*) 1·25 of its full-load value.

Neglect voltage drop at the brushes and armature reaction.

***85. Discuss the factors which must be given consideration in the design of a d.c. compound motor in which a 1:2 variation of speed at constant power output is to be obtained by variation of the shunt field.**

The characteristics, at 525 V, of a compound traction motor with full shunt and series field are—

Armature current A	25	100	150	200
Speed m.p.h. . .	12·8	11·3	10·6	10·1
Tractive effort lb .	435	1,950	3,100	4,240

Calculate the speed and tractive effort at 525 V and 200 A with the series field winding only. The resistances of the windings are: armature 0·1 ohm, series 0·05 ohm, shunt 350 ohms. The number of turns in each shunt coil is 1,200 and in each series coil 12.

CHAPTER X

86. How may the speed of a d.c. shunt motor be controlled? Explain in detail the happenings consequent upon reducing the flux of a shunt motor by 10 per cent, taking the case of a 220-volt motor having a resistance for its armature, etc., of 0·2 ohm carrying a current of 50 A prior to weakening the field. The load torque may be assumed to remain constant, and iron and friction losses may be neglected.

87. A shunt motor has to work over a wide range of speed, obtained by field control. When working with a very weak field, the motor shows a tendency to "flash over" between brushes. Explain fully the reason for this.

88. Explain in detail the method of "series-parallel" control of a pair of similar motors.

89. Deduce an expression for the loss of energy that takes place when a pair of motors are run up to speed, using series-parallel control, assuming that the current in each motor armature remains constant during the whole of the starting period.

90. A large mill-motor is controlled by the Ward-Leonard method. Explain this method and indicate clearly the advantages gained by the use of such a method.

*91. Sketch and describe a starter with no-voltage and over-load release suitable for a 10 h.p., 200-volt shunt motor. Explain the factors that influence the choice of the number of steps, and show how the resistance should be divided between the various steps.

92. Calculate the values of the resistance elements for the starter of a shunt motor to conform to—

Maximum current at starting = 100 A.
Supply voltage, 500 volts. Motor resistance, 0·2 ohm.
Number of steps, 11; i.e. 10 resistance elements.

What will be the minimum value of the current?

93. As in the above example, but the minimum current to be 100 A. What will be the maximum value of the current in this case?

*94. A d.c. shunt motor whose internal resistance is 0·5 ohm, is to be started by a controller which keeps the current between the limits of 75 and 100 A, the supply voltage being 500. Calculate the number and values of the resistance steps; prove the theory of the method of calculation you employ.

95. The motor in Ex. 94 is started by means of a resistance of 5 ohms divided into 10 equal steps. Notching-up takes place when the current has fallen to 80 A. Determine the value of the current rush on each notch—neglecting inductance effects.

96. Show that if notching-up takes place when the current is I_1, and the current then attained is the value I_2 on each notch, the values of the resistance elements for the starter of a series motor form a geometric progression.

97. Fig. 182 represents the magnetization curve for a series motor having a resistance of 0·12 ohm. The maximum current at starting is to be 300 A,

and there are to be 5 resistance elements, i.e. 6 steps. Determine the values of the resistance elements.

98. Give a graphical construction for the solution of the above example, Ex. 97.

99. The full-load current in the armature of a shunt motor is 120 A, the line voltage being 440, the resistance of the armature circuit 0·2 ohm, and the speed 800 rev/min. What will be the speed if the total torque on the motor is reduced to 60 per cent of the full-load value, and a resistance of 1·5 ohm is included in the armature circuit, the field strength being unaltered?

100. A series motor working off a 500-volt system runs at a speed of 800 rev/min, when the load is such that the current is 120 A. The resistance of the motor is 0·15 ohm, of which 0·04 ohm is the resistance of the main field circuit.

Assuming that the flux is directly proportional to the armature current, calculate—

(*a*) The speed when the armature current is 60 A and the supply voltage 550.

(*b*) The speed when the torque is one-half of the full-load value, and the field winding is connected in parallel with a diverting resistance of 0·08 ohm.

(*c*) The speed when a resistance of 1 ohm is connected in series with the motor, and the torque is 75 per cent of full-load value (in (*b*) and (*c*) the supply voltage is 500).

101. Two series motors have slightly different air gaps. When joined to a 500-volt supply and taking 100 A, the one runs at 600 rev/min and the other at 660 rev/min. The resistance of each motor is 0·3 ohm. At what speed will the motors run if mechanically coupled and connected in series to a 500-volt supply, when the load current is 100 A? What is the value of the voltage between the terminals in the case of the machine having the larger air gap under these conditions?

102. Discuss the manner in which (*a*) shunt motors, and (*b*) series motors share the load if mechanically coupled to the same load and working off the same supply system in parallel.

Two series motors are connected to the same supply system, and coupled to the same load. The one when running at 500 rev/min normally takes a current of 50 A; the other, when taking 50 A, runs at 1,000 rev/min, in each case the supply voltage being 500. Neglecting the effects of saturation, at what speed will the motors run when the total current is 100 A? Each motor has a resistance of 0·8 ohm. Calculate also the value of the current in each motor under these conditions.

103. A series motor when working at full load off a 500-volt supply takes a current of 100 A and runs at 800 rev/min. The resistance of the field winding is 0·05 ohm, and of the armature and interpole windings, 0·25 ohm total.

It is required to develop one-half of full-load torque at a speed of 900 rev/min, and this is effected by connecting a resistance in parallel with the armature. Determine the value of this resistance, and of the current in the field winding and armature circuit respectively. (The magnetization curve may be assumed to be a straight line and iron losses neglected.)

*104. Explain the principle of the current-limit method of controlling the operation of a contactor-type starter for a d.c. motor.

Such a starter for a 15-h.p., 460-V shunt motor has four sections of starting resistors.

Determine the resistance of each section if the initial starting current is equal to the full-load current and the upper limit of current on the second and following steps is $1\frac{3}{4}$ times the full-load current. The resistance of the armature circuit of the motor is 0·7 ohm and that of the shunt field circuit is 460 ohm. The efficiency at full load is 88 per cent.

If the speed at full load and normal voltage is 800 rev/min calculate the speed at the instant of cutting out the first section of the starting resistor.

†105. A 4-pole, 460-V, d.c. shunt motor has a wave-wound armature having 1,160 conductors and a resistance between brushes of 1·2 ohms. The motor

runs at a no-load speed of 1,000 rev/min, taking an armature current of 1·1 A, and the field resistance is 328 ohms. The magnetization curve for the motor is given by—

Field current A	0·4	0·6	0·8	1·0	1·2	1·4
Flux per pole Wb	0·0075	0·0095	0·0105	0·0112	0·0116	0·0119

Determine the resistance to be added to the field circuit in order that the motor may run at 1,200 rev/min when giving an output of 10 h.p.

Assume that the bearing friction and windage losses remain independent of speed.

CHAPTERS XI AND XII

*106. Enumerate the losses occurring in a shunt motor, and state how each loss will vary when the load on the machine is varied. Describe a method of determining the efficiency of a shunt motor at various loads, without actually putting a load on the motor. Criticize the assumptions made in the method described.

107. The following readings are made on a shunt motor—

Light load, armature input 4 A at 200 volts.
Resistance of field winding, hot, 150 ohms.
Resistance of armature circuit, hot, 0·10 ohm.

Determine the output and efficiency of the motor when the armature current is 50, 100, and 150 A respectively.

108. Describe one method of testing the efficiency of two similar d.c. generators at full load without using the full power of either. What are the chief sources of error in making this test?

109. Describe a method of carrying out a "back-to-back" efficiency test upon a pair of large railway motors, i.e. series motors. Show how the efficiency of each machine can be deduced from readings taken during the test.

110. The iron losses in a d.c. generator, when driven at normal speed and at normal excitation, are 8 kW.

When driven at half speed, and the same excitation, the loss is reduced to 3 kW.

Determine the values of the eddy current and hysteresis loss at half speed and at full speed.

111. Discuss, analytically, the shape of the efficiency curve of a shunt motor. At what load does maximum efficiency occur, and at what current is the output a maximum?

112. Determine the brush friction loss for a commutator having 8 brush arms, 6 brushes per arm, each 1·9 × 3·1 cm. The peripheral speed of the commutator is 20 metres per sec, the brush pressure 100 g per cm², and the coefficient of friction 0·2.

Give a brief description of the properties of the more important types of brushes.

113. Explain what is meant by the "temperature time constant." Assuming that the temperature increases according to an exponential law, determine the time constant for the case when the temperature rise after 20 minutes' run is 50 per cent of the final value.

114. The temperature rise of a motor after working for 40 minutes on full load is 16·5° C; after a further 40 minutes the temperature rise becomes 27·5° C. Assuming that the temperature rise increases according to an exponential law, determine the final temperature rise and the time constant.

115. A direct current generator develops a fault whilst running on light load. Explain the symptoms that would arise for: (*a*) a short-circuited coil; (*b*) an open-circuited coil; and (*c*) an earthed coil. How can the fault be located and temporarily remedied?

116. The following data relate to a Hopkinson efficiency test on a pair of

similar shunt machines. Determine the output and efficiency of the machine working as a motor.

Resistance of each field winding, 200 ohms.

Resistance of each armature circuit, 0·1 ohm.

The field windings are connected as in Fig. 203, and take, respectively, 2 A and 1·8 A.

The current in the armature of the generating machine is 200 A, and the total current flowing from the supply system into the two armatures is 35 A. The supply pressure is 500 volts.

117. Two traction motors are tested by the modified Hopkinson test indicated in Fig. 205. Determine the output and efficiency of the motoring machine from the following data—

Resistance of each field winding, 0·04 ohm.
Resistance of each armature circuit, 0·12 ohm.
Total current from supply system, 20 A.
Total current from booster, 230 A.
Voltage between terminals of machine acting as motor, 750.
Voltage between terminals of booster, 60.

*118. What is the meaning of the thermal time-constant of a machine and on what factors does it depend?

A totally-enclosed motor is continuously rated at 10 h.p., its final temperature rise on full load being 40° C. Its thermal time-constant is 1·5 hours. Assuming that the final steady-state temperature rise is proportional to the square of the load current, determine the one-hour rating of the machine.

Derive any expression used.

*119. A 460-V, d.c. compound motor runs at a speed of 550 rev/min at no load, the armature current being 5·2 A and the shunt field current 1·6 A. The resistance of the armature and series windings is 0·13 ohm. The number of turns per pole of the shunt and series windings are 2,000 and 9 respectively.

Determine for an armature current of 130 A, (i) the speed, (ii) the output h.p., and (iii) the efficiency. Assume the core, friction and windage loss at this load to be 10 per cent greater than at no load. A portion of the no-load magnetization characteristic as a separately-excited generator running at a constant speed of 500 rev/min is—

M.M.F. per pole AT	3,000	3,500	4,000	4,500
Armature voltage V	407	428	448	465

*120. Explain the procedure of the tests required to determine the no-load iron loss of a 500-h.p., 550-V, 500-rev/min, d.c. motor.

A sample calculation should be given together with circuit diagrams showing the necessary instruments (with ranges) and suitable ratings of the test plant required.

†121. Explain the process of commutation in a d.c. generator and describe, with relevant theory, the use of compoles.

Describe tests which are required on a large machine involving bucking and boosting the compole currents to determine whether the compoles are operating satisfactorily. Draw diagrams of test circuits and sketch graphs of typical results showing how the graphs indicate the adjustments that should be made.

†122. A totally-enclosed motor has a temperature rise of 20° C after ½ hour, and 35° C after 1 hour, on full load. Sketch the form of the graph showing the rise of temperature with time and determine the temperature rise after 2 hours on full load.

If a fan is fitted driving cooling air over the external surface of the machine which reduces the final steady-state temperature rise of the machine to 80 per cent of the original value, determine the temperature rise from cold after 1 hour on full load.

†123. Explain the meaning of the term thermal time-constant as applied to electrical plant.

A totally-enclosed motor has a thermal time-constant of 2 hours and a final temperature rise of 10° C on no-load and 40° C on full load.

Determine the limits between which the temperature rise fluctuates when the motor operates on a load cycle consisting of alternate periods of 1 hour on full-load and 1 hour on no-load, steady-state conditions having been established.

CHAPTER XIII

124. It is required to supply a load at 80 volts; a 220-volt supply is available. Explain an economical method of doing this, and compare the economy of the method with that of employing simple motor-generator connexions.

125. A shunt generator works in parallel with a large secondary battery. Discuss in detail the manner in which the load is shared.

126. Explain the action of negative boosters as used for traction work.

*127. What is an automatic reversible booster and for what purpose is it employed? Make a diagrammatic sketch of the circuits of such a machine, showing how it should be connected in the supply system.

128. Explain in detail the action of a rotating balancer set as used for a d.c. three-wire system.

129. A static balancer is connected to the slip rings of a "three-wire" d.c. generator. Show that, for the case of two slip rings, the potential of the mid-point of the coil is midway between the potentials of the brushes on the commutator.

*130. Explain the principle of one type of rotating d.c. amplifier. Describe briefly the construction and sketch the arrangement of the field winding, giving reasons for any special features.

CHAPTERS XIV TO XIX

131. Deduce from first principles an expression connecting the main dimensions of the armature of a direct current generator with the electric and magnetic loading constants.

*132. Prove from first principles that for a dynamo-electric machine the output in volt-amperes $= KD^2Ln$, where $D =$ the diam, and L the length of the armature core, and n is the speed. Show fully how and why the output coefficient K changes with size and type of machine, and show that in all designs it approaches a fixed maximum value.

133. Assuming in each case that the output coefficient is 4, determine suitable diameters for—

(*a*) A 2,000-kW generator operating at 200 rev/min.
(*b*) A 500-kW generator operating at 250 rev/min.
(*c*) A 1,000-h.p. motor operating at 150 rev/min.

In each case the diameter is to be worked out for core lengths of 30, 35, and 40 cm respectively.

134. Discuss in detail the factors that influence the choice of the core length to adopt for a given design.

The maximum flux density in the air gap of a machine is 0·95 Wb/m². If the maximum voltage between commutator segments is not to exceed 20 on no-load, determine the maximum core length that can be used for—

(*a*) An armature 100 cm diameter running at 600 rev/min.
(*b*) An armature 150 cm diameter running at 350 rev/min.

135. Explain in detail the factors that determine the best number of poles to adopt for a direct current machine.

Determine the preliminary values of the diameter and length of the armature and the number of poles for—

136. A d.c. generator to develop 750 kW, 440 volts at 400 rev/min.

137. A 500-h.p. d.c. shunt motor to operate at 350 rev/min off a 500-volt supply.

138. A d.c. generator to develop 1,500 kW, 500 volts at 250 rev/min.

139. A d.c. shunt generator to develop 50 kW, 440 volts at 750 rev/min.

140. A d.c. generator to develop 1,000 kW, 220 volts at 350 rev/min.

141. Discuss in detail the factors that influence the choice of the type of armature winding to adopt for a direct current generator or motor.

Obtain suitable windings, giving details of the number of slots, number of conductors per slot, number of commutator segments, commutator pitch, and equalizing connexions, if any, for—

142. An 8-pole, 500-volt generator; flux per pole approx. 0·09 Wb, 400 rev/min.

143. A 12-pole, 440-volt generator; flux per pole approx. 0·1 Wb, 250 rev. min.

144. A 4-pole, 440-volt motor; flux per pole approx. 0·015 Wb, 800 rev/min.

145. A 4-pole, 220-volt motor; flux per pole approx. 0·014 Wb, 1,000 rev/min.

146. A 16-pole, 500-volt generator; diameter 200 cm, length 40 cm, 200 rev/min.

147. A 6-pole, 440-volt motor; diameter 65 cm, length 30 cm, 350 rev/min.

148. A 2-pole, 220-volt motor; diameter 14 cm, length 9 cm, 1,500 rev/min.

*149. Design the shunt field coils of a direct current generator to conform to the following data; 18,500 ampere-turns per pair of poles; length of inside turn, 120 cm; number of poles, 8; voltage between terminals of field winding, 400 volts.

Assume a specific resistance, hot, of 2 micro-ohms per cm cube, that 1 kW can be dissipated per square metre of external cylindrical surface, and that insulation increases the diameter of the wire by 0·36 mm.

150. The shunt winding of a 250-volt, 10-pole generator consists of 350 turns per pole; the mean length of turn is 160 cm; round conductor 3·64 mm diameter is used.

Determine the maximum ampere-turns and loss of power in the winding (*a*) when cold, and (*b*) when hot, if there is no regulating resistance included in the circuit. Determine also the ampere-turns developed per pole when the voltage between field terminals is 190 volts, with the winding hot.

151. Determine the value of the ampere-turns per pole developed by the shunt coils of a bipolar motor, the coils of which consist of 5,200 turns per pole of an area of 0·0507 mm², and 3,400 turns per pole of an area of 0·078 mm². The mean length of turn is 50 cm, and the p.d. between field terminals 400 volts.

152. The shunt field coils of a 400-volt, 8-pole generator are composed partly of conductor 2 mm diameter, and partly of conductor 2·5 mm diameter, in the ratio of 3 turns of 2 mm diam to 2 turns of 2·5 mm diam. The mean length of turn is 130 cm. Determine the ampere-turns per pole developed by the winding when hot, with full voltage between field terminals.

153. Determine the average value of the reactance voltage, at full load, for the complete slot group of conductors of the 1,000-kW generator designed on page 405 *et seq.* Assume that the current changes according to a straight-line law, and that $\lambda = 4{\cdot}5 \times 10^{-6}$ Wb/m.

154. Determine the average value of the reactance voltage of the central coils in the slot of the machine in Ex. 153.

155. Calculate the brush friction loss for the 1,500-kW generator, data of which are given on page 469; take $\mu = 0{\cdot}15$, and $p = 120$ g/cm².

156. Determine the value of the field ampere-turns on open circuit for the 350-kW generator, of which full data are given on pages 463–8. The leakage coefficient to be taken as 1·15 at normal e.m.f. of 440 volts, and the value of the ampere-turns calculated for 90, 100, 110, and 115 per cent of normal voltage. Estimate the value of the leakage coefficient at the various values of the voltage.

157. Work out completely the performance of the 350-kW machine on page 463.

158. Obtain suitable dimensions for the commutator and brushes of a

1,200-kW, 440-volt generator. The external diameter of the armature is 170 cm, and the machine has 12 poles and 540 armature coils.

159. Work out completely the performance of the 1,500-kW generator, of which full data are given on page 469.

*160. Discuss the factors which influence the choice of the specific electric and magnetic loadings to be used in the design of a d.c. machine.

Determine the diameter and length of the armature core for a 55-kW, 110-V, 1,000-rev/min, 4-pole, shunt generator, assuming the specific electric and magnetic loadings as 26,000 ampere-conductors per m and 0·5 Wb/m² (5,000 gauss) respectively. The pole arc should be about 70 per cent of the pole pitch and the length of the core about 1·1 times the pole arc. Allow 10 A for the field current at full load and assume the voltage drop in the armature and brushes at full load to be $3\frac{1}{2}$ per cent of the rated voltage.

Determine also suitable values for the number of armature conductors and the number of slots.

*161. A 350-kW, 600-V, 8-pole, d.c. generator has an armature 80 cm in diameter. The specific electric loading at full load is approximately 28,000 ampere-conductors per metre.

Determine, for a suitable armature winding which has symmetrical 3-phase tappings connected to slip-rings,

(a) the number of commutator segments,
(b) the number of slots and the slot-pitch,
(c) the number of conductors,
(d) the commutator segments to be connected to the slip rings.

*162. Define the terms specific electric loading and specific magnetic loading, as applied to an electromagnetic machine. Discuss the factors which influence the values of these quantities to be adopted in the design of a d.c. generator.

A 250-kW, 460-V, 600 rev/min, 6-pole, d.c. generator is built with an armature diameter of 72 cm and a net core length of 27 cm. The lap armature winding has 660 conductors. Using data obtained from this machine, determine preliminary dimensions for the armature core and the number of armature conductors, slots and commutator segments for a 350-kW, 500-V, 725-rev/min, 6-pole d.c. generator. Assume a square pole face with the pole arc 70 per cent of the pole pitch.

*163. Discuss the distribution of temperature within a field coil of a d.c. machine and explain the factors which determine this distribution.

A 10-pole, 500-V, d.c. shunt generator requires 6,000 ampere-turns per pole. The poles are of rectangular cross-section 15 × 25 cm, and the available winding cross-section is 15 × 3 cm. A conductor of circular cross-section is to be used, resistivity $2{\cdot}0 \times 10^{-6}$ ohm-cm (hot) and insulation increases the diameter by 0·02 cm.

Determine (a) the cross-sectional area of the conductor,
(b) the number of turns per coil,
(c) the dissipation in W/cm² based on the area of the outside of the coil and the two end surfaces.

*164. Calculate the size of the conductor and number of turns for the field coil of a 6-pole, 460-V, d.c. shunt motor. The coil is to supply 4,000 AT at the working temperature (the resistivity at this temperature being 2×10^{-6} ohm-cm); the length of the inside turns is 74 cm, the length available for the windings is 13 cm, the space factor of the winding is 0·52, and the permissible W/cm² of external surface (excluding the ends) is 0·12. Solution should not be attempted by assuming a numerical value for the depth of the winding.

†165. The armature of a 200-kW, 500-V, 6-pole, d.c. generator is 75 cm in diameter and the specific electrical loading at full load is to be approximately 27,500 ampere-conductors per metre. It has tappings brought out to slip rings for a 3-phase static balancer.

Design a suitable winding, quoting (a) the number of slots, (b) the number of conductors, (c) the number of commutator segments, and (d) data showing the connexions between the coils and slip-rings.

†166. An 8-pole, 500-V, d.c. shunt generator with all field coils connected in series, requires 5,000 AT per pole. The cross-section of the poles, including surface insulation, is rectangular (12 cm × 20 cm) and the available cross-section on the spools for the winding is 10 cm × 2·5 cm.

Determine (*a*) the cross-sectional area of the conductor to be used, (*b*) the number of turns per spool, and (*c*) the dissipation in watts per square centimetre based on the area of the outside surface together with the two end surfaces of the coil.

The conductor to be used is circular in cross-section, of resistivity $2{\cdot}0 \times 10^{-6}$ ohm-cm, insulation increasing its diameter by 0·02 cm. Allow for a voltage drop in the field regulator of 50 V.

†167. Define the terms specific electric loading and specific magnetic load ing, of a d.c. machine.

Determine suitable dimensions of the armature core, the number of armature conductors, the number of slots, and the number of commutator segments for a 250-kW, 500-V. 500-rev/min 6-pole d.c. generator. Assume a mean flux density over the pole-pitch to be 0·67 Wb/m², a specific electric loading of 280 ampere-conductors per cm, and a ratio core-length to pole pitch of approximately 0·8.

ANSWERS TO THE NUMERICAL EXAMPLES

1. 7·17 volts.
2. 4·8 volts.
3. 1,066 conductors.
4. 630 volts.
5. 15·35 lb.
6. 159,000 ft lb.
7. 218 lb ft.
8. 357 volts.
9. 183 volts.
11. 1·222.
12. 2·225 Wb/m².
16. (a) $y_C = 64$ or 65. (c) $y_C = 32$.
    (b) $y_C = 43$. (d) $y_C = 70$.
18. $y_C = 61$ or 62; coil to span 10 teeth.
19. 18 conductors per slot; 6 coil sides per slot; 3 turn coils; either 41 slots and 123 commutator segments with $y_C = 61$ or 62; or 43 slots and 129 commutator segments with $y_C = 64$ or 65.
21. 6 conductors per slot; 128 slots; 384 segments; 12 equalizers, connexions to ring 1: 1–97–193–289; ring 2 connected: 9–105–201–297, etc.
22. 74 slots; 8 conductors per slot; 296 segments; $y_C = 98$; two separate windings.
23. 118 slots; 6 conductors per slot; 354 segments; $y_C = 89$.
    Equalizers connected: 1–178; 16–193; 31–208; 46–223; 60–237; 75–252; 90–267; 105–282; 119–296; 134–311; 149–326; 164–431.
24. 110 slots; 6 conductor sper slot; 330 segments; $y_C = 83$; coil to span 13 teeth.
25. 178 slots; 4 conductors per slot; 356 segments; $y_C = 59$.
    Equalizers connected; 1–179; 16–194; 31–209; 46–224; 61–239; 75–253; 90–268; 105–283; 120–298; 135–313; 150–328; 164–342.
26. 177 slots; 6 conductors per slot; 531 segments; $y_C = 88$, or 89.
    Equalizers connected to coils: 1–178–355; 17–194–371; 33–210–387; 49–226–403: 65–242–419; 81–258–435; 97–274–451; 113–290–467; 129–306–483; 145–322–499; 161–338–515.
27. 1·29.
29. 780; 65; 195.
34. 0·314, 0·754, 1·194 Wb/m².
36. 1,613 and 9,600 ampere-turns per pole.
37. 9,100 ampere-turns.
38. 2·5 microhenry ; 0·5 volt.
39. $i = 5(1-\varepsilon^{-1.125t})$.
41. 211 watt secs.
42. 2,770 watt secs.
44. $i = -0{\cdot}555$ A.
45. 0·986 secs.
49.

t/T_c	0	0·1	0·2	0·3	0·4	0·5	0·6	0·7	0·8	0·9	1·0
a' %	100	109	115	117	113	100	80	60	40	20	0
a'' %	0	20	40	60	80	100	113	117	115	109	100

57. 11.
58. 9½ (or 19 with two parallel circuits).
59. 108 volts.
60. 0·8 ohm.

65. (*a*) 1,500–2,500 ohms. (*c*) 40–60 ohms.
(*b*) 60–100 ohms. (*d*) 8–12 ohms.
66. (*a*) 1–1·5 ohms. (*b*) 0·04–0·06 ohm.
75. 645; (*a*) a fall of 110 V; (*b*) a fall of 71 V.
76. 7; 518 V.
79.

I, A	10	20	30	40	50	60	70	80
Torque, lb ft.	15·3	56·5	108	164	223	283	344	405
Speed, rev/min	2510	1340	1040	905	825	770	730	703

80. 925 and 900 rev/min; 76·7 and 175 lb ft.
84. (*a*) torque 0·447, speed 1·14, (*b*) torque 1·31, speed 0·945.
85. 11·7 m.p.h.; 3,660 lb.
92. 72·45 A; 1·375; 0·995; 0·72; 0·524; 0·381; 0·277; 0·201; 0·145; 0·105; 0·077 ohm.
93. 134 A; 0·95; 0·705; 0·853; 0·395; 0·29; 0·22; 0·16; 0·12; 0·09; 0·06 ohm.
94. 9 steps; 1·25; 0·94; 0·70; 0·53; 0·39; 0·295; 0·22; 0·165 ohm.
95. 88; 88·9; 90; 91·5; 93·3; 96; 1001; 06·7; 120 and 160 A.
97. 0·23; 0·20; 0·17; 0·15; and 0·13 ohm.
99. 610 rev/min.
100. (*a*) 1,800 rev/min. (*b*) 1,400 rev/min. (*c*) 728 rev/min.
101. 295 rev/min; 240 V.
102. 743 rev/min; 34·6 and 65·4 A.
103. 13·5 ohms; 91·4 and 54·7 A.
104. 9·5, 4·24, 1·9, 0·86 ohm; 170 rev/min.
105. 440 ohms.
107. 12 h.p., 87·1 per cent; 24·4 h.p., 89·8 per cent; 36·1 h.p., 89·05 per cent.
110. At half speed, hysteresis loss 2 kW, eddy loss 1 kW: at full speed, hysteresis loss 4 kW, eddy loss 4 kW.
112. 1,109 watts.
113. 28·9 minutes.
114. 49·5° C; 99 minutes approx.
116. Motor output: 108 kW, efficiency 91·2 per cent.
117. Motor output: 172·4 kW, efficiency 92 per cent.
118. 14·3 h.p.
119. (i) 481 rev/min; (ii) 73·7 h.p. (iii) 90.8 per cent.
122. 54·7° C; 32·8° C.
123. 21·2° C, 28·6° C.
133. (*a*) 288, 267, and 250 cm.
(*b*) 129, 120, and 112 cm.
(*c*) 204, 189, and 176 cm.
134. (*a*) 33·5 cm. (*b*) 38·3 cm.
136. Suitable preliminary values are: diam, 115 cm; length, 34 cm; number of poles, 8.
137. Suitable preliminary values are: diam, 90 cm; length 35 cm; number of poles, 6.
138. Suitable preliminary values are: diam, 190 cm; length 36 cm; number of poles, 12.
139. Suitable preliminary values, with a ventilating fan, are: diam, 35 cm; length, 22 cm; 4 poles.
140. Suitable preliminary values are: diam, 145 cm; length, 32 cm; 12 poles.
142. 144 slots; 6 conductors per slot; 432 segments; $y_c = 1$; 12 equalizers connected: ring 1 to coils 1–109–217–325; ring 2 to coils 10–118–226–324, etc.; simple lap winding.
143. 180 slots; 6 conductors per slot; 540 segments; $y_c = 1$; 10 equalizers connected: 1–91–181–271–361–451; 10–100–190–280–370–460, etc.; simple lap winding.
144. 37 slots; 30 conductors per slot; 5 turn coils; 111 segments; $y_c = 55$. Coil span 9 teeth; simple wave winding.
145. 39 slots; 12 conductors per slot; 2 turns coil; 117 segments; $y_c = 58$. Coil span 10 teeth; simple wave winding.

146. 180 slots; 6 conductors per slot; 540 segments; equalizers as in No. 142; simple lap winding.
147. 65 slots; 6 conductors per slot; 194 segments; 1 dummy coil; simple wave winding; $y_c = 65$.
148. 21 slots; 36 conductors per slot; 63 segments; 6 turn coils; simple lap winding.
149. 1,250 turns per pole of conductor 2·5 mm diam; arranged in 14 layers.
150. 9,600, 8,000, and 6,100 ampere-turns per pole approx; 6·85 and 5·72 kW.
151. 1,180 ampere-turns per pole.
152. 7,100 ampere-turns.
153. 3·15 volts.
154. 4.95 volts.
155. 6 kW approx.
156. 4,110, 5,110, 7,070, and 9,230 ampere-turns. Leakage coefficient = 1·13, 1·15, 1·185, and 1·215 approx.
158. Commutator diameter may be taken 110 cm; length, excluding riser 50 cm; 12 brushes per spindle, each ¾ in. thick by 1¼ in. wide.
160. 36 cm; 21 cm; 240; 40.
161. (*a*) 480, (*b*) 96, 2·6 cm, (*c*) 960,
    (*d*) Ring 1: 1, 121, 241, 361.
    Ring 2: 41, 161, 281, 401.
    Ring 3: 81, 201, 321, 441.
162. diam 76 cm, length 28 cm; 540; 90; 270.
163. (*a*) 0·0215 cm², (*b*) 1,310, (*c*) 0·114 W/cm².
164. diam 0·105 cm, area 0·00862 cm²; 2,157.
165. Simple lap winding with two-turn coils—
    (*a*) 972, (*b*) 81, (*c*) 243,
    (*d*) Ring 1: 1 82 163
    Ring 2: 28 109 190
    Ring 3: 55 136 217
166. (*a*) 0·0128 cm², (*b*) 1,150, (*c*) 0·213 W/cm².
167. diam 73 cm, length 30·5 cm; 780; 78; 390.

INDEX